T0276153

CAMBRIDGE LIBRARY COLLECTION

Books of enduring scholarly value

Physical Sciences

From ancient times, humans have tried to understand the workings of the world around them. The roots of modern physical science go back to the very earliest mechanical devices such as levers and rollers, the mixing of paints and dyes, and the importance of the heavenly bodies in early religious observance and navigation. The physical sciences as we know them today began to emerge as independent academic subjects during the early modern period, in the work of Newton and other 'natural philosophers', and numerous sub-disciplines developed during the centuries that followed. This part of the Cambridge Library Collection is devoted to landmark publications in this area which will be of interest to historians of science concerned with individual scientists, particular discoveries, and advances in scientific method, or with the establishment and development of scientific institutions around the world.

A History of the Theory of Elasticity and of the Strength of Materials

A distinguished mathematician and notable university teacher, Isaac Todhunter (1820–84) became known for the successful textbooks he produced as well as for a work ethic that was extraordinary, even by Victorian standards. A scholar who read all the major European languages, Todhunter was an open-minded man who admired George Boole and helped introduce the moral science examination at Cambridge. His many gifts enabled him to produce the histories of mathematical subjects which form his lasting memorial. First published between 1886 and 1893, the present work was the last of these. Edited and completed after Todhunter's death by Karl Pearson (1857–1936), another extraordinary man who pioneered modern statistics, these volumes trace the mathematical understanding of elasticity from the seventeenth to the late nineteenth century. Volume 2 (1893) was split into two parts. Part 2 covers the work of Neumann, Kirchhoff, Clebsch, Boussinesq, and Lord Kelvin.

Cambridge University Press has long been a pioneer in the reissuing of out-of-print titles from its own backlist, producing digital reprints of books that are still sought after by scholars and students but could not be reprinted economically using traditional technology. The Cambridge Library Collection extends this activity to a wider range of books which are still of importance to researchers and professionals, either for the source material they contain, or as landmarks in the history of their academic discipline.

Drawing from the world-renowned collections in the Cambridge University Library and other partner libraries, and guided by the advice of experts in each subject area, Cambridge University Press is using state-of-the-art scanning machines in its own Printing House to capture the content of each book selected for inclusion. The files are processed to give a consistently clear, crisp image, and the books finished to the high quality standard for which the Press is recognised around the world. The latest print-on-demand technology ensures that the books will remain available indefinitely, and that orders for single or multiple copies can quickly be supplied.

The Cambridge Library Collection brings back to life books of enduring scholarly value (including out-of-copyright works originally issued by other publishers) across a wide range of disciplines in the humanities and social sciences and in science and technology.

A History of the Theory of Elasticity
and of
the Strength of Materials

VOLUME 2: PART 2
SAINT-VENANT TO LORD KELVIN (2)

ISAAC TODHUNTER
EDITED BY KARL PEARSON

CAMBRIDGE
UNIVERSITY PRESS

CAMBRIDGE
UNIVERSITY PRESS

University Printing House, Cambridge, CB2 8BS, United Kingdom

Published in the United States of America by Cambridge University Press, New York

Cambridge University Press is part of the University of Cambridge.

It furthers the University's mission by disseminating knowledge in the pursuit of education, learning and research at the highest international levels of excellence.

www.cambridge.org
Information on this title: www.cambridge.org/9781108070447

© in this compilation Cambridge University Press 2014

This edition first published 1893
This digitally printed version 2014

ISBN 978-1-108-07044-7 Paperback

This book reproduces the text of the original edition. The content and language reflect the beliefs, practices and terminology of their time, and have not been updated.

Cambridge University Press wishes to make clear that the book, unless originally published by Cambridge, is not being republished by, in association or collaboration with, or with the endorsement or approval of, the original publisher or its successors in title.

A.

B.

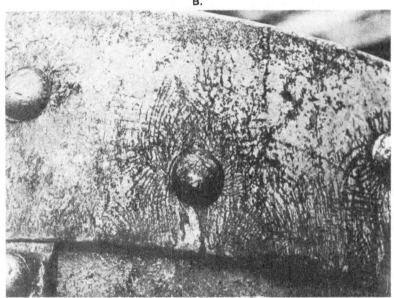

C.

LÜDERS CURVES IN STEEL.

A.—Square punch hole, Weardale Steel.
B.—Bent and punched butt strip of Dredger Bucket, Shelton Steel.
C.—Portion sheared off end of bar of Bush Steel ($2\frac{1}{2}'' \times \frac{3}{8}''$).
Specimens due to Mr. J. B. HUNTER: *see Art.* 1190.

A HISTORY OF
THE THEORY OF ELASTICITY

AND OF

THE STRENGTH OF MATERIALS

FROM GALILEI TO THE PRESENT TIME.

BY THE LATE

ISAAC TODHUNTER, D.Sc., F.R.S.

EDITED AND COMPLETED

FOR THE SYNDICS OF THE UNIVERSITY PRESS

BY

KARL PEARSON, M.A.

PROFESSOR OF APPLIED MATHEMATICS, UNIVERSITY COLLEGE, LONDON,
FORMERLY FELLOW OF KING'S COLLEGE, CAMBRIDGE.

VOL. II. SAINT-VENANT TO LORD KELVIN.

PART II.

CAMBRIDGE:
AT THE UNIVERSITY PRESS.

1893.

[*All Rights reserved.*]

Cambridge:

PRINTED BY C. J. CLAY, M.A. AND SONS,

AT THE UNIVERSITY PRESS.

ERRATA.

PART II.

Frontispiece, Specimen *B, for* butt strip, *read* cutting hip.
p. 282, *l.* 20 *for* Arts. 207—11 *read* Arts. 207*—11*.
p. 286, *l.* 4 from bottom *for* Davier *read* Darier.
p. 341, *l.* 4 from bottom *for* Art. 1863 *read* Art. 1563.

CHAPTER XII.

THE OLDER GERMAN ELASTICIANS: F. NEUMANN, KIRCHHOFF AND CLEBSCH.

SECTION I.

Franz Neumann.

[1192.] WE have already had occasion to deal with three important memoirs of F. Neumann's, which fall into the period occupied by our first volume, and we have now to turn to a work of his which, if only published in 1885, still in substance mainly belongs to the years 1857–8. To Franz Neumann's teaching in Königsberg is due much of the impulse which mathematical physics received in the fifties in Germany; the most distinguished German physicists of the past forty years have been nearly all pupils of Neumann's, and this remark is specially true in the field of elasticity. Of those who attended his lectures on this subject and received probably from him their first stimulus to original investigations, we may name Kirchhoff, Strehlke, Clebsch, Borchardt, Carl Neumann and Voigt as among the more important[1]. Franz Neumann's lectures on elasticity were given in Königsberg at different times from 1857 to 1874, and in 1885 were published under the supervision of O. E. Meyer of Breslau with the title: *Vorlesungen über die Theorie der Elasticität der festen Körper*

---

[1] O. E. Meyer includes in the list Von der Mühll, Minnigerode, Zöppritz, Gehring, Saalschütz, Wangerin and Baumgarten: see preface to the *Vorlesungen*, S. viii.

4 F. NEUMANN. [1193

und des Lichtäthers. The volume contains xiii + 374 pages, and
is based on the notebooks of the brothers L. and O. E. Meyer for
the years 1857–60, and those of Baumgarten and W. Voigt for
the years 1869–74. According to the Editor the work contains
all that was of importance in Neumann's lectures. The exact
amount of originality in the several investigations I shall endeavour
to point out in the course of my analysis, and I content myself
here with the following remarks from the preface :

> Zu den Gebieten, mit welchen Professor Neumann sich in jüngeren und
> späteren Jahren mit besonderer Vorliebe beschäftigt hat, gehört auch die
> Theorie der Elasticität ; es konnte daher nicht fehlen, dass seine Vorlesungen
> über diesen Gegenstand häufig eigene Arbeiten betrafen. Seinem ausgespro-
> chenen Wunsche, dass alle in verschiedenen Semestern vorgetragenen eigenen
> Untersuchungen in dieses Werk aufgenommen werden sollten, bin ich gern
> soweit nachgekommen, als es mir zu erreichen möglich war (S. v–vi).

The work is divided into twenty-one sections of which we note
the important points in the following articles.

[1193.] In Section 1, *Einleitung* (S. 1–7), we have first some remarks
on the origin of the theory of elasticity. Neumann attributes it not so
much to a development from the isolated problems of Bernoulli and
Euler as to the impulse given by Fresnel's new theory of light. He
says :

> Die exacte Beurtheilung seiner Beobachtungen führte Fresnel zu That-
> sachen, welche im geraden Widerspruch standen zu den anerkannten
> Principien der Wellenbewegung in elastischen Medien. In der Schallwelle
> ist die Bewegung der Theilchen parallel dem Strahl, die Welle eine longitudi-
> nale ; Fresnel fand, dass in der Lichtwelle jene Bewegung senkrecht gegen
> den Strahl gerichtet, die Welle also eine transversale ist, und doch soll der
> Unterschied der Eigenschaften beider Medien, der Luft und des Lichtäthers,
> nur quantitativ, nicht qualitativ sein. Die Mechaniker jener Zeit läugneten
> die Möglichkeit einer solchen Bewegung, weil sie unverträglich sei mit den
> hydrodynamischen Grundgleichungen, welche auf elastische Flüssigkeiten, auf
> Luft angewandt nur longitudinale Wellen kennen lehren. Fresnel, sich
> vertheidigend, machte darauf aufmerksam, dass möglicherweise in diesen
> Gleichungen nicht alle Kräfte berücksichtigt sein möchten, welche in
> elastischen Medien zur Wirkung kommen können. Er fand in der That,
> dass in den hydrodynamischen Gleichungen nur solche inneren Kräfte
> enthalten sind, welche aus einer Verdünnung oder Verdichtung des Mediums
> entstehen und welche wiederum eine Aenderung der Dichtigkeit hervorbrin-
> gen. Er stellte sich daher die Frage, ob es in einem elastischen Medium
> keine anderen Kräfte gebe, ob in einem solchen System, wie es die Theilchen
> eines elastischen Körpers bilden, nicht auch Kräfte entstehen können aus
> einer Verschiebung der Theilchen, durch welche die Dichtigkeit nicht geändert
> wird. Wie jetzt die Sachen liegen, ist es leicht, den Standpunkt, auf den
> Fresnel sich stellte, klar zu machen (S. 1–2).

This account of the origin of the theory of elasticity, attributing it
to the inability of the hydrodynamical equations to offer any explanation

of the phenomena of light, has been accepted by several writers (see the review of our first volume in the *Bulletin des sciences mathématiques* T. 12, p. 38, 1888), but it must be distinctly borne in mind that the first propounder of the theory was Navier, an elastician of the old, or Bernoulli-Eulerian school, who both in theory and practice had frequently dealt with elastic stresses by the old methods, and whose memoir of 1827 was preceded not by optical investigations but by researches on the elasticity of rods and plates.

Neumann after briefly referring to the labours of Navier, Poisson and Cauchy concludes his first section by defining stress on their lines, i.e. by supposing inter-molecular force central and a function only of the central distance.

[1194.] The second section is entitled : *Allgemeine Lehrsätze über die Druckkräfte* (S. 8–25) and develops the usual stress equations without regard to any molecular hypothesis. The third section (S. 26–36) discusses Cauchy's and Lamé's ellipsoids of stress and the principal tractions without reference, however, to those writers : see our Arts. 610*, (iv), and 1059*. The fourth section entitled : *Das System der Dilatationen* (S. 37–51) deals with the geometry of small strains, and discusses the ellipsoids of strain and the principal stretches. The fifth section is entitled : *Beziehungen zwischen den Druckkräften und den Verrückungen* (S. 52–9). It deals only with uncrystalline and presumably homogeneous and isotropic bodies. Neumann remarks that experiment shows us that stress and strain vanish and arise coevally ; hence he argues that one must be capable of being mathematically expressed as a function of the other. He then states that there can be no doubt that in uncrystalline bodies the axes of principal stretch and principal traction must coincide, and he continues :

Aus unserer Annahme, dass die Dilatationen kleine Grössen seien, folgt, dass die Druckkräfte, welche wir als Functionen jener anzusehen haben, in der Gestalt einer Entwickelung nach Potenzen der Dilatationen dargestellt werden können. Da ferner nach unserer Annahme die Dilatationen so kleine Grössen sind, dass wir nur ihre erste Potenz zu berücksichtigen brauchen, so müssen die Hauptdruckkräfte lineare Functionen der Dilatationen sein ; und zwar werden sie, da sie mit jenen zugleich verschwinden, ohne Hinzufügung eines constanten Gliedes ihnen einfach proportional zu setzen sein (S. 52-3).

Obviously here Neumann falls into the same *non-sequitur* as Cauchy in his memoir of 1827 (see our Art. 614*), as Maxwell in 1850 (see our Art. 1536*), or Lamé in 1852 (see our Art. 1051*). Neumann then obtains by transformation the ordinary stress-strain relations and the body-shift equations for an isotropic elastic solid. He employs Δ for our θ, $A - B$ for our 2μ, and B for our λ. Further he uses pressures not tractions throughout his work.

The Sections 2–5 of Neumann's work form an elementary theory of elasticity, at least so far as isotropic bodies are concerned. They do not possess any particular advantages in the present state of our science.

[1195.] The sixth section of the work (S. 60–6) is entitled *Navier's Differentialgleichungen*. It deduces the body-shift equations directly by Navier's method (see our Art. 266*); this method leads to uni-constant isotropy and avoids all introduction of the stresses. In starting with Navier's investigation Neumann adopts the historical plan. He points out the objections to Navier's process (S. 66 : see our Arts. 531*–2*), and then turns to Poisson's and Cauchy's treatment of the problem in his seventh section entitled : *Poisson's Ableitung der allgemeinen Gleichungen* (S. 67–79). Neumann's investigation follows fairly closely Poisson's of 1828. He deduces the shift-equations for the cases of isotropy and of three rectangular axes of elastic symmetry. The latter system he speaks of as crystalline, although it is often produced by working in bodies without crystalline structure. He says :

Zu diesen Krystallen, deren Zahl sehr gross ist, gehören alle Formen des regulären, viergliedrigen zwei- und zweigliedrigen und sechsgliedrigen Systems mit Ausnahme gewisser, hemiëdrischer Formen, bei denen die parallelen Krystallflächen fehlen, z. B. beim regulären Tetraëder. Wir nennen diese Formen die geneigtflächigen Hemiëder. Ferner findet eine solche symmetrische Vertheilung nicht mehr statt bei allen Krystallen des zwei- und eingliedrigen und des ein- und eingliedrigen Systems (S. 75).

The resulting equations involving six independent constants agree with those which would be obtained by substituting the stress-strain relations of our Art. 117 (a) with the rari-constant conditions $d = d'$, $e = e'$, $f = f'$, in the usual body stress-equations.

The seven sections with which we have already dealt belong to the 1857–8 notebooks. Section 8 is taken from a notebook of 1859–60, and is entitled : *Entwickelung der Gleichungen aus dem Princip der virtuellen Geschwindigkeit* (S. 80–106). This is a reproduction of the method of Carl Neumann's memoir of 1860 : see our Art. 667. F. Neumann, I think, supposes the first application of the principle of virtual moments to the theory of elasticity to have been made in the above memoir, but this is hardly correct : see our Arts. 268* and 759*. The method of the *Vorlesungen* is somewhat clearer and briefer than that of C. Neumann; it is also applied to bodies with three axes of elastic symmetry.

[1196.] Section 9 (S. 107–20), taken from a notebook of 1857–8, deals with the thermo-elastic equations in the method previously adopted by Duhamel and Neumann himself. We have seen that Neumann in 1841 (see our Art. 1196*) claimed priority in the deduction of these equations, and the Editor of the *Vorlesungen* (S. vi) apparently looks upon this section as an original part of the present work. The results do not seem to be more general than those of Duhamel (1838, see our Arts. 868* and 877*) and in all cases of doubt, priority of publication must be decisive.

Neumann like Duhamel limits his equations to the range in which extension is proportional to rise in temperature. His body-stress-

equations involving thermal effect (2) and (3), S. 113, are equivalent to
Equations (2) of our Art. 883* ; his surface stress-equations (1) and
(2), S. 114, to Equation (3) of the same article ; his remarks on the
relations between temperature and normal pressure, and between the
thermo-elastic-constant, the stretch-modulus and the thermal stretch-
coefficient are equivalent to those of Duhamel in our Arts. 875* and
888*.

[1197.] § 58 (S. 115–8) is entitled : *Krystallinische Körper.* In
it Neumann questions whether the thermo-elastic constant is in crystalline bodies the same for all directions. He suggests equations of the
form (see our Art. 883*) :

$$\rho \left(\frac{d^2u}{dt^2} - X \right) = \frac{d\widehat{xx}}{dx} + \frac{d\widehat{xy}}{dy} + \frac{d\widehat{zx}}{dz} - \beta_x \frac{dq}{dx},$$

$$\rho \left(\frac{d^2v}{dt^2} - Y \right) = \frac{d\widehat{xy}}{dx} + \frac{d\widehat{yy}}{dy} + \frac{d\widehat{yz}}{dz} - \beta_y \frac{dq}{dy},$$

$$\rho \left(\frac{d^2w}{dt^2} - Z \right) = \frac{d\widehat{zx}}{dx} + \frac{d\widehat{yz}}{dy} + \frac{d\widehat{zz}}{dz} - \beta_z \frac{dq}{dz},$$

in which he assumes, I suppose, the body to have three rectangular
axes of elastic symmetry, coinciding with the thermal axes. The
surface stress-equations will now be given by :

$$X' = (\widehat{xx} - \beta_x q) \cos l + \widehat{xy} \cos m + \widehat{zx} \cos n,$$

$$Y' = \widehat{xy} \cos l + (\widehat{yy} - \beta_y q) \cos m + \widehat{yz} \cos n,$$

$$Z' = \widehat{zx} \cos l + \widehat{yz} \cos m + (\widehat{zz} - \beta_z q) \cos n,$$

so that it is obvious that a rise of temperature is no longer equivalent
to a uniform surface traction : see our Arts. 684–5.

Hierauf beruht die Entscheidung durch die Beobachtung. Man bestimmt
durch directe Messung die Aenderung der Winkel, wenn der Druck auf die
Oberfläche des Krystalls geändert wird, wenn man ihn z. B. aus dem Drucke
einer Atmosphäre in den von 10 Atmosphären oder in den luftleeren Raum
bringt. Auf dieselbe Weise misst man die Winkeländerung, welche durch
eine Erhöhung der Temperatur, z. B. von 0° auf 100°, hervorgebracht wird.
Erhält man beide Male ein entsprechendes System von Winkeländerungen,
so sind alle drei Werthe von β unter sich gleich ; befolgen die Aenderungen
verschiedene Gesetze, so sind sie verschieden (S. 116–7).

Neumann then describes a method of making the needful measurements. He cites some experiments of Mitscherlich's (*Abhandlungen der
Berliner Akademie,* 1825, S. 212) upon calcspar. This material expands
in the direction of its axis owing to a rise of temperature and contracts
perpendicular to the axis. The stretch for 100° C. increase of temperature was found to be ·00286 and the squeeze − ·00056. Thus the
dilatation was ·00174. A similar result was exhibited by gypsum
which in three different directions had different stretches or squeezes.

Neumann does not cite any experiments to determine how far the
thermal results for these crystals are in accordance with those which
would be produced by uniform surface tractions. He merely remarks
that rods might be cut in certain directions from such crystals so that
they would not change their length with change of temperature :

Hier löst also eine krystallinische Substanz ein Problem, dessen Lösung
oft sehr gewünscht wird (S. 118).

The section concludes with a paragraph deducing the amplified form
of Fourier's differential equation for the conduction of heat. This is in
accord with Duhamel's results cited in our Art. 883*, Equation (i).

[1198.] The tenth section of the *Vorlesungen* is entitled
Kirchhoffs allgemeine Lehrsätze (S. 121–32). Of this section § 60
reproduces Kirchhoff's proof of the uniqueness of the solution of
the equations for the equilibrium of an elastic solid: see our
Art. 1255 : § 61 (S. 125–8) extends the proof of the uniqueness of
the solution to the case of vibrations. This, I think, had not
been done by either Kirchhoff or Clebsch and is original[1]
Neumann, as in the previous paragraph, supposes isotropy. We
will indicate his method of proof. If there be two solutions, then
their difference, given say by the shifts U, V, W, must satisfy
the body- and surface-equations with abstraction of body-force
and surface-load.

Consider the quadruple integral

$$\iiiint dt\,dx\,dy\,dz \left\{ \left(\rho\frac{d^2U}{dt^2} + \frac{d\widehat{xx}}{dx} + \frac{d\widehat{xy}}{dy} + \frac{d\widehat{zx}}{dz} \right) \frac{dU}{dt} \right.$$
$$+ \left(\rho\frac{d^2V}{dt^2} + \frac{d\widehat{xy}}{dx} + \frac{d\widehat{yy}}{dy} + \frac{d\widehat{yz}}{dz} \right) \frac{dV}{dt}$$
$$\left. + \left(\rho\frac{d^2W}{dt^2} + \frac{d\widehat{zx}}{dx} + \frac{d\widehat{yz}}{dy} + \frac{d\widehat{zz}}{dz} \right) \frac{dW}{dt} \right\},$$

which is zero owing to the body stress-equations. Integrate the stress
terms by parts ; the surface integrals then vanish owing to the surface
stress-equations. Substitute for the stresses from the stress-strain relations, and the whole will be found a complete differential with regard
to the time. Integrating out with regard to the time we find :

[1] The whole of this section is due to the lectures of 1859–60, and thus precedes
Clebsch's *Treatise*. Kirchhoff's investigation was first given in the memoir of 1858 :
see our Art. 1255.

$$0 = \iiint dx\, dy\, dz \left\{ \left(\frac{dU}{dt}\right)^2 + \left(\frac{dV}{dt}\right)^2 + \left(\frac{dW}{dt}\right)^2 \right\}$$

$$+ \iiint dx\, dy\, dz \left[2\mu \left\{ \left(\frac{dU}{dx}\right)^2 + \left(\frac{dV}{dy}\right)^2 + \left(\frac{dW}{dz}\right)^2 \right\} + \lambda\theta^2 \right.$$

$$\left. + \mu \left\{ \left(\frac{dV}{dz} + \frac{dW}{dy}\right)^2 + \left(\frac{dW}{dx} + \frac{dU}{dy}\right)^2 + \left(\frac{dU}{dy} + \frac{dV}{dx}\right)^2 \right\} \right].$$

Hence it follows that all the squared terms must separately vanish at all points of the body. We see then that U, V, W are not functions of the time and that they can only express a translation and rotation of the body as a whole.

[1199.] § 62 of the *Vorlesungen* is entitled : *Verallgemeinerung des Beweises für Krystalle.* It is a not very satisfactory extension of the proof of the preceding section to bodies for which the stress-strain relations are of the form :

$$\widehat{xx} = as_x + fs_y + es_z, \qquad \widehat{yz} = d\sigma_{yz},$$

$$\widehat{yy} = fs_x + bs_y + ds_z, \qquad \widehat{zx} = e\sigma_{zx},$$

$$\widehat{zz} = es_x + ds_y + cs_z, \qquad \widehat{xy} = f\sigma_{xy};$$

i.e. to bodies for which *we can assume rari-constancy and which possess three rectangular axes of elastic symmetry.* Even if we suppose rari-constancy, such bodies are by no means the only existing type of crystal. Further Neumann's proof depends on the conditions that

$$a > e + f, \qquad b > f + d, \qquad c > d + e \ldots\ldots\ldots\ldots(i).$$

Neumann demonstrates this as follows. Crystals, he states, do not according to experiment differ widely from isotropic bodies, hence we must have :

$$3\lambda = a - \kappa_1 = b - \kappa_2 = c - \kappa_3,$$

$$\lambda = d - \varpi_1 = e - \varpi_2 = f - \varpi_3,$$

where κ_1, κ_2, κ_3, ϖ_1, ϖ_2, ϖ_3 are very small quantities as compared with λ. Hence it follows that the relations (i) above must be true. This supposes again the limit to be uni-constant isotropy. Now the objection to this sort of proof is that relations akin to (i) may hold, and certainly the uniqueness of the solution must hold, for wood and other materials, in which there is no approach to isotropy at all. Neumann's concluding words would seem to suggest that he considered the proposition proved for *all* bodies which occur in nature. In a footnote the remark is made that the laws of double refraction require that in the case of the ether we should have

$$a = 3(e + f - d), \quad b = 3(f + d - e), \quad c = 3(d + e - f) \ldots\ldots(ii),$$

and that since d, e, f differ only slightly, relations (i) must also be satisfied for the ether. That relations (ii) are *not* absolutely necessary on the elastic jelly theory of the ether has been indicated in our Art. 148. A more complete proof of the uniqueness of the solution of the equations of elasticity is given in Kirchhoff's *Vorlesungen*[1]: see our Arts. 1240, 1255 and 1278.

[1200.] § 63 (S. 129–32) belongs to the lectures of 1873–4. It is an investigation of the elastic energy of the stresses for an isotropic solid; it is so far more general than that to be found in the usual text-books, in that it regards possible changes of temperature due to the strain.

Let X, Y, Z be the body-forces at the point x, y, z of the solid, and X', Y', Z' the surface-load at the element dS of the surface. Then we can deduce from the thermo-elastic equations (see our Art. 1197) the following relation:

$$\frac{1}{2}\frac{d}{dt}\iiint \rho \left\{ \left(\frac{du}{dt}\right)^2 + \left(\frac{dv}{dt}\right)^2 + \left(\frac{dw}{dt}\right)^2 \right\} dx\,dy\,dz$$

$$= \frac{d}{dt}\iiint \rho (Xu + Yv + Zw)\, dx\,dy\,dz$$

$$+ \iint \left(X'\frac{du}{dt} + Y'\frac{dv}{dt} + Z'\frac{dw}{dt} \right) dS$$

$$- \frac{1}{2}\frac{d}{dt}\iiint \{\lambda\theta^2 + 2\mu\,(s_x{}^2 + s_y{}^2 + s_z{}^2) + \mu\,(\sigma_{yz}{}^2 + \sigma_{zx}{}^2 + \sigma_{xy}{}^2)\}\, dx\,dy\,dz$$

$$+ \iiint \beta q\, \frac{d\theta}{dt}\, dx\,dy\,dz \ldots \ldots \ldots \ldots \ldots (\text{iii}),$$

where

$$\frac{dq}{dt} = \frac{k}{c_v \rho}\, \nabla^2 q - \frac{\gamma - 1}{\delta}\, \frac{d\theta}{dt} \ldots \ldots \ldots \ldots (\text{iv}),$$

(see our Art. 885*).

Now if X', Y', Z' are independent of t, i.e. *if the surface load be always the same*, we may integrate the whole of this with regard to t except the last term of the last line. This last can be integrated easily in two cases:

(i) Steady temperature, or q no function of t. We have:

$$\frac{1}{2}\iiint \rho \left\{ \left(\frac{du}{dt}\right)^2 + \left(\frac{dv}{dt}\right)^2 + \left(\frac{dw}{dt}\right)^2 \right\} dx\,dy\,dz + \text{constant}$$

$$= \iiint \rho\,(Xu + Yv + Zw)\, dx\,dy\,dz + \iint (X'u + Y'v + Z'w)\, dS$$

$$- \tfrac{1}{2}\iiint \{\lambda\theta^2 + 2\mu\,(s_x{}^2 + s_y{}^2 + s_z{}^2) + \mu\,(\sigma_{yz}{}^2 + \sigma_{zx}{}^2 + \sigma_{xy}{}^2)\}\, dx\,dy\,dz$$

$$+ \iiint \beta q\theta\, dx\,dy\,dz \ldots \ldots \ldots \ldots \ldots (\text{v}).$$

[1] The importance of this proposition lies in the result, that if any *particular* solution be found which satisfies all the conditions of an elastic problem, this solution is the only admissible one.

(ii)   Suppose we neglect the first term on the right-hand-side of equation (iv), as for example in Newton's hypothesis as to the velocity of sound, then we have :

$$\frac{1}{2} \iiint \rho \left\{ \left(\frac{du}{dt}\right)^2 + \left(\frac{dv}{dt}\right)^2 + \left(\frac{dw}{dt}\right)^2 \right\} dx\,dy\,dz + \text{constant}$$

$$= \iiint \rho \,(Xu + Yv + Zw)\, dx\,dy\,dz + \iint (X'u + Y'v + Z'w)\, dS$$

$$- \tfrac{1}{2} \iiint \{\lambda\theta^2 + 2\mu\,(s_x^2 + s_y^2 + s_z^2) + \mu\,(\sigma_{yz}^2 + \sigma_{zx}^2 + \sigma_{xy}^2)\}\, dx\,dy\,dz$$

$$- \frac{1}{2} \iiint \frac{\delta}{\gamma - 1}\, \beta q^2\, dx\,dy\,dz \dots\dots\dots\dots (vi).$$

[1201.]   The eleventh section (S. 133–163) is entitled: *Anwendungen auf unkrystallinische Körper*, and is occupied with the application of the equations of bi-constant isotropic elasticity to certain simple problems.   The object of this section, we are told, is to clear up the doubtful points of those theories which starting from the molecular hypothesis reach uni-constant isotropy.   Neumann here, however, does not seem to lay sufficient stress on the possibility of various distributions of elastic homogeneity in the rods, wires, hollow cylinders and spheres of which he treats.   We may note one or two points.

(*a*)   He refers (S. 136–8) to the experiments of Cagniard de la Tour, Regnault, Wertheim and himself on the magnitude of the stretchsqueeze ratio : see our Arts. 368*, 1321*, 1358* and 736.   He himself had found that for iron-wire $\eta = 1/4$ nearly, but that it was nearer $1/3$ for other substances, which he unfortunately does not specify.

(*b*)   On S. 141–2 Neumann gives a theory of Wertheim's cylinder method of determining η : see our Art. 802.   He remarks on the extreme importance of ascertaining the value of η for truly isotropic bodies, as the development of the molecular theory depends so entirely upon it.   In investigating on S. 144–5 the stress in a hollow cylinder due to internal pressure, Neumann takes a stress-limit of strength and applies the theory of elasticity to rupture.   Both steps seem to me unjustifiable : see our Arts. 5 (*a*) and (*c*), 169 (*c*) and 320–1.

(*c*)   S. 146–153 deal with the oft-considered problem of the hollow spherical shell.   Neumann discusses Oersted's theory of the piezometer, and shows how Colladon and Sturm were correct in supposing that a hollow spherical shell with equal internal and external pressures contracts as a solid sphere would do under the same external pressure : see our Arts. 686*–690*.   He applies the theory to thermometer bulbs, and in particular shows how the reading of the thermometer is lower with the tube in a vertical than with the tube in a horizontal position owing to the internal pressure of the quicksilver on the bulb

12   F. NEUMANN.   [1202

being greater in the former case. He shows by a numerical example
that the difference of the reading in the two positions might amount
to ·2° C.

Bei Thermometern, welche Cylinder statt Kugeln haben, ist dieser Fehler
nicht so bedeutend, weil sie in der Regel eine stärkere Wand besitzen.
Hierin liegt einer der Vorzüge der Cylinderthermometer (S. 151).

The consideration (S. 151–3) of the strength of an isotropic
spherical shell and its comparison with the strength of a cylindrical
one, is for reasons we have frequently referred to, very questionable
when applied to glass vessels : see our Arts. 1358* and 119.

(d) § 74 (S. 153–5) deals with the problem of an isotropic solid
elastic sphere surrounded by a shell of different isotropic elastic
material, to the outer surface of which is applied a uniform pressure.
Neumann finds that the solid core will contract more or less than it
would do, if the external pressure were directly applied to it, according
as $3\lambda + 2\mu$ for the core is greater or less than it is for the shell.

(e) §§ 75–76 (S. 155–61) are introduced by the Editor, and give
methods of determining the elastic constants by torsion and *uniform
flexure* (i.e. flexure by a couple). These are practically the methods
adopted by Kirchhoff and Okatow to determine η : see our Arts. 1271–3.
The final paragraphs of this section (S. 161–3) entitled : *Beobachtungen
zur Bestimmung des Verhältnisses der beiden Elasticitätsconstanten* are
also mainly due to the Editor and give a short résumé of the various
experimental determinations of η due to Cornu, Mallock, Kirchhoff,
Okatow, Schneebeli, Kohlrausch, Loomis, Baumeister, Röntgen,
Amagat, W. Voigt, Littmann, and Everett. Accounts of the researches of these writers will be found under their names in our
index, and the results of later researches under the title *stretch-squeeze
ratio*. We can only remark here, that several of them still leave open
to question the true isotropy of the materials experimented on, and
they cannot thus be said to have finally settled the elastic constant
controversy : see our Arts. 925*, 932*, 192, 800, and 1271.

[1202.] The twelfth section is entitled: *Elasticität krystallinischer Stoffe* and occupies S. 164–202. This section is taken
from lecture notes of the years 1873–4. Neumann here rejects
the rari-constant equations for crystals with three axes of elastic
symmetry such as he had previously adopted in his work, and on
S. 165 expresses the stresses in terms of the strains by linear
relations involving 36 constants. Thus he writes:

Diese 36 Elasticitätsconstanten lassen sich im Allgemeinen nicht auf eine
geringere Anzahl zurückführen. Jedoch verringert sich in den allermeisten
Fällen ihre Zahl sehr erheblich, wenn der Krystall in Bezug auf eine oder
mehrere Ebenen symmetrisch gebildet ist. Nur in den seltener vorkommen-

den Fällen des ein- und eingliedrigen Systems, wie z. B. beim Kupfervitriol, liegt kein theoretischer Grund für eine Verminderung ihrer Anzahl vor (S. 165).

This statement of course is hardly true, for the principle of work leads us at once to the reduction of the 36 constants to 21. Notwithstanding this necessary modification (S. 179 ftn.), the section contains a good deal of valuable and, till 1885, unpublished work of Neumann. The results should be compared with those of Rankine : see our Arts. 450-1.

[1203.] We reproduce briefly Neumann's stress-strain relations in our own notation and with the additional relations between the constants due to Green's principle : see Arts. 78, 117, etc.

(a) *Crystal with one plane of elastic symmetry*, taken as that of zx. (*Zwei- und eingliedrigen oder monoklinische Krystalle.* S. 168.)

$$\widehat{xx} = as_x + f's_y + e's_z + h_1\sigma_{zx},$$

$$\widehat{yy} = f''s_x + bs_y + d's_z + h_2\sigma_{zx}, \qquad \widehat{xy} = f\sigma_{xy} + k'\sigma_{yz},$$

$$\widehat{zz} = e''s_x + d''s_y + cs_z + h_3\sigma_{yx}, \qquad \widehat{yz} = k''\sigma_{xy} + d\sigma_{yz},$$

$$\widehat{zx} = h_1's_x + h_2's_y + h_3's_z + e'\sigma_{zx},$$

Neumann has thus twenty constants, but we ought to put

$$e'', f'', d'', k'' = e', f', d', k',$$

and $h_1', h_2', h_3' = h_1, h_2, h_3$ respectively, or leave only thirteen constants.

(b) *Crystal with two planes of elastic symmetry at right angles*, taken as zx and yz. (*Zwei- und zweigliedrige Krystalle.* S. 169.)

$$\widehat{xx} = as_x + f's_y + e's_z, \qquad \widehat{yz} = d\sigma_{yz},$$

$$\widehat{yy} = f''s_x + bs_y + d's_z, \qquad \widehat{zx} = e\sigma_{zx},$$

$$\widehat{zz} = e''s_x + d''s_y + cs_z, \qquad \widehat{xy} = f\sigma_{xy}.$$

These equations also hold for crystals with *three* planes of elastic symmetry, and have *twelve* constants according to Neumann, but we ought to put $d'' = d'$, $e'' = e'$ and $f'' = f'$, which leaves only *nine* constants.

(c) *Crystal with two equal axes*, taken as those of x and y. (*Krystalle des viergliedrigen Systems.*)

$$\widehat{xx} = as_x + f's_y + e's_z, \qquad \widehat{yz} = d\sigma_{yz},$$

$$\widehat{yy} = f's_x + as_y + e's_z, \qquad \widehat{zx} = d\sigma_{zx},$$

$$\widehat{zz} = e''s_x + e''s_y + cs_z, \qquad \widehat{xy} = f\sigma_{xy}.$$

These equations according to Neumann have *seven* constants, but we ought to put $e'' = e'$ so that we have only *six* constants.

(d) *Regular Crystals*, or those having three rectangular equal axes taken as those of x, y, z.

$$\widehat{xx} = (a - f')\,s_x + f'\theta, \qquad \widehat{yz} = d\sigma_{yz},$$
$$\widehat{yy} = (a - f')\,s_y + f'\theta, \qquad \widehat{zx} = d\sigma_{zx},$$
$$\widehat{zz} = (a - f')\,s_z + f'\theta, \qquad \widehat{xy} = d\sigma_{xy}.$$

These have *three* independent constants.

[1204.] Neumann now passes to hexagonal and rhombohedral crystals.

(e) *Hexagonal Crystals*. (*Sechsgliedrige Krystalle*.)

Of these Neumann writes :

Es bleiben noch die Krystalle des hexagonalen Systems zu untersuchen übrig, deren Grundform die auf einem regulären Sechseck stehende gleichseitige Doppelpyramide ist. Die drei Diagonalen dieses Sechsecks bilden die drei gleichwerthigen Axen der Krystallform, deren vierte Axe von jenen verschieden ist. Um die Gesetze dieser Art von Symmetrie auf ein rechtwinkliges Coordinatensystem zu beziehen, benutzen wir die Formeln [Art. 1203 (*b*)], welche gültig sind, da die beschriebene Krystallform durch drei auf einander rechtwinklig stehende Ebenen symmetrisch theilbar ist. Dazu kommt als zweite Art der Symmetrie, dass eine Drehung um 60° zu einer von der ursprünglichen nicht unterschiedenen Stellung führt (S. 174).

Turning the axes of x, y through 60° round z, calculating the corresponding stresses and strains and causing them to have relations of the same form as in Art. 1203 (*b*), we find that we must have :

$$f'' = f', \quad a = b, \quad e = d, \quad 2f = a - f', \quad d'' = e'' \quad \text{and} \quad d = e.$$

We thus obtain the system :

$$\widehat{xx} = (2f + f')\,s_x + f's_y + e's_z, \qquad \widehat{yz} = e\sigma_{yz},$$
$$\widehat{yy} = f's_x + (2f + f')\,s_y + e's_z, \qquad \widehat{zx} = e\sigma_{zx},$$
$$\widehat{zz} = e''s_x + e''s_y + cs_z, \qquad \widehat{xy} = f\sigma_{xy}.$$

According to Neumann there are thus *six* constants : but Green's principle tells us that $e'' = e'$ also, or leaves only *five* constants.

(f) *Rhombohedral Crystals*.

Neumann remarks that a similar process to that of (e) enables us to obtain formulae for a rhombohedral crystalline system

dessen Grundform als eine doppelte dreiseitige Pyramide aufzufassen ist, jedoch mit einer solchen Bestimmung über das Gesetz der Symmetrie, dass einer Fläche der oberen Pyramide nicht eine gleiche der unteren entspricht, sondern dass eine Fläche der oberen mit einer Kante der unteren Pyramide auf derselben Seite des Krystalles liegt, und umgekehrt. Ein Rhomboëder ist also nicht durch mehrere auf einander rechtwinklige Ebenen symmetrisch theilbar (S. 176).

Neumann takes the chief axis of the crystal for axis of z and the plane of zx perpendicular to the face of one of the pyramids and so that an edge of the second pyramid also lies in it. Then zx is the only coordinate plane which is one of symmetry, and the formulae (a) hold for this case. A rotation, however, of 120° round the axis of z cannot affect the form of the stress-strain relations. This leads to the reduction of (a) to the types:

$$\widehat{xx} = (2f+f')\,s_x + f's_y + e's_z - h\sigma_{zx}, \qquad \widehat{yz} = h''\sigma_{xy} + d\sigma_{yz},$$

$$\widehat{yy} = f's_x + (2f+f')\,s_y + e's_z + h\sigma_{zx}, \qquad \widehat{zx} = h''(s_y - s_x) + d\sigma_{zx},$$

$$\widehat{zz} = e''s_x + e''s_y + cs_z, \qquad \widehat{xy} = f\sigma_{xy} + h\sigma_{yz}.$$

Here Neumann has *eight* constants, but Green's principle shows that $e'' = e'$, and $h'' = h$ or leaves only *six*. If we put $h = h'' = 0$ we obtain the hexagonal system as a particular case.

[1205.] Neumann now turns to some interesting problems on crystals involving the above formulae. These problems have been the starting-point of several important experimental investigations by Voigt, Baumgarten, Coromilas and others, and therefore deserve careful study.

§ 85 (S. 179–81) is entitled: *Zusammendrückung eines Krystalls durch allseitigen Druck.* Let p be the uniform pressure applied to the surface of a crystal, then the surface-stress equations will be satisfied if we take:

$$\widehat{xx} = \widehat{yy} = \widehat{zz} = -p, \qquad \widehat{yz} = \widehat{zx} = \widehat{xy} = 0,$$

and these will obviously satisfy also the body-stress equations. Hence a possible and therefore the only solution is to suppose the shifts linear functions of the coordinates x, y, z. Suppose we take:

$$u = Mx, \qquad v = Ny, \qquad w = Pz,$$

then from Art. 1203, (d) we find for a regular crystal:

$$M = N = P = -p/(a + 2f').$$

Thus the effect of uniform pressure on a regular crystal is only to change its boundary to a similar form.

Suppose we take:

$$u = Mx, \qquad v = My, \qquad w = Pz,$$

and apply these to the equations of Art. 1204 (f) for a rhombohedral crystal, we have:

$$-p = 2(f+f')M + e'P,$$

$$-p = 2e'M + cP,$$

whence:

$$\frac{M}{c-e'} = \frac{P}{2(f'+f-e')} = -\frac{p}{2\{(f'+f)c - e'^2\}}.$$

Thus the contraction of a rhombohedral crystal (or of a hexagonal, since the result does not involve h: see our Art. 1204 (e) and (f)) is different for different directions. A spherical surface becomes an ellipsoid of revolution. It is also possible, Neumann thinks, that M and P may be of opposite sign. Since f and f' are probably not very different from each other and from e' ($f = f'$ on the rari-constant hypothesis), this would seem to involve $c < e'$, or a given stretch s_z would have more effect in producing lateral than longitudinal stresses. Neumann adds :

Dann würde der allseitig gepresste Krystall sich in einer Richtung zusammenziehen, während er sich in einer andern ausdehnt, analog der schönen von Eilhard Mitscherlich gemachten Entdeckung, dass ein Krystall durch Erwärmung sich nicht allein ungleichmässig ausdehnt, sondern sogar in gewissen Richtungen sich zusammenziehen kann (S. 181).

[1206.] Neumann next turns to the still more interesting problem of a crystalline prism in the shape of a right six-face under uniform tractive load on a pair of parallel faces.

Suppose a rectangular coordinate system x, y, z to have relation to the axes of the crystal, and a second ξ, η, ζ to give the directions of the sides of the prism, so that the tractive load T is applied to the faces parallel to $\eta\zeta$ or in the direction of ξ. Then it will be found that the body-stress equations and the surface-stress equations can all be satisfied by taking the stresses equal to the constants as follows :

$$\begin{aligned} \widehat{xx} &= T\cos^2(\xi, x), & \widehat{yz} &= T\cos(\xi, y)\cos(\xi, z), \\ \widehat{yy} &= T\cos^2(\xi, y), & \widehat{zx} &= T\cos(\xi, z)\cos(\xi, x), \\ \widehat{zz} &= T\cos^2(\xi, z), & \widehat{xy} &= T\cos(\xi, x)\cos(\xi, y) \end{aligned} \right\} \dots\dots \text{(i)}.$$

The shifts are thus linear and of the form :

$$\begin{aligned} u &= Mx + p'y + nz \\ &= Mx + \tfrac{1}{2}(p+p')y + \tfrac{1}{2}(n+n')z - \tfrac{1}{2}(p-p')y + \tfrac{1}{2}(n-n')z, \\ v &= px + Ny + m'z \\ &= \tfrac{1}{2}(p+p')x + Ny + \tfrac{1}{2}(m+m')z - \tfrac{1}{2}(m-m')z + \tfrac{1}{2}(p-p')x, \\ w &= n'x + my + Pz \\ &= \tfrac{1}{2}(n+n')x + \tfrac{1}{2}(m+m')y + Pz - \tfrac{1}{2}(n-n')x + \tfrac{1}{2}(m-m')y \end{aligned} \right\} \dots\text{(ii)}.$$

The second method of writing these equations shows that we can only hope to determine the six quantities M, N, P, $\tfrac{1}{2}(m+m')$, $\tfrac{1}{2}(n+n')$, $\tfrac{1}{2}(p+p')$ by substituting in (i) ; for the terms with

$$\tfrac{1}{2}(m-m'), \quad \tfrac{1}{2}(n-n'), \quad \tfrac{1}{2}(p-p')$$

denote merely a rotation of the prism as a whole.

Neumann first considers the case of a regular crystal. We have at once from Art. 1203 (d):

$$T\cos^2(\xi, x) = (a - f')\,M + f'\theta, \qquad d\,(m + m') = T\cos(\xi, y)\cos(\xi, z),$$

$$T\cos^2(\xi, y) = (a - f')\,N + f'\theta, \qquad d\,(n + n') = T\cos(\xi, z)\cos(\xi, x),$$

$$T\cos^2(\xi, z) = (a - f')\,P + f'\theta, \qquad d\,(p + p') = T\cos(\xi, x)\cos(\xi, y),$$

whence

$$M = \frac{T}{a - f'}\left\{\cos^2(\xi, x) - \frac{f'}{a + 2f'}\right\},$$

$$N = \frac{T}{a - f'}\left\{\cos^2(\xi, y) - \frac{f'}{a + 2f'}\right\},$$

$$P = \frac{T}{a - f'}\left\{\cos^2(\xi, z) - \frac{f'}{a + 2f'}\right\}.$$

Let s_r be the stretch in the direction r having direction cosines l_1, m_1, n_1, with regard to x, y, z, then we easily find from our Art. 54*:

$$s_r = \frac{T}{a - f'}\left\{l_1^2\cos^2(\xi, x) + m_1^2\cos^2(\xi, y) + n_1^2\cos^2(\xi, z) - \frac{f'}{a + 2f'}\right\}$$

$$+ \frac{T}{d}\Big\{m_1 n_1\cos(\xi, y)\cos(\xi, z) + n_1 l_1\cos(\xi, z)\cos(\xi, x) + l_1 m_1\cos(\xi, x)\cos(\xi, y)\Big\}\text{(iii)}.$$

This may be compared with Neumann's investigation of 1834: see our Arts. 795*–9* and *Corrigenda* to Vol. I., p. 3, and compare our Art. 309.

Suppose we wish to find s_ξ, then, $l_1 = \cos(\xi, x)$, $m_1 = \cos(\xi, y)$, $n_1 = \cos(\xi, z)$; and after slight reductions we have:

$$s_\xi = T\left\{\left(\frac{1}{a - f'} - \frac{1}{2d}\right)\left(\cos^4(\xi, x) + \cos^4(\xi, y) + \cos^4(\xi, z)\right)\right.$$

$$\left. - \left(\frac{f'}{(a - f')(a + 2f')} - \frac{1}{2d}\right)\right\}\ldots\text{(iv)}.$$

This result was published by W. Voigt as from Neumann's lectures (in *Poggendorffs Annalen, Ergänzungs-Band* VII., S. 5, 1876) and has been experimentally verified for alum by Beckenkamp: *Zeitschrift für Krystallographie*, Bd. 10, S. 41, 1885. From the above equation we find at once Neumann's biquadratic surface for the stretch-modulus $E_\xi\,(= T/s_\xi)$: see our Art. 799*.

In the case in which the prism is cut parallel to an axis of the crystal, we have for the stretch-modulus E_0:

$$E_0 = \frac{(a + 2f')(a - f')}{a + f'},$$

which value of E_0 stands in a simple relation to the dilatation-modulus F (Vol. I. p. 885), deduced from the Equations (iii) of this article as:

$$F = \frac{a + 2f'}{3}.$$

[1207.] Neumann next investigates the directions in which the stretch-modulus as given by (iv) takes maximum or minimum values. To obtain these values Neumann transfers to polar coordinates with the axis of x as polar-axis, or he takes :

$$\cos(\xi, x) = \cos a, \quad \cos(\xi, y) = \sin a \cos \phi, \quad \cos(\xi, z) = \sin a \sin \phi.$$

He then obtains the following directions in which the stretch-modulus is a maximum or minimum, namely the normals to the faces of the following geometrical crystalline forms :

(i) The cube

$$\left.\begin{array}{l} \sin a = 0, \\ \cos a = 0, \quad \cos \phi = 0, \\ \cos a = 0, \quad \sin \phi = 0, \end{array}\right\} \quad \frac{1}{E_C} = \frac{a + f'}{(a - f')(a + 2f')};$$

(ii) The octahedron

$$\left.\begin{array}{l} \tan^2 a = 2, \quad \tan^2 \phi = 1, \end{array}\right\} \quad \frac{1}{E_O} = \frac{1}{3(a + 2f')} + \frac{1}{3d};$$

(iii) The rhombic dodecahedron

$$\left.\begin{array}{l} \tan^2 a = 1, \quad \cos \phi = 0, \\ \tan^2 a = 1, \quad \sin \phi = 0, \\ \cos a = 0, \quad \tan^2 \phi = 1, \end{array}\right\} \quad \frac{1}{E_D} = \frac{a}{2(a - f')(a + 2f')} + \frac{1}{4d}.$$

We easily find :

$$3\left(\frac{1}{E_C} - \frac{1}{E_O}\right) = 4\left(\frac{1}{E_C} - \frac{1}{E_D}\right) = 12\left(\frac{1}{E_D} - \frac{1}{E_O}\right) = \frac{2}{a - f'} - \frac{1}{d},$$

whence the relative magnitudes of E_C, E_O, E_D may be determined according as $2d$ is $>$ or $< a - f'$. Since we have $1/E_C + 3/E_O = 4/E_D$, we cannot determine the three elastic constants a, f', d by ascertaining the values of the stretch-moduli E_C, E_O, E_D.

[1208.] In the following paragraph (§ 89, S. 188—90) Neumann investigates the lateral squeeze which accompanies a longitudinal traction. In this case l_1, m_1, n_1 of our Art. 1206 are subject to the condition :

$$l_1 \cos(\xi, x) + m_1 \cos(\xi, y) + n_1 \cos(\xi, z) = 0\ldots\ldots\ldots(\text{v}),$$

and the formula (iii) of that article can then be thrown into the form :

$$s' = T\left\{\left(\frac{1}{a - f'} - \frac{1}{2d}\right)\left(l_1^2 \cos^2(\xi, x) + m_1^2 \cos^2(\xi, y) + n_1^2 \cos^2(\xi, z)\right)\right.$$
$$\left. - \frac{1}{a - f'}\frac{f'}{a + 2f'}\right\} \ldots(\text{vi}),$$

s' denoting the lateral stretch, which is here a squeeze.

(a) Suppose the prism cut in any way parallel to an axis of the crystal, then two out of the three l_1, m_1, n_1 are zero and we have from (v) and (vi):

$$s' = -\frac{T}{a-f'}\frac{f'}{a+2f'}.$$

Hence the stretch-squeeze ratio is in this case *constant* in *all* directions perpendicular to the traction, and by aid of the value of E_0 in Art. 1206 we see that it is given by:

$$\eta = f'/(a +f').$$

(*b*) Suppose the traction in the direction of a normal to the octahedron, then

$$\cos^2(\xi, x) = \cos^2(\xi, y) = \cos^2(\xi, z) = 1/3,$$

and therefore from (iv):

$$s' = -T\left(\frac{1}{2d} - \frac{1}{a+2f'}\right).$$

This result differs from that given by Neumann on S. 189. The stretch-squeeze modulus is here more complex, but is still constant for all directions in the plane of the cross-section.

(*c*) Suppose the traction in the direction of a normal to the rhombic dodecahedron, then $\cos^2(\xi, x) = \cos^2(\xi, y) = \frac{1}{2}$, and $\cos(\xi, z) = 0$, and if χ be the angle the direction of the squeeze makes with the axis of z we have:

$$s'_\chi = -T\left\{\frac{1}{a-f'}\frac{f'}{a+2f'} - \frac{1}{2}\left(\frac{1}{a-f'} - \frac{1}{2d}\right)\sin^2\chi\right\}.$$

Thus the squeeze varies with the direction in the cross-section.

[1209.] § 90 (S. 190–5) entitled: *Aenderung der Winkel eines regulären Krystalls durch Druck* deduces expressions in terms of the elastic constants for the changes in the angles of a crystalline prism under uniform longitudinal traction. By simple optical methods, which are indicated by Neumann, these changes can be easily measured and we thus have a further means of ascertaining the elastic constants.

Consider a plane whose direction-cosines with regard to x, y, z in the unstrained condition are given by $\cos a_1$, $\cos\beta_1$, $\cos\gamma_1$, and let these after strain become $\cos(a_1 + \delta a_1)$, $\cos(\beta_1 + \delta\beta_1)$, $\cos(\gamma_1 + \delta\gamma_1)$. Then with the notation of our Art. 1206 Neumann easily shows that:

$$\begin{aligned}\cos(a_1 + \delta a_1) &= (\cos a_1 - M\cos a_1 - p\cos\beta_1 - n'\cos\gamma_1)\,q_1,\\ \cos(\beta_1 + \delta\beta_1) &= (\cos\beta_1 - p'\cos a_1 - N\cos\beta_1 - m\cos\gamma_1)\,q_1,\\ \cos(\gamma_1 + \delta\gamma_1) &= (\cos\gamma_1 - n\cos a_1 - m'\cos\beta_1 - P\cos\gamma_1)\,q_1\end{aligned}\right\}\dots\text{(i),}$$

where q_1 is found by squaring these expressions, adding, neglecting the squares of small quantities and taking the root, to be :

$$\begin{aligned}q_1 = 1 &+ M\cos^2 a_1 + N\cos^2\beta_1 + P\cos^2\gamma_1 + (m + m')\cos\beta_1\cos\gamma_1\\ &+ (n + n')\cos\gamma_1\cos a_1 + (p + p')\cos a_1\cos\beta_1\dots\text{(ii).}\end{aligned}$$

Let a second plane be given by $\cos a_2$, $\cos \beta_2$, $\cos \gamma_2$ in the unstrained and $\cos (a_2 + \delta a_2)$, $\cos (\beta_2 + \delta \beta_2)$, $\cos (\gamma_2 + \delta \gamma_2)$ in the strained position, then we have, if σ, σ_0 be the angles between the two planes after and before strain, from equations of the type (i):

$$\cos \sigma = \{\cos \sigma_0 - 2 \, (M \cos a_1 \cos a_2 + N \cos \beta_1 \cos \beta_2 + P \cos \gamma_1 \cos \gamma_2)$$
$$- (m + m') \, (\cos \beta_1 \cos \gamma_2 + \cos \beta_2 \cos \gamma_1)$$
$$- (n + n') \, (\cos \gamma_1 \cos a_2 + \cos \gamma_2 \cos a_1)$$
$$- (p + p') \, (\cos a_1 \cos \beta_2 + \cos a_2 \cos \beta_1)\} \, q_1 q_2 \dots\dots\dots (iii),$$

where q_2 is an expression similar to q_1 in (ii) but involving a_2, β_2, γ_2. Neumann takes only the special case when the planes are originally at right angles and therefore $\sigma_0 = 90°$, $\cos \sigma_0 = 0$. Hence, if $\sigma = \sigma_0 + \delta\sigma$, we may replace $\cos \sigma$ by $-\delta\sigma$, and substituting the values of the constants given in our Art. 1206 we reach the result:

$$\delta\sigma = \frac{2T}{a-f'} \{\cos^2 (\xi, x) \cos a_1 \cos a_2 + \cos^2 (\xi, y) \cos \beta_1 \cos \beta_2$$
$$+ \cos^2 (\xi, z) \cos \gamma_1 \cos \gamma_2\}$$
$$+ \frac{T}{d} \{\cos (\xi, y) \cos (\xi, z) (\cos \beta_1 \cos \gamma_2 + \cos \beta_2 \cos \gamma_1)$$
$$+ \cos (\xi, z) \cos (\xi, x) (\cos \gamma_1 \cos a_2 + \cos \gamma_2 \cos a_1)$$
$$+ \cos (\xi, x) \cos (\xi, y) (\cos a_1 \cos \beta_2 + \cos a_2 \cos \beta_1)\} \quad \dots\dots (iv).$$

Neumann takes two special cases of this:

(i) *Change in angle between the two rectangular faces of a prism which are parallel to the direction of the traction.*

Here:

$$\cos (\xi, x) \begin{Bmatrix} \cos a_1 \\ \cos a_2 \end{Bmatrix} + \cos (\xi, y) \begin{Bmatrix} \cos \beta_1 \\ \cos \beta_2 \end{Bmatrix} + \cos (\xi, z) \begin{Bmatrix} \cos \gamma_1 \\ \cos \gamma_2 \end{Bmatrix} = 0,$$

hence multiplying these together, we have by (iv):

$$\delta\sigma = T \left(\frac{2}{a-f'} - \frac{1}{d} \right) \{\cos^2 (\xi, x) \cos a_1 \cos a_2$$
$$+ \cos^2 (\xi, y) \cos \beta_1 \cos \beta_2 + \cos^2 (\xi, z) \cos \gamma_1 \cos \gamma_2\} \dots\dots\dots (v).$$

If the traction be in the direction of an axis of the crystal or of a normal to the octahedron, it is easy to shew that $\delta\sigma = 0$; if in the direction of a normal to the rhombic dodecahedron we have:

$$\delta\sigma = T \left\{ \frac{1}{2d} - \frac{1}{a-f'} \right\} \cos a_1 \cos a_2,$$

where $a_1 = 90° - a_2$.

(ii) *Change in angle between a loaded face of the prism and a free face.*

Here a_1, β_1, γ_1 are (ξ, x), (ξ, y), (ξ, z) respectively, hence

$$\left.\begin{matrix}1\\0\end{matrix}\right\} = \cos(\xi, x) \begin{Bmatrix}\cos a_1\\\cos a_2\end{Bmatrix} + \cos(\xi, y) \begin{Bmatrix}\cos \beta_1\\\cos \beta_2\end{Bmatrix} + \cos(\xi, z) \begin{Bmatrix}\cos \gamma_1\\\cos \gamma_2\end{Bmatrix},$$

and by multiplying these and using (iv) we find :

$$\delta\sigma = T\left\{\frac{2}{a - f'} - \frac{1}{d}\right\}\{\cos^3(\xi, x)\cos a_2$$

$$+ \cos^3(\xi, y)\cos \beta_2 + \cos^3(\xi, z)\cos \gamma_2\}\ldots\ldots\ldots\ldots(vi).$$

Various special cases are deduced from this general formula, S. 194-5.

[1210.] Neumann next indicates methods of dealing with the like problems in the case of prisms cut from rhombohedral crystals. He takes in § 91 (S. 195-9) the case of a uniform longitudinal tractive load applied to such a prism. We have now to solve equations like (i) of our Art. 1206, when the values of the shifts (ii) in that article are substituted in the stress-strain relations (f) of our Art. 1204. Neumann gives the values of M, N, P, $m + m'$, $n + n'$, $p + p'$ on S. 195, and taking the chief axis of the rhombohedral crystal as polar axis, so that

$$\cos(\xi, z) = \cos\gamma, \quad \cos(\xi, x) = \sin\gamma\cos\phi, \quad \cos(\xi, y) = \sin\gamma\sin\phi,$$

he obtains for the stretch s_ξ in the direction ξ of the traction T as in Art. 1206 :

$$s_\xi = T\left\{\frac{1}{4}\left(\frac{c}{\kappa_1} + \frac{d}{\kappa_2}\right)\sin^4\gamma + \frac{f' + f}{\kappa_1}\cos^4\gamma - \left(\frac{e'}{\kappa_1} - \frac{f}{\kappa_2}\right)\sin^2\gamma\cos^2\gamma\right.$$

$$\left. + \frac{h'}{2\kappa_2}\sin^3\gamma\cos\gamma\cos 3\phi\right\}\ldots\ldots\ldots\ldots\ldots(vii),$$

where $\kappa_1 = c\,(f + f') - e'^2, \quad \kappa_2 = fd - h^2.$

Thus the reciprocal of the stretch-modulus $1/E_\xi\ (= s_\xi/T)$ is given for every direction. Putting $1/E_\xi$ proportional to $1/r^4$, where r is a radius-vector we have a biquadratic surface, the properties of which Neumann discusses at some length (S. 196-9). Perpendicular to the chief axis (z) the equatorial section is a circle ; the section by a plane through the axis of z making an angle of 30° with the axis of x and that by the plane yz are alike and are oval curves of the type

$$1/r^4 = H\sin^4\gamma + I\cos^4\gamma - K\sin^2\gamma\cos^2\gamma.$$

Maxima or minima of r are given by $\phi = 0°$, 60° and 120°; and for $\phi = 0°$ (or for the plane xz) these are investigated by Neumann. It is found that in general there are in that plane three directions of maximum or minimum r. Experiments of Baumgarten on calcspar (*Poggendorff's Annalen*, Bd. 152, S. 369, 1874) and Coromilas on gypsum and mica (*Zeitschrift für Krystallographie*, Bd. I., S. 407) appear to some extent to confirm Neumann's theoretical results. We note from equation (vii), however, that only *four* relations between the six constants of a rhombohedral crystal can be found by pure tractive experiments.

[1211.] The next problem dealt with is that of a rhombohedral crystal under uniform surface pressure (§ 92, S. 199–200). Substitute the values (ii) of Art. 1206 in (f) of Art. 1204 equating the tractions to $-p$ and the shears to zero, we find :

$$-p = (2f+f')\, M + f'N + e'P - h\,(n+n'), \qquad 0 = h\,(p+p') + d\,(m+m'),$$

$$-p = f'M + (2f+f')\, N + e'P + h\,(n+n'), \qquad 0 = h\,(N-M) + d\,(n+n'),$$

$$-p = e'M + e'N + cP, \qquad\qquad\qquad 0 = f\,(p+p') + h\,(m+m').$$

Whence we see that $m + m' = n + n' = p + p' = N - M = 0$ and

$$\frac{M}{c-e'} = \frac{P}{2\,(f+f'-e')} = \frac{-p}{2\,\{(f+f')\,c - e'^2\}} \dots\dots(viii),$$

a result agreeing with that in our Art. 1205.

Further the plane $\dfrac{x}{A} + \dfrac{y}{B} + \dfrac{z}{C} = 1$ is converted by the uniform pres-

sure into the plane $\dfrac{x}{A\,(1+M)} + \dfrac{y}{B\,(1+M)} + \dfrac{z}{C\,(1+P)} = 1$, whence we can easily ascertain the change of angle between any two planes as in our Art. 1209. The dilatation will give the value of $2M + P$, the change in angle can be so taken as to give $M - P$, whence it follows that M and P can be found. These are not functions of the coefficients which occur in (vii) of our Art. 1210. Thus we obtain two further relations to determine the six elastic constants. Neumann, who has *eight* and not *six* constants, does not shew how the remaining two relations are to be found. He concludes this section with the words:

Eine experimentelle Untersuchung dieser Verhältnisse würde auch für die Beantwortung der Frage von Bedeutung sein, ob die Ausdehnung eines Krystalls durch Wärme und seine Zusammenziehung durch Abkühlung denselben Gesetzen folgt, wie seine Formveränderung durch Verminderung oder Steigerung des Druckes. Findet man, dass beide Vorgänge in gleicher Weise vor sich gehen, und dass das Verhältniss von M zu P bei Erwärmung denselben constanten Werth annimmt, welchen es bei Zusammendrückung besitzt, so würde daraus folgen, dass die in §. 58 (see our Art. 1197) zur Definition des thermischen Druckes eingeführten Constanten β_x, β_y, β_z auch in einem Krystalle für drei Axenrichtungen den gleichen Werth besitzen (S. 200).

[1212.] This section of the lectures concludes with a paragraph (§ 93, S. 201–2) added by the Editor and entitled: *Neuere Untersuchungen über die Elasticität der Krystalle.* The Editor remarks that Neumann, besides the problems on crystals considered in the present lecture (see our Arts. 1202–11), has also dealt with the two important problems of the flexure and torsion of small prisms cut in any direction from a crystal: see our Art. 1230. The results of his researches

have been published by W. Voigt, who has himself written many
valuable memoirs based upon Neumann's investigations[1], which fall into
a later period than that of our present volume.

For rock salt Voigt has found (last memoir cited in our footnote) in
terms of the constants for regular crystals given in our Art. 1203, (d):

$$a = 4753, \qquad f' = 1313, \qquad d = 1292,$$

where a stress of one kilogramme per square millimetre is taken as
the unit. Here we have almost. exactly $f' = d$,—i.e. $|xxyy| = |xyxy|$ in the
general constant notation,—thus the additional condition of rari-constancy is nearly satisfied: see our Art. 116, ftn. It is very doubtful
if this holds for all regular crystals. Voigt found for fluorspar:

$$a = 14550, \qquad f' = 2290, \qquad d = 3380,$$

while for the same material Klang with an inferior theory and method
gave (*Wiedemanns Annalen*, Bd. 12, S. 331, 1881):

$$a = 13200, \qquad f' = 4250, \qquad d = 3300.$$

Both observers therefore agree in the inequality of f' and d, but in
opposite senses. Although numerous other regular crystals have had
their elasticity investigated, there is still scarcely material enough for
us to consider the multi- or rari-constancy of crystalline structures
as finally determined[2].

[1213.] The next five sections of Neumann's lectures (S.
203–299) are entitled: *Gesetze für die Fortpflanzung ebener Wellen*,
and belong properly to the History of the Undulatory Theory of
Light. We shall therefore here only briefly refer to their
contents. Neumann proceeds in his usual semi-historical method
and with his characteristic clearness; hence these hundred pages,
accompanied as they are by editorial references to later work, are
most instructive, and the student is hardly likely to find a better
introduction to the elastic jelly theory of the ether.

[1214.] § 13 (S. 203–40) is entitled: *Theorie der Wellenbewegungen
auf Grund der Molekularhypothese* (Lecture Notes of 1857–8). This
deals with the laws of polarisation and double-refraction of light as
previously investigated by Cauchy (*Mémoires de l'Académie*, T. x p. 293.
Paris, 1831) and Neumann himself (*Poggendorffs Annalen*, Bd. 25,
S. 418, 1832) on the *rari-constant* hypothesis. Neumann explains

[1] *Poggendorffs Annalen*, Ergänzungs-Band VII. S. 1 u. 177, 1876; *Wiedemanns
Annalen*, Bd. 16, S. 273, 398 u. 416, 1882; *Sitzungsberichte d. Berliner Akademie*,
Zweiter Halbband, 1884, S. 989–1004, and many others of later date.

[2] An important *assumption* is indeed made by Neumann and others : namely
that crystals really fall crystallographically and elastically into the *same* classes.
For example, is it *a priori* certain that a regular crystallographic crystal is a regular
" elastic crystal "?

in the following words why he starts from this narrow basis, instead of the more general crystalline equations which he has given in the previous section :

Wir thun das nicht nur in Rücksicht auf die geschichtliche Entwickelung der Theorien, welche uns jetzt beschäftigen werden, sondern auch deshalb, weil es nicht nothwendig ist anzunehmen, dass die Bewegungen des Lichtäthers in Krystallen genau denselben Kräften und Gesetzen unterliegen, wie die wägbare Substanz des Körpers selbst (S. 203).

Neumann starts from elastic equations involving only *six* constants, or from the equations of the stresses as given in our Art. 1203 (*b*), where on the rari-constant hypothesis all the accents are to be removed, i.e. $e = e' = e''$, $f = f' = f''$ and $d = d' = d''$. He determines in the usual manner three pairs of waves, each pair having a different velocity but its members consisting of waves propagated with the same velocity in opposite directions. He shows (S. 207–8) how the arbitrary functions may be determined in terms of the initial disturbance; and further, how for each pair of waves the direction of the shift is different, but for the same pair is the same at all places and for all times, and is independent of the initial disturbance (S. 210–11). He then investigates (S. 211–13) the ellipsoid of wave-propagation (*Fortpflanzungsellipsoid*), and discusses some interesting general problems of wave-motion (S. 215–23), concluding this part of the section with a determination of the wave-velocities (S. 225). He next turns to the more purely optical applications of his formulae, especially to Fresnel's laws of double-refraction and of the polarisation of light in crystals. He remarks that (on the *rari-constant* hypothesis) the formulae of our Art. 1203 (*b*) must for various optical media be thus simplified :

(*a*) Uncrystalline medium $a = b = c = 3d = 3e = 3f$,⎫   ⎧$e = e' = e''$,
(*b*) Regular crystal $a = b = c$, $d = e = f$,   ⎬ and ⎨$f = f' = f''$,
(*c*) Uniaxial crystal $a = b$, $d = e$,   ⎭   ⎩$d = d' = d''$
(*d*) Biaxial crystal $e = e' = e''$, $f = f' = f''$, $d = d' = d''$.

After an investigation of wave-motion in uncrystalline media (S. 227–8), Neumann deduces Fresnel's laws for the velocity of propagation in biaxial crystals, *provided the plane of polarisation be defined as the plane through the direction of the wave and the direction of the vibration.* The plane of polarisation is thus perpendicular to that given by Fresnel's definition, and the above is usually spoken of in Germany as Neumann's definition (*Neumann'sche Definition*) of the plane of polarisation. The deduction of Fresnel's laws even with this definition requires the following relations to hold among the six elastic constants (see our Art. 148) :

$$(c - d)(b - d) = 4d^2, \quad (a - e)(c - e) = 4e^2, \quad (b - f)(a - f) = 4f^2;$$

which, neglecting the squares of the differences of d, e, f, may be replaced by :

$$a = 3(e + f - d), \quad b = 3(f + d - e), \quad c = 3(d + e - f),$$

see our Art. 1199. Neumann remarks that the differences of d, e, f are
small as they depend on the differences of the refractive indices of the
two rays[1], and neglecting the differences of d, e, f he demonstrates
the transversality of two of the waves (S. 229–240). Of the third or
longitudinal wave he gives no physical account in this section.

[1215.] In Section 14 entitled : *Theorie der Lichtwellen im incom-
pressiblen Aether* (S. 241–56; Lecture Notes of 1859–60) F. Neumann
follows Carl Neumann[2] in supposing the ether incompressible and so
disposing of the longitudinal wave. He remarks :

Die strenge Durchführung der auf der Hypothese der Incompressibilität
des Lichtäthers beruhenden Rechnung wird uns nur auf transversale Wellen
führen und Resultate liefern, die sich mit den Resultaten der Beobachtung in
vollkommener Coincidenz befinden (S. 241–2).

This seems too strong a statement.

Carl Neumann's equations are of the type :

$$\rho \frac{d^2u}{dt^2} = a \frac{d^2u}{dx^2} + f \frac{d^2u}{dy^2} + e \frac{d^2u}{dz^2} + 2f \frac{d^2v}{dxdy} + 2e \frac{d^2w}{dxdz} - \frac{dp}{dx},$$

subject to the condition

$$\frac{du}{dx} + \frac{dv}{dy} + \frac{dw}{dz} = 0.$$

Besides involving this condition, the theory introduces the terms

$$\frac{dp}{dx}, \frac{dp}{dy}, \frac{dp}{dz}$$

into the body-shift equations of the previous section. F. Neumann
remarks of these equations :

Die neu eingeführte Grösse p hat eine bestimmte physikalische Bedeutung.
Sie ist der durch die Bewegung entstehende *hydrostatische Druck*. Denn die
verbesserten Differentialgleichungen sind im Grunde nur die hydrodyna-
mischen, in welche ausser den äusseren Kräften noch die inneren Molekular-
kräfte eingeführt sind, während die in jenen Gleichungen vorkommenden
Quadrate der Geschwindigkeiten dem Grundprincipe der Elasticitätstheorie
entsprechend fortfallen (S. 245).

In the course of the section it is shown that :

$$p = \text{a constant} + 3 \left(d \frac{du}{dx} + e \frac{dv}{dy} + f \frac{dw}{dz} \right),$$

[1] These differences do not always seem small, e.g. in the case of Iceland Spar.
Neumann's reasoning here does not seem by any means conclusive.
[2] The Editor (p. vii) distinctly states that Section 14 is taken from the Lecture-
Notes of 1859–60. In the section itself there is reference to Carl Neumann as the
propounder of the theory of an incompressible ether, but the earliest reference
(S. 241) to a paper by him is 1863 (*Die magnetische Drehung der Polarisationsebene
des Lichtes*, Halle, 1863). Had Carl Neumann communicated the idea to
F. Neumann before 1859?

so that for a non-crystalline medium $p = a$ constant :

und dieser constante Werth muss, $p=0$, gleich dem Werthe im Welten-
raume sein, da wir uns den letzteren nicht als zusammengedrückt vorstellen
können (S. 253).

It does not, however, seem more difficult to suppose the ether in
space under high pressure, than to suppose it rigidly fixed at an
infinite distance as required by the recent theory of Sir William
Thomson (*Phil. Mag.* November, 1888).

F. Neumann deduces the laws of Fresnel from the above type of
body-shift equations, provided the same relations as in the previous
section hold among the elastic constants (see our Art. 1214) and
provided the plane of polarisation is parallel to the vibrations. The
advance made by C. Neumann's hypothesis is confined to the disappearance of the longitudinal wave and to the exact transversality
of the other two waves. F. Neumann (see his S. 252) seems especially
satisfied with the hypothesis of an incompressible ether and he remarks
with regard to the coincidence of the planes of polarisation and vibration :

Hierauf ist besonderes Gewicht zu legen, weil die entgegengesetzte Ansicht
noch sehr verbreitet ist. Alle strengen Durchführungen der Theorie aber
führen zu dem von der gewöhnlichen Meinung abweichenden Resultate.

Some judicious remarks, due I think to Neumann's Editor, occur on
S. 256. We ought not to expect the same relations necessarily to hold
for the elastic constants of the ether in a crystal as hold for the elastic
constants of the crystalline material itself. The vibrations of the
material and of the ether may follow quite different laws. For example, an *optically* uniaxial crystal possesses the same optical elasticity
for *all* directions perpendicular to the optic axis. Hence for the ether
in such a crystal, Neumann's Editor considers that the relations $e = f$,
$b = c = 3d$, must hold. But such relations do not hold for the elastic
material of the crystal itself, for were its elasticity the same in all
directions round the axis, its crystalline form could only be cylindrical.

[1216.] Section 15 is entitled : *Theorie transversaler Wellen in
Krystallen* (S. 257-75). Neumann here deals with Lamé's theory of
the ether in crystals[1]. He remarks that the theories we have previously considered have been theoretically deduced only for media
symmetrical about three rectangular planes, but experience shows that
Fresnel's laws are true also for unsymmetrical crystalline forms. Such
crystals may have three rectangular axes, in relation to which the
medium is optically symmetrical. Hence a generalisation of our theory
is very desirable. Neumann gives Lamé's theory with considerable

There is a mere footnote reference to Green's memoir of 1839: see our
Art. 917*.

modifications and I think improvements. The hypotheses from which he starts are the following: (a) the medium possesses the property of propagating *plane* waves; (b) accurately transverse vibrations are possible in it; (c) there are stresses in it which arise and disappear with the strains; the stresses obey the general laws of statical equilibrium; and (d) they are assumed to be linear functions of the strains.

These hypotheses enable us to reduce the 36 constants to 12, and Neumann puts the body-shift equations into the form :

$$\rho \frac{d^2u}{dt^2} = A \frac{d\theta}{dx} + C_1 \frac{d\theta}{dy} + B_1 \frac{d\theta}{dz} + \frac{dV}{dz} - \frac{dW}{dy},$$

$$\rho \frac{d^2v}{dt^2} = C_1 \frac{d\theta}{dx} + B \frac{d\theta}{dy} + A_1 \frac{d\theta}{dz} + \frac{dW}{dx} - \frac{dU}{dz},$$

$$\rho \frac{d^2w}{dt^2} = B_1 \frac{d\theta}{dx} + A_1 \frac{d\theta}{dy} + C \frac{d\theta}{dz} + \frac{dU}{dy} - \frac{dV}{dx},$$ ....(i),

where
$$U = \alpha \tau_{yz} - \gamma \tau_{zx} - \beta \tau_{xy},$$
$$V = - \gamma \tau_{yz} + b \tau_{zx} - \alpha \tau_{xy},$$
$$W = - \beta \tau_{yz} - \alpha \tau_{zx} + c \tau_{xy}$$

τ_{yz},...... expressing the twists, $\frac{1}{2} \left(\frac{dw}{dy} - \frac{dv}{dz} \right)$,...... and θ the dilatation as usual (S. 261). Neumann apparently treats these equations as if we had 12 independent constants, but if we apply Green's principle to the stresses on S. 261 we find :

$$A = B = C, \quad A_1 = B_1 = C_1 = 0 ....(ii),$$

or, we have only *seven* independent constants.

Neumann (S. 262–7) shows that by a transformation of the coordinate axes we can get rid either of the coefficients A_1, B_1, C_1 or of the coefficients α, β, γ, and that the axes for which these groups respectively vanish are not necessarily the same. He remarks :

Unsere Betrachtung führt uns also auf eine Doppelnatur des Mediums, insofern nämlich, als Eigenschaften verschiedener Art sich auf verschiedene Axensysteme beziehen können, ein Punkt, auf den wir zurückkommen (S. 266).

Neumann later in his work makes a considerable point of this double system of axes, but it seems to me that if the principle of the conservation of energy applies to the ether in a crystal, then (ii) must hold and so $A_1 = B_1 = C_1 = 0$ for *all* axes.

The type of body-shift equations, when we transform them so that α, β, γ vanish, is given by :

$$\rho \frac{d^2u}{dt^2} = A \frac{d\theta}{dx} + C_1 \frac{d\theta}{dy} + B_1 \frac{d\theta}{dz} + b \frac{d\tau_{zc}}{dz} - c \frac{d\tau_{xy}}{dy} ....(iii).$$

[1217.] In § 122 (S. 267–8) Neumann deals with the 'longitudinal wave'; it is not exactly longitudinal unless (ii) holds, but is marked by the existence of a dilatation θ; it would thus be better termed the *pressural wave*. In § 123 (S. 269–71) the problem of transverse waves is dealt with. The wave surface deduced is accurately Fresnel's, and his laws are shown to be absolutely correct *provided we accept Neumann's definition of the plane of polarisation*. In § 124 (S. 272) Neumann supposes the ether incompressible and puts $\theta = 0$; he then introduces terms into equations (iii) corresponding to a hydrostatic pressure and writes them in the form:

$$\rho \frac{d^2u}{dt^2} = b \frac{d\tau_{zx}}{dz} - c \frac{d\tau_{xy}}{dy} - \frac{dp}{dx} \dots\dots\dots\dots\dots(iv).$$

These practically agree in form with Carl Neumann's equations, only p has a slightly different meaning. F. Neumann seems to see in such equations a completely satisfactory system giving only two waves and these with purely optical properties:

welche in jeder Hinsicht den durch die Erfahrung gelieferten Gesetzen entsprechen (S. 272).

[1218.] We have considered somewhat at length Neumann's treatment of Lamé's theory because in § 125 (S. 272–4) he takes the double system of axes of Art. 1216 as the basis for some important considerations with regard to the different kinds of axes in crystals. He remarks that the properties of the longitudinal and transverse waves seem to depend upon different systems of rectangular axes, and hence he argues that *the different physical properties of a crystal can be distributed symmetrically about different systems of rectangular planes.* We have seen (Art. 1206) that so far as this argument is based on there being different sets of axes for longitudinal and transverse waves, it is only valid if we suppose the ether in a crystal *not to obey Green's Principle.* Neumann thinks that these systems of axes will fall together only when the material of the crystal is symmetrical about three rectangular planes, in which case, he adds, experience shows that the optical elastic axes[1] coincide with those of other kinds of physical symmetry.

If this triple-plane symmetry does not exist in the crystal, it is still theoretically possible that the various systems of axes may coincide, but this is not the result of experiment, so far as concerns at least the optical and thermal axes (axes of greatest and least stretch by heat). Neumann here refers to his memoir of 1833: see our Art. 788*, in which he had shown that the difference between the thermal and optical axes was for gypsum not sufficiently great to be measurable, but he

---

[1] I use the term 'optical axes' for the three rectangular axes about which the ether in a crystal is optically, or, on the elastic theory, elastically symmetrical. They are not necessarily the elastic axes of the crystalline material, and must not be confused with the 'optic axes' or normals to the circular sections of the optical 'ellipsoid of elasticity.' The optical axes are the axes of this ellipsoid.

had in a later research (*Poggendorffs Annalen*, Bd. 35, S. 81–95[1] and S. 203–5, 1835) convinced himself that this result could only be an approximation to the truth:

Bei einer genaueren Untersuchung der von Mitscherlich entdeckten Thatsache, dass die optischen Axen eines zweiaxigen Krystalls ihre Lage gegen einander bei Erwärmung oder Abkühlung verändern, bemerkte der Verfasser, dass die beiden Axen sich mit ungleicher Geschwindigkeit bewegen, und machte damit die Entdeckung, dass nicht bloss die beiden Richtungen einfacher Lichtbrechung, sondern auch das rechtwinklige Axensystem, von welchem die optischen Eigenschaften abhängen, eine mit der Temperatur veränderliche Lage im Krystall hat. Hieraus folgt, dass die optischen Elasticitätsaxen nicht bei jeder Temperatur mit den thermischen zusammenfallen können. Es giebt also in der That in derartigen Krystallen zwei Axensysteme verschiedener Richtung (S. 273).

Our statement therefore of Neumann's results in Art. 792* must be corrected in the sense of this later conclusion of the same scientist.

[1219.] Neumann then briefly refers to the other systems of crystalline axes which have been investigated. Plücker found that the diamagnetic phenomena in crystals depend upon their optical axes (see *Poggendorffs Annalen*, Bd. 72, S. 315–50, 1847, and Plücker u. Beer: Bd. 81, S. 115–62, 1850, Bd. 82, S. 42–74, 1851), but Ångström has shown that the chief axes of conduction of heat (directions of maximum and minimum capacity for propagation of heat), which also form a rectangular system, do not coincide with the optical axes (see our Art. 685). Similar results were obtained by Sénarmont (*Comptes rendus*, T. xxv. pp. 459–61, 1847, *Annales de chimie*, T. xxi. pp. 457–470, 1847 and T. xxii. pp. 179–211, 1848[2]). The same want of coincidence appears to be true for the axes of electrical conduction (Wiedemann: *Poggendorffs Annalen*, Bd. 76, S. 404–12, 1849, Bd. 77, S. 534–7, 1849; Sénarmont: *Annales de chimie*, T. 28, pp. 257–78, 1850), of distribution of hardness (see our Art. 685) and of atmospheric disintegration (Pape: *Poggendorffs Annalen*, Bd. 124, S. 329–36, 1865, Bd. 125, S. 513–63, 1865, Bd. 133, S. 364–99, 1868, and Bd. 135, S. 1–29, 1868), which have all relation to differently situated systems of axes.

Bei dieser Verwickelung der Verhältnisse erscheint es als das Wahrscheinlichste, dass die veränderliche Lage aller dieser Axensysteme von einem anderen festen Axensystem abhängt, falls nicht schon eins der genannten jenes vermuthete feste ist (S. 274).

[1] This paper is entitled: *Ueber die optischen Eigenschaften der hemiprismatischen oder zwei- und eingliedrigen Krystalle* and Neumann shows in it that there is a 'dispersion' of the optic axes of elasticity (in Fresnel's sense of the word). Thus each colour has its own axes not only in magnitude but in position (gypsum). Further these axes change with the temperature and each differently (gypsum, borax, adularia).

[2] Sénarmont shows that in gypsum and crystals of the unsymmetrical prismatic system there is no simple relation between the position and magnitude of the thermal axes and the axes of optical elasticity.

[1220.] The final paragraph of this Section (S. 274–5) is entitled :
Ueber die Aenderung der optischen Axen mit der Temperatur. This is
an attempt to explain the alteration of the optical axes with the
temperature on the basis of Lamé's theory as Neumann has developed
it. It assumes not only that the formulae of Lamé hold for the motion
of the ether in crystals, but also that equations of the same form hold
for the elastic deformation of the crystalline material. Further it
supposes a change of temperature to have the same effect as a uniform
surface pressure : see our Arts. 875* and 1196. Suppose p to be this
pressure equivalent in effect to the temperature change. Neumann
holds that the axes for which the a, β, γ of our Art. 1216 vanish are
the optical axes. Then we must have stress relations corresponding to
the body-shift equations of type (iii) in our Art. 1216 ; these give us :

$$-p = A\theta - cs_y - bs_z, \qquad 0 = A_1\theta + \tfrac{1}{2}a\sigma_{yz},$$

$$-p = B\theta - as_z - cs_x, \qquad 0 = B_1\theta + \tfrac{1}{2}b\sigma_{zx},$$

$$-p = C\theta - bs_x - as_y, \qquad 0 = C_1\theta + \tfrac{1}{2}c\sigma_{xy}.$$

To satisfy these take

$$u = H_1 x + h_3 y + h_2 z,$$

$$v = h_3 x + H_2 y + h_1 z,$$

$$w = h_2 x + h_1 y + H_3 z.$$

We easily find :

$$\theta = \frac{(a^2 + b^2 + c^2 - 2ab - 2bc - 2ca)\,p}{aA\,(b + c - a) + bB\,(c + a - b) + cC\,(a + b - c) - 2abc},$$

$$h_1 = -\frac{A_1\theta}{a}, \qquad h_2 = -\frac{B_1\theta}{b}, \qquad h_3 = -\frac{C_1\theta}{c}.$$

Hence, Neumann remarks, since h_1, h_2, h_3 do not in general vanish,
the axes of the stretch ellipsoid corresponding to u, v, w do not in
general coincide with the axes of coordinates, i.e. the optical axes.
Now the axes of this stretch-ellipsoid are the only lines which do not
alter their position with the dilatation, so that it follows :

dass durch einen allseitigen Druck auf die Oberfläche eines Krystalls die
optischen Hauptaxen desselben und also auch die optischen Farbenaxen ihre
Lage ändern werden (S. 275).

Thus on the assumption of identity in effect between pressure and
temperature, the proposition appears proved. We observe, however,
that the principle of energy seems to require, $A_1 = B_1 = C_1 = 0$, and that
then the optical and crystalline elastic axes would *coincide*. As they
do not, it seems probable that the assumption that the formulae of
Lamé hold both for the elastic deformation of a crystalline material
and for its bound ether is incorrect.

[1221.] Neumann next turns to the problem of *dispersion*. In Section 16 (S. 276–89) he gives very clearly and concisely Cauchy's explanation of the dispersion of light. He remarks, however; that Cauchy's theory would lead us to expect dispersion as well in gases and in space itself as in solid and fluid bodies, since the dispersion depends only on the action of the more distant particles of ether and not on that of the particles of matter. Neumann himself in 1841 (*Die Gesetze der Doppelbrechung des Lichtes...*, *Abhandlungen der Berliner Akademie d. Wissenschaft*, 1841, *Zweiter Theil* (Footnote), S. 28–32) was among the first to attribute dispersion to the influence of the ponderable particles on the particles of ether. (O'Brien, as Neumann's Editor remarks, had reached almost simultaneously the same explanation : see his *On the Propagation of Luminous Waves in the interior of Transparent Bodies : Cambridge Philosophical Transactions*, Vol. VII. p. 397, 1842.) Accordingly Neumann in Section 17 (S. 290–9) develops his own theory of dispersion, as depending on the action of the ponderable particles. He considers only the case of an uncrystalline medium. The general remarks § 136 (S. 296–7) are given just as they were delivered in the lectures of 1857–8, and at that time they were full of suggestion for further researches in dispersion based upon Neumann's theory. Such researches, inspired doubtless by Neumann's work, have been made by Ketteler, Sellmeier, Lommel, Voigt and others (S. 137), but these fall far beyond our limits and we must refer the reader for their discussion to the *Report on Optical Theories* by Glazebrook published in the *British Association Report* for 1885.

In concluding my brief analysis of Neumann's application of the theory of elasticity to light, I must again express my sense of its value and clearness. As an introduction to elastico-optic theories for the use of students the lectures of 1857–9 seem to me still unequalled.

[1222.] Sections 18–21 (S. 300–74) are entitled: *Gesetze der Bewegungen dünner Körper* and deal with strings, membranes and rods. Section 18 (S. 300–17) deals with the vibrations of strings. We may note several points in this section.

(*a*) Neumann obtains for a perfectly flexible string—defined as a body of prismatic form, which is so thin that we can take at every point of one and the same cross-section the same value of the molecular forces —the following equations (S. 303):

$$\omega\rho\left(\frac{d^2u}{dt^2} - X\right) = \frac{d}{ds}\left(\omega \cdot \widehat{xx}\,\frac{ds}{dx}\right),$$

$$\omega\rho\left(\frac{d^2v}{dt^2} - Y\right) = \frac{d}{ds}\left(\omega \cdot \widehat{xx}\,\frac{ds}{dx}\frac{dy}{dx}\right), \left.\rule{0pt}{60pt}\right\} \quad\ldots\ldots\ldots\ldots(i).$$

$$\omega\rho\left(\frac{d^2w}{dt^2} - Z\right) = \frac{d}{ds}\left(\omega \cdot \widehat{xx}\,\frac{ds}{dx}\frac{dz}{dx}\right)$$

Here ω is the cross-section, ρ the density, ds an element of length of the string, and X, Y, Z the components of body-force per unit mass upon it. Among the stresses Neumann finds the following relations to hold:

$$\frac{\widehat{xx}}{(dx)^2} = \frac{\widehat{yy}}{(dy)^2} = \frac{\widehat{zz}}{(dz)^2} = \frac{\widehat{yz}}{dydz} = \frac{\widehat{zx}}{dzdx} = \frac{\widehat{xy}}{dxdy} \quad\ldots\ldots\ldots\ldots\text{(ii)}.$$

(b) For the special case of a string without body-forces, stretched in the direction of the axis of x, we may neglect for small oscillations the squares of dy/dx and dz/dx. Hence it follows that $\widehat{yy} = \widehat{zz} = 0$ approximately, and

$$\widehat{xx} = E\, du/dx\,;$$

whence we reach for a uniform cross-section the equations (S. 304):

$$\left.\begin{array}{l}
\rho\,\dfrac{d^2u}{dt^2} = E\,\dfrac{d^2u}{dx^2}, \quad \rho\,\dfrac{d^2v}{dt^2} = E\,\dfrac{d}{dx}\left(\dfrac{du}{dx}\dfrac{dv}{dx}\right), \\[3mm]
\rho\,\dfrac{d^2w}{dt^2} = E\,\dfrac{d}{dx}\left(\dfrac{du}{dx}\dfrac{dw}{dx}\right)
\end{array}\right\} \quad\ldots\ldots\ldots\text{(iii)}.$$

There is a good deal that seems original and valuable about Neumann's deduction of these equations.

(c) S. 305–311 are occupied with a discussion of the wave-motion involved in an equation of the type $d^2u/dt^2 = a^2\,d^2u/dx^2$. Neumann then passes to the vibrations of a stretched string consisting of two diverse pieces, and deals with the problems of the wave reflection and 'refraction' which occur at the junction. The work is clear but does not present anything of special note.

[1223.] The nineteenth section (S. 318–31) treats of the vibrations of a stretched membrane. Neumann's deduction of the equations is very similar to Lamé's: see our Arts. 1072*–6*. He deals with several simple problems and then discusses the nodal lines of square membranes. His treatment here again corresponds closely to Lamé's: see also our Art. 825 (e) and Lord Rayleigh's *Theory of Sound*, Vol. I. pp. 250–92.

[1224.] The twentieth section is entitled: *Theorie des geraden Stosses cylindrischer Stäbe* (S. 332–50). Neumann remarks that the ordinary theory of impact between elastic bodies is given in mechanical text-books as if it had a simple and correct basis. He reproduces it but without the Newtonian modification to account for the loss of energy: see our Arts. 35* and 217. Neumann remarks that such loss of energy generally does take place and that investigations based on the theory of elasticity lead to results often in direct contradiction with those of the ordinary Newtonian theory. Neumann then proceeds to investigate the longitudinal impact of two right-circular cylinders. He does not at first reduce the problem to the simple case of the impact of two thin rods, the particular problem which was later dealt with by Saint-Venant : see our Arts. 203–20.

Let $r\chi$ be the radial shift at a point distant r from the axis and x from one end of the cylinder, u the corresponding longitudinal shift, then we have the following equations to determine χ and u for one cylinder :

$$\left.\begin{aligned}
\rho\frac{d^2u}{dt^2} &= (\lambda + 2\mu)\frac{d^2u}{dx^2} + \mu\left(\frac{d^2u}{dr^2} + \frac{1}{r}\frac{du}{dr}\right) + (\lambda + \mu)\left(r\frac{d^2\chi}{dxdr} + 2\frac{d\chi}{dx}\right), \\
\rho\frac{d^2\chi}{dt^2} &= \mu\frac{d^2\chi}{dx^2} + (\lambda + 2\mu)\left(\frac{d^2\chi}{dr^2} + \frac{3}{r}\frac{d\chi}{dr}\right) + (\lambda + \mu)\frac{1}{r}\frac{d^2u}{dxdr}
\end{aligned}\right\}\ \dots\dots(\text{i}).$$

A similar pair with different dilatation-coefficient and slide-modulus hold for the second cylinder.

Further at the curved surfaces of the cylinders we must have conditions of the type :

$$\widehat{xr} = \mu\left(\frac{du}{dr} + r\frac{d\chi}{dx}\right) = 0, \quad \widehat{rr} = (\lambda + 2\mu)\,r\frac{d\chi}{dr} + 2\,(\lambda + \mu)\,\chi + \lambda\frac{du}{dx} = 0,\dots(\text{ii}),$$

while at the terminal cross-sections of both cylinders we must have the stresses of type :

$$\left.\begin{aligned}
\widehat{xx} &= (\lambda + 2\mu)\frac{du}{dx} + \lambda\left(r\frac{d\chi}{dr} + 2\chi\right), \\
\widehat{xr} &= \mu\left(\frac{du}{dr} + r\frac{d\chi}{dx}\right)
\end{aligned}\right\}\ \dots\dots\dots\dots\dots(\text{iii}),$$

either zero, or equal for the two cylinders at their common surface.

This is the most general statement of the problem, u and χ and their time fluxions being supposed initially given.

[1225.] Neumann does not solve the problem in all its generality. He assumes first :

$$\left.\begin{aligned}
u &= u_0 + u_1r + u_2r^2 + \dots\dots + u_nr^n + \dots\dots, \\
\chi &= \chi_0 + \chi_1r + \chi_2r^2 + \dots\dots + \chi_nr^n + \dots\dots
\end{aligned}\right\}\ \dots\dots\dots\dots(\text{iv}),$$

where $u_0,\ u_1,\ u_2,\dots\chi_0,\ \chi_1,\ \chi_2\dots$ are functions only of x and the time. The substitution of (iv) in (i), shows that all the coefficients of odd

powers of r must vanish. Neumann then contents himself with values
of the form :

$$\left.\begin{array}{l} u = u_0 + u_2 r^2, \\ \chi = \chi_0 + \chi_2 r^2 \end{array}\right\} \quad\ldots\ldots\ldots\ldots\ldots\ldots\ldots(v),$$

which he seems to think will be approximately true if the cylinders be
thin enough. There seems to me to be exactly the same strong
objections to this method of treatment as Saint-Venant has raised
against Cauchy's method of dealing with the problem of torsion : see
our Arts. 661*, 29 and 395.

Neumann says the terms in r^2 will give a first approximation if the
cylinder be thin, but he does not *justify* this statement. Why should
not u_4 or u_6 be large as compared with u_2? This case is what actually
occurs in Cauchy's attempt to investigate the torsion of a rectangular
prism by an expansion of this kind (Saint-Venant, *Leçons de Navier*,
pp. 621–6, footnote). We have *a priori* nothing to show that the
arbitrary functions u_0, u_2..., χ_0, χ_2... do not vary *inversely* as the
dimensions of the cross-section. Clearly the ratio u_{2m}/u_{2m-2} as to its
order is the inverse square of a line, but for aught Neumann says to
the contrary this line may be the radius of the cylinder and not its
length.

If we substitute (v) in (i) and (ii) and *now neglect terms in* r^2
we find :

$$\left.\begin{array}{l} \rho \dfrac{d^2 u_0}{dt^2} = (\lambda + 2\mu) \dfrac{d^2 u_0}{dx^2} + 4\mu u_2 + 2(\lambda + \mu) \dfrac{d\chi_0}{dx}, \\[2mm] \rho \dfrac{d^2 \chi_0}{dt^2} = \mu \dfrac{d^2 \chi_0}{dx^2} + 8(\lambda + 2\mu)\chi_2 + 2(\lambda + \mu)\dfrac{du_2}{dx}, \\[2mm] 0 = 2u_2 + \dfrac{d\chi_0}{dx}, \\[2mm] 0 = 2(\lambda + \mu)\chi_0 + \lambda \dfrac{du_0}{dx} \end{array}\right\} \quad\ldots\ldots(vi).$$

Eliminating χ_0, u_2, we have from the first of (vi) :

$$\rho \frac{d^2 u_0}{dt^2} = E \frac{d^2 u_0}{dx^2} \quad\ldots\ldots\ldots\ldots\ldots\ldots(vii),$$

the ordinary equation for the longitudinal vibrations of a thin rod.

Further : $\widehat{xr} = \mu \left(2u_2 + \dfrac{d\chi_0}{dx}\right)r, = 0$ by (vi),

and $\widehat{rr} = \left\{(4\lambda + 6\mu)\chi_2 + \lambda \dfrac{du_2}{dx}\right\} r^2.$

Thus $\widehat{rr} = 0$, at the surface, if we again neglect the square of the
external radius in the stresses, which Neumann appears to think we
may do to the required degree of approximation. The relation between

χ_2 and u_2 cannot be so chosen as to make the terms in r^2 vanish. Neumann says (S. 340):

dass auch den Bedingungen für die Cylinderfläche insoweit genügt wird, als es bei dem erstrebten Grade der Annäherung erforderlich ist.

Thus his degree of approximation is not really to r^2.
Further we have:

$$\widehat{xx} = E\frac{du_0}{dx} + r^2\left\{(\lambda + 2\mu)\frac{du_2}{dx} + 4\lambda\chi_2\right\},$$

or, again neglecting the terms in r^2:

$$\widehat{xx} = E\frac{du_0}{dx} \dots\dots\dots\dots\dots\dots\dots\dots(\text{viii}).$$

Thus Neumann reaches in (vii) and (viii) the ordinary equations for the longitudinal vibrations of thin rods, but I do not see that his process gives these equations with any greater accuracy or any less degree of assumption than the usual one. Proceeding from these equations he deals on S. 340-9 with the longitudinal impact of two free rods and of one fixed and one free rod. This section is taken from the Lecture Notes of 1857-8, and thus Neumann's discussion of the problem precedes Saint-Venant's by ten years; but although he reaches some of Saint-Venant's results, his processes, analytical and graphical, are far less complete, and his discussion more special. In view of the excellence of Saint-Venant's work and the space we have devoted to it, we pass by Neumann's pages with the mere recognition of his priority. A reference to experimental work in this field added by his Editor, does not cover much more ground than our Arts. 203-4, 210 and 214.

[1226.] The twenty-first and last section of Neumann's work (S. 351-74) deals with the elasticity of thin rods. It belongs to the Lecture Notes of the years 1859-60. The Editor considers this portion of Neumann's work original (S. vi), but I think it corresponds very closely to the methods adopted by Poisson and Cauchy: see our Arts. 466* and 620*. Neumann supposes the thin rod of uniform cross-section and with its axis initially in the axis of x. *He then supposes that the shifts can be expanded in ascending powers of the assumed small linear dimensions of the cross-section.* We have already noted (see our Arts. 661* and 75) the objections to such an assumption and seen to what erroneous results it leads in the case of torsion: see our Arts. 805, 1225 and the references there.

Neumann obtains for a rod of circular cross-section of radius R, acted upon by body-forces X, Y, Z, the relations:

$$\left. \begin{array}{l} -\tfrac{1}{4}R^2 \dfrac{d^3\widehat{xx}}{dx^2dy} = \rho Y + \tfrac{1}{8}\rho R^2 \left\{ \dfrac{d^2Y}{dy^2} + \dfrac{d^2Y}{dz^2} + 2\dfrac{d^2X}{dydx} \right\}, \\[3mm] -\tfrac{1}{4}R^2 \dfrac{d^3\widehat{xx}}{dx^2dz} = \rho Z + \tfrac{1}{8}\rho R^2 \left\{ \dfrac{d^2Z}{dy^2} + \dfrac{d^2Z}{dz^2} + 2\dfrac{d^2X}{dxdz} \right\} \end{array} \right\} \dots\dots (i).$$

These equations are to hold only for $y = z = 0$, or at the axis of the rod, but they are true for all manner of elastic distributions.

[1227.] Neumann then treats especially the case of isotropy. He neglects in the stresses all the terms multiplied by R^2 and so finds for $y = z = 0$:

$$\widehat{xy} = \widehat{yz} = \widehat{zx} = \widehat{yy} = \widehat{zz} = 0 \dots\dots\dots\dots\dots (ii),$$

and further:

$$\frac{d\widehat{xx}}{dy} = E\frac{d^2v}{dx^2}, \quad \frac{d\widehat{xx}}{dz} = E\frac{d^2w}{dx^2} \dots\dots\dots\dots\dots (iii),$$

whence he obtains also for $y = z = 0$:

$$\left. \begin{array}{l} -\rho\dfrac{d^2v}{dt^2} + \tfrac{1}{4}\rho R^2 \dfrac{d^4v}{dx^2dt^2} = \tfrac{1}{4}R^2 E \dfrac{d^4v}{dx^4}, \\[3mm] -\rho\dfrac{d^2w}{dt^2} + \tfrac{1}{4}\rho R^2 \dfrac{d^4w}{dx^2dt^2} = \tfrac{1}{4}R^2 E \dfrac{d^4w}{dx^4} \end{array} \right\} \dots\dots\dots\dots (iv).$$

The reader will find that Neumann's reasoning is almost identical with that of Poisson and Cauchy (see our Arts. 467 and 620*), but it is by no means sufficient. The results (ii) only hold under certain very narrow limitations, and equations (iv) require at least a discussion like that of Kirchhoff or of Clebsch to justify their adoption: see our Art. 1251. Thus this portion of Neumann's work seems neither original nor valid. On S. 362–4 a similar process for a rod of rectangular cross-section is given; this again corresponds to Cauchy's work: see our Arts. 618*–624*.

According to Neumann's Editor (S. 355 ftn.), Neumann had also obtained by a similar method the equations for a crystalline rod, and his results have been published by Baumgarten and Voigt in the memoirs referred to in our Arts. 1210 and 1212. But I see no reason why his method should be more satisfactory in the complex than in the simple case.

[1228.] On S. 364–8 we have the simple case of a doubly-supported bar with an isolated load, and on S. 368–73, the transverse vibrations, tones and nodes of a thin rod discussed,—without, however, anything of novelty. The volume of *Vorlesungen* concludes with a brief note by Neumann's Editor of earlier and of more recent work on the theory of rods.

The general impression left on my mind after the perusal of Neumann's lectures is that they form the best *elementary* treatise on elasticity and its relation to light that I have met with in the German tongue. They contain a good deal of original matter and are without the difficulties of Clebsch's analysis, or the monotony of Lamé's isotropic solids.

[1229.] Of some other memoirs of Neumann's bearing on elasticity we have treated in our first volume, namely in Arts. 788*-801* and Arts. 1185*-1213*. Further memoirs belonging essentially to the theory of light, but appealing to that of elasticity, are the following:

(a) *Theorie der doppelten Strahlenbrechung: Poggendorffs Annalen*, Bd. 25, 1832, S. 418-454. This deduces the laws of double refraction from the equations of elasticity.

(b) *Theoretische Untersuchung der Gesetze nach welchen das Licht an der Grenze zweier vollkommen durchsichtigen Medien reflectirt und gebrochen wird: Abhandlungen der Berliner Akademie*, 1835, *Mathematische Klasse*, S. 1-160. Experimental results bearing on this theory were published by Neumann in *Poggendorffs Annalen*, Bd. 42, 1837, S. 1-29.

(c) *Reproduction der Fresnel'schen Formeln über totale Reflexion. Poggendorffs Annalen*, Bd. 40, 1837, S. 497-514. This deduces the laws of reflection and refraction including the case of total reflection from the theory of elasticity. Experimental results are given.

(d) The memoir entitled: *Ueber die optischen Eigenschaften der hemiprismatischen Crystalle* in *Poggendorffs Annalen*, Bd. 35, 1835, S. 81-94, and S. 203-5, should be taken as modifying the results of the memoir of 1834 stated in our Arts. 789*-93*. It announces the discovery of the dispersion of the optical axes in gypsum and the dependence of their position on the temperature. See our Art. 1218, or Neumann's *Vorlesungen über die Theorie der Elasticität*, S. 273.

(e) Neumann's *Vorlesungen über theoretische Optik* edited by E. Dorn, Leipzig, 1885, contribute nothing to the *elastic* theory of light; the brief application of that theory on S. 275 *et seq.* is due to Voigt.

A criticism of Neumann's elastic theories of light will be found in Glazebrook's *Report on Optical Theories*, especially in the parts of that *Report* referring to MacCullagh, whose theories are closely allied to Neumann's.

[1230.]   Voigt in a paper entitled: *Bestimmung der Elastici-tätsconstanten des Steinsalzes* published in *Poggendorffs Annalen, Ergänzungsband* VII., 1876, S. 1–53, and S. 177–214, gives two results due to Neumann.   The first (S. 5) is given in Neumann's *Vorlesungen*, S. 185 (see our Art. 1206), and gives the stretch-modulus for a prism cut in any direction from a crystal of the regular system with equal axes.   The second result is for the angle of torsion τ per unit length of a prism cut from a like crystal. It is given without proof.   If l be the length of the prism, $\alpha \times \beta$ its rectangular cross-section, M the applied couple, the formula is:

$$\tau = \frac{3M}{\alpha^3\beta^3} \left\{ \frac{\alpha^2 + \beta^2}{d} - 4\left(\frac{1}{2d} - \frac{1}{a - f'}\right) \right.$$
$$\times \left[\cos^2(l, x)\left(\alpha^2 \cos^2(\alpha, x) + \beta^2 \cos^2(\beta, x)\right)\right.$$
$$+ \cos^2(l, y)\left(\alpha^2 \cos^2(\alpha, y) + \beta^2 \cos^2(\beta, y)\right)$$
$$\left.\left. + \cos^2(l, z)\left(\alpha^2 \cos^2(\alpha, z) + \beta^2 \cos^2(\beta, z)\right)\right]\right\}$$

where x, y, z are the directions of the crystalline axes, and a, f', d the constants of our Art. 1203, (d).

Take the axis of the prism in the direction of x, and we have

$$\tau = \frac{3M}{\alpha^3\beta^3}\frac{\alpha^2 + \beta^2}{d}, \text{ or for the square, } = \frac{6M}{a^4 d} \ldots\ldots\ldots(\text{i}).$$

Saint-Venant's formula cited in our Art. 30 gives

$$\tau = \frac{6M}{\cdot 843 a^4 d} \ldots\ldots\ldots\ldots\ldots\ldots\ldots\ldots\ldots(\text{ii}),$$

so that this would give an error of about 18 p.c. in Neumann's formula.   I suspect Neumann deduced his formula from, or by a method similar to, Cauchy's erroneous investigation of the torsion of a rectangular prism, which leads to a result like (i).   Anyhow the formula is I think incorrect, and so probably are all the numerical determinations based upon it.

SECTION II.

Kirchhoff [1].

[1231.] The contributions of Kirchhoff to our subject consist of five or six memoirs published in various journals from 1848–79 and nearly all reprinted on S. 237–339 of the *Gesammelte Abhandlungen* (hereinafter referred to as *G. A.*) edited by Kirchhoff himself, Leipzig, 1882, of three or four memoirs (1882–4) reprinted in Boltzmann's *Nachtrag* to the *Abhandlungen*, Leipzig, 1891, and of five lectures in the *Vorlesungen über mathematische Physik : Mechanik*, of which a first edition appeared in Leipzig, January, 1876 and a second in the November of the same year. The *Vorlesungen* contain a good deal of the material of the earlier memoirs in an improved form, but it must be confessed that Kirchhoff's methods seem, at least to the Editor of the present work, frequently obscure and occasionally wanting in strictness. His contributions, however, to the theory of elastic wires and of thin plates are of such importance as to give him a permanent place in the history of elasticity.

[1232.] We give the titles only of the two earliest elastic papers by our author :
Note relative à la théorie de l'équilibre et du mouvement d'une plaque élastique : Comptes rendus, T. 27, pp. 394–7. Paris, 1848.
Note sur les vibrations d'une plaque circulaire : Comptes rendus, T. 29, pp. 753–6. Paris, 1849.
The substance of these papers, which are not free from misprints, is embodied in the memoir of 1850, considered in our next article.

[1233.] *Ueber das Gleichgewicht und die Bewegung einer elastischen Scheibe : Crelles Journal*, Bd. 40, S. 51–88. Berlin, 1850. (*G. A. S.* 237–79.) The author was at this time *Privat-docent* in the Berlin University.
The aim of the memoir is twofold : (i) to obtain the correct

---

[1] Some account of Kirchhoff's life and labours will be found in a *Necrologue* by Hofmann in the *Berichte der chemischen Gesellschaft zu Berlin*, Jahrg. 20, Juli—December, 1887, S. 2771–7 and in a somewhat florid *Festrede* delivered at Graz on November 15, 1887 by Ludwig Boltzmann entitled : *Gustav Robert Kirchhoff*, and published at Leipzig, 1888. Kirchhoff died October 17, 1887.

equations, especially those at the boundary, for the equilibrium
and motion of an elastic plate: (ii) to determine if possible
from a comparison of the theory with experiments on the nodes
and notes of vibrating plates whether Poisson's or Wertheim's
value of the stretch-squeeze ratio η is the correct one. The
memoir consists of five sections preceded by a short historical
introduction.

[1234.] In the introduction Kirchhoff refers to the memoirs
of Sophie Germain and notes that Lagrange first gave the correct
body-shift equation for a thin plate: see our Arts. 283*–306*.
He notes the errors into which Sophie Germain fell and demonstrates them by applying her equations to a particular case: see
his S. 51–4 (*G. A. S.* 237–40). The theory of Poisson (see our
Arts. 474*–93*) is then referred to and Kirchhoff remarks:

Aber auch diese Theorie bedarf einer Berichtigung, und dieselbe zu
geben, ist eben meine Absicht. Poisson gelangt, indem er seine allgemeinen Gleichungen des Gleichgewichts elastischer Körper auf den Fall
einer Scheibe anwendet, zu derselben partiellen Differentialgleichung,
zu welcher die Hypothese von Sophie Germain geführt hat, aber zu
andern Grenzbedingungen, und zwar zu drei Grenzbedingungen. Ich
werde beweisen, dass im Allgemeinen diesen nicht gleichzeitig genügt
werden kann (*sic!*); woraus dann folgt, dass auch nach der Poisson'schen
Theorie eine Platte im Allgemeinen keine Gleichgewichtslage haben
müsste (S. 54; *G. A. S.* 240–1).

Kirchhoff certainly emphasizes Poisson's error a little too
strongly considering he does not indicate any mistake in Poisson's
process. The real difficulty lies of course in the exact amount
of 'thinness' to be attributed to the plate, and we have already
pointed out how Thomson and Tait have practically reconciled
Poisson and Kirchhoff, while the researches of Saint-Venant and
Boussinesq have put the whole matter into a clearer light: see our
Art. 394 and Chapter XIII. The points raised, however, by Kirchhoff's investigation have been extremely valuable and important,
and have led to much good work. Like problems with regard to
the boundary conditions for thin shells have recently been discussed
in instructive memoirs by Love, Lamb and Basset.

[1235.] The first section of the memoir occupies S. 54–60 (*G. A.*
S. 241–7) and deals with the general equations of elasticity. Kirchhoff

shows that a single variational equation contains in itself the six body- and surface-equations of elasticity.

Let δU denote the virtual moment of the applied forces during strain, $dxdydz$ an element of volume of the elastic body, then this equation is:

$$\delta U - \mu\delta \iiint \left\{ s_1{}^2 + s_2{}^2 + s_3{}^2 + \frac{\lambda}{2\mu}(s_1 + s_2 + s_3)^2 \right\} dxdydz = 0 \ldots\ldots(i),$$

where s_1, s_2, s_3 are the principal stretches[1], the body is supposed isotropic, and the integration taken over its whole volume.

By means of the discriminating cubic Kirchhoff expresses

$$\left\{ s_1{}^2 + s_2{}^2 + s_3{}^2 + \frac{\lambda}{2\mu}(s_1 + s_2 + s_3)^2 \right\} = \Omega, \text{ say,}$$

in terms of the three stretches and three slides for any set of rectangular axes, and then shows that the development of the variations leads to the ordinary six equations of elasticity. He remarks that Green had already given equation (i), without, however, using the principal stretches: see Kirchhoff's footnote S. 56 (*G. A. S.* 243) and our Art. 918*.

Kirchhoff having deduced the elastic equations proceeds to a proof of equation (i):

Ich werde jetzt eine Ableitung der Gleichung (i) geben, aus welcher hervorgehen wird, dass sie eine allgemeinere Gültigkeit hat, als die Gleichungen (6) [i.e. the six equations of elasticity]. Betrachtungen, die denen, welche hier folgen, ganz ähnlich sind, hat Lagrange mehrmals in seiner Mechanik, z. B. bei der Herleitung der Gleichgewichtsbedingung eines elastischen Stabes, angestellt (S. 589; *G. A. S.* 246).

The proof does not seem to me very convincing. Kirchhoff practically *assumes* that the virtual moment of the stresses on $dxdydz$ must be of the form

$$- dxdydz\,(S_1\delta s_1 + S_2\delta s_2 + S_3\delta s_3),$$

and further that S_1, S_2, S_3 must be symmetrical functions of the type

$$S_1 = as_1 + b\,(s_2 + s_3),$$

a and b being elastic constants. I do not think the proof can be considered rigid.

[1236.] In the second section, which occupies S. 60–63 (*G. A. S.* 247–251), Kirchhoff deals with the problem of an infinitely thin plate bounded by parallel planes and any cylindrical surface whose generators are perpendicular to these planes. The plate is

---

[1] Kirchhoff uses K for our μ, and θ for our $\lambda/(2\mu)$, while he takes $q =$ our $E = 2K\dfrac{1+3\theta}{1+2\theta}$, but he uses q in another sense also on his S. 61.

supposed strained by body-forces, and by surface-forces on the edge only, the plane faces having no load. The strains are supposed infinitely small but the shifts are not necessarily so. The plate is supposed isotropic. In order to apply equation (i) Kirchhoff makes two assumptions which, he says, are to be regarded as results of experiment and which correspond exactly with those which James Bernoulli made in regard to an elastic rod (S. 60; *G. A. S.* 248 and our Art. 19*).

These assumptions are the following:

(i) Every straight line in the plate which was originally perpendicular to the plate-surfaces, remains straight after the strain and perpendicular to the surfaces which were originally planes parallel to the plate-surfaces.

(ii) All elements of the mid-surface (i.e. that surface which in the unstrained condition of the plate was plane, parallel to the plate-surfaces and half-way between them) remain after strain without stretch.

Kirchhoff makes no appeal to any *definite* experiment as confirming these assumptions, and the reference to James Bernoulli is distinctly unfortunate. It is true that the Bernoulli-Eulerian hypothesis leads to an equation, which Saint-Venant has shown is really true for the flexure of long bars, but the assumptions by which that equation is reached are not true, and it seems unadvisable to make assumptions, which, even if true for certain types of strain, need not be true for all types which lead to Lagrange's plate equation: see our Arts. 70 and 79. The assumptions which it is really needful to make and the arguments in favour of them have been dealt with by Boussinesq and Saint-Venant: see the memoirs on plates of the former discussed in our Chapter XIII., and our Arts. 385, 388 and 394.

Kirchhoff's treatment must therefore be looked upon as interesting and suggestive, but not as rigid or final.

[1237.] On S. 61–2 (*G. A. S.* 248–9) the values of the principal stretches are deduced, and equation (i) of our Art. 1235 is reduced to the form:

$$\delta U - \mu \delta \iiint d\omega dz \left\{ \left(\frac{dq}{dz}\right)^2 + \left(\frac{z}{\rho_1}\right)^2 + \left(\frac{z}{\rho_2}\right)^2 + \frac{\lambda}{2\mu}\left(\frac{dq}{dz} + \frac{z}{\rho_1} + \frac{z}{\rho_2}\right)^2 \right\} \dots \text{(ii)}.$$

Here $d\omega$ is an element of the mid-surface, and the axis of z is taken perpendicular to this, ρ_1, ρ_2 are the principal radii of curvature of the mid-surface at $d\omega$, and dq/dz is really the stretch in the direction z at the point distant z from $d\omega$. Kirchhoff goes through a rather long investigation on S. 62–3 (*G. A. S.* 249–51), which I do not find very clear, to prove that

$$(\lambda + 2\mu)\,\frac{dq}{dz} + \lambda \left(\frac{z}{\rho_1} + \frac{z}{\rho_2}\right) = 0 \quad \ldots\ldots\ldots\ldots\ldots\text{(iii)}.$$

The physical meaning of equation (iii) is, however, that the stress $\widehat{zz}$, perpendicular to the plate faces, is to vanish at every point of the plate. Since the plate is supposed infinitely thin and to have no load on its surfaces, this seems, at any rate as an approximation, a reasonable conclusion.

By the aid of equation (iii) and integration with regard to z, Kirchhoff reduces (ii) to the form:

$$\delta U - \tfrac{2}{3}\epsilon^3\mu\delta \iint d\omega \left\{\frac{1}{\rho_1{}^2} + \frac{1}{\rho_2{}^2} + \frac{\lambda}{\lambda + 2\mu}\left(\frac{1}{\rho_1} + \frac{1}{\rho_2}\right)^2\right\} = 0 \quad \ldots\ldots\text{(iv)},$$

where 2ϵ is the thickness of the plate.

This is I believe the first occasion on which the work done in bending a thin isotropic elastic plate to curvatures $1/\rho_1$, $1/\rho_2$ at any point was expressed in terms of those curvatures, and this is one of the merits of Kirchhoff's memoir.

[1238.] The third section of the memoir (S. 63–70; *G. A. S.* 251–9) deduces by variation of equation (iv) the equation for the transverse shift at any point of the mid-surface and the boundary or edge conditions of the plate. Kirchhoff deals only with the case treated by Poisson, namely when the mid-surface shift is very small. He obtains the two edge conditions and the shift-equation in the manner which is now to be found in several text-books: see Lord Rayleigh's *Theory of Sound*, Vol. I. pp. 293–300, and compare Thomson and Tait's *Natural Philosophy*, Part II. pp. 181–90. The equations agree with those obtained by Saint-Venant and Boussinesq much later indeed, but by what seems to me very much more conclusive reasoning: see our Arts. 383–8 and 394.

[1239.] In the last pages of this section (S. 67–70; *G. A.* S. 258–9) Kirchhoff shows that the *two* boundary conditions and the shift-equation are sufficient to determine completely (the translation of the plate as a whole excepted) the value of the transverse shift, and he thence argues that Poisson's equations can only be satisfied in special cases, as they involve an additional equation. For the exact meaning of Poisson's boundary conditions, see our Arts. 488* and 394.

[1240.] Kirchhoff's proof of the uniqueness of the solution of the plate equations is of interest, as it is, I think, the first appearance of a method afterwards extended by himself and then by Clebsch: see our Art. 1255. In general terms it may be indicated as follows: Consider the double integral

$$I = \iint \left\{ \left(\frac{d^2w}{dx^2}\right)^2 + \left(\frac{d^2w}{dxdy}\right)^2 + 2\left(\frac{d^2w}{dy^2}\right)^2 + \frac{\lambda}{\lambda+2\mu}\left(\frac{d^2w}{dx^2} + \frac{d^2w}{dy^2}\right)^2 \right\} d\omega$$

over the mid-surface of the plate. If ds be any element of the edge, and dn an element of its normal measured inwards, ϕ the angle between the normal at ds and the positive direction of the axis of x, then this integral may by partial integration be expressed in the form (S. 70; G. A. 258):

$$I = \frac{2(\lambda+\mu)}{\lambda+2\mu}\iint w\left\{\frac{d^4w}{dx^4} + 2\frac{d^4w}{dx^2dy^2} + \frac{d^4w}{dy^4}\right\} d\omega$$

$$+ \int\left\{\frac{2(\lambda+\mu)}{\lambda+2\mu}\left[\left(\frac{d^3w}{dx^3} + \frac{d^3w}{dxdy^2}\right)\cos\phi + \left(\frac{d^3w}{dx^2dy} + \frac{d^3w}{dy^3}\right)\sin\phi\right]\right.$$

$$\left. - \frac{d}{ds}\left[\frac{d^2w}{dxdy}(\cos^2\phi - \sin^2\phi) + \left(\frac{d^2w}{dy^2} - \frac{d^2w}{dx^2}\right)\cos\phi\sin\phi\right]\right\} w\,ds$$

$$- \int\left\{\frac{\lambda}{\lambda+2\mu}\left(\frac{d^2w}{dx^2} + \frac{d^2w}{dy^2}\right) + \frac{d^2w}{dx^2}\cos^2\phi + 2\frac{d^2w}{dxdy}\cos\phi\sin\phi\right.$$

$$\left. + \frac{d^2w}{dy^2}\sin^2\phi\right\}\frac{dw}{dn}\,ds.$$

Now suppose two solutions w_1 and w_2 of the plate equations possible and let $w_1 - w_2 = w$. Then if we subtract the body shift-equations for w_1 and w_2 from each other, and each of the corresponding boundary-conditions likewise, the applied forces vanish and we obtain, with constant multipliers, exactly the three expressions in the curled brackets in the value of I equated to zero. Hence I must be zero, whence it follows that throughout the plate

$$\frac{d^2w}{dx^2} = 0, \quad \frac{d^2w}{dy^2} = 0, \quad \frac{d^2w}{dxdy} = 0,$$

or, $w_1 - w_2 = C_1 x + C_2 y$, C_1 and C_2 being constants; that is to say the shift difference corresponds to a translation of the plate as a whole. The reader may convince himself that the expressions in curled brackets are really identical with the body and boundary equations by reference to our Arts. 392-4.

[1241.] The fourth section of the memoir occupies S. 70–81 (*G. A. S.* 259–71). It deals with the vibrations of a *free* circular plate, without surface-load or body-force. The solution is more general than Poisson's, as it does not suppose the vibrations to be the same along all radii. The initial shift and shift-velocity at any point of the plate are supposed given in terms of the radius-vector and the vectorial angle. For a complete plate the solution is expressed in a doubly-infinite series of functions akin to Bessel's functions, there being a constant to be determined as one of the roots of an algebraic equation of infinite order. The analysis is too lengthy to be reproduced here, but it possesses considerable interest. I may note especially the manner in which the equation (15) on S. 74 (*G. A. S.* 263) is transformed on S. 74–7 (*G. A. S.* 264–6) to one proceeding by ascending powers of the variable. Physically the most valuable part of the memoir lies in the discovery of the equations for the frequencies of the notes and the positions of nodal lines.

The form of the transverse shift for a single tone is given by

$$w = \{(A \cos n\psi + B \sin n\psi) \cos (4\lambda^2_{nm}at)$$
$$+ (C \cos n\psi + D \sin n\psi) \sin (4\lambda^2_{nm}at)\} U_{nm} \ldots\ldots\ldots\ldots(i),$$

where: A, B, C, D are constants ultimately depending on the initial conditions, t is the time from the epoch, ψ the radial angle, n a positive integer, $a^2 = \frac{4}{3} \frac{\mu(\lambda + \mu)}{\lambda + 2\mu} \frac{\epsilon^2}{\rho} \left(= \frac{H\epsilon^2}{3\rho}\right)$, H being the plate-modulus of our Art. 323, ρ the density and 2ϵ the thickness of the plate), and

$$U_{nm} = X^{(n)} . \left\{ (n^2 - 4\gamma R^2) Y^{(n)} - R \frac{dY^{(n)}}{dR} \right\}_{(R=\lambda_{nm}b)}$$
$$- Y^{(n)} \left\{ (n^2 + 4\gamma R^2) X^{(n)} - R \frac{dX^{(n)}}{dR} \right\}_{(R=\lambda_{nm}b)}, \Bigg\} \ldots\ldots\ldots(ii),$$

where
$$\gamma = \frac{2(\lambda + \mu)}{\lambda + 2\mu} = \frac{H}{2\mu}, \quad R = \lambda_{nm}r$$

r being any radius vector, and b the radius of the plate; further,

$$\binom{X^{(n)}}{Y^{(n)}} = \frac{R^n}{n!} \left[1 \pm \frac{R^2}{1.(n+1)} + \frac{R^4}{1.2.(n+1)(n+2)} \right.$$
$$\left. \pm \frac{R^6}{1.2.3.(n+1)(n+2)(n+3)} + \text{etc.} \right] \ldots\ldots\ldots(iii);$$

and finally λ_{nm} is the mth root of

$$0 = (4\gamma - 1)\, n^2\, (n-1) + \sum_{k=1}^{k=\infty} (-1)^k\, \frac{N_k}{D_k}\, R^{4k},$$

where $R = \lambda_{nm}b$,

$N_k = -n^2(n^2-1) + 4\gamma(n+2k)(n+2k+1)\{n(n-1) - 2k + 4\gamma k(n+k)\},$

$D_k = 1 . 2 \ldots k . \overline{n+1} . \overline{n+2} \ldots \overline{n+k} . \overline{n+1} . \overline{n+2} \ldots \overline{n+2k+1}$

$\left.\vphantom{\begin{array}{c}1\\1\\1\\1\end{array}}\right\}$ ...(iv).

[1242.] The fifth section of the memoir which occupies S. 81–88 (G. A. S. 271–9) deals with the numerical solution of the equations we have given in the previous article, and implies a very great amount of laborious calculation. The tones are compared with those obtained by Chladni and the nodal lines with those obtained by Strehlke with two circular glass plates. We will cite some of the results of Kirchhoff's investigations.

(a) First with regard to the *notes*, their periods are obviously given by $\pi/(2\lambda^2_{nm}a)$. Kirchhoff calculates (S. 84–5; G. A. S. 275–7) the values of $\log_{10}(\lambda_{nm}b)^4$ from equation (iv) and finds for these values:

	$m=0$		$m=1$		$m=2$	
	$\lambda=\mu$	$\lambda=2\mu$	$\lambda=\mu$	$\lambda=2\mu$	$\lambda=\mu$	$\lambda=2\mu$
$n=0$	—	—	0·693 67	0·711 68	1·963 08	1·967 12
$n=1$	—	—	1·415 53	1·420 12	2·348 29	2·350 22
$n=2$	0·278 37	0·236 38	1·891 17	1·889 97		
$n=3$	1·006 51	0·970 14	2·246 93	2·242 98		

Further Kirchhoff has shown S. 83–4 (G. A. S. 273–5) that when $\lambda_{nm}b$ is great its value is very approximately given by

$$\lambda_{nm}b = \tfrac{1}{4}\pi\,(n + 2m)\ldots\ldots\ldots\ldots\ldots\ldots\ldots(v).$$

This practically covers the values of $\lambda_{nm}b$ not given by the logarithms of $(\lambda_{mn}b)^4$ in the above table.

Hence for a plate of known elasticity all the notes may be calculated, or the frequencies of all the sub-tones may be found in terms of that of the fundamental tone. These results are compared with Chladni's experiments.

Chladni found by experiment that the frequencies of vibration $(2\lambda^2_{nm}a/\pi)$ in the tones which had in their nodal figures the same number of diameters (i.e. those tones which correspond to the same value of n) were, with the exception of the lowest, nearly as the squares of successive even or uneven numbers according as the number of nodal diameters was even or odd : see S. 82–3 of the memoir (G. A. S. 273).

The frequencies of the high tones vary as λ^2_{nm}, or as $(n + 2m)^2$, and thus Chladni's experiments so far agree with Kirchhoff's theory. Kirchhoff then compares Chladni's results for the lower tones with those calculated first on Wertheim's hypothesis $(\lambda = 2\mu)$ and then on Poisson's $(\lambda = \mu)$ for the case of a plate whose lowest tone is taken as C. He assumes that Chladni's results were all obtained for the same constant temperature. The results of Poisson's hypothesis are closer than those of Wertheim's to Chladni's observations, but the divergences are so great in both cases that no conclusions can be drawn with regard to the relative value of the hypotheses. The frequency of the tone, especially for large values of m and n, varies so little with the value of λ/μ, that experiments on plates can hardly be crucial between the two hypotheses.

(b) Secondly with regard to the *nodal lines*. These are given by the values of r and ψ for which, independently of the time, $w = 0$. Clearly from equation (i) we have the radii of the nodal circles given by the values of r $(< b)$ for which $U_{nm} = 0$, and also the nodal diameters given by the values of ψ for which

$$A \cos n\psi + B \sin n\psi = 0, \qquad C \cos n\psi + D \sin n\psi = 0.$$

Thus there can be no nodal diameters unless the plate be so disturbed that $A : B :: C : D$. If this equation holds we have n nodal diameters, each adjacent pair separated by the angle π/n. n and m may thus be considered as giving the number of nodal diameters and nodal circles respectively which can occur in connection with the tone defined by λ_{nm}.

Kirchhoff says with regard to this:

Diese allgemeinen Resultate der Theorie sind im Wesentlichen mit der Erfahrung in Uebereinstimmung. Der Versuch zeigt, dass die Knotenlinien aus Kreisen bestehen, die mit der Peripherie der Scheibe concentrisch sind, und aus Durchmessern, die diese in gleiche Theile theilen, wenn man von gewissen Verzerrungen absieht, die diese Linien erleiden und die, wie mir scheint, hauptsächlich darin ihren Grund haben, dass die Scheibe nicht vollkommen frei ist, wie die Theorie sie voraussetzt[1]. Der Versuch zeigt aber auch, dass bei einem Tone, bei dem zuweilen Durchmesser als Knotenlinien vorkommen, die Durchmesser zuweilen fehlen. Fehlen sie, so ordnet sich der auf die Scheibe gestreute Sand zwar auch in Durchmessern an : diese bleiben aber nicht fest während der Bewegung der Scheibe, sondern oscilliren. (S. 82 ; G. A. S. 272-3.)

To this general experimental confirmation of the theory Kirchhoff adds a comparison of the radii of the nodal circles calculated on Poisson's and Wertheim's hypotheses with those obtained experimentally by Strehlke from two circular glass plates (Art. 359*), and by Savart on three such plates (Art. 320*).

---

[1] It may be also questioned whether any such perfectly *isotropic* and *homogeneous* plate as is supposed in the theory can be really prepared.

I place below his comparison of experimental and theoretical numbers
for the ratios of the radii of the nodal circles to the radius of the plate :

	Experiment.					Theory.	
	Strehlke. Plates.		Savart. Plates.			Poisson, $\lambda = \mu$.	Wertheim, $\lambda = 2\mu$.
	1	2	1	2	3		
$n=0,$ $m=1$	0·6792	0·6782	0·6819	0·6798	0·6812	0·68062	0·67941
$n=1,$ $m=1$	0·7811	0·7802	—	—	—	0·78136	0·78088

Kirchhoff says of these results :

Die aus der Wertheim'schen Annahme abgeleiteten Resultate weichen
von den aus der Poisson'schen abgeleiteten nur wenig ab ; mit den Strehlke'schen Beobachtungen stimmen jene noch besser überein als diese. Wie mir
scheint, spricht dieses aber nicht gegen die Poisson'sche Annahme, denn eine
vollkommene Uebereinstimmung zwischen der Theorie und dem Versuche darf
man nicht erwarten, weil die dem Versuche unterworfenen Scheiben nicht
die Eigenschaften in aller Strenge besitzen, welche in der Theorie ihnen
beigelegt werden. (S. 87 ; *G. A.* S. 278-9.)

The correspondence between experiment and theory is not by any
means so remarkable as the fact that such different hypotheses as those
of Poisson and Wertheim give such very similar results. The nodal
lines of vibrating circles obviously afford no crucial test of the truth of
uni-constancy.

Kirchhoff on S. 88 (*G. A.* S. 279) gives the results of further
experiments of Strehlke's on less perfect plates, and also the calculated
values of the radii of the nodal circles for $m = 1$, $n = 1$, 2, 3, and for
$m = 2$, $n = 1$, on both Wertheim's and Poisson's hypotheses. See our
Arts. 512*-520*, and 1344*-1348*.

[1243.] *Über die Schwingungen einer kreisförmigen elastischen
Scheibe : Poggendorffs Annalen,* Bd. 81, 1850, S. 258-264. (*G.
A.* S. 279-85.) This is a résumé of the memoir in *Crelles Journal*
just discussed ; see our Arts. 1233-42. It contains, however, more
detailed numerical results and still further theoretical calculations
of the frequencies of the notes and the position of the nodal lines.

(*a*) We may draw attention especially to the *numerical* calculation
of the frequencies on S. 261 (*G. A.* S. 282). The fundamental note
being that in which the nodal figure consists of two perpendicular

diameters, the period of a single corresponding vibration is taken as the unit of time, and Kirchhoff finds the numbers of vibrations corresponding to the sub-tones which take place in this unit of time. The numbers thus obtained are the same for all plates whatever their substance and dimensions, provided we assume any fixed relation between λ and μ. The sub-tones are more fully calculated on Poisson's hypothesis ($\lambda = \mu$) than on Wertheim's ($\lambda = 2\mu$), and they are given here for reference :

Ratios of Frequency of Sub-tone to that of fundamental Note.

$\lambda = \mu$	$n = 0$	$n = 1$	$n = 2$	$n = 3$	$n = 4$	$n = 5$
$m = 0$			1·0000	2·3124	4·0485	6·1982
$m = 1$	1·6131	3·7032	6·4033	9·6445	13·3937	17·6304
$m = 2$	6·9559	10·8383	15·3052	20·3249		
$m = 3$	15·9031					

These do not agree very closely with Chladni's results, also converted into numbers by Kirchhoff, who considers that more accurate observations of the frequencies would be of value.

(*b*) With regard to the radii of the nodal circles Kirchhoff also gives more complete results, especially for the hypothesis $\lambda = \mu$. The ratios of the radii of the nodal circles to the radius of the plate are given by the following table :

$\lambda = \mu$	$n = 0$	$n = 1$	$n = 2$	$n = 3$	$n = 4$	$n = 5$
$m = 1$	·68062	·78136	·82194	·84523	·86095	·87256
$m = 2$	$\begin{cases} ·39151 \\ ·84200 \end{cases}$	·49774 ·87057	·56043 ·88747	·60365 ·89894		
$m = 3$	$\begin{cases} ·25679 \\ ·59147 \\ ·89381 \end{cases}$					

The numbers on Wertheim's hypothesis are not carried as far, but they are in close accordance, so far as they go. The results are compared with Strehlke's measurement, on four glass and two metal discs, and there is close correspondence between Kirchhoff's theory and experiment : see S. 262–4 (*G. A. S.* 283–5).

Kirchhoff gives the following expressions on S. 262 (*G. A. S.* 283) for the number N of vibrations in unit time corresponding to the fundamental note of a circular plate of radius b and small thickness 2ϵ:

$$N = 1 \cdot 04604 \, \frac{\epsilon}{b^2} \sqrt{\frac{E}{\rho}}, \text{ for } \lambda = \mu, \text{ and } = 1 \cdot 02357 \, \frac{\epsilon}{b^2} \sqrt{\frac{E}{\rho}}, \text{ for } \lambda = 2\mu$$

(see our Arts. 511* and 518*). He remarks that, so far as he is aware, no experiments have as yet been made to test these results.

[1244.] *Über die Gleichungen des Gleichgewichtes eines elastischen Körpers bei nicht unendlich kleinen Verschiebungen seiner Theile. Sitzungsberichte der mathem.-naturwiss. Classe der k. Akademie der Wissenschaften*, Bd. IX. S. 762–773. Wien, 1852. Kirchhoff did not republish this in his *Gesammelte Abhandlungen*, and therefore was possibly dissatisfied with its method and results. He commences his memoir by referring to the paper of Saint-Venant discussed in our Art. 1617* (I.) *et seq.* Saint-Venant had briefly indicated a method of finding the equations of elasticity when the shifts are not infinitely small. Kirchhoff remarks:

Diese Gleichungen habe ich auf zwei verschiedenen Wegen abgeleitet, von denen der erste im Wesentlichen mit dem von St. Venant angedeuteten übereinzukommen scheint, der zweite auf der Entwickelung einer früher von mir (Crelles Journ. XL. [see our Art. 1235]) aufgestellten Formel beruht (S. 762).

[1245.] Kirchhoff takes as his variables not the shifts u, v, w of the point x, y, z but the coordinates of the point x, y, z, after shift, or $\xi = x + u$, $\eta = y + v$, $\zeta = z + w$. He then states rather than proves that body- and surface-stress equations of the usual types, namely[1]:

$$\rho X + \frac{d\widehat{xx}}{dx} + \frac{d\widehat{xy}}{dy} + \frac{d\widehat{xz}}{dz} = 0,$$

$$X_0 = l\widehat{xx} + m\widehat{xy} + n\widehat{xz},$$

hold, where, however, the stress symbols have not their usual meaning. They denote stresses parallel to the coordinate axes across planes originally but no longer parallel to the coordinate planes. Thus relations of the type

$$\widehat{xy} = \widehat{yx}$$

will no longer be true. (*Diese neun Drucke sind im Allgemeinen schief gegen die Ebenen gerichtet, gegen die sie wirken, und es sind nicht drei von ihnen dreien anderen gleich*, S. 763.)

---

[1] It should be noted that we use tensions where he uses *pressures*.

[1246.]   The next step is to express these nine stresses in terms of the three principal tractions. This occupies S. 764–7. In the course of the investigation the following process occurs. Let r be the direction of an element of a line in the unstrained state which takes the direction r' in the strained state; let e be an element of length of r defined by its terminal coordinates x, y, z and $x + \delta x$, $y + \delta y$, $z + \delta z$, then if e becomes ϵ with terminals given by ξ, η, ζ, $\xi + \delta \xi$, $\eta + \delta \eta$, $\zeta + \delta \zeta$, we have :

$$\delta \xi = \frac{d\xi}{dx}\, \delta x + \frac{d\xi}{dy}\, \delta y + \frac{d\xi}{dz}\, \delta z$$

or, $\epsilon \cos (r', x) = e \left\{ \dfrac{d\xi}{dx} \cos (r, x) + \dfrac{d\xi}{dy} \cos (r, y) + \dfrac{d\xi}{dz} \cos (r, z) \right\}$,

with similar equations for $\epsilon \cos (r', y)$ and $\epsilon \cos (r', z)$. Kirchhoff cancels ϵ and e on either side, which is allowable he says "*wenn wir berück-sichtigen, dass ϵ von e nur unendlich wenig verschieden ist*" (S. 765). Now it is not shown that the terms Kirchhoff is thus neglecting are not of the order of the quantities he proposes to retain. In fact, if r be taken to coincide with x, he finds the cosine of the angle between the strained and unstrained directions of x to be $d\xi/dx = 1 + u_x$, which is quite incorrect. If we keep e/ϵ in, we should have it as a factor of the right-hand sides of equations (6) of S. 765. Thus in Kirchhoff's expressions on S. 766 for the stresses, we must read for his principal pressures P_1, P_2, P_3 the quantities

$$P_1 e_1/\epsilon_1, \;\; P_2 e_2/\epsilon_2, \;\; P_3 e_3/\epsilon_3,$$

or, if s_1, s_2, s_3 be the stretches in the directions of the principal pressures :

$$P_1/(1 + s_1), \;\; P_2/(1 + s_2), \;\; P_3/(1 + s_3)$$

respectively. (Kirchhoff uses λ_1, λ_2, λ_3 for our s_1, s_2, s_3.)

[1247.]   Kirchhoff next assumes that the principal pressures will be linear functions of the principal stretches, or that

$$P_1 = - 2\mu' \left\{ s_1 + \frac{\lambda'}{2\mu'} (s_1 + s_2 + s_3) \right\}.$$

He writes K for μ' above, and θ for $\lambda'/(2\mu')$, using the same letters K and θ for these elastic constants as he had used in the memoir of 1850 (see our Art. 1235). He is justified in doing this because he neglects the square of the strain. If we retain the square of the strain, and still assume the principal pressures linear functions of the principal stretches, then λ' and μ' will not be the λ and μ of our ordinary notation. Thus Sir W. Thomson in his memoir of 1862 (*Phil. Trans.* 1863, p. 612, or *Treatise on Natural Philosophy*, Part II. p. 464) remarks :

And it may be useful to observe that for all values of the variables A, B, C, a, b, c [the strain energy] must therefore be expressible in the same form, with varying coefficients, each of which is always finite, for all values of the variables.

Here $A-1$, $B-1$, $C-1$, a, b, c are the generalised components of strain, and it has just been noted that if these are infinitely small the strain-energy may be expressed as a homogeneous quadratic function of them with *constant* coefficients. Hence Sir W. Thomson considers that the coefficients of elasticity vary as the strain increases in magnitude and thus for finite strain may no longer be represented by λ and μ.

[1248.] To be more general then than Kirchhoff, that is to deal with any magnitude of strain, we ought to replace in the equations (7) and (8) of Kirchhoff's S. 767, the quantities P_1, P_2, P_3 by expressions of the type

$$P_1 = -2\mu' \left\{ s_1 + \frac{\lambda'}{2\mu'} (s_1 + s_2 + s_3) \right\} \Big/ (1 + s_1).$$

These will agree with Kirchhoff's values if the strains are so small that the products of the principal stretches may be *neglected*.

Neglecting the square of s_1, etc. Kirchhoff finds values for s_1, s_2, s_3 in terms of quantities which he denotes by the letters L, M, N, l, m, n. These quantities are related in the following manner to Thomson's A, B, C, a, b, c and to the ϵ_x, ϵ_y, ϵ_z, η_{yz}, η_{zx}, η_{xy} of our Art. 1619*:

$$2L = A - 1 = 2\epsilon_x, \quad 2M = B - 1 = 2\epsilon_y, \quad 2N = C - 1 = 2\epsilon_z,$$

$$2l = a = \eta_{yz}, \quad\quad 2m = b = \eta_{zx}, \quad\quad 2n = c = \eta_{xy}.$$

But it must be noted that while all these quantities are generalised components of strain, Kirchhoff's expressions for s_1, s_2, s_3 in terms of L, M, N and therefore his expressions for the stress in terms of these strain-components are true only for *infinitely small strains*.

[1249.] Expressed in the notation of our work we have according to Kirchhoff the following expressions for the stress-symbols as defined in our Art. 1245:

$$\widehat{xx} = 2\mu \left\{ (1 + u_x) \left(\epsilon_x + \frac{\lambda}{2\mu} (\epsilon_x + \epsilon_y + \epsilon_z) \right) + u_y \frac{\eta_{xy}}{2} + u_z \frac{\eta_{zx}}{2} \right\},$$

$$\widehat{xy} = 2\mu \left\{ (1 + u_x) \frac{\eta_{xy}}{2} + u_y \left(\epsilon_y + \frac{\lambda}{2\mu} (\epsilon_x + \epsilon_y + \epsilon_z) \right) + u_z \frac{\eta_{yz}}{2} \right\},$$

with others written down by proper cyclical interchanges. These results, as we have seen, are obtained on the assumption that the square of the strain may be neglected. Now Kirchhoff's last set of equations on S. 769 shows that s_1, s_2, s_3 are of the same order as ϵ_x, ϵ_y, ϵ_z, and therefore these latter quantities are also small; but $\epsilon_x = u_x + \frac{1}{2}(u_x{}^2 + v_x{}^2 + w_x{}^2)$, and therefore if u_x be positive, u_x and ϵ_x must be practically of the same order, hence it is difficult to see how as a rule we can neglect $s_1{}^2$ and retain products like $u_x \epsilon_x$. But if we do not reject $s_1{}^2$ Kirchhoff's investigation is invalid. Thus it does not seem that much importance can be attributed to the expressions given above for the stress-symbols in terms of the generalised components of strain

[1250.] More weight is I think to be laid on Kirchhoff's second method of investigation, which at any rate, till it assumes the strain-energy to be a quadratic function of the principal stretches, does not suppose the strains necessarily small.

Let W be the strain-energy, then in our notation Kirchhoff finds for the values of the stress-symbols as defined in our Art. 1245 (S. 772):

$$\widehat{xx} = \frac{dW}{du_x}, \qquad \widehat{xy} = \frac{dW}{du_y}, \qquad \widehat{xz} = \frac{dW}{du_z},$$

$$\widehat{yx} = \frac{dW}{dv_x}, \qquad \widehat{yy} = \frac{dW}{dv_y}, \qquad \widehat{yz} = \frac{dW}{dv_z},$$

$$\widehat{zx} = \frac{dW}{dw_x}, \qquad \widehat{zy} = \frac{dW}{dw_y}, \qquad \widehat{zz} = \frac{dW}{dw_z}.$$

But $W =$ a function of ϵ_x, ϵ_y, ϵ_z, η_{yz}, η_{zx}, η_{xy}, where these generalised strain-components have the values given in our Art. 1619*.

Whence it follows that:

$$\frac{dW}{du_x} = \frac{dW}{d\epsilon_x}\frac{d\epsilon_x}{du_x} + \frac{dW}{d\eta_{zx}}\frac{d\eta_{zx}}{du_x} + \frac{dW}{d\eta_{xy}}\frac{d\eta_{xy}}{du_x},$$

or

$$\widehat{xx} = \frac{dW}{d\epsilon_x}(1+u_x) + \frac{dW}{d\eta_{zx}} u_z + \frac{dW}{d\eta_{xy}} u_y.$$

Similarly

$$\widehat{xy} = \frac{dW}{d\epsilon_y} u_y + \frac{dW}{d\eta_{yz}} u_z + \frac{dW}{d\eta_{xy}}(1+u_x),$$

$$\widehat{xz} = \frac{dW}{d\epsilon_z} u_z + \frac{dW}{d\eta_{yz}} u_y + \frac{dW}{d\eta_{zx}}(1+u_x).$$

Substitute these expressions in the body-stress equations of Art. 1245 and we have precisely the generalised equations given by C. Neumann in 1860 (see our Art. 670) and by Thomson in 1862 (*Phil. Trans.* 1863, p. 611, *Nat. Phil.* Part II. p. 463). These equations are thus involved in Kirchhoff's results on S. 772 and 789, although he passes them by to express the value of W in the doubtful form:

$$W = \mu\left(\epsilon_x^2 + \epsilon_y^2 + \epsilon_z^2\right) + \frac{\mu}{2}\left(\eta_{yz}^2 + \eta_{zx}^2 + \eta_{xy}^2\right) + \frac{\lambda}{2}\left(\epsilon_x + \epsilon_y + \epsilon_z\right)^2,$$

on the assumption that the squares of the strains may be neglected.

[1251.] *Über das Gleichgewicht und die Bewegung eines unendlich dünnen elastischen Stabes: Crelles Journal*, Bd. 56, S. 285-313. Berlin, 1858 (*G. A.* S. 285-316).

This memoir is substantially reproduced in the twenty-eighth *Vorlesung* of Kirchhoff's *Mechanik*, S. 407-428, with some modifications and improvements. Kirchhoff's theory in both places is owing to its brevity and generality rather hard reading. It is given in a somewhat simpler and clearer fashion by Clebsch in his *Elasticität*, S. 192 *et seq.* It belongs to a branch of our subject that Kirchhoff was among the first to treat with any exactness, namely the equilibrium and motion of elastic bodies having one or two dimensions infinitely small, i.e. thin rods, wires, plates and shells. The subject is a difficult one, and it is only the confirmation, which the results reached receive when we approach them as limiting cases of bodies of finite dimensions (as, for example, has been done for certain cases by Clebsch), that enables us to set aside the doubts raised by some of the processes adopted.

[1252.] The memoir opens with the following historical account of its object :

Poisson hat in seinem *Traité de mécanique* eine Theorie der endlichen Formänderungen entwickelt, die ein unendlich dünner, ursprünglich gerader oder krummer, elastischer Stab durch Kräfte, die theils auf sein Inneres, theils auf seine Enden wirken, erfährt. De Saint-Venant hat jedoch nachgewiesen, dass die Voraussetzungen, von denen Poisson dort ausgegangen ist, theilweise unrichtig sind, und hat zum ersten Male die Torsion und Biegung eines unendlich dünnen Stabes von beliebigem Querschnitt, von den Grundgleichungen der Theorie der Elasticität ausgehend, mit Strenge untersucht. De Saint-Venant hat dabei aber nur den Fall behandelt, dass der Stab ursprünglich cylindrisch ist, dass die Formänderungen unendlich klein sind, und dass die Axe des Stabes eine Axe der Elasticität ist. In der vorliegenden Abhandlung untersuche ich, von den Gleichungen der Theorie der Elasticität ausgehend, die Formänderungen eines unendlich dünnen Stabes von überall gleichem Querschnitt ohne diese beschränkenden Annahmen. S. 285 (*G. A.* S. 285-6.)

[1253.] The first section of the memoir occupies S. 286–93 (*G. A.* S. 286–95) and relates to certain general principles which are afterwards applied to the special problem of the thin rod. Kirchhoff first proves a principle which Clebsch has termed *Kirchhoff's Principle* and which he has thus stated in his *Theorie der Elasticität*, S. 191 :

Die innern Verschiebungen eines sehr kleinen Körpers sind nur abhängig von den Kräften, welche auf seine Oberfläche wirken, nicht aber von denjenigen, welche auf sein Inneres wirken, vorausgesetzt, dass die letzteren nicht gegen die erstern ausserordentlich gross sind.

Kirchhoff's demonstration of this principle is given in analytical form on S. 286–90 of his memoir (*G. A.* S. 286–91) and is repeated with slight variations on S. 407–9 of the *Vorlesungen*.

After studying both demonstrations I am obliged to confess that
they carry no conviction to my mind. Clebsch after citing the
principle as due to Kirchhoff adds:

von dessen Richtigkeit man sich leicht von vorn herein überzeugt (S. 191).

Clebsch's statement of the proof is as follows:

Man sieht diesen Satz sofort ein, wenn man folgende Erwägung anstellt.
Nehmen wir an, dass die Grösse der auf das Aeussere wirkenden Kräfte, bezogen
auf die Flächeneinheit, und die Grösse der auf das Innere wirkenden Kräfte,
bezogen auf die Volumeneinheit, entweder vergleichbar seien, oder die erstere
sehr gross gegen letztere; nur der umgekehrte Fall sei ausgeschlossen. Dann
ist die Grösse der wirklich auf die Oberfläche des kleinen Körpers wirkenden
Kraft der ganzen Oberfläche oder einem Theil desselben proportional, erhält also
jedenfalls einen Faktor, welcher von der Ordnung der Grösse dieser Oberfläche
ist. Die absolute Grösse der auf das Innere wirkenden Kraft hingegen wird
proportional mit seinem Volumen. Sind nun die Dimensionen des kleinen
Körpers kleine Grössen erster Ordnung, so ist seine Fläche von der zweiten
Ordnung, sein Volumen von der dritten; der Faktor also, mit welchem die auf
das Aeussere wirkenden Kräfte behaftet sind, ist um eine Ordnung niedriger, als
derjenige, mit welchem die auf das Innere wirkenden Kräfte behaftet sind.
Sind also nur die letzten nicht an sich gegen die erstern sehr gross, so wird ihre
Wirkung sehr klein gegen letztere und ist somit zu vernachlässigen. Ich
bemerke dass genau dasselbe Princip bereits im Anfang unserer Untersuchung
benutzt wurde, indem man die innern Verschiebungen eines Elements nur von
den auf seine Oberfläche wirkenden Spannungen, nicht aber von den auf sein
Inneres wirkenden Kräften abhängig machte (S. 191-2).

This statement of Clebsch's appears to contain all the arguments of
Kirchhoff's analysis. But I do not see any reason why exactly the same
argument should not be applied to the elementary right six-face from
which we deduce our fundamental elastic equations; indeed the last
words of Clebsch seem to indicate that in some fashion we do apply it.
The reasoning does not seem to me to clearly explain why for a body of
infinitesimal dimensions we may neglect the right-hand side of the
typical equation :

$$\frac{d\widehat{xx}}{dx} + \frac{d\widehat{xy}}{dy} + \frac{d\widehat{xz}}{dz} = \rho\left(\frac{d^2u}{dt^2} - X\right) \dots\dots\dots (i),$$

but not the right-hand side of the typical equation for the surface-load :

$$l\widehat{xx} + m\widehat{xy} + n\widehat{xz} = X_0 \dots\dots\dots (ii).$$

I am indeed doubtful whether if *all* the dimensions of the body are made
infinitesimal the principle has any real meaning. If we are dealing,
however, with a wire or thin plate, it is the shifts of points on the axis
of the wire or the mid-plane of the plate that we are anxious to discover,
and these shifts depend upon the *resultant* body and *resultant* surface
forces over elements of the wire or plate. In the case of a wire the
dimensions of the cross-section of which are ϵ, and of which δs is an
element of length, the resultant body force is of order $\epsilon^2\delta s\,(\rho X)$ and the

resultant surface force of order $\epsilon \delta s X_0$; hence, if ρX be not very great as compared with X_0, the former term vanishes as compared with the latter when ϵ is extremely small. In the case of the plate, if τ be its thickness and $\delta \omega$ an element of its surface, these resultant forces are of the order $\tau \delta \omega (\rho X)$ and $\delta \omega X_0$ respectively, and, if ρX be not very great as compared with X_0, the former vanishes as compared with the latter, if τ be extremely small. Thus for wires and thin plates we may put the right-hand side of equation (i) zero, if we are merely seeking the shifts of points on the central axis or mid-plane; but if we were to suppose these bodies to have a sensible, if very small cross-section, then it seems to me that for the relative shifts of points on the same cross-section there is no reason why the body- and surface-forces should not have like effect. On the whole the method by which Boussinesq and Saint-Venant approach kindred problems seems to me slightly more convincing than the somewhat vague reasoning of Kirchhoff and Clebsch: see our Arts. 384–94 and Chapter XIII.

[1254.] The next general principle considered by Kirchhoff is similar to that of his memoir on plates. He states that the six stresses expressed as linear functions of the six strains would involve 36 constants, but that 15 of these are equal to 15 others because the expression

$$\widehat{xx}ds_x + \widehat{yy}ds_y + \widehat{zz}ds_z + \widehat{yz}d\sigma_{yz} + \widehat{zx}d\sigma_{zx} + \widehat{xy}d\sigma_{xy}$$

must be the complete differential of a homogeneous function F of the six strains. He remarks in a footnote that this follows easily from the mechanical theory of heat and explains why this is so, concluding with the words:

Diese Betrachtung ist, wie ich glaube, schon von W. Thomson im Quarterly Mathematical Journal (April, 1855[1]) angestellt; ich habe die citirte Stelle nicht einsehen können (S. 290; *G. A. S.* 291).

The strain-energy leads Kirchhoff to the equation of variation:

$$\delta U - \delta \iiint F dx\, dy\, dz = 0 \ldots\ldots\ldots\ldots\ldots\ldots\ldots(i),$$

where δU is the virtual moment of the external forces. A similar form of this equation occurs in the memoir on plates (see our Art. 1235) and had already·been given by Green and others.

[1255.] From a certain property of the function F, Kirchhoff proceeds to show that the above equation, or the general equations of elasticity, determine uniquely the values of the shifts u, v, w, the

[1] *Mathematical and Physical Papers*, Vol. I. pp. 300–5.

translation or rotation of the body as a whole being neglected. This general proof of the uniqueness of the solution of the equations of elasticity has been adopted by Clebsch and Boussinesq (see our Art. 1331 and Chapter XIII.), and was probably suggested by Saint-Venant's memoir on *Torsion* : see our Arts. 6 and 10.

Suppose there are two solutions of the equilibrium equations of elasticity. Substitute the shifts in the three body- and three surface-equations and subtract the corresponding equations for either system of solutions, then there must be values of u, v, w differing from zero (i.e. the difference of the two systems of shifts) for which the right-hand sides of the six equations vanish, or which satisfy equations of the type

$$\frac{d\widehat{xx}}{dx} + \frac{d\widehat{xy}}{dy} + \frac{d\widehat{zx}}{dz} = 0,$$

$$l\widehat{xx} + m\widehat{xy} + n\widehat{xz} = 0.$$

Multiply the first of these equations by $u\,dx\,dy\,dz$, and the corresponding equations by $v\,dx\,dy\,dz$ and $w\,dx\,dy\,dz$ respectively; add and integrate by parts over the whole volume of the solid. Then by means of the second or surface set of equations we easily find

$$\iiint \left(\widehat{xx}s_x + \widehat{yy}s_y + \widehat{zz}s_z + \widehat{yz}\sigma_{yz} + \widehat{zx}\sigma_{zx} + \widehat{xy}\sigma_{xy} \right) dx\,dy\,dz = 0,$$

or $$\iiint F dx\,dy\,dz = 0 \dots\dots\dots\dots\dots\dots\dots(ii).$$

Now for an isotropic body

$$F = \mu \left(s_x^2 + s_y^2 + s_z^2 \right) + \tfrac{1}{2}\mu \left(\sigma_{yz}^2 + \sigma_{zx}^2 + \sigma_{xy}^2 \right) + \tfrac{1}{2}\lambda \left(s_x + s_y + s_z \right)^2.$$

Hence for an isotropic body we must have :

$$s_x = s_y = s_z = \sigma_{yz} = \sigma_{zx} = \sigma_{xy} = 0,$$

or the strains all zero. Thus the two systems of shifts can only differ by a translation or rotation of the body as a whole.

Kirchhoff adds to this proof for *isotropic* bodies :

da bei denjenigen Körpern, welche in verschiedenen Richtungen eine verschiedene Elasticität besitzen, die Unterschiede der Elasticität nur klein sind, so wird man annehmen dürfen, dass bei allen in der Natur vorkommenden Körpern F dieselbe Eigenschaft hat (S. 291; *G. A.* S. 293).

The *Eigenschaft* in question is that of never being negative and only vanishing when the six strains are each separately zero. That bodies with aeolotropic elasticity (e.g. wood) have in fact only 'small differences in their elasticity' seems more than doubtful, but Kirchhoff gives no experimental data. Clebsch in his *Treatise* (S. 68–70) deals with the same problem of the unique solution, and asserts without further proof that F must be a positive quantity and that its vanishing involves the vanishing of the six strains individually. Clebsch may only be thinking of the form of F for isotropic elastic solids; its form for aeolotropic solids requires some further discussion. At any rate Kirchhoff's argument from nearly equal elasticities does not seem conclusive. A modified proof is given by Kirchhoff on S. 394–5 of his *Vorlesungen*, which does not exclude the case of aeolotropic bodies, although any reference to them is omitted. He states however that for a compressible, frictionless fluid, F will take the form given by $\mu = 0$ and λ finite, in which case the vanishing of F does not involve $u = v = w = 0$ for the case of no motion of the fluid as a whole, i.e. the slides may be finite.

[1256.] Kirchhoff concludes the first section of his memoir by throwing equation (i) into a form suitable for a body in which the shifts are not very small, but the strains in each elementary portion are small. We have only to sum F for all these elementary portions, and we have :

$$\delta U - \delta \Sigma \iiint F \, dx \, dy \, dz = 0 \dots\dots\dots\dots\dots\dots\dots (iii).$$

If the body be in motion and T be its kinetic energy this equation becomes " *durch ein bekanntes Prinzip der Mechanik* " :

$$\int dt \{\delta T + \delta U - \delta \Sigma \iiint F \, dx \, dy \, dz\} = 0 \dots\dots\dots\dots (iv).$$

The application of these equations to the case of a thin rod or wire is made in the following sections.

[1257.] Kirchhoff's second section occupies S. 293–302 (*G. A.* S. 295–304) and is substantially reproduced on S. 410–19 of the *Vorlesungen*. Clebsch on S. 190–202 of his *Treatise* deals with the same matter, but soon forsakes Kirchhoff's processes for deductions based on his own solution of Saint-Venant's problem. We shall return to Clebsch's work later (Art. 1359), but may remark here that it is in some respects more, in others less, satisfactory than Kirchhoff's original investigation of the problem.

Kirchhoff supposes the rod to be initially right-cylindrical, and in this initial state takes a rectangular system of axes at the centroid P of any cross-section consisting of the axis of the rod (1) and the principal axes of the cross-section (2, 3). Let x, y, z be the coordinates of any point of the rod relative to these axes before strain and $x + u$, $y + v$, $z + w$ be the coordinates after strain relative to rectangular axes x, y, z, of which the axis of x is the strained position of 1, and the axis of z is perpendicular to the plane through x and 2. Now if x, y, z be supposed to receive only values of the order of the linear dimensions of the cross-section, then x, y, z, u, v, w are quantities which fulfil the conditions required for the equation (iii) to hold. Let ξ, η, ζ be the coordinates of P after strain referred to any rectangular axes in space, and let the former set (x, y, z) make the system of angles whose direction-cosines are given by

$$a_0, \ \beta_0, \ \gamma_0$$
$$a_1, \ \beta_1, \ \gamma_1$$
$$a_2, \ \beta_2, \ \gamma_2$$

with the axes ξ, η, ζ.

Then the coordinates of the point x, y, z after strain with regard to ξ, η, ζ are given by three equations of the type

$$\xi + a_0 (x + u) + a_1 (y + v) + a_2 (z + w) \dots\dots\dots\dots\dots (\text{v}).$$

If s be the distance of the point P from an end of the rod in its unstrained condition, quantities like (v) must be functions of $s + x$, or their partial differentials with regard to s and x must be equal. Since ξ, η, ζ and the direction-cosines are not functions of x we find :

$$\begin{Bmatrix} a_0 \\ \beta_0 \\ \gamma_0 \end{Bmatrix} \left(1 + \frac{du}{dx} \right) + \begin{Bmatrix} a_1 \\ \beta_1 \\ \gamma_1 \end{Bmatrix} \frac{dv}{dx} + \begin{Bmatrix} a_2 \\ \beta_2 \\ \gamma_2 \end{Bmatrix} \frac{dw}{dx}$$

$$= \begin{Bmatrix} d\xi/ds \\ d\eta/ds \\ d\zeta/ds \end{Bmatrix} + \begin{Bmatrix} da_0/ds \\ d\beta_0/ds \\ d\gamma_0/ds \end{Bmatrix} (x + u) + \begin{Bmatrix} da_1/ds \\ d\beta_1/ds \\ d\gamma_1/ds \end{Bmatrix} (y + v) + \begin{Bmatrix} da_2/ds \\ d\beta_2/ds \\ d\gamma_2/ds \end{Bmatrix} (z + w)$$

$$+ \begin{Bmatrix} a_0 \\ \beta_0 \\ \gamma_0 \end{Bmatrix} \frac{du}{ds} + \begin{Bmatrix} a_1 \\ \beta_1 \\ \gamma_1 \end{Bmatrix} \frac{dv}{ds} + \begin{Bmatrix} a_2 \\ \beta_2 \\ \gamma_2 \end{Bmatrix} \frac{dw}{ds} \dots\dots\dots\dots\dots\dots (\text{vi}).$$

Multiply these equations respectively by a_0, β_0, γ_0, then by a_1, β_1, γ_1 and then by a_2, β_2, γ_2 and add in each case, and we find after certain reductions

$$\left. \begin{aligned} \frac{du}{dx} &= \frac{du}{ds} + r (y + v) - q (z + w) + \epsilon, \\[2mm] \frac{dv}{dx} &= \frac{dv}{ds} + p (z + w) - r (x + u), \\[2mm] \frac{dw}{dx} &= \frac{dw}{ds} + q (x + u) - p (y + v) \end{aligned} \right\} \dots\dots\dots\dots (\text{vii}),$$

where

$$\epsilon = \sqrt{\left(\frac{d\xi}{ds} \right)^2 + \left(\frac{d\eta}{ds} \right)^2 + \left(\frac{d\zeta}{ds} \right)^2} - 1.$$

Clearly ϵ is the stretch in ds, and the following relations must hold :

$$\frac{d\xi}{ds} = a_0 (1 + \epsilon), \ \frac{d\eta}{ds} = \beta_0 (1 + \epsilon), \ \frac{d\zeta}{ds} = \gamma_0 (1 + \epsilon) \dots\dots (\text{vii}) \ bis.$$

Further p, q, r are given by the following expressions:

$$p = a_1 \frac{da_2}{ds} + \beta_1 \frac{d\beta_2}{ds} + \gamma_1 \frac{d\gamma_2}{ds},$$

$$q = a_2 \frac{da_0}{ds} + \beta_2 \frac{d\beta_0}{ds} + \gamma_2 \frac{d\gamma_0}{ds}, \quad \Bigg\} \dots\dots\dots\dots\text{(viii)}.$$

$$r = a_0 \frac{da_1}{ds} + \beta_0 \frac{d\beta_1}{ds} + \gamma_0 \frac{d\gamma_1}{ds}$$

[1258.] Kirchhoff now remarks that $\dfrac{du}{dx}$, $\dfrac{dv}{dx}$, $\dfrac{dw}{dx}$ are infinitely great as compared with u, v, w, if we only give to x values of the order of the linear dimensions of the cross-section; further, if $\dfrac{du}{ds}$, $\dfrac{dv}{ds}$, $\dfrac{dw}{ds}$ are not infinitely great as compared with u, v, w, these differentials with regard to s will be infinitely small as compared to those with regard to x. Thus by neglecting infinitely small quantities of the higher order we have:

$$\frac{du}{dx} = ry - qz + \epsilon,$$

$$\frac{dv}{dx} = pz - rx, \quad \Bigg\} \dots\dots\dots\dots\dots\text{(ix)}.$$

$$\frac{dw}{dx} = qx - py$$

For the proof of these assertions Kirchhoff refers rather vaguely to his first paragraph, and there is a similar reference in the *Vorlesungen*, S. 412 (*Gestützt auf die am Ende des vorigen § gemachte Bemerkung*). Clebsch in his *Treatise*, S. 202, puts the matter thus:

Bemerken wir nun, dass bei der Differentiation nach x sich die Grössen u, v, w immer um eine Ordnung unendlich kleiner Grössen erniedrigen, was bei der Differentiation nach s im Allgemeinen nicht geschehen wird, und dass u, v, w klein gegen x, y, z, so reduciren sich diese Gleichungen sich auf (ix).

The argument does not seem to me by any means clear, and I think equations (ix) would be incorrect if there were an appreciable longitudinal or buckling load.

[1259.] By integrating (ix) we find

$$u = u_0 + (ry - qz + \epsilon)\, x,$$
$$v = v_0 + pzx - \tfrac{1}{2}rx^2, \quad \Bigg\} \dots\dots\dots\dots\text{(x)},$$
$$w = w_0 + \tfrac{1}{2}qx^2 - pxy$$

where u_0, v_0, w_0 are quantities independent of x.

By forming the expressions for the strains it will be found that they are all independent of x, so that the body-stress equations reduce to :

$$\frac{d\widehat{xy}}{dy} + \frac{d\widehat{zx}}{dz} = 0,$$

$$\frac{d\widehat{yy}}{dy} + \frac{d\widehat{yz}}{dz} = 0,$$ ....(xi),

$$\frac{d\widehat{yz}}{dy} + \frac{d\widehat{zz}}{dz} = 0$$

and the surface-stress equations at the curved surface to

$$\widehat{xy}\frac{dg}{dy} + \widehat{zx}\frac{dg}{dz} = 0,$$

$$\widehat{yy}\frac{dg}{dy} + \widehat{yz}\frac{dg}{dz} = 0,$$ ....(xii),

$$\widehat{yz}\frac{dg}{dy} + \widehat{zz}\frac{dg}{dz} = 0$$

where $g = 0$ is the equation to the contour of a cross-section, and therefore g is a function of z and y only. Further (xii) supposes no forces to act on the surface of the rod except at the terminal cross-sections.

The arbitrary constants in the values of u, v, w may be determined by the conditions that for $y = z = 0$,

$$u_0 = 0, \quad v_0 = 0, \quad w_0 = 0, \frac{dw_0}{dy} = 0 \dots\dots\dots\dots\text{(xiii)}.$$

[1260.] We easily find for the strains

$$s_x = ry - qz + \epsilon, \quad s_y = \frac{dv_0}{dy}, \quad s_z = \frac{dw_0}{dz},$$

$$\sigma_{yz} = \frac{dv_0}{dz} + \frac{dw_0}{dy}, \quad \sigma_{zx} = \frac{du_0}{dz} - py, \quad \sigma_{xy} = \frac{du_0}{dy} + pz \quad \Bigg\} \dots\dots\text{(xiv)}.$$

The stresses are given as linear functions in terms of these strains, the form of the functions depending on the elastic nature of the rod. If the axis of the rod be parallel to an axis of elasticity we have formulae of the following type, which Kirchhoff cites from an account of a memoir by Rankine (Art. 418) in the *Fortschritte der Physik*, 1850–1, S. 244–9 :

$$\widehat{xx} = |xxxx|\, s_x + |xxyy|\, s_y + |xxzz|\, s_z + |xxyz|\, \sigma_{yz},$$
$$\widehat{yy} = |yyxx|\, s_x + |yyyy|\, s_y + |yyzz|\, s_z + |yyyz|\, \sigma_{yz},$$
$$\widehat{zz} = |zzxx|\, s_x + |zzyy|\, s_y + |zzzz|\, s_z + |zzyz|\, \sigma_{yz}, \quad \Bigg\} \dots\dots\dots\text{(xv)},$$
$$\widehat{yz} = |yzxx|\, s_x + |yzyy|\, s_y + |yzzz|\, s_z + |yzyz|\, \sigma_{yz},$$
$$\widehat{zx} = |zxzx|\, \sigma_{zx} + |zxxy|\, \sigma_{xy},$$
$$\widehat{xy} = |xyzx|\, \sigma_{zx} + |xyxy|\, \sigma_{xy}$$

where the constants have the usual meanings and inter-constant relations : see our Art. 78, p. 77, footnote, and Vol. I. p. 885.

The first body-stress and first surface-stress equations, (xi) and (xii), easily give us :

$$|xxzx| \frac{d^2u_0}{dz^2} + 2\,|zxxy| \frac{d^2u_0}{dydz} + |xyxy| \frac{d^2u_0}{dy^2} = 0 \quad \ldots\ldots\ldots\text{(xvi)},$$

and

$$\left\{ |xxxx| \left(\frac{du_0}{dz} - py \right) + |zxxy| \left(\frac{du_0}{dy} + pz \right) \right\} \frac{dg}{dz}$$

$$+ \left\{ |zxxy| \left(\frac{du_0}{dz} - py \right) + |xyxy| \left(\frac{du_0}{dy} + pz \right) \right\} \frac{dg}{dy} = 0 \ldots\ldots\text{(xvii)}.$$

These equations with the first of (xiii) determine fully u_0, and the other equations of (xi), (xii) and (xiii) determine v_0 and w_0. They should be compared with those obtained for the case of a rod of finite cross-section by Saint-Venant and later by Clebsch : see our Arts. 17, 83 and 1334.

Even if the axis of the rod be not parallel to an elastic axis and (xv) do not hold, (xi), (xii) and (xiii) determine u_0, v_0, w_0 uniquely and as linear homogeneous functions of p, q, r, ϵ : see Kirchhoff's S. 297–8 ($G.\ A.\ S.$ 299). If the values of u_0, v_0, w_0 thus found be substituted in (x) we have u, v, w as linear homogeneous functions of p, q, r, ϵ. The coefficients of these quantities will be independent of s and thus if dp/ds, dq/ds, dr/ds, $d\epsilon/ds$ are not infinitely great as compared with p, q, r, ϵ respectively, equations (x) satisfy the hypothesis we have made in Art. 1258 with regard to du/ds, dv/ds, dw/ds.

[1261.]   The strains will be given by (xiv) as linear homogeneous functions of p, q, r, ϵ also. If these functions be substituted in the value of the strain-energy F, we obtain F as a quadratic function of these quantities, which is independent of x. Integrate this over the cross-section and suppose $\int\int F dy dz = f$, then we may write for equations (iii) and (iv) respectively :

$$\delta U - \delta \int f ds = 0 \ldots\ldots\ldots\ldots\ldots\ldots\ldots\text{(xviii)},$$

$$\int dt\,\{\delta T + \delta U - \delta \int f ds\} = 0 \ldots\ldots\ldots\ldots\ldots\text{(xix)}.$$

It may be noted that Thomson and Tait start from f as a quadratic function of p, q, r, ϵ : see their *Natural Philosophy*, Part II. §§ 592–5. Kirchhoff describes a general method of calculating the values of the coefficients of this function in terms of the usual elastic constants, but it is one which it would not be easy to apply except to special cases.

[1262.]   Kirchhoff remarks on S. 299 ($G.\ A.\ S.$ 301) that the equations (xi), (xii) and (xiii) can be satisfied by the hypothesis made by Saint-Venant in his memoirs on *Torsion* and *Flexure*, namely :

$$\widehat{yy} = \widehat{zz} = \widehat{yz} = 0.$$

See our Arts. 77 (ii), 316–8 and 1334.

He shews, indeed, that this hypothesis gives a possible solution, but he does not prove that it is the only one. His discussion does not bring very much confirmation to Saint-Venant's theory and hardly justifies the note on p. 616 of Moigno's *Statique*; still it is of value as shewing the relation between the two investigations—a relation which has been still more clearly brought out by the researches of Clebsch: see our Arts. 1334–7[1].

[1263.] Kirchhoff next investigates an expression for the kinetic energy T. After some analysis which involves a rather difficult consideration of the relative magnitude of various quantities, he finds :

$$T = \tfrac{1}{2}\rho \iint \left\{ \left(\frac{d\xi}{dt}\right)^2 + \left(\frac{d\eta}{dt}\right)^2 + \left(\frac{d\zeta}{dt}\right)^2 + K^2 P^2 \right\} \omega ds \ \ldots\ldots(\text{xx}),$$

where $\qquad \omega K^2 = \iint (y^2 + z^2)\, d\omega,$

and $\qquad P = a_1 \dfrac{da_2}{dt} + \beta_1 \dfrac{d\beta_2}{dt} + \gamma_1 \dfrac{d\gamma_2}{dt}.$

This might I think have been deduced from general dynamical principles rather more briefly than by Kirchhoff's analysis: see his S. 299–301 (*G. A. S.* 301–3).

[1264.] The second section of the memoir concludes with the extension of the previous results to rods whose unstrained form is curved, the cross-section, however, being the same throughout :

Unter dieser Bedingung wird der Stab durch passende, auf sein Inneres wirkende Kräfte cylindrisch gemacht werden können ; dabei werden seine Theile unendlich kleine Dilatationen erleiden ; bezieht man die Grössen x, y, z und u, v, w auf den Zustand, in dem der Stab sich dann befindet, statt auf seinen natürlichen Zustand, und bezeichnet durch u', v', w' die Werthe, die u, v, w annehmen, wenn man den Stab in seinen natürlichen Zustand und in eine beliebige Lage übergehen lässt, so werden die Gleichungen (iii) und (iv) richtig, wenn man in F statt u, v, w setzt : $u - u'$, $v - v'$, $w - w'$. Daher werden die Gleichungen (xviii)-(xx) auch jetzt gelten, wenn man in f für p, q, r, ϵ gesetzt hat : $p - p'$, $q - q'$, $r - r'$, $\epsilon - \epsilon'$, wo p', q', r', ϵ' die Werthe bedeuten, die p, q, r, ϵ annehmen, wenn man den Stab in seinen natürlichen Zustand und in eine beliebige Lage übergehen lässt. Es sind nämlich in diesem Falle $u - u'$, $v - v'$, $w - w'$ dieselben linearen Funktionen von $p - p'$, $q - q'$, $r - r'$, $\epsilon - \epsilon'$, wie in dem früheren u, v, w von p, q, r, ϵ (S. 302 ; *G. A.* S. 304).

The process here is a very general extension of that by which we deduce the bending-moment at any point of a plane curved rod to be $E\omega\kappa^2 (1/\rho - 1/\rho_0)$ from the value $E\omega\kappa^2/\rho$ in the case of a straight rod : see our Art. 257* and compare Arts. 619–20.

[1] A good deal of Kirchhoff's later work depends upon the supposition that $\widehat{yy} = \widehat{zz} = \widehat{zy} = 0$ is true for rods. Kirchhoff's method of reaching this result has been legitimately criticised by Saint-Venant: see his *Clebsch* pp. 178–81, especially § 7, and our Art. 316.

[1265.] The third section of the memoir further develops equation (xviii) on the assumption that the only external forces are those acting on the terminal cross-sections (S. 302–8, *G. A. S.* 304–11). We have to seek by the processes of the Calculus of Variations four functions p, q, r, ϵ of s, but these quantities are defined by differential coefficients of ξ, η, ζ, a_0, β_0, γ_0, a_1, β_1, γ_1, a_2, β_2, γ_2, between which certain relations hold. Kirchhoff adopts the method of indeterminate multipliers and uses A, B, C, M_0, M_1, M_2 to denote respectively the multipliers of the three relations (vii) *bis* and the three relations (viii). He finds by the ordinary processes of the Calculus and by the elimination of the other multipliers the following sets of equations :

$$\frac{df}{dp}=M_0, \qquad \frac{df}{dq}=M_1, \qquad \frac{df}{dr}=M_2 \ \ldots\ldots\ldots\ldots\ldots(\text{xxi}),$$

$$\frac{df}{d\epsilon}=Aa_0+B\beta_0+C\gamma_0 \quad (=S,\text{ say}) \ \ldots\ldots\ldots\ldots\ldots(\text{xxii}),$$

$$\frac{dA}{ds}=0, \qquad \frac{dB}{ds}=0, \qquad \frac{dC}{ds}=0 \ \ldots\ldots\ldots\ldots\ldots\ldots(\text{xxiii}),$$

$$\left.\begin{aligned}\frac{dM_0}{ds}&=M_2q-M_1r,\\[4pt]\frac{dM_1}{ds}&=M_0r-M_2p-(Aa_2+B\beta_2+C\gamma_2),\\[4pt]\frac{dM_2}{ds}&=M_1p-M_0q+(Aa_1+B\beta_1+C\gamma_1)\end{aligned}\right\}\ \ldots\ldots\ldots(\text{xxiv}).$$

Kirchhoff then deduces the following simple meanings of the quantities A, B, C, M_0, M_1, M_2 :

A, B, C are the sums of the components, parallel to the axes of ξ, η, ζ respectively, of the elastic stresses which act upon the cross-section determined by s, from the side of that portion of the rod which corresponds to greater values of s; M_0, M_1, M_2 are the moments of the same stresses about the axes of x, y, z respectively; these moments are positive when they correspond to a right-handed screw motion round the corresponding axis, such a motion round the x-axis turning a point on the z-axis into the y-axis.

In the *Vorlesungen*, S. 419–21, Kirchhoff starts with these meanings of A, B, C, M_0, M_1, M_2[1] and deduces from statical considerations equations (xxiv) and then equations (xxi) and (xxii). The former set is given more easily by the statical process, the latter by the Calculus of Variations; both processes are instructive especially when compared. Still a third process, more symmetrical and, perhaps, simpler than either of Kirchhoff's, is given by Clebsch in his *Treatise* S. 204–9.

---

[1] M_0, M_1, M_2 are replaced by M_x, M_y, M_z respectively in that work.

[1266.]   Since f is a quadratic function of $p-p'$, $q-q'$, $r-r'$, and $\epsilon - \epsilon'$, it follows from equations (xxi) and (xxii) that M_0, M_1, M_2 and S can be expressed as linear functions of those quantities.   Kirchhoff uses the following system of coefficients:

	$p-p'$	$q-q'$	$r-r'$	$\epsilon-\epsilon'$
M_0	a_{00}	a_{01}	a_{02}	a_{03}
M_1	a_{10}	a_{11}	a_{12}	a_{13} ...(xxv),
M_2	a_{20}	a_{21}	a_{22}	a_{23}
S	a_{30}	a_{31}	a_{32}	a_{33}

where $a_{ij} = a_{ji}$.

Kirchhoff remarks that these a's are not all of equal order since $\epsilon - \epsilon'$ is a mere number, but $p-p'$, $q-q'$, $r-r'$ are the reciprocals of a length.   Hence the a's involving one 3 as subscript must be one linear dimension lower than those containing no subscript 3 and one linear dimension higher than that containing two subscripts 3.   The linear factor can, moreover, only be a linear dimension of the cross-section of the rod, and so an infinitely small quantity.   Thus coefficients with one subscript 3 are infinitely small as compared with a_{33} and infinitely great as compared with those with no subscript 3.   Thus we cannot neglect the terms in $\epsilon - \epsilon'$ in the expressions (xxv), for, although $\epsilon - \epsilon'$ may be very small as compared with $p - p'$, $q - q'$, $r - r'$, still its coefficients are infinitely greater than the others.

From the value of S indicated in (xxv) we find :

$$\epsilon - \epsilon' = - \frac{a_{30}(p-p') + a_{31}(q-q') + a_{32}(r-r') - S}{a_{33}},$$

and if this value of $\epsilon - \epsilon'$ be substituted in the first three expressions we see that unless S is infinitely great as compared with M_0, M_1, M_2 we may neglect the terms in S, thus we find expressions of the form :

$$\begin{aligned} M_0 &= b_{00}(p-p') + b_{01}(q-q') + b_{02}(r-r'), \\ M_1 &= b_{10}(p-p') + b_{11}(q-q') + b_{12}(r-r'), \\ M_2 &= b_{20}(p-p') + b_{21}(q-q') + b_{22}(r-r') \end{aligned} \quad \text{.........(xxvi),}$$

where $b_{ij} = b_{ji}$, and the b's are easily expressed as functions of the a's.

Kirchhoff shows on S. 307 ($G.$ $A.$ S. 310) that S is infinitely great as compared with M_0, M_1, M_2, only when the direction of the resultant of the constant forces A, B, C differs everywhere infinitely little from that of the tangent to the axis of the rod.

Equations [i.e. (xxi)–(xxvi)] theoretically sufficient to fully solve the problem have now been found.

[1267.]   In the last paragraph of this section Kirchhoff points out a very interesting elastico-kinetic analogy (S. 307-8; $G.$ $A.$ S. 310).   Suppose the rod in its unstrained condition straight, or that $p' = q' = r' = 0$.   Then if we substitute the values of the M's from (xxvi) in (xxiv) we obtain the same differential equations as those for the rotation of a heavy body about a fixed point. The symbols used in our elastic investigations must be interpreted in the following manner for the rotating body:

The axes ξ, η, ζ are axes fixed in space, the axes x, y, z are axes fixed in the body at time s, the origin of the latter system is the fixed point of the body and the axis of x passes through its centroid; $-A$, $-B$, $-C$ are the components of the weight of the body parallel to the axes of ξ, η, ζ, multiplied by the x-coordinate of the centroid; finally if m be an element of the mass of the body which has x, y, z for its coordinates, then we must have:

$$b_{00} = \Sigma m\,(y^2 + z^2), \qquad b_{12} = -\Sigma myz,$$
$$b_{11} = \Sigma m\,(z^2 + x^2), \qquad b_{20} = -\Sigma mzx,$$
$$b_{22} = \Sigma m\,(x^2 + y^2), \qquad b_{01} = -\Sigma mxy.$$

To determine the form of the elastic rod, when the corresponding problem of the rotating body is solved, requires us only to perform the three integrations which give the coordinates of a point on the axis of the rod, namely:

$$\xi = \int a_0 ds, \qquad \eta = \int \beta_0 ds, \qquad \zeta = \int \gamma_0 ds.$$

Here the longitudinal stretch ϵ is neglected.

Kirchhoff's elastico-kinetic analogy has been discussed by several later writers: see Thomson and Tait, *Natural Philosophy* Vol. II. §§ 609–13; Hess, *Mathematische Annalen*, Bd. 23, S. 181–212 and Bd. 25, S. 1–38, 1884–5; Greenhill, *Proceedings of the London Mathematical Society*, Vol. XVIII. p. 278, 1888.

[1268.] The fourth and last section of Kirchhoff's memoir is devoted to the following special case: the rod in its original unstrained state is a wire of circular cross-section and its axis has the form of a helix. The rod is supposed to be of homogeneous and isotropic elasticity. See S. 308–13 (*G. A. S.* 311–316). Kirchhoff easily deduces the following expression for the f of our Art. 1261 where the notation of the elastic constants is that of the present work[1]:

$$f = \frac{\omega}{2}\left\{\mu K^2 p^2 + E\left[\tfrac{1}{2}K^2\,(q^2 + r^2) + \epsilon^2\right]\right\} \dots\dots\dots(\text{xxvii}),$$

where $\omega K^2 = \int\int(y^2 + z^2)\,d\omega = 2\int\int y^2 d\omega.$

[1] Our ω, μ, E, ωK^2, stand for the λ, K, $2\dfrac{1+3\theta}{1+2\theta}K$, μ of Kirchhoff's memoir.

Equations (xxvi) take the form :

$$\begin{aligned}
M_0 &= \mu\omega K^2 \,(p - p'), \\
M_1 &= \tfrac{1}{2} E\omega K^2 \,(q - q'), \\
M_2 &= \tfrac{1}{2} E\omega K^2 \,(r - r')
\end{aligned} \Big\} \quad \dots\dots\dots\dots\text{(xxviii)}.$$

[1269.] Kirchhoff now takes θ' for the angle a tangent to the helix makes with the axis, in the unstrained condition, $\dfrac{1}{n'}\sin\theta'$ for the radius of the cylinder on which it lies, and for the unstrained coordinates he puts :

$$\xi' = s\cos\theta', \quad \eta' = \frac{1}{n'}\sin\theta'\sin n's, \quad \zeta' = -\frac{1}{n'}\sin\theta'\cos n's,$$

whence we obtain for a_0', β_0', γ_0' the values :

$$a_0' = \cos\theta', \quad \beta_0' = \sin\theta'\cos n's, \quad \gamma_0' = \sin\theta'\sin n's.$$

Since the cross-section is *circular*, one of the six quantities a_1', β_1', γ_1', a_2', β_2', γ_2' may be assumed to be an arbitrary function of s. Kirchhoff takes :

$$a_1' = \sin\theta'\cos l's,$$

where l' is an arbitrary constant. Hence, after some analysis, he deduces :

$$\begin{aligned}
p' &= l' - n'\cos\theta', \qquad q' = -n'\sin\theta'\cos l's, \\
r' &= -n'\sin\theta'\sin l's
\end{aligned} \Big\} \quad \dots\dots\dots\text{(xxix)}.$$

These equations might have been deduced by other considerations. Now assume ξ, η, ζ, a_0, β_0, γ_0, a_1, β_1, γ_1, a_2, β_2, γ_2 equal to the expressions for the same quantities with dashes, only replacing the constants θ', n', l' by new *constants* θ, n, l.

It will be found that all the equations of the problem are satisfied except (xxiv) whatever be the values of θ, n, l. Further using (xxviii) it will be found that (xxiv) can be satisfied if we take :

$$\begin{aligned}
l &= l', \\
A &= \frac{n}{\sin\theta}\{L\,(n\cos\theta - n'\cos\theta')\sin\theta - N(n\sin\theta - n'\sin\theta')\cos\theta\}, \\
B &= C = 0
\end{aligned} \right\} \;\dots\text{(xxx)},$$

where $\qquad\qquad L = \mu\omega K^2, \qquad N = \tfrac{1}{2} E\omega K^2.$

The condition $B = C = 0$ denotes that *the force acting at the end of the helical wire must have the direction of the axis of the helix.* This is one of the conditions that the values of ξ, η, ζ shall be those assumed, or that the helix shall be strained into a second helix.

Equations (xxviii) give us

$$\begin{aligned}
M_0 &= -L\,(n\cos\theta - n'\cos\theta'), \\
M_1 &= -N\,(n\sin\theta - n'\sin\theta')\cos l's, \\
M_2 &= -N\,(n\sin\theta - n'\sin\theta')\sin l's
\end{aligned} \Big\} \quad \dots\dots\dots\text{(xxxi)};$$

whence if M_ξ, M_η, M_ζ be the couples which act upon the end of the helix with respect to the three axes, we easily find :

$$M_\xi = -\{L(n\cos\theta - n'\cos\theta')\cos\theta + N(n\sin\theta - n'\sin\theta')\sin\theta\},$$
$$M_\eta = A\zeta, \qquad\qquad\qquad M_\zeta = -A\eta \qquad\qquad\qquad\Big\}\dots(\text{xxxii}),$$

where A is given by (xxx) and η, ζ refer to the end of the wire.

The last two equations of (xxxii) evidently give a second condition for the preservation of the helical form, namely : *that the couples* M_η *and* M_ζ *must be exactly equal to those couples which the force* A *would produce round axes, parallel to* η *and* ζ *through the end of the wire, if* A's *point of action were a point of the axis of the helix rigidly united to the end of the wire.*

Thus the helical wire remains helical in form only when the system of force applied at one terminal consists of a force A in the axis of the helix and a couple M_ξ about this axis. If A and M_ξ are given, equations (xxx) and (xxxii) give the values of the constants n and θ which occur in the values of ξ, η, ζ. Finally we note that the elongation of the axis of the helix is given by $s(\cos\theta - \cos\theta')$, and the rotation of the terminal round the axis by $s(n-n')$, where s equals the total length of the helix.

The whole of this investigation deserves careful comparison with the methods of Giulio, J. Thomson and Saint-Venant: see our Arts. 1219*–1223*, 1382*–1384*, 1593*–1595* and 1608*. Kirchhoff remarks that J. Thomson has considered the case in which $M_\zeta = 0$:

aber die Betrachtungen, die er über denselben anstellt, sind nicht strenge, und das Resultat, zu dem er gelangt, ist nicht genau. (S. 313; *G. A.* S. 316.)

[1270.] Special examples of Kirchhoff's method have been given by himself in the *Vorlesungen*: see our Art. 1283, by Clebsch: see his *Treatise* §§ 51–3, and by numerous other writers. Thomson and Tait, after referring to the elastico-kinetic analogy as a beautiful theorem due to Kirchhoff, continue : " to whom also the first thoroughly general investigation of the equations of equilibrium and motion of an elastic wire is due." See *Natural Philosophy*, Part II. § 609.

The present memoir of Kirchhoff's has been made the basis of much of Clebsch's work and has suggested the methods of several later writers. As the most important of Kirchhoff's elastic papers, we have given it fuller treatment than, perhaps, the space at our disposal warranted.

[1271.] *Ueber das Verhältniss der Quercontraction zur Längendilatation bei Stäben von federhartem Stahl: Poggendorffs Annalen,* Bd. 108, S. 369–392 (*G. A. S.* 316–39). Leipzig, 1859.

This memoir is an attempt to settle by direct experiment the problem of uni-constancy. Kirchhoff states the object of his experiments as follows :

Nach theoretischen Betrachtungen von Poisson sollte das Verhältniss der Quercontraction zur Längendilatation immer 1/4 sein; Wertheim schloss aus seinen Versuchen, dass dasselbe 1/3 ist; nach einer mehrfach ausgesprochenen Ansicht hat es weder den einen noch den andern Werth und ist verschieden bei verschiedenen Substanzen. Bei den meisten Körpern, bei denen man eine gleiche Elasticität in verschiedenen Richtungen annehmen kann, stellt sich der experimentellen Bestimmung dieses Verhältnisses der Umstand hindernd in den Weg, dass bei ihnen, auch bei sehr kleinen Formänderungen, bleibende Dehnung und elastische Nachwirkung in erheblichem Grade sich zeigen. Es ist dieses der Fall bei ausgeglühten Metalldrähten und Glassstäben. Bei hart gezogenen Metalldrähten ist eine bleibende Dehnung und eine elastische Nachwirkung viel weniger bemerklich; aber bei ihnen ist sicher die Elasticität in verschiedenen Richtungen verschieden. Bei gehärteten Stahlstäben dagegen kann man wohl mit Wahrscheinlichkeit eine Gleichheit der Elasticität in verschiedenen Richtungen voraussetzen; und da diese überdiess mehr noch als hart gezogene Drähte einem idealen elastischen Körper ähnlich sind, so erscheinen sie vorzugsweise geeignet zu Versuchen über den Werth jenes Verhältnisses. S. 369 (*G. A. S.* 316–7.)

These words of Kirchhoff appreciate so fully the real difficulties of settling the constant-controversy by experiment, that we have reproduced them. It is a pity that they have not been always sufficiently regarded by the many elasticians at home and abroad who have sought to solve this moot-point by experiments on *wires.* Kirchhoff's own rods of 'federhart' steel were, however, portions of drawn wire, and there may indeed be a suspicion as to whether they can be considered to represent accurately enough the ideal isotropic elastic body; even Kirchhoff himself seems to have had doubts on this point : see our Art. 1273.

[1272.] I cannot in this History enter at length into a description of Kirchhoff's experimental methods. They are ingenious and every precaution seems to have been taken to eliminate experimental sources of error. Kirchhoff uses a

method of combined torsion and flexure. He supposes that his
rods are not truly circular but elliptic, and that the square of
the eccentricity may be neglected. He does not, however, take
into account the distortion of the cross-section, and it seems to
me that this might possibly introduce errors whose magnitude
is as great as those Kirchhoff so ingeniously seeks to eliminate.

For three steel rods he finds for the stretch-squeeze ratio η

$$\eta = \cdot293, \quad \cdot295 \text{ and } \cdot294,$$

respectively : or the mean, $\eta = \cdot294$.

This is almost a mean between Wertheim's and Poisson's
values of η (i.e. 1/3 and 1/4).

For a hard drawn brass rod Kirchhoff found $\eta = \cdot387$, but he
remarks that no great stress can be laid on this result, as the
elasticity of such a rod is certainly different in the direction of the
axis and in the plane of the cross-section.

[1273.]   Of the results for the steel rods Kirchhoff writes :

Es wäre von Interesse zu prüfen, ob bei Stahlstäben von anderem
Querschnitte, als die hier untersuchten ihn haben, das genannte
Verhältniss sich eben so gross findet. Wäre das der Fall, so würde
dadurch die hier gemachte Annahme bestätigt werden, dass ein
gehärteter Stahlstab als homogen und von gleicher Elasticität in
verschiedenen Richtungen betrachtet werden darf. Gegen diese Annahme lassen sich Bedenken erheben; in der That kann man sich
vorstellen, dass bei der Härtung, bei der die Wärme von der Axe nach
der Peripherie hin abfliesst, die Elasticität in der Richtung der Axe
eine andere wird, als in den auf dieser senkrechten Richtungen, und
dass die Molecüle in den äusseren Schichten eine andere Anordnung
annehmen, als in den der Axe näheren. Findet dieses statt, so findet
es aber aller Wahrscheinlichkeit nach in verschiedenem Grade statt
je nach der Dicke des Stabes, und es wird jenes Verhältniss anders bei
dicken als bei dünnen Stäben sich ergeben müssen. S. 391. (*G. A.*
S. 338.)

It will thus be seen that Kirchhoff himself doubted the
absolute isotropy of his steel bars, and as no further experiments
on rods of other cross-sections seem to have been made, those of
the present memoir do not allow us to form any really definite
conclusion as to the truth or falsehood of the uni-constant
hypothesis.

[1274.] We may note here a paper of Kirchhoff's which is more closely associated with the theory of light than with that of elasticity. It is entitled : *Ueber die Reflexion und Brechung des Lichtes an der Grenze krystallinischer Mittel* and was first published in the *Abhandlungen der Berliner Akademie* (1876, S. 57-84, *G. A.* S. 352-376). We may very briefly indicate its general object (compare Glazebrook's *Report on Optical Theories,* p. 180). F. Neumann was the first to attempt to apply the theory of elasticity to the reflection and refraction of waves of light at the common surface of two crystalline media : see *Poggendorff's Annalen,* Bd. 25, 1832, S. 418-54, and *Abhandlungen der Berliner Akademie,* 1835, S. 1-160. Neumann supposes no body-forces to act upon the elements of the ether, but he does suppose surface-forces to act upon all surfaces which are the boundaries of different media. In his theory the direction of vibration makes a small angle with the wave-face, but a slight modification of the theory as noted by Neumann himself allows of exact parallelism. MacCullagh proceeds from a totally different hypothesis ; he assumes a form for the potential of the forces acting on an element of the ether which does not arise from an exact elastic theory ; the vibrations in this case are exactly parallel with the wave-front. But an examination of MacCullagh's potential shows that considered with regard to a small portion of the ether in a homogeneous medium, it may be supposed due to surface-forces acting on the surface of this portion. Thus the theory of MacCullagh in reality rests upon the assumption that, besides the ordinary elastic stresses, no other forces act upon the ether except at the boundaries of different media. This is exactly Neumann's hypothesis and the object in both cases is the same, i.e. to get rid of the longitudinal waves. Kirchhoff holds that : *Die beiden genannten Theorien dürfen daher als vollkommen übereinstimmend angesehen werden,* S. 58-9 (*G. A.* S. 352-3).

Kirchhoff's own memoir is to be looked upon as a generalisation and simplification of Neumann's and MacCullagh's work. He obtains a system of eight waves, four in either crystalline medium. This system is dealt with for certain special cases, but not with much detail. He lays special stress on his method of defining a *ray*: see S. 69 (*G. A.* S. 362-3). In the course of his work he refers to the labours of Green (*Camb. Phil. Trans.,* Vol. VII. 1839, pp. 121-40, *Collected Papers,* pp. 291-311, and our Art. 917*) and Lamé (*Leçons sur la théorie...de l'élasticité,* pp. 231-4, and our Art. 1097*) ; he cites Green as deducing a form of elastic potential which leads to results agreeing with Fresnel's, and Lamé for a particular form of the elastic equations. He does not discuss the question of the plane of polarisation nor the points in which MacCullagh's and Neumann's theories are not wholly in agreement with experiment: see Glazebrook (*op. cit.* pp. 157-9, 186 and 193). To obtain his own results he puts the dilatation zero and introduces extraneous surface-forces at the boundary of the two media. On S. 64 (*G. A.* S. 358) he writes:

Bei allen Krystallen, die es giebt, ist die Doppelbrechung nur eine kleine ; hierauf gestützt, darf man annehmen, dass bei jedem Krystall die Constanten der Elasticität des Aethers nur wenig von den Werthen abweichen, die sie in einem isotropen Körper haben können, und dass daher von den drei Wellen, die in ihm in einer Richtung sich fortpflanzen, die eine nahezu longitudinal ist, die beiden andern nahezu transversal sind, und dass die letzteren die Lichtwellen ausmachen.

How far this near equality of the crystalline constants with those of isotropy is really needful, and how far the hypotheses of zero dilatation and extraneous surface-forces are legitimate, it is for those to judge who are better acquainted than the present writer with optical principles. Certain points of Kirchhoff's paper, not very fully noticed by Glaze-brook, have been here indicated as possibly of value to those physicists who still seek aid from the theory of elasticity in expounding the theory of light.

1275. *Vorlesungen über mathematische Physik von Dr Gustav Kirchhoff* (*Professor der Physik an der Universität der Berlin*), *Bd. I. Mechanik.*

This work is in large octavo, and consists of x + 466 pages. The volume was published in three parts, two of which appeared in 1874 and the third in 1876. In a prospectus dated February 1874 the title is given thus: *Vorlesungen über analytische Mechanik mit Einschluss der Hydrodynamik und der Theorie der Elastizität fester Körper.* Thus Elasticity is expressly included in the volume, and we may expect to find it treated with some detail: S. 96–124 and 389–466 relate to our subject. A second edition of the book appeared in November, 1876, a third in 1883.

1276. The tenth *Lecture* occupies S. 96–109. This is purely geo-metrical, and relates to changes in position of the particles of a body, without any reference to the forces which produce these changes. Let x, y, z be the coordinates of a particle of a body; suppose that after a certain time these coordinates become respectively

$$h_1 + a_1x + a_2y + a_3z, \qquad h_2 + b_1x + b_2y + b_3z, \qquad h_3 + c_1x + c_2y + c_3z,$$

where h_1, h_2, h_3, a_1... are functions of the time but independent of x, y, z: that is, suppose we give the body a *homogeneous strain*. The terms h_1, h_2, h_3 correspond to a displacement of the body as a whole. It is shown that the aggregate of the other terms amounts to stretching the body in three directions at right angles to each other, and to rotating the body as a whole round an axis. In fact we have thus nine quantities at our disposal which we can express in terms of the nine quantities a_1, a_2.... For there are *three* dilatations, there are *three* angles which fix the directions of the axes of dilatation, and there are *three* constants involved in rotation round an axis. This indicates the nature of the main subject of the lecture, but does not reproduce quite the method in which Kirchhoff treats it. [The treatment can hardly be considered as so luminous or suggestive as Thomson and Tait's method

of discussing a homogeneous strain : see their *Natural Philosophy*, Part
I. §§ 180–5.　Kirchhoff concludes this *Lecture* by demonstrating that in
the case of continuous motion the surface of any body always contains
the same material points, S. 108–9.]

[1277.]　The eleventh *Lecture* occupies S. 110–124.　It establishes
the equations for the equilibrium or the motion of any body whose
parts are capable of relative motion.　Thus Kirchhoff is able to
deduce the equations both for a fluid and for an elastic solid from
the same investigation.　He deduces the principal properties of the
composition and resolution of stress, but he uses *pressures* instead of
tractions.　If T_1, T_2, T_3 be the principal tractions and l_1, m_1, n_1,
l_2, m_2, n_2, l_3, m_3, n_3 the cosines of the angles they make with the
coordinate axes, Kirchhoff deduces on S. 116 equations which in our
notation are of the type

$$\widehat{xx} = T_1 l_1^2 + T_2 l_2^2 + T_3 l_3^2,$$
$$\widehat{yz} = \widehat{zy} = T_1 m_1 n_1 + T_2 m_2 n_2 + T_3 m_3 n_3 \Bigg\} \dots\dots\dots\dots\dots(\mathrm{i}).$$

He also deals with the properties of the stress-ellipsoid.

On S. 116–9 it is shown that the Hamiltonian principle applied to
bodies whose parts are capable of relative but continuous motion leads
to an equation of the form :

$$0 = \int_{t_0}^{t_1} dt \, (\delta T + U' + F') \dots\dots\dots\dots\dots\dots(\mathrm{ii}),$$

where the integration is for the interval of time t_0 to t_1, while $T =$ the
kinetic energy of the body, $U' =$ the virtual moment of the applied forces,
and $F' = - \iiint dx\,dy\,dz\, (\widehat{xx}\delta s_x + \widehat{yy}\delta s_y + \widehat{zz}\delta s_z + \widehat{yz}\delta \sigma_{yz} + \widehat{zx}\delta \sigma_{zx} + \widehat{xy}\delta \sigma_{xy})$.　In
finding the value of F' in terms of the strains, Kirchhoff *assumes* that
the principal pressures (i.e. negative principal tractions) are for an
isotropic body in the same directions as the principal stretches, and are
linear, homogeneous functions of these stretches.　As in the memoirs he
uses K for our μ and θ for our $\lambda/2\mu$.　Kirchhoff's treatment of the
fundamental equations does not possess special advantages, but it leads
him fairly directly to the Hamiltonian equation (ii), which is the
starting-point for most of the physical investigations dealt with in
the *Vorlesungen*.

[1278.]　Kirchhoff's twenty-seventh *Lecture* occupies S. 389–406.
There is little to remark upon in his general treatment of the elastic
equations or of the strain-energy.　The reader must, however, be careful
to note that Kirchhoff's f in this *Lecture* is the expression

$$-\tfrac{1}{2}(\widehat{xx}s_x + \dots + \dots + \widehat{yz}\sigma_{yz} + \dots + \dots).$$

He uses pressures where we use tractions.　Hence it is equal but
opposite in sign to the F of our Arts. 1254 and 1256.　To the proof of

the uniqueness of the solution of the equations of elasticity which
occurs on S. 392-5 we have already referred : see our Art. 1255.
What is substantially added to the former proof is this: the elastic
solid is supposed to be in *stable* equilibrium when there is no body- or
surface-load.   From this it follows that

$$\int\int\int f\,dx\,dy\,dz$$

must be a maximum when the shifts u, v, w are all zero, that is, when
the strains vanish.   This maximum must also occur for zero values of
the strains when s_x, s_y, s_z, σ_{yz}, σ_{zx}, σ_{xy} are treated as variables independent of u, v, w.

Da nun f eine homogene Function zweiten Grades der genannten Argumente ist, so ist dieser Ausspruch gleichbedeutend mit dem, dass f *nie positiv
ist und nur verschwindet, wenn jedes seiner Argumente verschwindet* (S. 395).

This proof that the strain-energy (i.e. $-f$ in Kirchhoff's notation) is
always *positive* failed in the memoir of 1858 so far as applies to
aeolotropic bodies.   The proof here appears perfectly general : see, however, our Art. 6.

[1279.]   On S. 396-9 Kirchhoff investigates the dilatation-modulus
and the stretch-modulus : see our Art. 1065*.   There is no novelty to
note.   On S. 397-9, he deduces the six conditions of *compatibility*.
These had already been given by Saint-Venant and proved by Boussinesq : see our Art. 112.

[1280.]   Kirchhoff, having obtained the six equations just noticed,
proceeds (S. 399-403) some way in the solution of *Saint-Venant's
Problem* (see our Arts. 2 and 1333) by a method which while investigating flexure and torsion at the same time, is still somewhat
briefer than that of Clebsch.   He only finds, however, expressions for
the three finite stresses, and does not determine the shifts.   The equations (22) which he arrives at on S. 401 for $\widehat{xz}$ and $\widehat{yz}$ agree with
Clebsch's on S. 79 of his *Treatise* (see our Art. 1336).   We must
note that Kirchhoff's Ω differs from Clebsch's by a term of the form
$c_1(x^3 - 3xy^2) + c_2(y^3 - 3yx^2)$; their agreement will then be seen on substituting Kirchhoff's (22) in his (23) and comparing the result with
that given by Clebsch as (67) in his *Treatise*, S. 80.   Kirchhoff's
investigation was evidently suggested by Clebsch's, and we must refer
to our Arts. 1334-45 for a fuller consideration of the subject.   He
applies his results (S. 403-4) to calculate the stress in a right-circular
cylinder under combined flexure and torsion.

[1281.]   On his S. 405-6 Kirchhoff takes the simple example of
a hollow sphere subjected to uniform internal and external pressures.
This had already been dealt with in slightly different methods by various
writers : see our Arts. 1016*, 1094*, 123 and 1201 (c).

1282. The twenty-eighth *Lecture* occupies S. 407–428. This relates to rods having an indefinitely small transverse section. Clebsch says on S. 190 of his *Treatise* that Kirchhoff was the first who gave a rigorous theory of the subject: Clebsch's S. 190–222 correspond with this part of Kirchhoff's work, which is founded on the memoir in Crelle's *Journal*, Bd. 56 (see our Arts. 1251–70), but is in some respects improved. It is however still difficult; it will be necessary to compare it with the discussion given by Clebsch, to notice what is obscure, and to point out its merit as contrasted with what had been given by Poisson and others. Observe that Kirchhoff passes in the next *Lecture* to the case in which the shifts are very small: see his S. 429. Clebsch adopts a similar course for the problem; see his S. 233: and also for the problem of an elastic plate: see his S. 264. Kirchhoff refers on his S. 456 for the case of the finite shifts of an elastic plate to Clebsch, who was the first to treat of them: see our Art. 1350.

[1283.] The differences between the memoir and the lecture may be noted. The first two sections of the latter agree almost entirely with the memoir except for some changes of notation.

(a) The third section (S. 415–7) opens with an example which does not appear in the memoir, but is practically suggested by Saint-Venant's work, namely, the determination of the u_0, v_0, w_0 of our Art. 1259 when the cross-section is the ellipse

$$g = 1 - \frac{y^2}{a^2} - \frac{z^2}{b^2} = 0,$$

where I preserve the notation of that article.
The system of stresses

$$\widehat{zz} = \widehat{yy} = \widehat{yz} = 0,$$

$$\widehat{xy} = c\,\frac{z}{b^2}, \quad \widehat{zx} = -c\,\frac{y}{a^2},$$

where c is an arbitrary constant, and the stretch

$$s_x = ry - qz + \epsilon,$$

will be found to satisfy the equations of Art. 1259 and lead to the result:

$$\frac{d\sigma_{zx}}{dy} - \frac{d\sigma_{xy}}{dz} = 2p,$$

for the determination of c.
Kirchhoff indicates in general terms how the values of u_0, v_0, w_0 may then be found.

(b) S. 417–21 are practically reproductions of the memoir, but on S. 422 a slight modification is introduced. Equations (xxi) of our Art. 1265 show us that M_0, M_1, M_2 are differentials of a function f of p, q, r, ϵ. Equations (xxvi) give us, however, values of M_0, M_1, M_2

from which the ϵ which appears in (xxv) has disappeared. Kirchhoff now supposes G to be the function of p, q, r which f becomes when we eliminate ϵ by means of the fourth expression of (xxv). Then:

$$\frac{dG}{dp} = \frac{df}{dp} + \frac{df}{d\epsilon}\frac{d\epsilon}{dp}.$$

But since we may as a rule (see our Art. 1266) neglect S in the values of the M's, we may put $df/d\epsilon = S = 0$ in the above equation, whence it follows that

$$dG/dp = df/dp = M_0.$$

Similarly, $\qquad\qquad dG/dq = M_1, \quad dG/dr = M_2,$
or, the M's are given by the differentials with regard to p, q, r of the function G.

The equations (xxiv) of our Art. 1265 may then be written:

$$\frac{d}{ds}\left(\frac{dG}{dp}\right) = q\,\frac{dG}{dr} - r\,\frac{dG}{dq},$$

$$\frac{d}{ds}\left(\frac{dG}{dq}\right) = r\,\frac{dG}{dp} - p\,\frac{dG}{dr} - (A a_2 + B \beta_2 + C \gamma_2),$$

$$\frac{d}{ds}\left(\frac{dG}{dr}\right) = p\,\frac{dG}{dq} - q\,\frac{dG}{dp} + (A a_1 + B \beta_1 + C \gamma_1).$$

Kirchhoff now deduces the elastico-kinetic analogy from these equations by taking G as the *kinetic energy* of the rotating body: see our Art. 1267.

(c)   On S. 423 Kirchhoff notes that the problem of the heavy body rotating about a fixed point is not always solvable; but that it is solvable, when the weight is negligible, or again when the body is a solid of revolution and the fixed point about which it rotates is a point on its axis of revolution. Kirchhoff then demonstrates that the elastic problem analogous to the solid of revolution is that of an isotropic rod of circular cross-section.

In the latter case he really falls back on the early part of the treatment of the helix in the memoir: see our Arts. 1268-9. He obtains the value of f we have given in equation (xxvii) of Art. 1268, and the corresponding value G is then given by[1]

$$G = \tfrac{1}{2}\omega K^2 \{\mu p^2 + \tfrac{1}{2}E\,(q^2 + r^2)\}.$$

A special case of the rotation problem is now taken, which had been worked out in the fifth *Lecture*. Kirchhoff assumes the axis of the solid of revolution to describe a right cone about a vertical line. In

---

[1] Note that the f and F of the *Vorlesungen* are interchanged with the F and f of the memoir. Further their signs are reversed. In our discussion of Kirchhoff F is used for the strain-energy per unit volume ($= -f$ of the *Vorlesungen* and F of the memoir) and f is used for the total strain-energy per unit length of the rod ($= -F$ of the *Vorlesungen* and f of the memoir).

this case $q^2 + r^2$ and p^2 are both constants, and the elastico-kinetic analogy is that of a straight rod of circular cross-section bent into a helical shape. He gives as Γ in equation (43), S. 425, and as $M_{y'}$, S. 426, the values of the force and couple which will suffice to bend a straight rod or wire of circular cross-section into a helix of any required pitch and radius (equations (45) and (46), S. 426). If θ be the angle between the thread of the helix and its axis, a the radius of the cylinder upon which it lies, Kirchhoff's results may be expressed as follows:

The force parallel to the axis of the cylinder on which the helix lies

$$= \tfrac{1}{2} \frac{E\omega K^2}{a^2} \cos\theta \sin^2\theta - \frac{\mu\omega K^2}{a} p \sin\theta,$$

and the couple about this axis

$$= -\tfrac{1}{2} \frac{E\omega K^2}{a} \sin^3\theta - \mu\omega K^2 p \cos\theta,$$

where p remains an undetermined constant. Since $(K/a)^2$ is generally extremely small we have at once J. Thomson's theorem that helical springs act chiefly through torsion: see our Art. 1382*, and compare the results of our Arts. 1220* and 1608*.

Kirchhoff takes a special case in which p is chosen equal to $(\cos\theta\sin\theta)/a$: see his S. 426-7. The remainder of the chapter treats a problem similar to that of our Arts. 1268-9 but with a different notation and method.

[1284.] The twenty-ninth *Lecture* deals with the equations for the equilibrium and motion of an infinitely thin rod originally cylindrical, when the shifts are extremely small. It occupies S. 429-449 and contains a number of interesting points. The equations obtained for various special cases had all been previously considered, but not so directly from the general equations of elasticity, i.e. as a rule only from the Bernoulli-Eulerian hypothesis. We proceed to note its contents.

[1285.] In § 1 Kirchhoff deals with the problem of the equilibrium of an initially straight rod of uniform section, when the load is not infinitely nearly in the direction of its axis; no forces are supposed to act except on the terminals of the rod.

In this case p, q, r will be very small quantities, and the equations of our Art. 1283, (b) become:

$$\frac{d}{ds}\left(\frac{dG}{dp}\right) = 0, \quad \frac{d}{ds}\left(\frac{dG}{dq}\right) = -A_1, \quad \frac{d}{ds}\left(\frac{dG}{dr}\right) = A_2 \quad\ldots\ldots\ldots(i),$$

where A_1, A_2 may be looked upon as constants, since the direction of the force makes an angle with the axis of the rod, which varies only infinitesimally. Equation (i) gives M_0, M_1, M_2 as linear functions of s, and the arbitrary constants may be determined by the values of

$$M_0 (= dG/dp), \qquad M_1 (= dG/dq), \qquad M_2 (= dG/dr)$$

at a terminal of the rod.

If axes ξ, η, ζ in space be taken so that the axes x, y, z at each point of the rod differ infinitely little from them, we have

$$a_0 = 1, \ \beta_0 = \gamma_0 = 0, \qquad a_1 = \gamma_1 = 0, \ \beta_1 = 1, \qquad a_2 = \beta_2 = 0, \ \gamma_2 = 1,$$

nearly. Hence we find by Art. 1257 :

$$p = \frac{d\beta_2}{ds}, \qquad q = \frac{d\gamma_0}{ds}, \qquad r = \frac{da_1}{ds} = -\frac{d\beta_0}{ds} \ \dots\dots\dots\text{(ii)}^1.$$

Hence by equations (vii) *bis* of our Art. 1257 we have, neglecting small quantities of the second order, and writing $\beta_2 = \psi$:

$$q = \frac{d^2\zeta}{ds^2}, \qquad r = -\frac{d^2\eta}{ds^2}, \qquad p = \frac{d\psi}{ds} \ \dots\dots\dots\text{(iii)}.$$

Taking y and z for the principal axes of the cross-section and writing

$$\iint y^2 dy dz = \omega\kappa_1{}^2, \qquad \iint z^2 dy dz = \omega\kappa_2{}^2, \qquad \iint dy dz = \omega,$$

we have by *assuming* $\widehat{yy} = \widehat{zz} = \widehat{yz} = 0$ and supposing isotropy :

$$s_y = s_z = -\tfrac{1}{2} \frac{\lambda}{\lambda + \mu} s_x, \qquad \sigma_{yz} = 0 \ \dots\dots\dots\dots\text{(iv)}.$$

[1286.] That this *assumption* is made is not very clear from Kirchhoff's text. He merely refers in vague terms to § 6 of the previous lecture (*Eine Betrachtung, die ähnlich der im Anfange des* § 6 *der vorigen Vorlesung durchgeführten ist, lehrt* u. s. w.). § 6 appeals again without any further qualification to an equation (20ₐ) of § 3. Now at (20ₐ), S. 416 we are merely told that $\widehat{yy} = \widehat{zz} = \widehat{yz} = 0$ satisfies the equations. This passage corresponds to S. 299 of the memoir (see our Art. 1262) where there is a reference to Saint-Venant and there is a more hypothetical statement of these conditions as a *possible* solution. That they give the *only* possible solution is not shown by Kirchhoff and the difficulty is nowhere dealt with by him. This seems to me to form a very weak point in his theory. The matter will be found further discussed in our Arts. 316–8 and Chapter XIII.

[1287.] *Assuming* (iv) to hold, equations (xvi) and (xvii) of our Art. 1260 give us for isotropy :

$$\frac{d^2 u_0}{dz^2} + \frac{d^2 u_0}{dy^2} = 0, \qquad \left(\frac{du_0}{dz} - py\right)\frac{dg}{dz} + \left(\frac{du_0}{dy} + pz\right)\frac{dg}{dy} = 0.$$

---

[1] The last result follows from differentiating the identity $a_0 a_1 + \beta_0\beta_1 + \gamma_0\gamma_1 = 0$.

These are Saint-Venant's torsion equations, and from them we learn that u_0 must contain p as a factor: see our Art. 17. Thus from (iv) of Art. 1285 and (xiv) of Art. 1260 we find

$$s_x = ry - qz + \epsilon, \qquad s_y = s_z = -\tfrac{1}{2}\frac{\lambda}{\lambda + \mu}s_x, \qquad \sigma_{yz} = 0, \left.\begin{array}{c}\\\\\end{array}\right\} \quad \ldots\ldots(v),$$
$$\sigma_{zx} = c_1 p, \qquad\qquad \sigma_{xy} = c_2 p$$

c_1 and c_2 being functions of y and z.

Hence forming the expression for the strain-energy, we have :

$$F = \tfrac{1}{2}E\,(ry - qz + \epsilon)^2 + \tfrac{1}{2}\mu p^2\,(c_1{}^2 + c_2{}^2).$$

Integrating over the cross-section we find :

$$f = \iint F d\omega = \tfrac{1}{2}E\omega\,(\kappa_1{}^2 r^2 + \kappa_2{}^2 q^2 + \chi p^2 + \epsilon^2)\ \ldots\ldots\ldots\ldots(vi),$$

where

$$\chi = \frac{\mu}{E\omega}\iint (c_1{}^2 + c_2{}^2)\,d\omega.$$

Thus χ is the factor found for many sections by Saint-Venant. Kirchhoff merely indicates in the briefest language how f may be obtained. He uses a different notation : see our footnotes, pp. 826 and 836.

Now G is to be found as in our Art. 1283, (b) by putting $df/d\epsilon = 0$ and eliminating ϵ, whence we have :

$$G = \tfrac{1}{2}E\omega\,(\kappa_1{}^2 r^2 + \kappa_2{}^2 q^2 + \chi p^2)\ \ldots\ldots\ldots\ldots\ldots(vii).$$

Equations (i) now give us :

$$E\omega\kappa_1{}^2\frac{dr}{ds} = A_2, \qquad E\omega\kappa_2{}^2\frac{dq}{ds} = -A_1, \left.\begin{array}{c}\\\\\\\end{array}\right\}\ \ldots\ldots\ldots(viii).$$
$$\frac{dp}{ds} = 0$$

Whence, if the moments of the applied system of force at $s = l$ are M_0', M_1', M_2' about the axes of x, y, z respectively, and if we write $A_1 = Z'$, $A_2 = Y'$, we have by (iii) after integration :

$$E\omega\kappa_1{}^2\frac{d^2\eta}{ds^2} = (l - s)\,Y' + M_2',$$
$$E\omega\kappa_2{}^2\frac{d^2\zeta}{ds^2} = (l - s)\,Z' - M_1', \left.\begin{array}{c}\\\\\\\\\end{array}\right\}\ \ldots\ldots\ldots\ldots(ix).$$
$$E\omega\chi\frac{d\psi}{ds} = M_0'$$

The first two are the usual equations of flexure and the third that of torsion. The process by which they are obtained is more satisfactory than the Bernoulli-Eulerian method, but the assumption referred to in our Art. 1286 requires more consideration than is given to it by Kirchhoff.

[1288.] § 2 of this *Lecture* (S. 432-4) removes the restriction of the previous paragraph about the force not being nearly coincident with the direction of the axis of the rod. If it be nearly coincident, the expression we have found for f in (vi) is still true, only we must substitute, if $\xi = s + x$, for the stretch ϵ

$$\epsilon = \frac{dx}{ds} + \tfrac{1}{2} \left\{ \left(\frac{d\eta}{ds}\right)^2 + \left(\frac{d\zeta}{ds}\right)^2 \right\} \quad \ldots\ldots\ldots\ldots\ldots\ldots(x).$$

Kirchhoff then applies the principle of virtual moments to $\delta \int_0^l f ds$ and deduces

$$E\omega\epsilon = X' \ldots\ldots\ldots\ldots\ldots\ldots \ldots\ldots\ldots\ldots(xi),$$

where X' is the load-component in the direction of the axis of the rod at x. The equation of torsion remains the same as in (ix), but the type of flexure equation becomes

$$\left.\begin{array}{c} E\omega\kappa_1{}^2 \dfrac{d^4\eta}{ds^4} - X' \dfrac{d^2\eta}{ds^2} = 0, \\[2mm] \text{with the conditions for } s = l \text{ that} \\[2mm] E\omega\kappa_1{}^2 \dfrac{d^2\eta}{ds^2} = M_2', \\[2mm] E\omega\kappa_1{}^2 \dfrac{d^3\eta}{ds^3} - X' \dfrac{d\eta}{ds} = - Y' \end{array}\right\} \ldots\ldots\ldots\ldots\ldots(xii).$$

These equations agree with (ix), if we may put $X' = 0$.

[1289.] In § 3 Kirchhoff deals theoretically with a method for finding the stretch-modulus suggested by s'Gravesande. In this method a thin rod is stretched between two clamps and loaded in the middle, the stretch-modulus is then to be found from the observed central deflection.

At a clamped end of the rod there will act a couple, a shearing force and a tractive force. The shearing force, neglecting the weight of the rod, will be *one-half* the weight suspended from the centre. Kirchhoff appears to take it equal to the whole weight. Suppose the plane of the bent rod to be that of $\eta\xi$, then the problem is the same as if we took a cantilever of length $l = $ to half that of the rod and supposed the end $s = 0$ built-in, but to the free end $s = l$ applied a couple M_2', a traction X' and a shear $Y' = P/2$, where P is the applied central load.

The first equation of (xii) applies and we have:

$$\frac{d^4\eta}{ds^4} = h^2 \frac{d^2\eta}{ds^2} \quad \ldots\ldots\ldots\ldots\ldots\ldots\ldots\ldots(xiii),$$

where $h^2 = \epsilon/\kappa_1{}^2$ from (xi).

At $s = 0$, $\eta = 0$, and $d\eta/ds = 0$, hence the required form of solution is given by:

$$\eta = C_1 \left(e^{hs} - hs - 1\right) + C_2 \left(e^{-hs} + hs - 1\right).$$

Put $hl = 2p$, apply the latter two equations of (xii) and we get:

$$E\omega\kappa_1{}^2 \left(C_1 h^2 e^{2p} + C_2 h^2 e^{-2p}\right) = M_2' \dots(xiv),$$

$$E\omega\kappa_1{}^2 \left(C_1 h^3 e^{2p} - C_2 h^3 e^{-2p}\right) = -\tfrac{1}{2}P \dots(xv),$$

if we remember that $d\eta/ds = 0$ when $s = l$, which further involves:

$$C_1 \left(e^{2p} - 1\right) + C_2 \left(-e^{-2p} + 1\right) = 0 \dots(xvi).$$

For the central deflection η_l of the rod (corresponding to the deflection at $s = l$ of the cantilever) we have

$$\eta_l = C_1 \left(e^{2p} - 2p - 1\right) + C_2 \left(e^{-2p} + 2p - 1\right)\dots(xvii).$$

Whence after some reductions:

$$\eta_l = \frac{Pl^3}{8E\omega\kappa_1{}^2}\frac{1}{p^2}\left(1 - \frac{1}{p}\tanh p\right) \dots(xviii).$$

This gives η_l in terms of E, when p is known. Kirchhoff has 4 instead of 8 in the denominator of the right-hand of (xviii).

To find p we must return to (x) and note that $\int_0^l \frac{dx}{ds}\,ds$ is a known quantity, or if $2l'$ be the natural length of the rod

$$\int_0^l \frac{dx}{ds}\,ds = l - l' = \gamma l,$$

where γ is the uniform stretch of the rod between the two clamps before the mid load is put on. Hence we have

$$\frac{4p^2\kappa_1{}^2}{l} = \gamma l + \tfrac{1}{2}\int_0^l \left(\frac{d\eta}{ds}\right)^2 ds.$$

Whence Kirchhoff deduces:

$$4p^2\kappa_1{}^2 = \gamma l^2 + \frac{\eta_l^2}{8}\frac{2\cosh 2p + 4 - \dfrac{3}{p}\sinh 2p}{\left(\cosh p - \dfrac{1}{p}\sinh p\right)^2} \dots(xix),$$

and then shows that to a close approximation in the *special cases*, when $\kappa_1{}^2$ is very small as compared with either or both of γl^2 and η_l^2, or when p is very large, η_l and p are given by the equations:

$$\eta_l = \frac{Pl^3}{8E\omega\kappa_1{}^2}\frac{1}{p^2}\left(1 - \frac{1}{p}\right), \quad 4p^2\kappa_1{}^2 = \gamma l^2 + \frac{\eta_l^2}{2}\left(1 + \frac{1}{2p}\right) \dots(xx).$$

I have placed these results here as they seem to suggest a method of testing the stretch-modulus, which is not without its advantages. The equation (xiii) differs from that obtained by Poisson in his *Mécanique*, Vol. I. p. 607, who replaces the right-hand side by a term of the form $h^2\eta$.

[1290.] In § 4, S. 437–8 of this *Lecture*, Kirchhoff works out the case of a heavy rod stretched between two clamps. The method is precisely similar to that of our previous article, except that now the equations become:

$$\kappa_1^2 \frac{d^4\eta}{ds^4} - \epsilon \frac{d^2\eta}{ds^2} = \frac{g\rho}{E},$$

and, $d\epsilon/ds = 0,$

where ρ is the density and g gravitational acceleration.

Kirchhoff solves only the special cases in which κ_1^2 is infinitely great or little as compared with ϵl^2. It seems to me that in many practical cases they would probably be of about the same magnitude.

[1291.] In the remaining sections of this *Lecture* Kirchhoff deals with the vibrations of infinitely thin rods. Only the method by which he has obtained his equations seems to present novelty. I do not think any of his results are new. The following is a brief *résumé* of the contents.

(*a*) § 5 (S. 438–441). Discussion of the equations for the longitudinal and torsional vibrations of a cylindrical rod of infinitely small cross-section deduced from the Hamiltonian principle: see our Art. 1277, equation (ii).

(*b*) § 6 (S. 441–444). Discussion of the equation for the transverse vibrations of a similar rod. Reference is made to Strehlke for the calculation of the frequencies of the notes: see our Art. 356*.

(*c*) § 7 (S. 445–6). Deduction of the equation for the transverse vibrations of a stretched string, when the longitudinal traction due to the stretching is not immensely greater than that due to the transverse shift. Let γ be the permanent stretch of a string of length l, and η its shift at distance s from one terminal, then Kirchhoff gives an equation of the form:

$$\frac{\rho}{E} \frac{d^2\eta}{dt^2} = \left\{ \gamma + \frac{1}{2l} \int_0^l \left(\frac{d\eta}{ds}\right)^2 ds \right\} \frac{d^2\eta}{ds^2}.$$

As a particular solution assume

$$\eta = u \sin \frac{ns}{l}\pi,$$

then $u = b \cos am\, h\,(t - t_0),$ mod. $\kappa,$

if $$\kappa^2 = \frac{1}{2} \frac{n^2\pi^2b^2}{n^2\pi^2b^2 + 4l^2\gamma},$$

$$h^2 = \frac{n^2\pi^2}{4l^4}\frac{E}{\rho}(n^2\pi^2b^2 + 4l^2\gamma),$$

and u be an integer.

Thus we see that the period in this case is a function of the amplitude b of the vibration.

(*d*)   § 8 (S. 446–9).   Consideration of the equation for the transverse vibrations of a very tightly stretched string and the modes of solving it by Fourier's series or by arbitrary functions.

[1292.]   The thirtieth and last *Lecture* of Kirchhoff occupies S. 450–66.   It is devoted to a discussion of plates and membranes. The methods adopted by Kirchhoff are decidedly superior to those of his memoir on plates (see our Arts. 1233 and 1237–8), but the book was published after the *Treatise* of Clebsch and the memoir of Gehring: see our Arts. 1325 and 1411–15.   The first section closely resembles Gehring's work but Kirchhoff does not quote him. Gehring may of course have been much influenced by Kirchhoff's oral lectures, and the method naturally flows from that used in the memoir on thin rods: see our Art. 1251.   Kirchhoff himself remarks :

Aehnliche Betrachtungen, wie wir sie in Bezug auf einen unendlich dünnen, elastischen Stab in den letzten Vorlesungen durchgeführt haben, lassen sich auch in Bezug auf eine unendlich dünne elastische Platte anstellen.   Mit dem Gleichgewicht und der Bewegung einer solchen Platte wollen wir uns jetzt beschäftigen, dabei aber allein den Fall ins Auge fassen, dass dieselbe in ihrem natürlichen Zustande eben ist (S. 450).

[1293.]   In § 1 Kirchhoff obtains the equations for the *finite* shifts of an infinitely thin plate, each element of which is, however, subjected only to very small strain.   The method is similar to that of Clebsch's *Treatise*, S. 264 *et seq.*, where in a footnote its application to the *small* shifts of thin plates is attributed to Gehring, who, Clebsch remarks, followed up a hint given by Kirchhoff in a footnote to his memoir on rods (see *Crelles Journal*, Bd. 56, S. 308, or *G. A.* S. 311).   It is just possible that Kirchhoff practically gave the substance of the method in oral lectures before the appearance of Gehring's dissertation ; he certainly corrects Gehring's errors, and it seems therefore in place to indicate here the lines of the investigation.

Kirchhoff, after deducing an expression for the strain energy in the case of an infinitely thin plate with finite shifts, remarks :

Auf diesen Fall gehen wir nicht näher ein, sondern verweisen in

Bezug auf ihn auf die *Theorie der Elasticität fester Körper* von Clebsch, der zuerst die endlichen Formänderungen unendlich dünner Platten untersucht hat (S. 456).

When applied to finite shifts we may perhaps speak of it for convenience as the *Kirchhoff-Clebsch* method, and, when the equations for the small shifts of infinitely thin plates are deduced from it, as the *German* method, in order to distinguish it from the *French* method, or that due to Boussinesq and Saint-Venant : see our Arts. 384–8 and Chapter XIII.

[1294.] Let s_1, s_2 be the coordinates of a point P in the mid-plane of the plate referred to rectangular axes in that plane, when the plate is unstrained. At P consider in the unstrained state a system of rectangular axes 1, 2, 3 in the material of the plate, of which the first two are parallel to the axes s_1, s_2 and the third perpendicular to them and so to the mid-plane. After strain take a rectangular system x, y, z at P, so that x is a tangent to the strained position of the line 1 at P, y lies in the tangent plane to the mid-plane at P and z is perpendicular to this plane ; y and z will thus make small angles with 2 and 3. Let $x + u$, $y + v$, $z + w$ be the coordinates after strain of an element of the plate in the immediate neighbourhood of P referred to these axes, and so that x, y, z are the coordinates of this element when there is no strain, or the x, y, z axes coincide with 1, 2, 3. Further u, v, w are such functions of x, y, z that for $x = y = z = 0$

$$u = 0, \quad v = 0, \quad w = 0, \quad \frac{dv}{dx} = 0, \quad \frac{dw}{dx} = 0, \quad \frac{dw}{dy} = 0 \ldots\ldots\ldots\text{(i)}.$$

Let ξ, η, ζ be the coordinates of P after strain referred to any axes fixed in space. Let the cosines of the angles between the axes x, y, z after strain and ξ, η, ζ be given by the scheme :

	x	y	z
ξ	a_1	a_2	a_3
η	β_1	β_2	β_3
ζ	γ_1	γ_2	γ_3

Then the coordinates of what before strain was the point $s_1 + x$, $s_2 + y$, z will be given for the space axes by expressions of the type :

$$\xi + a_1 (x + u) + a_2 (y + v) + a_3 (z + w) \ldots\ldots\ldots\ldots\text{(ii)}.$$

These must be functions of $s_1 + x$ and $s_2 + y$, and hence as in the case of a rod (see our Art. 1257) it follows that the differentials with regard to

s_1 and x, and with regard to s_2 and y, must be equal each to each. Thus we have six equations of the types:

$$\left.\begin{array}{l} a_1\left(1+\dfrac{du}{dx}\right)+a_2\dfrac{dv}{dx}+a_3\dfrac{dw}{dx}=a_1\dfrac{du}{ds_1}+a_2\dfrac{dv}{ds_1}+a_3\dfrac{dw}{ds_1} \\[2mm] \qquad +\dfrac{d\xi}{ds_1}+\dfrac{da_1}{ds_1}\left(x+u\right)+\dfrac{da_2}{ds_1}\left(y+v\right)+\dfrac{da_3}{ds_1}\left(z+w\right), \\[3mm] a_1\dfrac{du}{dy}+a_2\left(1+\dfrac{dv}{dy}\right)+a_3\dfrac{dw}{dy}=a_1\dfrac{du}{ds_2}+a_2\dfrac{dv}{ds_2}+a_3\dfrac{dw}{ds_2} \\[2mm] \qquad +\dfrac{d\xi}{ds_2}+\dfrac{da_1}{ds_2}\left(x+u\right)+\dfrac{da_2}{ds_2}\left(y+v\right)+\dfrac{da_3}{ds_2}\left(z+w\right) \end{array}\right\} \ ..\ ...\text{(iii)}.$$

Here a, ξ may be changed into β, η or γ, ζ, without alteration of the subscripts. Kirchhoff writes:

$$\left.\begin{array}{l} 1+\sigma_1=\sqrt{\left(\dfrac{d\xi}{ds_1}\right)^2+\left(\dfrac{d\eta}{ds_1}\right)^2+\left(\dfrac{d\zeta}{ds_1}\right)^2}, \\[4mm] 1+\sigma_2=\sqrt{\left(\dfrac{d\xi}{ds_2}\right)^2+\left(\dfrac{d\eta}{ds_2}\right)^2+\left(\dfrac{d\zeta}{ds_2}\right)^2} \end{array}\right\} \ ....\text{(iv)},$$

and remarking that the axis of x coincides with the direction of the line 1 after strain we have:

$$\frac{d\xi}{ds_1}=a_1\left(1+\sigma_1\right), \quad \frac{d\eta}{ds_1}=\beta_1\left(1+\sigma_1\right), \quad \frac{d\zeta}{ds_1}=\gamma_1\left(1+\sigma_1\right)......\text{(v)}.$$

Further $\dfrac{d\xi}{ds_2}\dfrac{1}{1+\sigma_2}$ equals the cosine of the angle between the line 2 after strain and ξ, or $\cos\left(2,\ \xi\right)$. For the value of $\cos\left(2,\ \xi\right)$ Kirchhoff refers to some results of his tenth *Lecture*, but we easily find from projection that

$$\cos\left(2,\ \xi\right)=a_2+a_1\left(\frac{du}{dy}\right)_0=a_2+a_1\tau,\ \text{say},$$

where τ is the vanishingly small angle by which $(1, 2)$ differs from a right angle after strain. Thus we have equations of the form:

$$\left.\begin{array}{l} \dfrac{d\xi}{ds_2}=\left(a_2+a_1\tau\right)\left(1+\sigma_2\right), \quad \dfrac{d\eta}{ds_2}=\left(\beta_2+\beta_1\tau\right)\left(1+\sigma_2\right), \\[3mm] \qquad\qquad \dfrac{d\zeta}{ds_2}=\left(\gamma_2+\gamma_1\tau\right)\left(1+\sigma_2\right) \end{array}\right\} \ ......\text{(vi)}.$$

Further Kirchhoff writes:

$$\left.\begin{array}{l} p=a_3\dfrac{da_2}{ds}+\beta_3\dfrac{d\beta_2}{ds}+\gamma_3\dfrac{d\gamma_2}{ds}, \\[3mm] q=a_1\dfrac{da_3}{ds}+\beta_1\dfrac{d\beta_3}{ds}+\gamma_1\dfrac{d\gamma_3}{ds}, \\[3mm] r=a_2\dfrac{da_1}{ds}+\beta_2\dfrac{d\beta_1}{ds}+\gamma_2\dfrac{d\gamma_1}{ds} \end{array}\right\} \ ....\text{(vii)},$$

where the subscripts 1, 2 are to be attached to p, q, r according as they are attached to s.

By multiplying both types of equations (iii) first by a_1, β_1, γ_1, secondly by a_2, β_2, γ_2 and finally by a_3, β_3, γ_3, and adding in each case, we obtain the system :

$$du/dx = du/ds_1 + q_1 (z + w) - r_1 (y + v) + \sigma_1,$$
$$dv/dx = dv/ds_1 + r_1 (x + u) - p_1 (z + w),$$
$$dw/dx = dw/ds_1 + p_1 (y + v) - q_1 (x + u),$$
$$du/dy = du/ds_2 + q_2 (z + w) - r_2 (y + v) + \tau (1 + \sigma_2),$$
$$dv/dy = dv/ds_2 + r_2 (x + u) - p_2 (z + w) + \sigma_2,$$
$$dw/dy = dw/ds_2 + p_2 (y + v) - q_2 (x + u).$$

Neglecting terms of the second order of infinitely small quantities as in the case of the rod[1] (see our Art. 1258), and remembering that we must have $d^2u/dxdy$, etc. the same whichever system we derive them from, we find ultimately that $r_1 = r_2 = 0$, $p_1 + q_2 = 0$ and :

$$
\left.
\begin{aligned}
du/dx &= q_1 z + \sigma_1, & du/dy &= -p_1 z + \tau, \\
dv/dx &= -p_1 z, & dv/dy &= -p_2 z + \sigma_2, \\
dw/dx &= p_1 y - q_1 x, & dw/dy &= p_2 y + p_1 x
\end{aligned}
\right\} \quad \ldots\ldots(viii).
$$

Whence by integration

$$
\left.
\begin{aligned}
u &= u_0 - p_1 yz + q_1 zx + \sigma_1 x + \tau y, \\
v &= v_0 - p_2 yz - p_1 zx + \sigma_2 y, \\
w &= w_0 - \tfrac{1}{2} q_1 x^2 + p_1 xy + \tfrac{1}{2} p_2 y^2
\end{aligned}
\right\} \quad \ldots\ldots\ldots(ix),
$$

where u_0, v_0, w_0 are the values of u, v, w for $x = y = 0$.

The strains are easily seen to be given by the following expressions, which are independent of x and y :

$$
\left.
\begin{aligned}
s_x &= q_1 z + \sigma_1, & s_y &= -p_2 z + \sigma_2, & s_z &= dw_0/dz, \\
\sigma_{yz} &= dv_0/dz, & \sigma_{xx} &= du_0/dz, & \sigma_{xy} &= -2p_1 z + \tau
\end{aligned}
\right\} \quad \ldots\ldots(x).
$$

The body stress equations now reduce to :

$$d (\widehat{xz}, \quad \widehat{yz}, \quad \widehat{zz})/dz = 0 \ldots\ldots\ldots\ldots\ldots\ldots\ldots(xi).$$

[1295.] From equations (xi), which should be compared with the equations of our Art. 388 obtained by the French method, Kirchhoff argues as follows :

Nun wollen wir annehmen, dass auf die beiden Oberflächen der Platte Druckkräfte von solcher Grössenordnung wirken, dass sie bei einem Körper, dessen Dimensionen alle von gleicher Ordnung sind, nur Dilatationen erzeugen würden, die unendlich klein sind gegen die Dilatationen, die in der Platte stattfinden. Man darf dann, zunächst für die Oberflächen der Platte, und dann in Folge der abgeleiteten Gleichungen allgemein

$$\widehat{xx} = \widehat{yz} = \widehat{zz} = 0 \ldots\ldots\ldots\ldots\ldots\ldots\ldots\ldots\ldots(xii)$$

---

[1] As in the case of the rod so here I do not follow Kirchhoff's reasoning. A. E. H. Love in a *Note on Kirchhoff's theory of the deformation of elastic plates* (*Cambridge Philosophical Society, Proceedings*, Vol. vi. pp. 144–55, 1889), has endeavoured to strengthen Kirchhoff's process, but I think he leaves it still open to question.

setzen; man vernachlässigt dabei in den Dilatationen und in dem Ausdrucke des Potentials der durch diese erzeugten Kräfte, den wir zu bilden haben werden, nur Glieder, welche unendlich klein sind gegen die beibehaltenen (S. 454).

This reasoning is more complete than that by which equations similar to (xii) were dealt with in the case of a rod: see our Art. 1262. It is not, however, quite clear what the nature of the surface-forces are which will fulfil the condition imposed by Kirchhoff[1], and both this matter and that of the approximation in the preceding article require further consideration than is given to them in the *Vorlesungen*: see our Arts. 1262 and Chapter XIII.

[1296.] Equations (xii) and (i) suffice to determine u_0, v_0, w_0. If the material of the plate be isotropic we have

$$\sigma_{xz} = 0, \quad \sigma_{yz} = 0, \quad s_z + \frac{\lambda}{\lambda + 2\mu}(s_x + s_y) = 0,$$

whence:

$$du_0/dz = 0, \quad dv_0/dz = 0,$$

$$dw_0/dz = \frac{\lambda}{\lambda + 2\mu}\left\{(p_2 - q_1)z - \sigma_1 - \sigma_2\right\} \Bigg\} \dots\dots\dots(xiii).$$

Substituting from (xiii) in (x), and then the values of (x) in the expression F given in our Art. 1255 for the strain-energy of an isotropic solid, we find:

$$F = \mu \left\{ (q_1 z + \sigma_1)^2 + (p_2 z - \sigma_2)^2 + \tfrac{1}{2}(2p_1 z - \tau)^2 \right.$$

$$\left. + \frac{\lambda}{\lambda + 2\mu}\left((p_2 - q_1)z - \sigma_1 - \sigma_2\right)^2 \right\}.$$

Integrating this for the thickness $(2h)$ of the plate from $z = -h$ to h, we have finally for f:

$$f = \tfrac{2}{3}\mu h^3 \left(q_1{}^2 + p_2{}^2 + 2p_1{}^2 + \frac{\lambda}{\lambda + 2\mu}(q_1 - p_2)^2\right) \Bigg\} \dots\dots(xiv).$$

$$+ 2\mu h \left(\sigma_1{}^2 + \sigma_2{}^2 + \tfrac{1}{2}\tau^2 + \frac{\lambda}{\lambda + 2\mu}(\sigma_1 + \sigma_2)^2\right) \Bigg\}$$

The integral $\iint f ds_1 ds_2$ taken over the whole mid-plane gives the entire strain energy of the plate[2].

The six quantities σ_1, σ_2, τ, p_1, p_2, q_1 are all functions of s_1, s_2 and can be expressed in terms of the differentials with regard to s_1 and s_2 of

---

[1] No doubt $\widehat{xz}$, $\widehat{yz}$ and $\widehat{zz}$ are small as compared with the maximum values of $\widehat{xx}$, $\widehat{yy}$ and $\widehat{xy}$, but not necessarily as compared with *all* values of the latter. This at least is the conclusion I have drawn from considering the exact magnitude of the stresses neglected in the similar case of *rods*: *Quarterly Journal of Mathematics*, Vol. xxiv. pp. 63—110. London, 1890.

[2] It should be observed that we have interchanged Kirchhoff's f and F to preserve the notation of the memoir; further Kirchhoff in his *Vorlesungen* uses potential-energy and *not* strain-energy, so that he has $-f$ and $-F$ for our F and f. Compare our footnote p. 76.

ξ, η, ζ. Since τ is the angle by which 2 has approached 1 owing to the strain, we clearly have by (v), etc.:

$$(1 + \sigma_1)(1 + \sigma_2)\,\tau = \frac{d\xi}{ds_1}\frac{d\xi}{ds_2} + \frac{d\eta}{ds_1}\frac{d\eta}{ds_2} + \frac{d\zeta}{ds_1}\frac{d\zeta}{ds_2} \quad\ldots\ldots\ldots\text{(xv)}.$$

[1296 *bis*.] A result of the same form as (xiv) has been obtained by A. E. H. Love for the strain-energy of a thin shell (*The Small Free Vibrations and Deformation of a Thin Elastic Shell. Phil. Trans.*, Vol. 179, A. pp. 491–546, 1888. See p. 505). His result has been called in question by A. B. Basset (*On the Extension and Flexure of Cylindrical and Spherical Thin Elastic Shells. Phil. Trans.*, Vol. 181, A. pp. 433–480, 1890. See p. 433), and the validity of the criticism has been admitted by Love (*Proceedings of the Royal Society*, Vol. 49, pp. 100–2. London, 1891). Basset gives on p. 443 of his memoir an expression for the strain-energy of a distorted cylindrical shell. In this expression there occur terms multiplied by h^3 involving not only the quantities by which the bending is specified but also products of the extensions and of quantities depending principally on the bending. We might therefore be inclined to question whether such terms may not arise in the case of the plate, that is whether (xiv) represents sufficiently closely the strain-energy of a thin plate. Without discussing at this point Basset's method of investigation (which is open to the same sort of criticism as the method of Cauchy and Neumann considered in our Arts. 805 and 1225), we may still ask whether the terms it adds to the strain-energy are of importance in the case of the plate. To do this, we have only to make the radius of Basset's cylindrical shell infinite. It will then be found that Basset's expression for the strain-energy gives the following additional terms to the expression for f in (xiv) of our Art. 1296:

$$\frac{2}{3}\mu h^3 \frac{\lambda}{\lambda + 2\mu}\left\{\sigma_1\frac{d^2(\sigma_1 + \sigma_2)}{dx^2} + \sigma_2\frac{d^2(\sigma_1 + \sigma_2)}{dy^2}\right.$$
$$\left. + \frac{\lambda}{\lambda + 2\mu}(\sigma_1 + \sigma_2)\left(\frac{d^2}{dx^2} + \frac{d^2}{dy^2}\right)(\sigma_1 + \sigma_2) + \tau\frac{d^2(\sigma_1 + \sigma_2)}{dxdy}\right\}.$$

These terms do not therefore in the case of the plate involve the products of extensions and quantities specifying the bending. They form only an addition to the 'membrane terms' in the second line of f, and one of the order h^3 and therefore negligible as compared with those terms. Hence Basset's correction of Love's extension to shells of Kirchhoff's formula does not appear to have any bearing on the correctness of Kirchhoff's results for plates.

[1297.] If the plate has finite bending, we may neglect σ_1, σ_2 and τ as infinitely small, or put:

$$\left. \begin{array}{l} (d\xi/ds_1)^2 + (d\eta/ds_1)^2 + (d\zeta/ds_1)^2 = 1, \\ (d\xi/ds_2)^2 + (d\eta/ds_2)^2 + (d\zeta/ds_2)^2 = 1, \\ \dfrac{d\xi}{ds_1}\dfrac{d\xi}{ds_2} + \dfrac{d\eta}{ds_1}\dfrac{d\eta}{ds_2} + \dfrac{d\zeta}{ds_1}\dfrac{d\zeta}{ds_2} = 0 \end{array} \right\} \quad \ldots\ldots\ldots\ldots (\text{xvi}),$$

instead of (iv) and (xv).

These equations express the condition that the mid-plane remains unstrained, or that it should be a developable surface. In this case the strain-energy contains only the first line of the right-hand side of (xiv). For Clebsch's discussion of this case of finite bending, see our Arts. 1375–8.

[1298.] In § 2 of this *Lecture* (S. 456–9) Kirchhoff proceeds to find an expression for the strain-energy f, when the plate is very slightly bent. In this case we cannot in general neglect σ_1, σ_2 and τ. Now however, x and y may be written for s_1 and s_2, and the system ξ, η, ζ may be chosen so that ξ and η differ infinitely little from x and y, while ζ is infinitely small; thus we may put $\xi = x + u$ and $\eta = y + v$.

Kirchhoff now supposes that u, v and ζ are infinitely small as compared with h:

eine Annahme, die deshalb eine wesentliche ist, weil von beiden Gliedern, aus denen f [see (xiv)] sich zusammensetzt, das eine den Factor h^3, das andere nur den Factor h hat. Bei dieser Annahme ist es ausreichend, in beiden Gliedern nur die ersten Potenzen der Differentialquotienten von u, v, ζ zu berücksichtigen (S. 457).

Equations (iv) and (xv) then give us:

$$\sigma_1 = du/dx, \quad \sigma_2 = dv/dy, \quad \tau = du/dy + dv/dx \ldots \ldots \ldots (\text{xvii}),$$

and equations (v) and (vi):

$$a_1 = \beta_2 = \gamma_3 = 1, \quad a_2 = -\beta_1 = -dv/dx, \quad a_3 = -\gamma_1 = -d\zeta/dx,$$

$$\beta_3 = -\gamma_2 = -d\zeta/dy;$$

whence:

$$p_1 = d^2\zeta/dxdy, \quad p_2 = d^2\zeta/dy^2, \quad q_1 = -d^2\zeta/dx^2 \ldots\ldots(\text{xviii}).$$

If we now make the assumption that u, v, ζ are infinitely small as compared with h, we can use these first approximation values for p_1, p_2, q_1, which occur only in the terms of f multiplied by h^3, but we must proceed to terms of a higher order in the values of σ_1, σ_2 and τ. We find, if we keep the products and squares of differentials of ζ:

$$\left.\begin{array}{c} \sigma_1 = \dfrac{du}{dx} + \tfrac{1}{2}\left(\dfrac{d\zeta}{dx}\right)^2, \quad \sigma_2 = \dfrac{dv}{dy} + \tfrac{1}{2}\left(\dfrac{d\zeta}{dy}\right)^2, \\[2mm] \tau = \dfrac{du}{dy} + \dfrac{dv}{dx} + \dfrac{d\zeta}{dx}\dfrac{d\zeta}{dy} \end{array}\right\} \quad \ldots\ldots\ldots\text{(xix)}.$$

If the values given in (xviii) and (xix) for p_1, p_2, q_1, σ_1, σ_2 and τ be substituted in f in (xiv), we shall be neglecting only those portions of f which are infinitely small as compared with those retained (S. 457).

[1299.] The terms of f depending on h^3 are then :

$$\tfrac{2}{3}\mu h^3\left\{\left(\dfrac{d^2\zeta}{dx^2}\right)^2 + 2\left(\dfrac{d^2\zeta}{dxdy}\right)^2 + \left(\dfrac{d^2\zeta}{dy^2}\right)^2 + \dfrac{\lambda}{\lambda+2\mu}\left(\dfrac{d^2\zeta}{dx^2} + \dfrac{d^2\zeta}{dy^2}\right)^2\right\},$$

an expression which agrees with that contained in the memoir of 1850, and of which Kirchhoff (S. 458–9) proceeds to take the variation in the same manner: see our Art. 1237, (iv). The variation of the second line of f in (xiv) is given on S. 459 without, however, the intermediate stages. For comparison with the results of Clebsch, of Boussinesq and Saint-Venant, I cite it here. dl is an element of the perimeter of the plate, ϕ is the angle between the axis of x and the normal, drawn inwards, to the perimeter; for brevity $\lambda/(\lambda+2\mu) = \eta/(1-\eta)$ is written ν. The required part of f is the following expression multiplied by $4\mu h$:

$$\iint dxdy\left(\dfrac{d\sigma_1}{dx} + \tfrac{1}{2}\dfrac{d\tau}{dy} + \nu\,\dfrac{d(\sigma_1+\sigma_2)}{dx}\right)\delta u$$

$$+ \int dl\left(\sigma_1\cos\phi + \tfrac{1}{2}\tau\sin\phi + \nu(\sigma_1+\sigma_2)\cos\phi\right)\delta u$$

$$+ \iint dxdy\left(\dfrac{d\sigma_2}{dy} + \tfrac{1}{2}\dfrac{d\tau}{dx} + \nu\,\dfrac{d(\sigma_1+\sigma_2)}{dy}\right)\delta v$$

$$+ \int dl\left(\sigma_2\sin\phi + \tfrac{1}{2}\tau\cos\phi + \nu(\sigma_1+\sigma_2)\sin\phi\right)\delta v$$

$$+ \iint dxdy\left\{\dfrac{d}{dx}\left(\dfrac{d\zeta}{dx}\sigma_1 + \tfrac{1}{2}\dfrac{d\zeta}{dy}\tau + \nu\dfrac{d\zeta}{dx}(\sigma_1+\sigma_2)\right)\right.$$

$$\left. + \dfrac{d}{dy}\left(\dfrac{d\zeta}{dy}\sigma_2 + \tfrac{1}{2}\dfrac{d\zeta}{dx}\tau + \nu\dfrac{d\zeta}{dy}(\sigma_1+\sigma_2)\right)\right\}\delta\zeta$$

$$+ \int dl\left\{\cos\phi\left(\dfrac{d\zeta}{dx}\sigma_1 + \tfrac{1}{2}\dfrac{d\zeta}{dy}\tau + \nu\dfrac{d\zeta}{dx}(\sigma_1+\sigma_2)\right)\right.$$

$$\left. + \sin\phi\left(\dfrac{d\zeta}{dy}\sigma_2 + \tfrac{1}{2}\dfrac{d\zeta}{dx}\tau + \nu\dfrac{d\zeta}{dy}(\sigma_1+\sigma_2)\right)\right\}\delta\zeta\ldots\ldots\ldots\text{(xx)}.$$

[1300.] In the following paragraphs Kirchhoff makes special applications of these expressions for the several parts of the variation of the strain-energy.

(*a*) § 3 (S. 459–60). A plate has no load on its faces but its edge is fixed, i.e. *u* and *v* are given there. The variational equations lead to $\zeta = 0$, and to the 'membrane' equations for *u*, *v*, which follow from the 1st and 3rd lines of (xx). These agree with those given by Cauchy and Lamé: see our Arts. 640*, 1072* and 389.

(*b*) § 4 (S. 460–65). This deals with the transverse vibrations of plates and gives briefly certain portions of the memoir of 1850: see our Arts. 1233 *et seq.*

(*c*) § 5 (S. 465–6). Kirchhoff concludes his *Lectures* by investigating the differential equation for the transverse vibrations of a membrane stretched in any manner. In this case *u*, *v* are any shifts which satisfy the differential equations for equilibrium of a stretched membrane given in our Arts. 389–391. If these shifts are considerable as compared with the thickness of the plate, we need only retain the portion of *f* indicated in (xx), putting therein $\delta v = \delta u = 0$ everywhere and $\delta \zeta = 0$ along the perimeter. If *u* and *v* are also so great compared with ζ that we can neglect the second approximation in (xix) and use (xvii), we have:

$$\rho \frac{d^2\zeta}{dt^2} = 2\mu \left\{ \frac{d}{dx}\left[\frac{du}{dx}\frac{d\zeta}{dx} + \tfrac{1}{2}\left(\frac{du}{dy} + \frac{dv}{dx}\right)\frac{d\zeta}{dy} + \nu\left(\frac{du}{dx} + \frac{dv}{dy}\right)\frac{d\zeta}{dx}\right] \right.$$
$$\left. + \frac{d}{dy}\left[\frac{dv}{dy}\frac{d\zeta}{dy} + \tfrac{1}{2}\left(\frac{du}{dy} + \frac{dv}{dx}\right)\frac{d\zeta}{dx} + \nu\left(\frac{du}{dx} + \frac{dv}{dy}\right)\frac{d\zeta}{dy}\right] \right\} \quad \ldots\ldots\ldots(\text{xxi}).$$

This for example is the proper equation for the small vibrations of a very tightly but irregularly stretched drum-head of any form. *u* and *v* are independent of the time and may be any of the numerous functions that satisfy the equations for the equilibrium of a membrane. Kirchhoff cites the special case of an uniformly stretched membrane for which $u = ax$, $v = ay$, *a* being a constant, and deduces the usual equation.

Kirchhoff's method should be carefully compared with that of Boussinesq: see our Chapter XIII.

[1301.] A second and posthumous volume of Kirchhoff's *Vorlesungen über mathematische Physik* entitled *Mathematische Optik* and edited by K. Hensel was published at Leipzig in 1891. In this volume Kirchhoff bases his theory of light upon the equations of an elastic medium. This naturally leads him to Neumann's hypothesis, i.e. that the vibrations take place in the plane of polarisation : see our Arts. 1214 and 1217.

Die Fresnel'sche Annahme ist aber nicht verträglich mit der Hypothese, welche wir an die Spitze unserer optischen Betrachtungen stellten und die sich durch ihre nicht zu übertreffende Einfachheit empfiehlt, mit der Hypothese nämlich, dass der Aether in den durchsichtigen Mitteln in Bezug auf die Lichtbewegung sich verhält wie ein elastischer fester Körper, auf dessen Theile keine anderen Kräfte wirken, als die durch die relativen Verschiebungen erzeugten (S. 141).

Kirchhoff does not discuss how far Neumann's hypothesis leads to results in accordance with experiment, nor does he consider the objections which have been raised to it on several sides : see Glazebrook's *Report on Optics*, pp. 169, 180 and our Art. 1274.

Valuable as many parts of these Lectures on Optics are, they do not, so far as the theory of elasticity is concerned, add much to the researches of F. Neumann : see our Arts. 1213–22.

[1302.] *Ueber die Transversalschwingungen eines Stabes von veränderlichem Querschnitt: Berliner Monatsberichte*, Jahrgang 1879, S. 815–28 (*G. A. S. 339–351*).

The type of rod which Kirchhoff proposes to deal with is defined in the following words :

Es werde zunächst ein Stab ins Auge gefasst, dessen Querschnitt in der Richtung der Länge beliebig, nur so variirt, dass alle Querschnitte unendlich klein sind, ihre Schwerpunkte in einer Geraden liegen und ihre Hauptaxen die gleichen Richtungen haben. Ein solcher Stab kann unendlich kleine Schwingungen ausführen, bei denen die Verschiebungen immer in einer dieser beiden Richtungen geschehen ; um solche Schwingungen soll es sich handeln; die Differentialgleichung derselben ist bekannt und leicht mit Hülfe des Hamilton'schen Principes abzuleiten (S. 815; *G. A. S.* 340).

Kirchhoff cites Lord Rayleigh's *Theory of Sound*, Vol. I. p. 240, as giving the equation for the vibrations. If ξ be the shift at time t of the centroid of the cross-section distant z from one end of the rod, the equation for the vibrations parallel to one system, x, of principal axes of the cross-sections is, in the usual notation of our work :

$$\omega\rho \frac{d^2\xi}{dt^2} + E \frac{d^2}{dz^2}\left(\omega\kappa^2 \frac{d^2\xi}{dz^2}\right) = 0 \dots\dots\dots\dots(i):$$

see our Art. 343, equation (i), putting in it ξ for u, $\rho\omega g$ for p, and $\mathfrak{g} = 0$. Taking a simple tone, or putting $\xi = u \sin pt$, p being a constant, we have

$$\omega\rho p^2 u = E \frac{d^2}{dz^2}\left(\omega\kappa^2 \frac{d^2u}{dz^2}\right) \dots\dots\dots\dots\dots(ii).$$

The conditions to be satisfied at a terminal of the rod, whether fixed or free, are :

$$\frac{d}{dz}\left(\omega\kappa^2 \frac{d^2u}{dz^2}\right)\delta u = 0, \quad \text{and} \quad \omega\kappa^2 \frac{d^2u}{dz^2}\delta\frac{du}{dz} = 0\dots\dots\dots(iii),$$

where δ is the usual symbol of variation.

[1303.] Kirchhoff remarks that equation (ii) can be solved in general terms when the coordinates x and y of the boundary of the cross-section are of the form :

$$x = z^m f_1(\chi), \qquad y = z^n f_2(\chi),$$

χ being any variable quantity and f_1, f_2 given functions of it. In this case we easily find, ξ being parallel to x, and κ the swing-radius about y,

$$\omega = \omega_0 z^{m+n}, \qquad \omega\kappa^2 = \omega_0\kappa_0^2 z^{3m+n},$$

where ω_0 and $\omega_0\kappa_0^2$ are the values of ω and $\omega\kappa^2$ for $z = 1$.

Equation (ii) now becomes :

$$\rho p^2 z^{m+n} u = E\kappa_0^2 \frac{d^2}{dz^2}\left(z^{3m+n}\frac{d^2u}{dz^2}\right)\dots\dots\dots\dots(iv),$$

and may be solved by a series of the form :

$$u = Az^h + A_1 z^{h+(4-2m)} + A_2 z^{h+2(4-2m)} + \dots\dots$$

provided h satisfies the equation

$$h(h-1)(h-2+3m+n)(h-3+3m+n) = 0.$$

See S. 816 (G. A. S. 341).

Kirchhoff discusses the relations between the constants A, A_1, A_2, etc., and the special cases which can arise according as m is $> =$ or < 2 (S. 817–8. G. A. S. 342). He does not, however, enter into special details except for two interesting cases, namely :

(a)   when $m = 1$, $n = 0$,

(b)   ...... $m = 1$, $n = 1$.

In both these cases the integrals admit of being expressed by Bessel's functions with real or imaginary arguments. We devote the following four articles to a consideration of Kirchhoff's results.

[1304.] *Case* (a). If $m = 1$, $n = 0$, and χ be a constant, then the cross-section is rectangular, and the rod is bounded by two parallel planes and a pair of planes perpendicular to these. If the latter meet at a very small angle, the rod may be looked upon as a *very thin wedge*.

Equation (iv) will now be found to be satisfied by either of the alternatives[1]:

$$\frac{1}{z}\frac{d}{dz}\left(z^3\frac{du}{dz}\right) = \pm up\sqrt{\frac{\rho}{E\kappa_0^2}},$$

or, if

$$\zeta = zp\sqrt{\frac{\rho}{E\kappa_0^2}},$$

by

$$\zeta\frac{d^2u}{d\zeta^2} + 2\frac{du}{d\zeta} = \pm u \dots \dots \dots (v).$$

The first forms of solutions, answering to the + and − signs respectively, are

$$u_1 = \frac{d\phi}{d\zeta}, \qquad u_2 = \frac{d\psi}{d\zeta},$$

where

$$\left.\begin{array}{c}\phi\\\psi\end{array}\right\} = 1 \pm \frac{\zeta}{(1\,!)^2} + \frac{\zeta^2}{(2\,!)^2} \pm \frac{\zeta^3}{(3\,!)^2} + \dots \dots \dots (vi).$$

The second forms of solution involve $\log\zeta$, and are thus unsuitable if the end $z = 0$ of the rod be free. Therefore u is of the form:

$$u = C_1\frac{d\phi}{d\zeta} + C_2\frac{d\psi}{d\zeta} \dots \dots \dots (vii).$$

This must satisfy (iii) at the free end $z = 0$, which requires:

$$\zeta^3\frac{d^2u}{d\zeta^2} = 0, \quad \text{and} \quad \frac{d}{d\zeta}\left(\zeta^3\frac{d^2u}{d\zeta^2}\right) = 0,$$

to be fulfilled.

At the base of the wedge, if we suppose it built-in, we must have:

$$u = 0, \quad \text{and} \quad \frac{du}{dz} = 0.$$

This leads to:

$$C_1\frac{d\phi}{d\zeta} + C_2\frac{d\psi}{d\zeta} = 0,$$

$$C_1\frac{d^2\phi}{d\zeta^2} + C_2\frac{d^2\psi}{d\zeta^2} = 0;$$

whence, by writing down the differential equations satisfied by ϕ and ψ, we find that for the base value of z

$$\frac{d}{d\zeta}(\phi\psi) = 0.$$

This is the equation from which the frequencies of the notes must be

---

[1] By taking $z' = 1/z$ the equation reduces to the form $\frac{d^2u}{dz'^2} = \pm\beta^2z'^{-3}u$, β^2 being a constant. This is a case of Riccati's equation and may be solved by Bessel's functions: see Forsyth's *Treatise on Differential Equations*, § 111.

deduced. Kirchhoff finds the value of $\phi\psi$ in a series exactly as he found a similar product in his memoir on plates (see our Art. 1241), namely by ascertaining the differential equation which the product must satisfy. Thus he obtains

$$\phi\psi = 1 - \frac{\zeta^2}{2!(1!)^2} + \frac{\zeta^4}{4!(2!)^2} - \frac{\zeta^6}{6!(3!)^2} + \ldots\ldots,$$

and hence for the frequencies we require the roots of :

$$1 - \frac{\zeta^2}{3!(2!)^2} + \frac{\zeta^4}{5!(3!)^2} - \frac{\zeta^6}{7!(4!)^2} + \ldots\ldots = 0 \ldots\ldots\ldots \text{(viii)}.$$

If l be the length of the wedge he deduces for the fundamental note,

$$\zeta = 5{\cdot}315 = lp \sqrt{\frac{\rho}{E\kappa_0{}^2}},$$

and if $2a$ be the depth, parallel to the direction of vibration, of the base of the wedge, we have

$$p = 5{\cdot}315 \frac{a}{l^2} \sqrt{\frac{E}{3\rho}} \ldots\ldots\ldots\ldots\ldots \text{(ix)},$$

since $\kappa_0{}^2 : \kappa^2$ (for $z = l$) :: $1 : l^2$.

For a prismatic rod of uniform rectangular cross-section of depth $2a$, we should have had :

$$p = 3{\cdot}516 \frac{a}{l^2} \sqrt{\frac{E}{3\rho}} \ldots\ldots\ldots\ldots\ldots \text{(x)},$$

supposing the material and the fixing the same. Hence the fundamental note of the wedge is higher than that of a rod of uniform rectangular cross-section equal to its base.

[1305.] Kirchhoff next proceeds to find how great the shift at the free end of the wedge may be without danger to its elasticity, when the wedge is vibrating solely with its note of lowest pitch. Let s_0 be the limiting safe stretch, then we must have the maximum stretch at every point of the wedge less than this. But this maximum stretch occurs at the contour of the cross-section, and for a cross-section distant z from the free end is $= \frac{az}{l}\frac{d^2\xi}{dz^2} = \frac{az}{l}\frac{d^2u}{dz^2} \sin pt$, or giving $\sin pt$ its maximum value and substituting for z in terms of ζ, we must have

the maximum of $\quad \frac{a\zeta}{l} p \sqrt{\frac{\rho}{E\kappa_0{}^2}} \frac{d^2u}{d\zeta^2} < s_0,$

or, substituting for p from (ix),

the maximum value of $\quad 5{\cdot}315 \frac{a}{l^2} \zeta \frac{d^2u}{d\zeta^2} < s_0.$

Kirchhoff calculates the maximum value of $\zeta \dfrac{d^2 u}{d\zeta^2}$ for the fundamental note (S. 823–4 ; *G. A.* S. 347) and shows that it equals

$$4{\cdot}992\,C,$$

where $U = 19{\cdot}563\,C$ is the maximum shift at the free end. The
maximum stretch occurs at the cross-section for which $\zeta = 3{\cdot}688$,
or $z = \dfrac{3{\cdot}688}{p}\sqrt{\dfrac{E\kappa_0^{\,2}}{\rho}}$, or substituting from (ix), at the point $z = {\cdot}694l$.

Thus with the wedge the greatest strain is *not* at the built-in
terminal, and further the position of the section of greatest strain
varies according to the note the wedge is sounding.

The safe shift of the free end is found for the fundamental note to
be given by

$$
\begin{cases}
U < {\cdot}737 s_0 \,\dfrac{l^2}{a}, \\[2ex]
\;\;\; < 3{\cdot}919 \,\dfrac{s_0}{p} \sqrt{\dfrac{E}{3\rho}}.
\end{cases}
$$

For a rod of uniform rectangular cross-section we have the corresponding expressions :

$$
\begin{cases}
U < {\cdot}284 s_0 \,\dfrac{l^2}{a}, \\[2ex]
U < \dfrac{s_0}{p} \sqrt{\dfrac{E}{3\rho}}.
\end{cases}
$$

Hence if we take a prismatic rod of the same material, of the same
length and on the same base as a wedge, the free end of the latter can
make, in the case when both swing with their fundamental notes,
oscillations of $2{\cdot}6$ times the amplitude of the former. If both be of the
same material and have the same fundamental note (i.e. p the same
for both) but be on bases of different size or shape, then the wedge can
safely receive oscillations at its free end of nearly four times the
amplitude of those of the prism.

These results seem of considerable interest and possibly possess some
practical application.

[1306.] *Case* (*b*). Kirchhoff next passes to the case: $m = n = 1$.
This corresponds to the rod having the form of a very sharp cone.

The differential equation (iv) now takes the form :

$$\frac{\rho p^2}{E\kappa_0^{\,2}}\, z^2 u = \frac{d^2}{dz^2}\left(z^4 \frac{d^2 u}{dz^2}\right),$$

or if $\zeta = zp\sqrt{\dfrac{\rho}{E\kappa_0^{\,2}}}$ we have to find solutions of :

$$\zeta \frac{d^2 u}{d\zeta^2} + 3\frac{du}{d\zeta} = \pm\, u \quad\ldots\ldots\ldots\ldots\ldots\ldots(\text{xi}).$$

Kirchhoff shows that the complete solution in this case is of the form

$$u = C_1 \frac{d^2\phi}{d\zeta^2} + C_2 \frac{d^2\psi}{d\zeta^2},$$

where ϕ and ψ have the values given in (vi).

The equation for the frequencies of the notes, the terminal conditions being as in the previous case, is:

$$\frac{d}{d\zeta}\left(\frac{d\phi}{d\zeta}\frac{d\psi}{d\zeta}\right) = 0,$$

or $$\frac{1}{2!\,3!} - \frac{\zeta^2}{1!\,3!\,5!} + \frac{\zeta^4}{2!\,4!\,7!} - \frac{\zeta^6}{3!\,5!\,9!} + \ldots\ldots = 0.$$

The least root of this is $\zeta_0 = 8\cdot718$, and we find for the fundamental note

$$p = 8\cdot718\frac{1}{l}\sqrt{\frac{E\kappa_0^2}{\rho}}.$$

If κ be the swing radius of the base, or for $z = l$, we have $\kappa : \kappa_0 :: l : 1$,

and $$p = 8\cdot718\frac{\kappa}{l^2}\sqrt{\frac{E}{\rho}} \ldots\ldots\ldots\ldots\ldots \ldots\ldots\text{(xii).}$$

For a cylindrical rod of the same material and base we should have had:

$$p = 3\cdot516\frac{\kappa}{l^2}\sqrt{\frac{E}{\rho}}.$$

Thus the frequencies of the fundamental notes of a sharp conical and of a cylindrical rod of the same length and on the same base are in the ratio of $8\cdot718 : 3\cdot516$.

[1307.]   Finally Kirchhoff proceeds, as in the corresponding case of our Art. 1305, to measure the safe amplitude for the fundamental vibration at the free end of the cone. He finds with the same notation as in that article, a now denoting the maximum distance of any point on the fixed base from the neutral axis:

$$\left\{\begin{array}{l} U < \cdot790s_0\dfrac{l^2}{a}, \\[2ex] \quad < 6\cdot889\dfrac{s_0}{p}\dfrac{\kappa}{a}\sqrt{\dfrac{E}{\rho}}. \end{array}\right.$$

For the cylindrical rod we have:

$$U < \cdot284s_0\frac{l^2}{a},$$

$$< \frac{s_0}{p}\frac{\kappa}{a}\sqrt{\frac{E}{\rho}}.$$

Hence we conclude that for a cylinder and a very sharp cone of the same material and length and on the same base, the cone can have at its free end amplitudes of 2·8 times the magnitude of those of the cylinder, when both vibrate with their fundamental note. If the bases be of the same shape but different size, and the fundamental notes be the same, then the cone can have at its free end amplitudes nearly 7 times those of the cylinder.

The fail-point of the cone for its fundamental note is on the cross-section given by

$$\zeta = 4\cdot464,$$

or,

$$z = \frac{4\cdot464}{p} \sqrt{\frac{E\kappa_0{}^2}{\rho}} = \cdot512\, l.$$

Thus, the fail-point of the cone is about its mid-section.

[1308.] *Bemerkungen zu dem Aufsatze des Herrn Voigt:* "*Theorie des leuchtenden Punktes.*" *Crelles Journal für die Mathematik*, Bd. 90, S. 34. Berlin, 1881. (*G. A. Nachtrag*, S. 17–22.) Voigt deals with an infinitely extended isotropic elastic medium surrounding a rigid sphere at the surface of which there is no slipping. Supposing the sphere to have an infinitely small oscillatory motion it is required to ascertain the vibrations of the medium. He applies the conclusions to be drawn from such a mechanism to the theory of an incandescent point. Obviously the most complex oscillatory motion can be constructed from: (*a*) an oscillatory rotation round a diameter, and (*b*) an oscillatory translation of the sphere as a whole. Kirchhoff shows that the solutions for these special cases can be obtained by an easier method than that of Voigt's memoir.

[1309.] The expressions for the shifts u, v, w at any point of the medium may be put into the form

$$u = \frac{dP}{dx} + \frac{dV}{dz} - \frac{dW}{dy},\; v = \frac{dP}{dy} + \frac{dW}{dx} - \frac{dU}{dz},\; w = \frac{dP}{dz} + \frac{dU}{dy} - \frac{dV}{dx},$$

where P is a solution of

$$\frac{d^2P}{dt^2} = a^2 \nabla^2 P,$$

and U, V, W are solutions of

$$\frac{d^2\phi}{dt^2} = b^2 \nabla^2 \phi,$$

a and b being the velocities with which longitudinal and transverse waves are propagated. These equations are attributed by Kirchhoff to

Clebsch, but they had been previously given by Lamé in his *Leçons sur l'élasticité*, pp. 144–6. In the notation of our work $a^2 = (\lambda + 2\mu)/\rho$ and $b^2 = \mu/\rho$: see our Arts. 1078* and 1394.

Kirchhoff now obtains a solution for Case (*a*) by taking

$$P = U = V = 0 \text{ and } W = \frac{1}{r} F (r - bt),$$

where r is the distance of any point of the medium before strain from the centre of the sphere, and F is an undetermined function. If R be the radius of the sphere and $f(t)$ the angle of rotation *from x to y* at time t, we easily find that for $r = R$, we must have :

$$\frac{dF}{dt} + \frac{b}{r} F + br^2 f(t) = 0.$$

The solution of this equation is given by :

$$\frac{F(r-bt)}{r} = W = -\frac{bR^2}{r} e^{-\frac{bt+R-r}{R}} \int_0^{t+\frac{R-r}{b}} f(t)\, e^{\frac{bt}{R}}\, dt.$$

If $f(t) = 0$, when $t < 0$, then the above solution supposes W and $dW/dt = 0$ for $t = 0$ and $r > R$, that is the medium is supposed to be at rest in its unstrained position before the vibration of the sphere begins at time $t = 0$.

[1310.] Let the motion in Case (*b*) be parallel to the axis of then a suitable solution will be obtained by taking

$$P = \frac{dQ}{dz}, \quad U = \frac{dS}{dy}, \quad V = -\frac{dS}{dx}, \quad W = 0,$$

where, F_1 and F_2 being undetermined functions :

$$Q = \frac{1}{r} F_1 (r - at), \quad S = \frac{1}{r} F_2 (r - bt).$$

If we put $r = R$, $u = v = 0$ and $w = f(t)$, we obtain after some analysis

$$F_1 (R - at) = 2 \frac{a^2}{b^2} F_2 (R - bt) + 3a^2 R \int_0^t dt \int_0^t f(t)\, dt,$$

$$F_2 (R - bt) = \frac{b^2 R}{\lambda_2 - \lambda_1} \left\{ e^{\lambda_1 t} \int_0^t \chi(t)\, e^{-\lambda_1 t}\, dt - e^{\lambda_2 t} \int_0^t \chi(t)\, e^{-\lambda_2 t}\, dt \right\},$$

where

$$\chi(t) - f(t) + \frac{3a}{R} \int_0^t f(t)\, dt + \frac{3a^2}{R^2} \int_0^t dt \int_0^t f(t)\, dt ;$$

and λ_1, λ_2 are the roots of the quadratic :

$$\lambda^2 + \frac{2a+b}{R} \lambda + \frac{2a^2 + b^2}{R^2} = 0.$$

By writing $t + (R - r)/a$ and $t + (R - r)/b$ for t in $F_1(R - at)$ and $F_2(R - bt)$ respectively, we obtain the values of Q and S from these results.

As before, if $f(t) = 0$ for $t < 0$, it will be found that these results suppose the medium at rest in its unstrained position before the sphere begins to oscillate.

Voigt's solution for this case may be obtained from Kirchhoff's by supposing a infinitely great.

[1311.] *Zur Theorie der Lichtstrahlen. Sitzungsberichte der k. Akademie d. Wissenschaften*, Jahrgang 1882, Zweiter Halbband, S. 641-69. Berlin, 1882. *Annalen der Physik*, Bd. 18, S. 663–95. Leipzig, 1883. (*G. A. Nachtrag*, S. 22-54.)

This memoir belongs properly to the theory of light. It starts, however, from the basis of an isotropic elastic medium, in which the dilatation θ is put zero. Kirchhoff remarks :

Die Schlüsse, durch welche man, hauptsachlich gestützt auf Betrach-tungen von Huyghens und Fresnel, die Bildung der Lichtstrahlen, ihre Reflexion und Brechung, sowie die Beugungserscheinungen zu erklären pflegt, entbehren in mehrfacher Beziehung der Strenge. Eine vollkom-men befriedigende Theorie dieser Gegenstände aus den Hypothesen der Undulationstheorie zu entwickeln, scheint auch heute noch nicht möglich zu sein; doch lässt sich jenen Schlüssen eine grössere Schärfe geben. Ich erlaube mir der Akademie Auseinandersetzungen vorzu-legen, welche hierauf abzielen, und deren wesentlichen Inhalt ich in meinen Universitätsvorlesungen seit einer Reihe von Jahren vorge-tragen habe. Das gleiche Ziel in Bezug auf die Beugungserscheinungen ist inzwischen in einigen veröffentlichten Abhandlungen von den Herren Fröhlich und Voigt verfolgt (*Berichte* S. 641; *G. A. Nachtrag*, S. 22).

[1312.] In the course of his work Kirchhoff gives a proof of a generalisation of *Huyghens' Principle* which was first stated by Helmholtz (*Journal für Mathematik*, Bd. 57, S. 1. Berlin, 1860.)

Let ϕ be a solution of the equation

$$\frac{d^2\phi}{dt^2} = b^2 \nabla^2 \phi,$$

on which the transverse vibrations of the medium can be made to depend, and let σ be a closed surface containing none of the points of disturbance; let dn be an element of the normal at $d\sigma$ measured inwards, and let r_0 be the distance of $d\sigma$ from a chosen point O inside σ. Lastly let $d\phi/dn = f(t)$.

Then Kirchhoff deduces from Green's Theorem (*Mathematical Papers*, p. 23) that

$$4\pi\phi_0(t) = \iint \Omega d\sigma,$$

where
$$\Omega = \frac{d}{dn}\left\{\frac{\phi\left(t - \frac{r_0}{b}\right)}{r_0}\right\} - \frac{f\left(t - \frac{r_0}{b}\right)}{r_0},$$

and ϕ_0 is the value of ϕ at O (S. 646; *G. A. Nachtrag*, S. 28).

Thus it is always possible to replace the system of disturbing points by a new distribution of disturbing points over the surface σ (supposed to contain none of the old points), provided we know the values of ϕ and $d\phi/dn$ due to the old system over this surface, and that the point O lies inside it.

Kirchhoff discusses further what modifications are introduced when a disturbing point lies inside the surface o.

The major portion of the memoir is too closely associated with the theory of light to be discussed here.

[1313.] *Ueber die Formänderung, die ein fester elastischer Körper erfährt, wenn er magnetisch oder diëlectrisch polarisirt wird.* Sitzungsberichte der k. Akademie d. Wissenschaften, Jahrgang 1884, Erster Halbband, S. 137–56. Berlin, 1884. *Annalen der Physik*, Bd. 24, S. 52–74. Leipzig, 1885. (*G. A. Nachtrag*, S. 91–113.)

Sir William Thomson and Clerk-Maxwell have both discussed the mechanical forces called into play in a body when placed in an electro-magnetic field, and Helmholtz has extended their results by introducing, besides the constant of induction, a second constant which is to be determined by the changes which result from a change of density in the medium. Kirchhoff proposes to still further generalise their conclusions by introducing a third constant to express the changes experienced by the induction owing to the existence of the most general form of strain, when the body is elastic. Lorberg in an article entitled: *Ueber Electrostriction* (*Annalen der Physik*, Bd. 21, S. 300–29, 1884) simultaneously reached by different considerations like results.

[1314.] Kirchhoff[1] conceives an elementary sphere of iron, which he supposes isotropic, to have undergone the uniform stretches s_1, s_2, s_3 in three rectangular directions. Then, if A_0, B_0, C_0 be the components

---

[1] I have partially changed Kirchhoff's notation to agree better with the customary English one of Maxwell. He uses $\lambda_1, \lambda_2, \lambda_3, \mu_1, \mu_2, \mu_3, k, a, \beta, \gamma, A, B, C, \bar{A}, \bar{B}, \bar{C}$ for our $s_1, s_2, s_3, A_0, B_0, C_0, \kappa, A, B, C, P_x, P_y, P_z, \bar{P}_x, \bar{P}_y, \bar{P}_z$, respectively.

of the magnetic intensity, due to constant magnetic forces J_1, J_2, J_2 acting in these directions, Kirchhoff takes, if $\theta = s_1 + s_2 + s_3$:

$$\left. \begin{aligned} A_0 &= (p - p'\theta - p''s_1)\,J_1, \\ B_0 &= (p - p'\theta - p''s_2)\,J_2, \\ C_0 &= (p - p'\theta - p''s_3)\,J_3 \end{aligned} \right\} \quad \dots\dots\dots\dots\dots\text{(i)}.$$

Next taking this sphere as an infinitely small part of a finite mass of iron, which has been magnetised by given external forces, he puts:

$$J_1 = \tfrac{4}{3}\pi A_0 - \frac{d\phi}{dv_1}, \quad J_2 = \tfrac{4}{3}\pi B_0 - \frac{d\phi}{dv_2}, \quad J_3 = \tfrac{4}{3}\pi C_0 - \frac{d\phi}{dv_3} \quad \dots\dots\text{(ii)},$$

where v_1, v_2 and v_3 are the directions of the principal stretches s_1, s_2, s_3, and ϕ is equal to the sum of V, the potential due to external magnetism, and Q, the potential of the whole magnetised mass of iron, at the element.

Substituting (ii) in (i) and neglecting the terms involving the squares of the strain we have three relations of the type:

$$A_0 = -(\kappa - k'\theta - k''s_1)\frac{d\phi}{dv_1} \quad \dots\dots\dots\dots\text{(iii)}.$$

Here κ, k' and k'' are constants, functions of p, p' and p'', and taken as depending solely on the nature of the iron. In a second paper (see our Art. 1319) Kirchhoff states that although the theory supposes κ a constant, it really varies immensely with the value of

$$\sqrt{\left(\frac{d\phi}{dv_1}\right)^2 + \left(\frac{d\phi}{dv_2}\right)^2 + \left(\frac{d\phi}{dv_3}\right)^2}$$

Calling this expression R we have by (iii) and (ii), supposing the strain terms zero or small as compared with κ, and taking $J = \sqrt{J_1^2 + J_2^2 + J_3^2}$,

$$R = J/(1 + \tfrac{4}{3}\pi\kappa).$$

In some experiments of Stoletow (*Annalen der Physik* Bd. 146, S. 461, 1872) cited by Kirchhoff in his second paper, κ for soft iron rises from 21·5 to 174 as R varies from ·43 to 3·2, and sinks to 42·1 as R increases further to 30·7. Ewing (*Phil. Trans.* 1885, p. 548) has shown that the fluctuations in the values of κ (Maxwell's 'coefficient of induced magnetisation') for soft iron largely exceed even those Kirchhoff cites from Stoletow. The bearing of this variation of κ on the fundamental differential equation is not considered by Kirchhoff in his memoir.

Further it is more than doubtful whether experiments in the case of soft iron, nickel or cobalt justify Kirchhoff's neglect of the terms involving the square of the strain. His equations and conclusions are therefore given here with every reservation.

[1315.] Reducing the above results for the principal stretch-axes of each element to general axes x, y, z in space, parallel to which the components of magnetic intensity at the point x, y, z are A, B, C, Kirchhoff finds :

$$A = -\left(\kappa - k'\theta - k''s_x\right)\frac{d\phi}{dx} + \tfrac{1}{2}k''\sigma_{xy}\frac{d\phi}{dy} + \tfrac{1}{2}k''\sigma_{zx}\frac{d\phi}{dz},$$

$$B = \tfrac{1}{2}k''\sigma_{xy}\frac{d\phi}{dx} - \left(\kappa - k'\theta - k''s_y\right)\frac{d\phi}{dy} + \tfrac{1}{2}k''\sigma_{yz}\frac{d\phi}{dz}, \quad \left.\right\} \ \ldots\ldots(\text{iv}).$$

$$C = \tfrac{1}{2}k''\sigma_{zx}\frac{d\phi}{dx} + \tfrac{1}{2}k''\sigma_{yz}\frac{d\phi}{dy} - \left(\kappa - k'\theta - k''s_z\right)\frac{d\phi}{dz}$$

He now proceeds to determine the general differential equation for ϕ. Since $\phi = V + Q$, and Q is given by :

$$Q = \iiint \left\{ A \frac{d}{dx}\left(\frac{1}{r}\right) + B \frac{d}{dy}\left(\frac{1}{r}\right) + C \frac{d}{dz}\left(\frac{1}{r}\right)\right\}\, d\varpi,$$

where $d\varpi$ is an element of the mass of iron, and r the distance of this element from the point at which Q is the potential, we have :

$$\frac{1}{4\pi}\nabla^2\phi - \frac{1}{4\pi}\nabla^2 Q = \frac{1}{4\pi}\nabla^2 V,$$

or, by integrating the expression for Q by parts :

$$\frac{1}{4\pi}\nabla^2\phi - \frac{dA}{dx} - \frac{dB}{dy} - \frac{dC}{dz} = \frac{1}{4\pi}\nabla^2 V. \quad\ldots\ldots\ldots\ldots(\text{v})^1$$

Here A, B, C must be given the values in (iv) above. Following an idea of Helmholtz's, Kirchhoff supposes the iron to change not abruptly but gradually to air, so that κ, k', k'' take values varying from those they have in iron to those for air, or zero, through a thin shell over the surface of the iron-mass, this shell being ultimately reduced to an infinite thinness (*Berichte* S. 140-1 ; *G. A. Nachtrag*, S. 95).

Kirchhoff shows how (v) may be replaced by an equation expressing that the variation of a certain integral vanishes, but to discuss this integral would carry us beyond our limits (S. 141-4 ; *G. A. Nachtrag*, S. 96-101).

[1316.] If P_x, P_y, P_z represent the terms that must be added to the body-forces ρX, ρY, ρZ in the body-stress equations of type :

$$\frac{d\widehat{xx}}{dx} + \frac{d\widehat{xy}}{dy} + \frac{d\widehat{zx}}{dz} + \rho X = 0,$$

to represent the effect of the magnetisation, and $\bar{P}_x$, $\bar{P}_y$, $\bar{P}_z$ the terms that

---

[1] For the iron mass itself $\nabla^2 V = 0$, and if k' and k'' were to be neglected as small compared with κ we should have the usual equation

$$\frac{d}{dx}(1+4\pi\kappa)\frac{d\phi}{dx} + \frac{d}{dy}(1+4\pi\kappa)\frac{d\phi}{dy} + \frac{d}{dz}(1+4\pi\kappa)\frac{d\phi}{dz} = 0.$$

must be added to the surface-stresses X_0, Y_0, Z_0 in the surface-stress equations of type

$$\widehat{xx} \cos (nx) + \widehat{xy} \cos (ny) + \widehat{xz} \cos (nz) = X_0,$$

where n is the normal to the surface measured *inwards*, then Kirchhoff shows (S. 146-8 ; *G. A. Nachtrag*, S. 102-4) that:

$$P_x = -\tfrac{1}{2}\frac{d\kappa}{dx}R^2 + \tfrac{1}{2}\frac{d}{dx}(k'R^2) + \tfrac{1}{2}\frac{d}{dx}\left(k''\left(\frac{d\phi}{dx}\right)^2\right) + \tfrac{1}{2}\frac{d}{dy}\left(k''\frac{d\phi}{dx}\frac{d\phi}{dy}\right)$$

$$+ \tfrac{1}{2}\frac{d}{dz}\left(k''\frac{d\phi}{dx}\frac{d\phi}{dz}\right), \quad \Biggr\} \ \ ...(vi),$$

and

$$\overline{P}_x = -2\pi\kappa^2\left(\frac{d\phi}{dn}\right)^2 \cos(nx) - \frac{\kappa - k'}{2}R^2 \cos(nx) + \frac{k''}{2}\frac{d\phi}{dx}\frac{d\phi}{dn}.$$

with similar values for P_y, P_z and $\overline{P}_y$, $\overline{P}_z$.

Here R^2 represents as before

$$\left(\frac{d\phi}{dx}\right)^2 + \left(\frac{d\phi}{dy}\right)^2 + \left(\frac{d\phi}{dz}\right)^2.-$$

These results agree with those of Helmholtz if k' and k'' be put zero.

[1317.] Kirchhoff remarks (S. 149 ; *G. A. Nachtrag*, S. 105):

Die in Bezug auf einen Eisenkörper angestellten Betrachtungen lassen sich auf ein Diëlectricum übertragen, wenn dieses an Stelle des Eisens und ein electrisirter Nichtleiter an Stelle des Magnets gesetzt wird. Der Nichtleiter kann aber auch durch Leiter ersetzt werden, da es für die Kräfte, die auf ein Element des Diëlectricums wirken, gleichgültig ist, ob die electrischen Flüssigkeiten, von denen diese Kräfte herrühren, soweit sie in endlicher Entfernung von dem Elemente liegen, in ihren Trägern beweglich sind, oder nicht.

On S. 150-2 (*G. A. Nachtrag*, S. 106-9) Kirchhoff points out how another method, which has been, indeed, adopted by Boltzmann, does not lead to the correct equations.

[1318.] Finally Kirchhoff works out the case of a spherical condenser of glass bounded by two concentric surfaces of radii r_1 and $r_2 (r_2 > r_1)$. These surfaces are provided with conducting coatings, the inner of which is maintained at potential ϕ_0 and the outer at potential zero, and pressures on these coatings are supposed to be at once transferred to the glass surfaces. Kirchhoff agreeing with Korteweg (*Annalen der Physik*, Bd. 9, S. 48-61, 1880) finds that the extension of the internal radius is given by:

$$u_1 = \frac{1}{2E}\frac{\phi_0^2}{(r_2 - r_1)^2}\frac{r_2^2}{r_1}\left(\frac{1}{4\pi} + \kappa - \frac{2\mu k' - \lambda k''}{2(\lambda + \mu)}\right),$$

where E is the stretch-modulus and λ and μ the usual elastic coefficients.

The κ, k' and k'' of this article are of course not those of Art. 1314, but constants of the dielectric, $1 + 4\pi\kappa$ being the K of Maxwell, or the specific inductive capacity. It is an analytical not a physical relation, which enables us to apply the results for magnetisation to the case of a dielectric : see our Art. 1317.

[1319.] *Ueber einige Anwendungen der Theorie der Formänderung, welche ein Körper erfährt, wenn er magnetisch oder diëlectrisch polarisirt wird. Sitzungsberichte d. k. Akademie der Wissenschaften,* Jahrgang 1884, Zweiter Halbband, S. 1155–70. Berlin, 1884. *Annalen der Physik,* Bd. 25, S. 601–17. Leipzig, 1885 (*G. A. Nachtrag,* S. 114–31).

The only portion of this memoir which concerns elastic solids is § 5, which deals with the change in form undergone by an isotropic iron sphere of radius r_0 when magnetised by a constant magnetic force of intensity J in the direction x.

In this case at a great distance from the sphere the centre being the origin, and the notation that of our Art. 1314 :

$$\phi = - Jx,$$

and inside the sphere :

$$\phi = -\frac{J}{1 + \frac{4}{3}\pi\kappa}\, x.$$

Whence from equations (vi) of our Art. 1316 we have to find the strains in an elastic medium subjected to no body-forces, for $P_x = P_y = P_z = 0$, but to the surface-stresses $\overline{P}_x$, $\overline{P}_y$, $\overline{P}_z$ given by :

$$\overline{P}_x/x = \frac{\beta}{r_0^3}\left(2\pi\kappa^2 x^2 + \frac{\kappa - k' - k''}{2}\, r_0^2\right),$$

$$\overline{P}_y/y = \overline{P}_z/z = \frac{\beta}{r_0^3}\left(2\pi\kappa^2 x^2 + \frac{\kappa - k'}{2}\, r_0^2\right),$$

where

$$\beta = J^2 \left/ \left(1 + \frac{4\pi}{3}\kappa\right)^2\right.$$

The surface-stresses consist therefore of :

(a)   A uniform surface traction $= \frac{1}{2}\beta(\kappa - k')$.

(b)   A variable surface traction $= 2\pi\beta\kappa^2 \cos^2 \psi$, where ψ is the angle the outwardly directed normal at any point makes with the direction of magnetisation.

(c)   A variable surface *pressure* parallel to the direction of magnetisation $= \frac{1}{2}\beta k'' \cos \psi$.

We can easily ascertain the corresponding shifts and strains :

(a) This corresponds to a uniform dilatation of the sphere and to a radial shift U at central distance r given by :

$$U = \tfrac{1}{2}\beta \left(\kappa - k' \right) \frac{r}{3\lambda + 2\mu},$$

and consequent dilatation :

$$\theta = \tfrac{3}{2}\beta \left(\kappa - k' \right) \frac{1}{3\lambda + 2\mu}.$$

(b) If ρ be the distance of a point from the axis of x, the shifts u, V, at x, ρ parallel and perpendicular to the axis of magnetisation are given by :

$$u = a_1 x^3 + b_1 \rho^2 x + c_1 r_0^2 x,$$
$$V = a_1' x^2 \rho + b_1' \rho^3 + c_1' r_0^2 \rho,$$

where Kirchhoff finds for the constants the values which in our notation are expressed by :

$$a_1 = - \frac{2\eta}{(7 + 5\eta)\, \mu r_0^2}\, 2\pi\beta\kappa^2, \quad b_1 = \frac{7 - 6\eta}{4\eta} a_1, \quad c_1 = - \frac{7 + 3\eta - 2\eta^2}{4\eta\,(1 + \eta)} a_1,$$

$$a_1' = - \frac{7 - 8\eta}{4\eta} a_1, \qquad b_1' = - \tfrac{1}{2} a_1, \qquad c_1' = \frac{3 + 2\eta}{2\,(1 + \eta)} a_1.$$

(c) This gives us shifts parallel and perpendicular to the axis of magnetisation measured by

$$u = - \frac{x}{E}\tfrac{1}{2}\beta k'', \qquad V = - \frac{\eta\rho}{E}\tfrac{1}{2}\beta k''.$$

The combination of these cases gives the total strain due to the magnetisation.

[1320.] If we suppose, that κ is immensely greater than k' and k'', we have only to consider the shifts given by (b).

Assuming the uni-constant isotropy of the sphere, or $\eta = \tfrac{1}{4}$, we have then by neglecting κ as compared with κ^2 :

$$u = \frac{3}{176\pi}\, \frac{J^2}{Er_0^2}\, \{- 10x^3 - 55\rho^2 x + 61 r_0^2 x\},$$

$$V = \frac{3}{176\pi}\, \frac{J^2}{Er_0^2}\, \{50x^2\rho + 5\rho^3 - 14 r_0^2 \rho\},$$

$$\theta = \frac{3}{176\pi}\, \frac{J^2}{Er_0^2}\, \{70x^2 - 35\rho^2 + 33 r_0^2\}.$$

The extension of the radius parallel to the magnetic force

$$= \frac{153}{176\pi}\, \frac{J^2 r_0}{E},$$

while the radii perpendicular to this undergo the compression

$$\frac{27}{176\pi} \frac{J^2 r_0}{E}.$$

[1321.] In the course of his discussion Kirchhoff refers to the great variations in the value of κ: see our Art. 1314. He points out that the uniform magnetic force might be obtained by placing the iron sphere in the axis of a coil, but that the extension of the radius in the direction of this axis would probably be far too small to be capable of measurement. Finally he refers to Joule's measurement in 1846 of the extension of an iron bar placed in such a coil : see our Art. 688.

Shelford Bidwell has, however, shown that an iron bar will shorten when the magnetising force is sufficiently increased, so that it is difficult to see the application of Joule's result to Kirchhoff's theory. Further J. J. Thomson (*Applications of Dynamics to Physics and Chemistry*, p. 54) has shown from Ewing's experiments on the relation of strain and magnetisation that Kirchhoff's results in the previous article sometimes give a very small part of the total strain in soft iron, the chief part being really due to the terms which connect the intensity of magnetisation with the strain (compare the k' and k'' of equation (iii) of our Art. 1314). It is not possible to discuss these matters here at length, but the reader is warned that Kirchhoff's results are not a complete representation of the relations between magnetism and strain brought to light by recent experimental researches.

Section III.

Clebsch[1]

[1322.] The first memoir due to Clebsch is entitled : *Ueber die Gleichgewichtsfigur eines biegsamen Fadens. Crelles Journal für die reine u. angewandte Mathematik.* Bd. 57, 1860, S. 93–110.

[1] For an account of Clebsch's life and work see the *Mathematische Annalen* founded by him, Bd. VI. S. 197–202 and Bd. VII. S. 1–55. Clebsch died Nov. 7, 1872, aged 39.

§§ 1–7 of this memoir are occupied with the equilibrium of an *inextensible* but flexible string; § 1 gives the general equations (S. 93–5); § 2 deals with a uniform heavy chain (S. 95); § 3 considers the equilibrium of a string under the action of 'centrifugal force' produced by rotation; a solution is obtained in terms of elliptic functions (S. 95–101); § 4 supposes the string constrained to remain on a given surface (S. 101–2), while §§ 5–7 take the special cases of any surface of revolution, a sphere, and a string on a sphere under the action of centrifugal force due to rotation respectively (S. 102–7). The equations are integrated by a process due to Jacobi.

[1323.] The remaining sections of the memoir deal more closely with our subject. § 8 is entitled: *Gleichgewicht dünner elastischer Faden* (S. 107–9). Clebsch supposes a force-function to exist and the cross-section to be so small that the string is perfectly flexible as well as elastic. He obtains his equations by making the integral

$$\int \left(\frac{Es^2}{2} - U \right) d\sigma$$

taken throughout the length σ of the string a minimum, E being the stretch-modulus, s the stretch in the element $d\sigma$, and U the corresponding force-function per unit length of $d\sigma$. Clebsch reduces the general solution to the discovery of a solution V of the partial differential equation :

$$\left(\frac{dV}{dx} \right)^2 + \left(\frac{dV}{dy} \right)^2 + \left(\frac{dV}{dz} \right)^2 = E^2 \left\{ \sqrt{ 1 - \frac{2}{E} \left(U + \frac{dV}{d\sigma} \right) } - 1 \right\}^2 .$$

The last section of the memoir is entitled: *Gleichgewicht eines dünnen elastischen Fadens unter dem Einfluss der Schwere* (S. 109–110). The statement is so brief that it is difficult to follow the reasoning of this last section.

[1324.] *Theorie der circularpolarisirenden Medien. Crelles Journal für reine u. angewandte Mathematik.* Bd. 57, 1860, S. 319–358. This memoir does not properly proceed from an elastic hypothesis, and the necessary optical terms are introduced into the equations by assuming a type of intermolecular force which has not received any physical explanation. Thus Clebsch's hypothesis is the following (S. 322–3) :

Nehmen wir an dass zwar in jedem Augenblick die Molecüle sich nach einer Function $f(r)$ der Entfernung anziehen, dass aber ausserdem durch die Bewegung selbst auf irgend eine Weise in denselben in jedem Augenblick eine (nicht wieder verschwindende) Kraft erregt wird, welche senkrecht gerichtet sein soll gegen eine der Verbindungslinie und der relativen Geschwindigkeit gleichzeitig parallele Ebene.

Further :

Dass die gedachte, in jedem Augenblick entstehende Kraft proportional ist einer Function der relativen Entfernung $F(r)$ und derjenigen Componenten

...der relativen Geschwindigkeit, welche gegen die Verbindungslinie senkrecht ist.

The resulting equations are not elastic equations, but similar to the optical equations of Cauchy, MacCullagh and Neumann, and the methods adopted are akin to those of the tractate on optics referred to in our Art. 1391. There is thus no need to consider the memoir at length under the history of elasticity.

1325. We have next to consider the work entitled: *Theorie der Elasticität der fester Körper von Dr A. Clebsch, Professor an der Polytechnischen Schule zu Carlsruhe.* This was published in large octavo at Leipzig in 1862, and contains xi. + 424 pages. The preface states briefly the object of the work; this may be said to be to furnish a sound basis for practical studies and applications. Accordingly the mathematical processes are kept as simple and elementary as possible; the general investigations given by Lamé and also any applications to the theory of light are omitted. On the other hand the researches of Saint-Venant on the Flexure and Torsion of Prisms, and those of Kirchhoff with respect to very slender rods, are fully considered. The work is divided into three parts; S. 1–189 treat of bodies having all their dimensions finite; S. 190–355 treat of bodies which have one dimension or two dimensions indefinitely small; S. 256–424 are devoted to applications. The work is subdivided into 92 sections.

[Notwithstanding Clebsch's preface and his position at Carlsruhe his book is certainly not suited for the technicist; the slightest comparison of his pages with those, for example, of his successor Grasshof will sufficiently demonstrate this fact. It is to the mathematical elastician that Clebsch in reality appeals, and the chief value of his book lies in the novelty of his analytical processes and his solutions of new elastic problems. Throughout the work Clebsch practically uses only the equations for isotropic materials, and this deprives the work of much physical and technical interest. In the French translation due to Saint-Venant and Flamant, suitable distributions of elasticity replace this isotropy of the original work. The copious notes of Saint-Venant and the correction of many of the innumerable errata of the original so increase the value of the translation, that it is safe to predict that for the future Clebsch will be chiefly read in the French edition (see our Arts. 298–400). For our present historical purposes, however, we follow the original, giving under the letters "*F. E.*" the corresponding pages of Saint-Venant's version.]

[1326.] The first seventeen sections of Clebsch (S. 1–50; *F. E.* pp. 1–113) contain a general theory of elasticity, which does not possess much novelty. The statements on S. 7 and 10 with regard to the numerical limits of the elastic constants[1] are only true for the isotropic materials of theory and not for the usual materials of construction : see our Arts. 169 (*d*) and 308 (*b*). The definition of the elastic limit, S. 4, requires modification, but the remark on S. 3 as to the fitness of excluding caoutchouc from the substances to which the theory of elasticity in its present form can be applied deserves notice. As a novelty we may refer to S. 23–7, where the reader will find Lamé's *ellipsoid of elasticity* and the *stress-director-quadric* (see our Arts. 1008* and 1059*) expressed in *tangential coordinates*; the analysis has probably more interest than the result practical value.

The linearity of the stress-strain relations is practically *assumed* by Clebsch, as he appeals to a mathematical process and not to experimental facts : see our Arts. 928*, 1051*, 1064*, and 299.

Clebsch terms the stresses *Spannungen* and the strains *Verschiebungen*[2], he uses *Zugkraft* also in the sense of *Spannung*, but it would I think be better to confine it to what in this volume we term *tractions*. He represents the stress system by t_{11}, t_{22}, t_{33}, t_{23}, t_{31}, t_{12} and the strains by α, β, γ, ϕ, χ, ψ.

[1327.] Clebsch next passes in § 18 to the special case of the equilibrium of a hollow spherical shell subjected to uniform surface-tractions. This has been fully considered by other writers (see our Arts. 1016*, 1093*, 123, and 1201 (*c*)), and their results should be compared with those of Clebsch, as there are misprints in his work. He uses also, here, as throughout his book, the maximum *stress* not the maximum *stretch* to suggest the *fail-limit* or condition of rupture : see our Arts. 4 (γ), 5, 169 (*c*) and 320–1.

The following section § 19 (S. 55–61) deals with the radial

---

[1] Clebsch uses μ for our η and F for our μ; he uses E, as we do, for the stretch-modulus.

[2] He also uses *Verschiebungen* occasionally for the *shifts*, e.g. S. 25.

vibrations of a sphere, and does not add much to Poisson's treatment of the like problem in his memoir of 1829 : see our Arts.
449*–463*.

These sections occupy S. 114–126 of *F. E.*

[1328.]   The next subject to which Clebsch turns is of more
interest.   § 20 (S. 62–67 ; *F. E.* pp. 126–132) is entitled : *Ueber
die Wurzeln der transscendenten Gleichungen, welche die Untersuchung von Schwingungen elastischer Körper mit sich führt,* and
its object is to show the reality of the roots, or the stability of
the small vibrational motions.   I reproduce the substance of
Clebsch's investigation here, as it appears to be original, and
is of considerable importance.

1329.   Let us suppose that for small vibrations the values of the
shifts u, v, w can be expressed in series of simple harmonic form.   Thus
we may put

$$u = u_1 \sin k_1 t + u_2 \sin k_2 t + u_3 \sin k_3 t + \ldots\ldots$$
$$+ u_1{}' \cos k_1 t + u_2{}' \cos k_2 t + u_3{}' \cos k_3 t + \ldots\ldots,$$

with similar expressions for v and w.   But as the treatment of the
terms which involve *cosines* is the same as that of the terms which
involve *sines*, we will omit the cosines entirely.   Hence we take

$$u = u_1 \sin k_1 t + u_2 \sin k_2 t + u_3 \sin k_3 t + \ldots\ldots$$
$$v = v_1 \sin k_1 t + v_2 \sin k_2 t + v_3 \sin k_3 t + \ldots\ldots$$
$$w = w_1 \sin k_1 t + w_2 \sin k_2 t + w_3 \sin k_3 t + \ldots\ldots$$

Now similarly each of the six elastic stresses will take the form of
such a series ; we will thus suppose that corresponding to

$$u_n \sin k_n t, \quad v_n \sin k_n t, \quad w_n \sin k_n t,$$

we have for the three tractions :

$$\nu_1 \sin k_n t, \quad \nu_2 \sin k_n t, \quad \nu_3 \sin k_n t,$$

and for the three shears :

$$\tau_1 \sin k_n t, \quad \tau_2 \sin k_n t, \quad \tau_3 \sin k_n t.$$

Then substituting in the body stress-equations, and supposing no
external forces to act, we have

$$\left.\begin{aligned}
- \rho k_n{}^2 u_n &= \frac{d\nu_1}{dx} + \frac{d\tau_3}{dy} + \frac{d\tau_2}{dz}, \\
- \rho k_n{}^2 v_n &= \frac{d\tau_3}{dx} + \frac{d\nu_2}{dy} + \frac{d\tau_1}{dz}, \\
- \rho k_n{}^2 w_n &= \frac{d\tau_2}{dx} + \frac{d\tau_1}{dy} + \frac{d\nu_3}{dz}
\end{aligned}\right\} \quad\ldots\ldots\ldots\ldots\ldots(1).$$

The values of ν_1, ν_2,...are connected with u_n, v_n, w_n by the same relations as the stresses are connected with the strains. Also we have the following equations holding at the bounding surfaces, supposing there to be no load and a, β, γ the direction-angles of the normal:

$$\left.\begin{array}{l} \nu_1\cos a + \tau_3\cos\beta + \tau_2\cos\gamma = 0, \\ \tau_3\cos a + \nu_2\cos\beta + \tau_1\cos\gamma = 0, \\ \tau_2\cos a + \tau_1\cos\beta + \nu_3\cos\gamma = 0 \end{array}\right\} \quad\ldots\ldots\ldots\ldots\ldots(2).$$

It will be observed that (1) and (2) are derived from the general body and surface stress-equations; each of these equations breaks up into sets obtained by considering separately the terms which involve

$$\sin k_1 t, \quad \sin k_2 t,\ldots\ldots \quad \sin k_n t,\ldots\ldots$$

Now let u_m, v_m, w_m correspond to $\sin k_m t$ in the values of u, v, w, where k_m is different from k_n. Let the corresponding forms of (1) and (2) be what we get by putting ν' instead of ν, and τ' instead of τ.

Consider the triple integral extended over the whole body

$$J = \iiint (u_m u_n + v_m v_n + w_m w_n)\, dx dy dz.$$

By means of (1) this gives at once, if ρ be constant:

$$-\rho k_m{}^2 J = \iiint \left\{ u_m \left(\frac{d\nu_1}{dx} + \frac{d\tau_3}{dy} + \frac{d\tau_2}{dz} \right) \right.$$
$$\left. + v_m \left(\frac{d\tau_3}{dx} + \frac{d\nu_2}{dy} + \frac{d\tau_1}{dz} \right) + w_m \left(\frac{d\tau_2}{dx} + \frac{d\tau_1}{dy} + \frac{d\nu_3}{dz} \right) \right\} dx dy dz.$$

Now by integration by parts, the triple integral can be transformed into a certain double integral extending over the boundaries, and a certain triple integral extending throughout the body; the double integral vanishes by (2); and we are thus left with the result

$$\rho k_m{}^2 J = \iiint \left\{ \nu_1 \frac{du_m}{dx} + \tau_1 \left(\frac{dv_m}{dz} + \frac{dw_m}{dy} \right) \right.$$
$$\left. + \nu_2 \frac{dv_m}{dy} + \tau_2 \left(\frac{dw_m}{dx} + \frac{du_m}{dz} \right) + \nu_3 \frac{dw_m}{dz} + \tau_3 \left(\frac{du_m}{dy} + \frac{dv_m}{dx} \right) \right\} dx dy dz.$$

Now in precisely the same way as this result has been obtained, by applying (1) with respect to u_m, v_m, w_m instead of with respect to u_n, v_n, w_n we obtain

$$\rho k_m{}^2 J = \iiint \left\{ \nu_1' \frac{du_n}{dx} + \tau_1' \left(\frac{dv_n}{dz} + \frac{dw_n}{dy} \right) \right.$$
$$\left. + \nu_2' \frac{dv_n}{dy} + \tau_2' \left(\frac{dw_n}{dx} + \frac{du_n}{dz} \right) + \nu_3' \frac{dw_n}{dz} + \tau_3' \left(\frac{du_n}{dy} + \frac{dv_n}{dx} \right) \right\} dx dy dz.$$

1330. But the right-hand members of the last two equations are the same. For we know that there is a certain homogeneous function of the second degree involving the six quantities

$$\frac{du_n}{dx}, \quad \frac{dv_n}{dz} + \frac{dw_n}{dy} \ \ldots\ldots;$$

such that its differential coefficient with respect to $\frac{du_n}{dx}$ is ν_1, its differential coefficient with respect to $\frac{dv_n}{dz} + \frac{dw_n}{dy}$ is τ_1, and so on; and in like manner similar considerations hold when we form the same function of the six quantities

$$\frac{du_m}{dx}, \quad \frac{dv_m}{dz} + \frac{dw_m}{dy} \ \ldots\ldots$$

Hence it will be found, as asserted, that the right-hand members of these equations are the same. Thus we have

$$\rho k_n{}^2 J = \rho k_m{}^2 J,$$

and since k_n and k_m are different it follows that J must be zero, that is

$$\iiint (u_m u_n + v_m v_n + w_m w_n)\, dx dy dz = 0 \ \ldots\ldots\ldots\ldots(3).$$

This result enables us to show that the quantities $k_1{}^2$, $k_2{}^2$,...... are all real. For suppose one of them were of the form $p + q\sqrt{-1}$, then another would be of the form $p - q\sqrt{-1}$; let then u_m, v_m, w_m correspond to the former, and u_n, v_n, w_n to the latter. Then u_m and u_n will be conjugate imaginary expressions of the forms $p_1 + q_1\sqrt{-1}$ and $p_1 - q_1\sqrt{-1}$; and so their product would be the sum of two squares. The like would hold for $v_m v_n$ and for $w_m w_n$; and thus the integral in (3) would be necessarily a positive quantity, and so could not vanish as it must by (3).

Clebsch then proceeds to show that $k_1{}^2$, $k_2{}^2$,......must all be *positive* quantities. For put

$$J' = \iiint (u_n{}^2 + v_n{}^2 + w_n{}^2)\, dx dy dz,$$

the integral extending throughout the whole body; thus J' denotes what J would become if we put u_n, v_n, w_n for u_m, v_m, w_m respectively. Then transforming this as we did J we get

$$\rho k_n{}^2 J' = \iiint \left\{ \nu_1 \frac{du_n}{dx} + \ldots\ldots \right\} dx dy dz$$

$$= 2 \iiint F dx dy dz \ \ldots\ldots\ldots\ldots \ \ldots\ldots\ldots\ldots\ldots(4),$$

where F is that homogeneous function of the six quantities

$$\frac{du_n}{dx}, \quad \frac{dv_n}{dz} + \frac{dw_n}{dy}, \ \ldots\ldots$$

to which we have already referred.

Now Clebsch says, in substance, that the right-hand member of (4) must be a positive quantity. It is really the strain-energy corresponding to certain small shifts; hence since J' is necessarily positive it follows from (4) that k_n^2 must be positive also. Clebsch should have referred to some standard treatise on Mechanics for the proposition which he here asserts [see our Art. 1278]; his own words after arriving at equation (4) are

Das negative Differential des dreifachen Integrals rechts bedeutete aber die Arbeit, welche die innern Kräfte bei einer kleinen Verschiebung leisten, daher stellt das Integral selbst, mit entgegengesetzten Zeichen genommen, die Arbeit dar, welche die innern Kräfte leisten, wenn der Körper aus seiner naturlichen Lage verschoben wird, bis er die Verschiebungen u_n, v_n, w_n erhält. Diese Arbeit ist ihrer Natur nach negativ, genau entgegengesetzt der ihr gleichen positiven Arbeit der äussern Kräfte, welche zu einer solchen Verschiebung nothwendig ist. Das dreifache Integral ist also nothwendig positiv; aber auch J', welches eine Summe positiver Glieder ist, daher muss denn auch k_n^2 nothwendig positiv sein, was zu beweisen war (S. 65-6).

Clebsch applies the results obtained to the problem of the vibrating sphere in order to justify an equation there assumed to be true.

1331. The twenty-first section consists of a demonstration that the problem of the equilibrium of an elastic body is *a determinate* problem. (S. 67–70; *F. E.* S. 132–6.) This is a modification of Kirchhoff's proof: see our Arts. 1255 and 1278.

Suppose if possible that such a problem admitted two solutions, one in which the shifts are u', v', w', and another in which the shifts are u'', v'', w''. Write down the body and surface shift-equations, first with respect to u', v', w', and next with respect to u'', v'', w''. Make subtractions of corresponding equations; for the result we obtain equations of elastic equilibrium, *with no applied forces whatever*, and where the displacements are denoted by $u'-u''$, $v'-v''$, $w'-w''$ respectively : it is our object to show that the displacements in this case must all be zero, that is

$$u' - u'' = 0, \quad v' - v'' = 0, \quad w' - w'' = 0.$$

If this be shown it amounts to establishing that there is only one solution of the problem of equilibrium.

Let us then suppose that no applied forces whatever act, and let us denote the shifts as usual, by u, v, w. If we proceed as in our last article we have results like those obtained there, provided we put k_m^2 and k_n^2 zero; for now as we are supposing no motion, these quantities do not occur. Thus corresponding to (4) of that article we have now

$$\iiint F\,dx\,dy\,dz = 0 \quad \ldots\ldots\ldots\ldots\ldots\ldots(1).$$

But this integral owing to its physical meaning has a positive value and cannot, therefore, be zero unless F be zero. This can only be the case when all the variables in F vanish, that is when :

$$\left.\begin{array}{l} \dfrac{du}{dx} = 0, \quad \dfrac{dv}{dy} = 0, \quad \dfrac{dw}{dz} = 0, \\[2mm] \dfrac{dv}{dz} + \dfrac{dw}{dy} = 0, \quad \dfrac{dw}{dx} + \dfrac{du}{dz} = 0, \quad \dfrac{du}{dy} + \dfrac{dv}{dx} = 0 \end{array}\right\} \dots\dots\dots\dots(2).$$

It seems to me that these equations are not obtained in a very convincing manner : compare our Art. 1278.

The values of u, v, w which satisfy these equations are of the following forms :

$$u = a + \gamma y - \beta z, \quad v = b + \alpha z - \gamma x, \quad w = c + \beta x - \alpha y \dots\dots\dots(3),$$

where a, b, c, α, β, γ are constants. These are easily shown to follow from (2). For from the first three of these we see that u cannot involve x, that v cannot involve y, and that w cannot involve z; then from the last of them du/dy cannot involve y, and from the fifth of them du/dz cannot involve z : in this way the assigned formulae are obtained.

Now the equations (3) exhibit only such motions as the body can take as a *whole*, and which consequently do not give rise to any *relative* shifts, and so do not call out any stresses. For a, b, c correspond to shifts parallel to the axes of x, y, z respectively ; α, β, γ correspond to a small rotation of the body round a straight line inclined to the axes at angles whose direction cosines are proportional to α, β, γ respectively. The conclusion is that any problem relating to the equilibrium of an elastic body becomes perfectly definite if we exclude all such shifts as the body could take as a whole.

[1332.] S. 70–148 of Clebsch's treatise are occupied with what he has termed *Saint-Venant's Problem*, that is to say with the torsion and flexure of prisms. This forms Chapter II. of the French edition (pp. 137–294). Clebsch's treatment is very instructive, as he combines in one investigation the general results of Saint-Venant's two classical memoirs : see our Arts. 1 and 69. At the same time the slight value of his book for technical students is well brought out by the fact that he passes over all the important practical examples (the elliptic cross-section alone excepted) which Saint-Venant has given of his theory (see our Arts. 18–49 and 87–97), and devotes himself especially to the case of a prism bounded by two confocal elliptic cylinders. The analysis is interesting, but the practical application is small. We have here a good example of how the love of original investigation may render it impossible even for a mathematician of genius to write a textbook especially suitable for a particular class of students. In this respect his very originality may handicap him, and Clebsch's treatise has never won for itself the same type of readers as those of Navier, Lamé or Grashof: see our Arts. 279*, 1043*

[1333.] Clebsch states Saint-Venant's Problem in the following manner (S. 72–3):

Welches sind die Gleichgewichtszustände eines cylindrischen Körpers, auf dessen cylindrische Oberfläche keine Kräfte wirken, und dessen Inneres keinen äussern Kräften unterworfen ist, bei welchen die den Körper zusammensetzenden Fasern keinerlei seitlichen Druck erleiden. Welches sind die Kräfte, welche auf die freie Endfläche wirken müssen, um dergleichen Zustände hervorzurufen.

1334. We will now indicate Clebsch's method of investigating this problem. To free the shifts from pure translational and rotational terms we may fix a point in the body and a linear and a planar element at that point. This Clebsch does in the following manner.

Suppose the body of any cylindrical form. Take the axis of z parallel to that of the cylinder, so that originally a section at right angles to the axis is parallel to the plane of xy. We shall suppose that the origin is a *fixed* point, so that we have $u = 0$, $v = 0$, $w = 0$ at this point, that is where x, y, z vanish. For a point in the plane of xy very near the origin the displacements parallel to the axes of x, y, z respectively may be denoted by

$$\left(\frac{du}{dx}\right)_0 dx + \left(\frac{du}{dy}\right)_0 dy, \quad \left(\frac{dv}{dx}\right)_0 dx + \left(\frac{dv}{dy}\right)_0 dy, \quad \left(\frac{dw}{dx}\right)_0 dx + \left(\frac{dw}{dy}\right)_0 dy.$$

Suppose then that $\left(\dfrac{dw}{dx}\right)_0 = 0$, and $\left(\dfrac{dw}{dy}\right)_0 = 0$; this amounts to assuming that an infinitesimal element originally in the plane of xy remains in that plane, or that there is no rotation round an axis in that plane. Let us further assume that $\left(\dfrac{dv}{dx}\right)_0 = 0$; then there is no motion parallel to the axis of y of any point of the infinitesimal element which is on the axis of x, and so there can be no rotation round an axis perpendicular to the plane of xy.

We take then these six conditions to hold when $x = 0$, $y = 0$, $z = 0$:

$$u = 0, \quad v = 0, \quad w = 0, \quad \frac{dw}{dx} = 0, \quad \frac{dw}{dy} = 0, \quad \frac{dv}{dx} = 0 \quad \ldots\ldots(1).$$

These conditions in fact make the six constants of (3) in our Art. 1331, namely a, b, c, α, β, γ, all vanish. The six conditions might be assumed differently; thus for instance, instead of $\dfrac{dv}{dx} = 0$ we might take $\dfrac{du}{dy} = 0$, keeping all the others : but we shall adhere to the form adopted in (1).

We assume that no body-force whatever acts; and that there is no load on the curved boundary of the cylinder, but only on the terminal cross-sections. The direct problem now would be to let given forces act at the terminals and then seek to determine u, v, w; but instead of this Clebsch follows Saint-Venant in an indirect course. He proposes to seek the conditions that must hold, and the forces that must act on the body, in order that throughout the body we may have

$$\widehat{xx} = 0, \quad \widehat{yy} = 0, \quad \widehat{xy} = 0.$$

The assumptions made that $\widehat{xx}$, $\widehat{yy}$, $\widehat{xy}$ shall all vanish amount to supposing the cylinder to consist of slender fibres, rectangular if we please, and that these fibres exercise on each other no stress perpendicular to their length. As no transversal stress exists on such a fibre we must have

$$\widehat{zz} = E \frac{dw}{dz} \quad \dotfill (2),$$

and

$$\frac{du}{dx} = \frac{dv}{dy} = -\eta \frac{dw}{dz} \quad \dotfill (3).$$

These relations flow at once from the conditions

$$\widehat{xx} = 0, \quad \widehat{yy} = 0,$$

as we see from Art. 78.

The condition that $\widehat{xy} = 0$ leads to

$$\frac{du}{dy} + \frac{dv}{dx} = 0 \dotfill (4).$$

The body-stress equations now take the form

$$\frac{d\widehat{zx}}{dz} = 0, \quad \frac{d\widehat{yz}}{dy} = 0, \quad \frac{d\widehat{zx}}{dx} + \frac{d\widehat{yz}}{dy} + \frac{d\widehat{zz}}{dz} = 0 \dotfill (5).$$

Substitute the values of $\widehat{yz}$, $\widehat{zx}$, $\widehat{zz}$, and these become, supposing the elasticity to have a planar distribution perpendicular to the axis of the prism :

$$\frac{d^2 u}{dz^2} + \frac{d^2 w}{dx dz} = 0, \quad \frac{d^2 v}{dz^2} + \frac{d^2 w}{dy dz} = 0 \dotfill (6),$$

$$\left(\frac{E}{\mu} - 2\eta \right) \frac{d^2 w}{dz^2} + \frac{d^2 w}{dx^2} + \frac{d^2 w}{dy^2} = 0 \dotfill (7),$$

where E, μ and η must be now regarded as *independent* elastic constants[1]: see our Arts. 310–3 and 321 (d).

---

[1] We have here followed Saint-Venant in extending Clebsch's results to a planar elastic distribution. Clebsch supposes the body isotropic and therefore has 2 for our $E/\mu - 2\eta$.

The conditions relative to the cylindrical surface reduce to

$$0 = \widehat{zx} \cos p + \widehat{yz} \sin p,$$

that is to $\quad\left(\dfrac{du}{dz} + \dfrac{dw}{dx}\right)\cos p + \left(\dfrac{dv}{dz} + \dfrac{dw}{dy}\right)\sin p = 0 \quad\ldots\ldots\ldots\ldots(8),$

where p is the angle that the outwardly directed normal makes with the axis of x.

1335. The twenty-third section proceeds to the solution of the equations just obtained.

The equations to be discussed are the following:

$$\frac{du}{dx} = \frac{dv}{dy} = -\eta\,\frac{dw}{dz} \quad\ldots\ldots\ldots\ldots\ldots\ldots\ldots(1),$$

$$\frac{du}{dy} + \frac{dv}{dx} = 0\ldots\ldots\ldots\ldots\ldots\ldots\ldots\ldots\ldots\ldots\ldots(2),$$

$$\frac{d^2u}{dz^2} + \frac{d^2w}{dxdz} = 0\ldots\ldots\ldots\ \ldots\ldots\ldots\ldots\ldots(3),$$

$$\frac{d^2v}{dz^2} + \frac{d^2w}{dydz} = 0\ldots\ldots\ldots\ldots\ldots\ldots\ldots\ldots\ldots\ldots(4),$$

$$\frac{d^2w}{dx^2} + \frac{d^2w}{dy^2} + \left(\frac{E}{\mu} - 2\eta\right)\frac{d^2w}{dz^2} = 0\ \ldots\ldots\ldots\ldots\ldots\ldots(5).$$

Differentiate (5) with respect to z; and subtract (3) differentiated with respect to x, and (4) differentiated with respect to y; thus

$$\left(\frac{E}{\mu} - 2\eta\right)\frac{d^3w}{dz^3} - \frac{d^3u}{dxdz^2} - \frac{d^3v}{dydz^2} = 0$$

Hence by (1) we find:

$$\frac{d^3w}{dz^3} = 0 \ \ldots\ldots\ldots\ldots\ldots\ldots\ldots\ldots\ldots\ldots(6).$$

Differentiate (3) with respect to y, and (4) with respect to x, and add: thus

$$\frac{d^3u}{dydz^2} + \frac{d^3v}{dxdz^2} + 2\frac{d^3w}{dxdydz} = 0\ ;$$

the sum of the first and second terms vanishes by (2); and thus

$$\frac{d^3w}{dxdydz} = 0.$$

Differentiate (5) with respect to z, and use (6); thus

$$\frac{d^3w}{dzdx^2} + \frac{d^3w}{dzdy^2} = 0\ldots\ldots\ldots\ldots\ldots\ldots\ldots\ldots(7).$$

Differentiate (3) with respect to x, and (4) with respect to y; the first terms are equal by (6), and we have

$$\frac{d^3w}{dx^2 dz} = \frac{d^3w}{dy^2 dz}$$

Comparing this with (7) we see that

$$\frac{d^3w}{dx^2 dz} = 0, \qquad \frac{d^3w}{dy^2 dz} = 0.$$

Thus we have shown that the following differential coefficients of $\dfrac{dw}{dz}$ must vanish :

$$\frac{d^2}{dz^2}\left(\frac{dw}{dz}\right), \quad \frac{d^2}{dx^2}\left(\frac{dw}{dz}\right), \quad \frac{d^2}{dy^2}\left(\frac{dw}{dz}\right), \quad \frac{d^2}{dx\,dy}\left(\frac{dw}{dz}\right).$$

On account of the first three of these dw/dz cannot contain x, y or z to a power higher than the first; and on account of the last dw/dz cannot contain xy : hence

$$\frac{dw}{dz} = a + a_1 x + a_2 y + z\,(b + b_1 x + b_2 y) \quad \dots\dots\dots\dots(8).$$

From this we have by (1)

$$\frac{du}{dx} = \frac{dv}{dy} = -\eta\,\{a + a_1 x + a_2 y + z\,(b + b_1 x + b_2 y)\}.$$

Integrating the last two equations we get

$$u = -\eta\left(ax + \frac{a_1 x^2}{2} + a_2 xy\right) - \eta z\left(bx + \frac{b_1 x^2}{2} + b_2 xy\right) + \phi\,(y, z),$$

$$v = -\eta\left(ay + a_1 xy + \frac{a_2 y^2}{2}\right) - \eta z\left(by + b_1 xy + \frac{b_2 y^2}{2}\right) + \psi\,(x, z),$$

where $\phi\,(y, z)$ denotes some function of y and z, and $\psi\,(x, z)$ some function of x and z : these must now be determined.

From (3) and (4) we have

$$\frac{d^2u}{dz^2} = -a_1 - b_1 z, \qquad \frac{d^2v}{dz^2} = -a_2 - b_2 z\,;$$

thus ϕ and ψ do not involve any power of z higher than the third; also the coefficients of z^2 and z^3 are constants in each.

It follows from (2), combined with the expressions obtained for u and v, that y in $\phi\,(y, z)$ and x in $\psi\,(x, z)$ cannot occur to a power higher than the second. Thus for the forms of ϕ and ψ we obtain

$$\phi\,(y, z) = a' + a_1' y + a_2' y^2 + z\,(b' + b_1' y + b_2' y^2) - \frac{a_1 z^2}{2} - \frac{b_1 z^3}{6},$$

$$\psi\,(x, z) = a'' + a_1'' x + a_2'' x^2 + z\,(b'' + b_1'' x + b_2'' x^2) - \frac{a_2 z^2}{2} - \frac{b_2 z^3}{6}$$

The equations (3) and (4) are now fully satisfied by the values of u and v which we have obtained. Substitute these values in (2), and we find that the following relations must hold among the constants:

$$a_2'' = \eta \frac{a_2}{2}, \qquad a_2' = \eta \frac{a_1}{2}, \qquad a_1' + a_1'' = 0,$$

$$b_2'' = \eta \frac{b_2}{2}, \qquad b_2' = \eta \frac{b_1}{2}, \qquad b_1' + b_1'' = 0.$$

Put $a_1' = -a_1'' = a_0$, $b_1' = -b_1'' = b_0$; and we obtain finally

$$\left.\begin{aligned}
u = -\eta &\left(ax + a_1\frac{x^2 - y^2}{2} + a_2 xy\right) - \eta z \left(bx + b_1\frac{x^2 - y^2}{2} + b_2 xy\right) \\
&+ a' + a_0 y + z\left(b' + b_0 y\right) - \frac{a_1 z^2}{2} - \frac{b_1 z^3}{6}, \\
v = -\eta &\left(ay + a_1 xy + a_2\frac{y^2 - x^2}{2}\right) - \eta z\left(by + b_1 xy + b_2\frac{y^2 - x^2}{2}\right) \\
&+ a'' - a_0 x + z\left(b'' - b_0 x\right) - \frac{a_2 z^2}{2} - \frac{b_2 z^3}{6}
\end{aligned}\right\} \quad ...(9)$$

1336. These formulae satisfy equations (1), (2), (3), (4), and they constitute the most general solution of them; it will be seen that they fully determine u and v, except that they each involve some arbitrary constants. We proceed to find w, which has to satisfy (5) and (8). By integrating (8) we get

$$w = z\left(a + a_1 x + a_2 y\right) + \frac{z^2}{2}\left(b + b_1 x + b_2 y\right) + F(x, y),$$

where $F(x, y)$ denotes some function of x and y. Substitute in (5), then we get

$$\frac{d^2 F}{dx^2} + \frac{d^2 F}{dy^2} + \left(\frac{E}{\mu} - 2\eta\right)(b + b_1 x + b_2 y) = 0.$$

Assume

$$F(x, y) = \Omega - \left(\frac{E}{2\mu} - \eta\right)\left\{\frac{b}{2}(x^2 + y^2) + b_1 xy^2 + b_2 x^2 y\right\} + c - b'x - b''y,$$

where Ω denotes a function of x and y; then the equation becomes

$$\frac{d^2\Omega}{dx^2} + \frac{d^2\Omega}{dy^2} = 0 \quad ....(10)$$

This equation will not fully determine Ω; as we shall see, the condition holding at the cylindrical surface will aid in this. Introduce now the expressions found for u, v, w in the stresses, and we have the following set of formulae:

u and v as given by (9),

$$
\begin{aligned}
w = z\left(a + a_1 x + a_2 y\right) + \frac{z^2}{2}\left(b + b_1 x + b_2 y\right) + \Omega \\
-\left(\frac{E}{2\mu} - \eta\right)\left\{b_1 x y^2 + b_2 x^2 y + \frac{b}{2}\left(x^2 + y^2\right)\right\} + c - b'x - b''y\,;
\end{aligned}
$$

$$\widehat{xx} = 0, \quad \widehat{yy} = 0, \quad \widehat{xy} = 0,$$

$$\widehat{zz} = E\left\{a + a_1 x + a_2 y + z\left(b + b_1 x + b_2 y\right)\right\},$$

$$\widehat{zx} = \mu\left\{b_0 y - b\frac{E}{2\mu} x - b_1 \frac{\eta x^2 + \left(\dfrac{E}{\mu} - 3\eta\right) y^2}{2} - \left(\frac{E}{\mu} - \eta\right) b_2 x y + \frac{d\Omega}{dx}\right\},$$

$$\widehat{yz} = \mu\left\{-b_0 x - b\frac{E}{2\mu} y - b_2 \frac{\eta y^2 + \left(\dfrac{E}{\mu} - 3\eta\right) x^2}{2} - \left(\frac{E}{\mu} - \eta\right) b_1 x y + \frac{d\Omega}{dy}\right\}$$

$$\left.\right\} \quad (11).$$

To determine Ω we have equation (10); while equation (8) of our Art. 1334 now becomes

$$
\cos p\left\{b_0 y - \frac{E}{2\mu} bx - b_1 \frac{\eta x^2 + \left(\dfrac{E}{\mu} - 3\eta\right) y^2}{2} - \left(\frac{E}{\mu} - \eta\right) b_2 x y + \frac{d\Omega}{dx}\right\}
$$

$$
+ \sin p\left\{-b_0 x - \frac{E}{2\mu} by - b_2 \frac{\eta y^2 + \left(\dfrac{E}{\mu} - 3\eta\right) x^2}{2} - \left(\frac{E}{\mu} - \eta\right) b_1 x y + \frac{d\Omega}{dy}\right\} = 0
$$

$$\ldots\ldots\ldots\ldots(12).$$

1337. The twenty-fourth section relates to the functions which have to be determined in the solution of Saint-Venant's problem.

The first thing to be shown is that Ω is fully determined by (10) and (12) If there were two different forms of Ω which satisfied these conditions, then their difference which we will denote by Θ would satisfy the two conditions:

$$\frac{d^2\Theta}{dx^2} + \frac{d^2\Theta}{dy^2} = 0, \text{ at every point of the cross-section} \ldots\ldots\ldots(13),$$

$$\cos p\frac{d\Theta}{dx} + \sin p\frac{d\Theta}{dy} = 0, \text{ at every point of its contour} \ldots(14)$$

Consider now the following integral T extended over the whole cross-section:

$$T = \iint\left\{\left(\frac{d\Theta}{dx}\right)^2 + \left(\frac{d\Theta}{dy}\right)^2\right\} dx\,dy.$$

By a process frequently exemplified, this can be transformed into

$$T = \int \Theta \left(\frac{d\Theta}{dx} \cos p + \frac{d\Theta}{dy} \sin p \right) ds - \iint \Theta \left(\frac{d^2\Theta}{dx^2} + \frac{d^2\Theta}{dy^2} \right) dx\,dy,$$

where ds denotes an element of length of the contour of the cross-section, the first integral being taken round the whole contour, and the second over the whole cross-section. But by (13) and (14) the right-hand side is zero, and therefore T is zero; but this cannot be unless $d\Theta/dx$ and $d\Theta/dy$ vanish at every point; so that Θ must be a *constant*. Hence it follows that the two values of Ω which will solve our problem can differ only by a constant; so that if we add the condition that Ω shall vanish at some point, as for instance at the origin, then Ω is fully determined. We can impose this condition on Ω without any loss of generality, because in passing from F to Ω in Art. 1336 we have introduced an arbitrary constant c.

1338. Thus since Ω is fully determinate any form which we can give to it so as to satisfy the conditions (10) and (12) of the preceding section may be taken as the necessary form. Assume then

$$\Omega = bB + b_0 B_0 + b_1 B_1 + b_2 B_2,$$

where B, B_0, B_1, B_2 all separately satisfy (10); and let us add the following special conditions round the contour, so that (12) may be satisfied:

$$\frac{dB}{dx} \cos p + \frac{dB}{dy} \sin p = \left(\frac{E}{2\mu} \right)(x \cos p + y \sin p),$$

$$\frac{dB_0}{dx} \cos p + \frac{dB_0}{dy} \sin p = x \sin p - y \cos p,$$

$$\frac{dB_1}{dx} \cos p + \frac{dB_1}{dy} \sin p = \frac{\eta x^2 + \left(\frac{E}{\mu} - 3\eta \right) y^2}{2} \cos p + \left(\frac{E}{\mu} - \eta \right) xy \sin p,$$

$$\frac{dB_2}{dx} \cos p - \frac{dB_2}{dy} \sin p = \frac{\eta y^2 + \left(\frac{E}{\mu} - 3\eta \right) x^2}{2} \sin p + \left(\frac{E}{\mu} - \eta \right) xy \cos p$$

$$\quad\quad\quad\dots(15).$$

Thus B, B_0, B_1, B_2 are fully determinate; for each has to satisfy the general differential equation (13), and each has to satisfy round the contour the appropriate equation from (15).

1339. Clebsch now shows that b must be zero. Consider the expression

$$\iint \left(\frac{d^2B}{dx^2} + \frac{d^2B}{dy^2} \right) dx\,dy, \text{ taken over the cross-section;}$$

this must be zero by virtue of (10), if b be not zero. Integrate the first term once with respect to x, and the second term once with respect to y ; then according to a very common process we have:

$$0 = \int \left\{ \frac{dB}{dx} \cos p + \frac{dB}{dy} \sin p \right\} ds.$$

By the first of (15) this leads to

$$0 = \int (x \cos p + y \sin p)\, ds.$$

But by such a process as we have just indicated, this can be deduced from

$$0 = \iint \left\{ \frac{d^2}{dx^2} \left(\frac{x^2 + y^2}{2} \right) + \frac{d^2}{dy^2} \left(\frac{x^2 + y^2}{2} \right) \right\} dx dy,$$

that is

$$0 = 2 \iint dx dy ;$$

but this is impossible, for the integral is obviously equal to double the area of the cross-section. Thus as the only escape from this contradiction we must have $b = 0$.

We learn then that the solution which is furnished by equations (9) and (11) involves only the constants a, a_1, a_2, a', a'', a_0, b_1, b_2, b', b'', b_0, c, which all enter in a linear form; and besides these there is nothing arbitrary. For the function Ω is expressed in the form of a linear function of three of these constants, namely b_0, b_1, b_2, and involves nothing else which is indeterminate for a given cross-section. But these *twelve* constants will reduce to *six*, if we make use of the six conditions contained in (1) of our Art. 1334. These conditions lead to

$$a' = 0, \quad a'' = 0, \quad c = 0, \quad a_0 = 0, \quad b' = \left(\frac{d\Omega}{dx} \right)_0, \quad b'' = \left(\frac{d\Omega}{dy} \right)_0,$$

where in the last two equations the subscript 0 indicates that we are to put x and y each zero after differentiation.

The values of u, v, w as furnished by equations (9) and (11) take then the following simpler forms:

$$
\left.
\begin{aligned}
u &= -\eta \left\{ ax + a_1 \frac{x^2 - y^2}{2} + a_2 xy \right\} - \eta z \left\{ b_1 \frac{x^2 - y^2}{2} + b_2 xy \right\} \\
&\qquad + b_0 yz + z \left(\frac{d\Omega}{dx} \right)_0 - \frac{a_1 z^2}{2} - \frac{b_1 z^3}{6}, \\
v &= -\eta \left\{ ay + a_2 \frac{y^2 - x^2}{2} + a_1 xy \right\} - \eta z \left\{ b_2 \frac{y^2 - x^2}{2} + b_1 xy \right\} \\
&\qquad - b_0 xz + z \left(\frac{d\Omega}{dy} \right)_0 - \frac{a_2 z^2}{2} - \frac{b_2 z^3}{6}, \\
w &= z (a + a_1 x + a_2 y) + \frac{z^2}{2} (b_1 x + b_2 y) - \left(\frac{E}{2\mu} - \eta \right) (b_1 xy^2 + b_2 yx^2) \\
&\qquad + \Omega - x \left(\frac{d\Omega}{dx} \right)_0 - y \left(\frac{d\Omega}{dy} \right)_0
\end{aligned}
\right\} \dots(16).
$$

1340. The twenty-fifth section (S. 85-7) proceeds to the discussion of the solution.

There are six constants in the solution; we may then suppose them all to vanish except one, and so obtain an idea of the meaning of this constant. This Clebsch proposes to do. There would then be apparently *six* cases to discuss, but by a slight modification of the process it is found that a smaller number of cases is sufficient.

When any system of shifts occurs in a rod there are two points which deserve especial attention. We may determine the form assumed by a 'fibre' which was originally a straight line parallel to the axis of z; and we may determine the form assumed by a section which was originally a plane at right angles to this axis. Suppose now that x', y', z', denote the coordinates of a point of which the original coordinates were x, y, z; then

$$x' = x + u, \quad y' = y + v, \quad z' = z + w \quad \dots\dots\dots\dots(1).$$

In order to determine the form assumed by a 'fibre' we treat x and y as constant, and eliminate z between these three equations. Neglecting quantities which are small in comparison with those which we retain, this amounts to putting z' for z in u and v; denote the results thus obtained by u' and v' respectively: then we obtain

$$x' = x + u', \quad y' = y + v' \quad\dots\dots\dots\dots\dots(2).$$

In order to determine the form assumed by a cross-section we treat z as constant, and eliminate x and y; this amounts approximately to putting x' and y' for x and y respectively in w: and the result may be expressed thus:

$$z' = z + w' \quad\dots\dots\dots\dots\dots(3).$$

1341. As the first case to be considered we will suppose that all the constants vanish except a; then equations (16) of Art. 1339 reduce to

$$u = -\eta ax, \quad v = -\eta ay, \quad w = az.$$

The stresses all vanish except $\widehat{zz}$, and this is equal to Ea. The result corresponds to a simple longitudinal traction. Every straight line parallel to the axis of z becomes $(1 + a)$ times its original length, while a transverse line is reduced to $(1 - \eta a)$ times its original length.

1342. The twenty-sixth section (S. 87–91) continues the discussion of the results, which was commenced in the twenty-fifth section.

Suppose that all the constants in equations (16) of our Art. 1339 vanish except a_1 and b_1. Then

$$u = -a_1 \frac{\eta\,(x^2 - y^2) + z^2}{2} - b_1 \left\{ \frac{z^3}{6} + \eta z\,\frac{x^2 - y^2}{2} - z \left(\frac{dB_1}{dx}\right)_0 \right\},$$

$$v = -\eta xy\,(a_1 + b_1 z) + zb_1 \left(\frac{dB_1}{dy}\right)_0 , \qquad\qquad \left.\begin{array}{c}\\ \\ \\ \end{array}\right\} \ldots(1).$$

$$w^1 = a_1 xz + b_1 \left\{ \frac{xz^2}{2} - \left(\frac{E}{2\mu} - \eta\right) xy^2 + B_1 - x \left(\frac{dB_1}{dx}\right)_0 - y \left(\frac{dB_1}{dy}\right)_0 \right\}$$

Further from (11) of our Art. 1336 :

$$\widehat{zz} = Ex\,(a_1 + b_1 z),$$

$$\widehat{zx} = \mu b_1 \left\{ -\frac{\eta x^2 + \left(\frac{E}{\mu} - 3\eta\right) y^2}{2} + \frac{dB_1}{dx} \right\}, \qquad \left.\begin{array}{c}\\ \\ \\ \end{array}\right\} \ldots\ldots\ldots(2)$$

$$\widehat{yz} = \mu b_1 \left\{ -\left(\frac{E}{\mu} - \eta\right) xy + \frac{dB_1}{dy} \right\}$$

The equations (2) of Art. 1336 then become

$$x' = x - a_1 \frac{\eta\,(x^2 - y^2) + z'^2}{2} - b_1 \left\{ \frac{z'^3}{6} + \eta z'\,\frac{x^2 - y^2}{2} - z' \left(\frac{dB_1}{dx}\right)_0 \right\}, \left.\begin{array}{c}\\ \\ \end{array}\right\} \ldots(3)$$

$$y' = y - \eta xy\,(a_1 + b_1 z') + z'b_1 \left(\frac{dB_1}{dy}\right)_0$$

The second of these equations represents a plane, so that a 'fibre' which was originally parallel to the axis of the prism remains in one plane ; the first of these equations is that to the projection on the plane of xz of the curve which the 'fibre' becomes ; the curve is one of the *third* degree, which reduces to the common parabola when b_1 is zero. The plane denoted by the second equation is parallel to the axis of x ; in a particular case this plane will also be parallel to the axis of z, namely when

$$\eta xy = \left(\frac{dB_1}{dy}\right)_0 \ \ldots\ldots\ldots\ldots\ldots\ldots\ldots\ldots\ldots(4),$$

for then the equation reduces to

$$y' = y\,(1 - \eta xa_1).$$

Thus the 'fibres' which remain after displacement in a plane parallel to the axis originally constituted a hyperbolic cylinder determined by (4).

1343. The amount of the bending may be estimated by the shift of the end of the fibre determined by $x = 0$, $y = 0$. Suppose l the

In the value of w Clebsch has $\frac{z^2}{2}$ instead of our $\frac{xz^2}{2} - \left(\frac{E}{2\mu} - \eta\right) xy^2$; this makes his dimensions in u and w different : the mistake prevails through his twenty-sixth section.

length of the cylinder[1], and u_l, v_l, the corresponding values of u and v; then

$$u_l = -\frac{a_1}{2} l^2 - b_1 \left\{ \frac{l^3}{6} - l \left(\frac{dB_1}{dx} \right)_0 \right\},$$

$$v_l = l b_1 \left(\frac{dB_1}{dy} \right)_0 \qquad \qquad \qquad \qquad \dots\dots\dots\dots(5)$$

Clebsch also deals (S. 88–91) with the distorted form of the cross-section.

If instead of a_1, b_1 we cause all the constants except a_2, b_2 to vanish we obtain precisely similar results except that the bending now takes place in the plane yz.

1344. The twenty-seventh section (S. 91–4) continues the discussion commenced in the twenty-fifth.

Suppose that all the constants in equations (16) of our Art. 1339 vanish except b_0. Then

$$u = b_0 z \left\{ y + \left(\frac{dB_0}{dx} \right)_0 \right\}, \quad v = -b_0 z \left\{ x - \left(\frac{dB_0}{dy} \right)_0 \right\},$$

$$w = b_0 \left\{ B_0 - x \left(\frac{dB_0}{dx} \right)_0 - y \left(\frac{dB_0}{dy} \right)_0 \right\} \dots(6).$$

Further from (11) of our Art. 1336:

$$\widehat{zx} = \mu b_0 \left(y + \frac{dB_0}{dx} \right), \quad \widehat{yz} = -\mu b_0 \left(x - \frac{dB_0}{dy} \right), \quad \widehat{zz} = 0 \ \dots\dots(7).$$

If we shift the origin of coordinates, and put y_1 for $y + \left(\dfrac{dB_0}{dx} \right)_0$, and x_1 for $x - \left(\dfrac{dB_0}{dy} \right)_0$, the values of u and v become

$$u = b_0 z y_1, \quad v = -b_0 z x_1,$$

and then we see that they correspond to a *torsion*. The angle which expresses the amount of twisting is denoted by $b_0 z$, and so it varies as z.

Clebsch shows that the 'fibres' which originally were on the curved surface of any right circular cylinder of radius r

$$(x - a)^2 + (y - \beta)^2 = r^2,$$

will after strain lie on a hyperboloid of one sheet.

He says with respect to this section and the two which precede it:

So sind denn bei der Discussion dieser Resultate die drei Hauptformen, unter welchen ein elastischer Stab sich darstellt, sofort zu Anschauung

---

[1] Clebsch uses l without stating what it means; and he seems to say on his S. 88 that the bending takes place *in the plane* of xz: that is he treats v_l as if it were zero.

gekommen : *Ausdehnung, Biegung* und *Torsion*. Zugleich ist für die annähernde Behandlung wirklicher Probleme ein sicherer Ausgangspunkt gewonnen, und damit die Basis gegeben, auf welche eine minder strenge Fortentwicklung sich stützen kann (S. 94).

[1345.] The whole of the above investigation is concise, clear, and instructive, especially from the mathematical standpoint. It gives us the most general solution of the differential equations of elasticity subject to certain conditions, in particular the vanishing of the stresses $\widehat{xx}$, $\widehat{yy}$ and $\widehat{xy}$. It thus embraces Saint-Venant's results both for flexure and torsion and throws light on their mutual relationship : see our Arts. 17 and 82. It does not bring out to the student, however, quite so clearly as Saint-Venant's treatment the reason for these assumptions as to the stresses, and requires therefore to be supplemented by such considerations as we have referred to in our Arts. 77, 80 and 316–8. See also Saint-Venant's *Clebsch*, pp. 174–190. Certain misprints of Clebsch's have been tacitly corrected in our reproduction.

[1346.] Clebsch's twenty-eighth section is entitled : *Angenäherte Anwendung auf wirkliche Probleme* (S. 94–8, F. E. pp. 169–174). The discussion in this section does not seem to me to bring out fully the relationship between the theoretical surface stresses and such loads as can be applied in practice. Namely it is almost impossible to apply in practice any distribution of force which can be exactly represented by theory, we can only hope to obtain statically equivalent systems of loading : see our Arts. 8, 9, 21 and 100.

Clebsch supposes a statical system given by the force-components A, B, C parallel to the axes of x, y, z (origin the terminal, $z = 0$) and a couple system A', B', C' about those axes, applied to the terminal cross-section $z = l$. He takes the axes of x and y to coincide with the principal axes of inertia of a cross-section, and we may write :

$$\iint dxdy = \omega, \quad \iint x^2 dxdy = \kappa_2^2\omega, \quad \iint y^2 dxdy = \kappa_1^2\omega.$$

By the aid of these we can express the undetermined constants a, b_1, b_2, u_1, u_2, b_0 in terms of A, B, C, A', B', C' as is done by Clebsch on ·S. 98. The equations he gives contain integrals involving differentials of Ω. But it is shown in the following or twenty-ninth section (S. 99–102, F. E. pp. 191–5) that although Ω may not have been determined these integrals can be determined with one exception in terms of the cross-section and independently of Ω. Thus Clebsch deduces the follow-ing values for his constants (S. 102, *F. E.* p. 194):

$$\left.\begin{aligned}
& a = C/(E\omega), \quad b_1 = A/(E\omega\kappa_2^2), \quad a_1 = -B'/(E\omega\kappa_2^2), \\
& \qquad\qquad b_2 = B/(E\omega\kappa_1^2), \quad a_2 = A'/(E\omega\kappa_1^2), \\
& -b_0(\kappa_1^2 + \kappa_2^2) + \frac{1}{\omega}\iint\left(x\frac{d\Omega}{dy} - y\frac{d\Omega}{dx}\right)d\omega \\
& \quad + \frac{b_1}{2\omega}\iint\left\{\left(\frac{E}{\mu} - 3\eta\right)y^3 - \left(2\frac{E}{\mu} - 3\eta\right)x^2y\right\}d\omega \\
& \quad - \frac{b_2}{2\omega}\iint\left\{\left(\frac{E}{\mu} - 3\eta\right)x^3 - \left(2\frac{E}{\mu} - 3\eta\right)xy^2\right\}d\omega = \frac{C'}{\mu\omega}
\end{aligned}\right\}\quad \ldots\ldots(8).$$

Thus a, b_1, a_1, b_2, a_2 are given each in terms of a single element of the load system, but b_0 is given in terms of three, namely C', A and B. Clebsch says: "nur die letzte Gleichung enthält dann noch sämmtliche Grössen, so dass b_0 sich durch alle mit Ausnahme von C ausdrückt." This seems to me incorrect, as b_0 does not involve A' or B'

[1347.] The thirtieth section is entitled: *Symmetrische Querschnitte*, and occupies S. 102–6 (*F. E.* pp. 198–202). Here Clebsch investi-gates how the equations of our previous article may be simplified if the cross-section be symmetrical about two rectangular axes. Here after some reductions and for the case of a single force P acting parallel to the axis of x at the centroid of the terminal cross-section, $z = l$, we have:

$$\left.\begin{aligned}
& u = \frac{P}{E\omega\kappa_2^2}\left\{\eta\frac{x^2 - y^2}{2}(l - z) + \frac{z^2 l}{2} - \frac{z^3}{6} + z\left(\frac{dB_1}{dx}\right)_0\right\}, \\
& v = \frac{P}{E\omega\kappa_2^2}\eta xy(l - z), \\
& w = \frac{P}{E\omega\kappa_2^2}\left\{-lxz + \frac{z^2}{2}x - \left(\frac{E}{2\mu} - \eta\right)xy^2 + B_1 - x\left(\frac{dB_1}{dx}\right)_0\right\}, \\
& \widehat{zx} = \frac{\mu P}{E\omega\kappa_2^2}\left\{\frac{dB_1}{dx} - \eta\frac{x^2}{2} - \left(\frac{E}{\mu} - 3\eta\right)\frac{y^2}{2}\right\}, \\
& \widehat{yz} = \frac{\mu P}{E\omega\kappa_2^2}\left\{\frac{dB_1}{dy} - \left(\frac{E}{\mu} - \eta\right)xy\right\}, \\
& \widehat{zz} = -\frac{Px(l - z)}{\omega\kappa_2^2}
\end{aligned}\right\}\quad \ldots(9)$$

These values should be compared with those given in our Arts. 17, 83 and 84. Clebsch has an erroneous value of w in his equations (88) and (89) on S. 105. The error arises from the wrong value of w, already referred to (Art. 1342, *ftn.*), given on S. 87 in equation (75 a), and its influence extends to S. 110 of the *Treatise*.

[1348.] The following seven sections may be dealt with more briefly. They occupy S. 107–138; *F. E.* pp. 202–252.

(a) § 31 treats the case of the prism of elliptic cross-section : see Saint-Venant's results in our Arts. 18 and 90. There are errors on S. 110.

(b) § 32. General remarks on case of a hollow prism with, I think, wrong equations for c, b', and b'' : see our Art. 49.

(c) § 33. This contains I believe the first introduction of what are really conjugate functions into Saint-Venant's problem. Clebsch transforms the equations for Ω, i.e. for the B's (see our Arts. 1336 and 1338), into curvilinear coordinates in the plane of the cross-section.

The investigation has since been more elegantly carried out by Thomson and Tait : see their *Treatise on Natural Philosophy* 2nd Edn., Part II. pp. 250–3, but the idea is due to Clebsch : see our Art. 285.

(d) § 34. This develops the transformation of the preceding section for the case of elliptic coordinates.

(e) § 35 applies the whole investigation to the case of the *pure torsion* of a hollow cylinder the section of which is bounded by two confocal ellipses.

If the confocal ellipses be given by

$$\frac{x^2}{m^2 + a_1} + \frac{y^2}{n^2 + a_1} = 1,$$

$$\frac{x^2}{m^2 + a_0} + \frac{y^2}{n^2 + a_0} = 1,$$

Clebsch[1] finds for the value of

$$J \equiv \iint \left(x \frac{dB_0}{dy} - y \frac{dB_0}{dx} \right) d\omega,$$

$$J = \frac{\pi (m^2 - n^2)^2}{4} \frac{(\sqrt{m^2 + a_1} + \sqrt{n^2 + a_1})^2 - (\sqrt{m^2 + a_0} + \sqrt{n^2 + a_0})^2}{(\sqrt{m^2 + a_1} + \sqrt{n^2 + a_1})^2 + (\sqrt{m^2 + a_0} + \sqrt{n^2 + a_0})^2}$$

Thus all the constants of the problem (see our Art. 1344) are determined, and b_0 the angle of torsion per unit length of cylinder is given by :

$$b_0 = - \frac{C'}{\mu \{ (\kappa_1^2 + \kappa_2^2) \, \omega - J \}}$$

The values of κ_1^2 and κ_2^2 are easily expressible in terms of the axes of the two ellipses.

This result may be compared with Saint-Venant's for a hollow prism bounded by similar and similarly situated elliptic cylinders. Clebsch's analysis is interesting, but to make the cross-section with confocal instead of similar elliptic boundaries possesses no particular

---

Clebsch (and Saint-Venant editing him, p. 239) have 8 instead of 4 in the denominator of J, but this appears to be an error. Clebsch further drops the π n the numerator.

practical advantages, and the theoretical results are far more complicated. See also Saint-Venant's note on the subject pp. 240–2 of his edition of Clebsch.

(f) An instructive conclusion can, however, be drawn from Clebsch's result as to the possibly delusive character of torsional experiments upon bars which are not absolutely free from flaws. Suppose the inner elliptic surface to reduce to a thin cavity almost coinciding with the plane area between the focal lines of the outer elliptic surface. We thus have theoretically a fair approximation to the case of the torsion of an elliptic bar with a flaw along its axis, or with a rotten core, a not infrequent case in castings. If $M'(=C')$ be the couple required to produce an angle of torsion $\tau\,(=b_0)$ per unit length of a bar with cross-section and semi-axes b and $c\,(=\sqrt{m^2+a_1}$ and $\sqrt{n^2+a_1})$, we easily find from the above results by putting $\sqrt{n^2+a_0}=0$ and $\sqrt{m^2+a_0}=\sqrt{b^2-c^2}$, that:

$$M' = \mu\tau\pi bc\,\frac{(3b^2-c^2)\,c^2}{4b^2}$$

If M be the couple producing the same torsional angle in a sound bar of the same dimensions and material we have by Art. 18:

$$M = \mu\tau\pi bc\,\frac{b^2c^2}{b^2+c^2}.$$

Thus : $\qquad M'/M = \dfrac{(3b^2-c^2)\,(b^2+c^2)}{4b^4}.$

We find that this ratio varies from 1 to ·75, i.e. :

$$b/c = 1, \qquad M = M',$$
$$b/c = 2, \qquad M = \text{·}86M',$$
$$b/c = 3, \qquad M = \text{·}80M',$$
$$b/c = \infty, \qquad M = \text{·}75M'.$$

It would thus appear that the determination of the slide-modulus from torsional experiments on cast bars may be liable to considerable error, if there be flaws, as so frequently happens, in the core of the bar.

The maximum slide σ might be calculated for this case from the formula :

$$\frac{\sigma^2}{b_0^2} = \text{max.}\left\{x^2+y^2+2\left(y\frac{dB_0}{dx}-x\frac{dB_0}{dy}\right)+\left(\frac{dB_0}{dx}\right)^2+\left(\frac{dB_0}{dy}\right)^2\right\},$$

and its value compared with that given for the case of a sound bar in Art. 18. The analysis would be somewhat lengthy, but it would be interesting to compare the result with Mr Larmor's conclusions: *Philosophical Magazine*, Vol. 33, p. 70, 1892.

(g) In § 36 we have a discussion of the *Elasticitätsellipsoid* for a case like the present when the three stresses $\widehat{xx}$, $\widehat{yy}$ and $\widehat{xy}$ are zero. The principal tractions are now

$$T''' = 0, \qquad T' = \frac{\widehat{zz}}{2} + \sqrt{\widehat{zx}^2 + \widehat{yz}^2 + \frac{\widehat{zz}^2}{4}},$$

$$T'' = \frac{\widehat{zz}}{2} - \sqrt{\widehat{zx}^2 + \widehat{yz}^2 + \frac{\widehat{zz}^2}{4}}$$

Obviously T' and T'' are always of opposite sign, or one principal traction is negative and the other positive.

Clebsch gives an elegant geometrical construction for determining the position of the ellipse to which the ellipsoid reduces and so the directions of the principal tractions. His consideration, however, of the spot at which the danger of rupture is greatest (S. 132) seems to me invalid as it is based on a maximum stress limit.

(h)　The same objection applies to his § 37 entitled : *Grenzen für die Grösse der äussern Kräfte.* The concluding paragraph of that section (S. 138, *F. E.* p. 252) contains several statements which do not seem in accord with experience, and a very loose conception of the limit of elasticity as well as of the different practical effects of pressure and traction is exhibited : see our Arts. 164, 321 and 709–10.

[1349.]　The next or thirty-eighth section of Clebsch's *Treatise* is entitled: *Vergleichung mit der gewöhnlichen Theorie. Grundlagen für weitere Anwendungen* (S. 139–48 ; *F. E.* pp. 283–94). This compares the theory just developed for flexure with the Bernoulli-Eulerian, and for torsion with the extension of Coulomb's theory to prisms of other than circular cross-section. Clebsch criticises with considerable severity the earlier theories. He remarks that even Saint-Venant's theory only covers the special case of flexure in which constant forces act upon a free end and continues :

Es wird eine weitere Aufgabe der strengen Theorie sein, ähnliche Gleichungen für allgemeinere Fälle aufzustellen. Da dies inzwischen bisher nicht gelungen ist[1], so wird man einstweilen jener Gleichungen sich auch fortfahren zu bedienen, wenn das Innere des Körpers durch Kräfte ergriffen wird, oder wenn an verschiedenen Stellen des Körpers Einzelkräfte angreifen. Man wird sich aber dabei den Mangel an Strenge nicht verhehlen dürfen. In einem spätern Abschnitt wird sich zeigen, dass für *sehr kleine* Querschnitte dies Verfahren allerdings zulässig ist (S. 142, *F. E.* p. 287).

---

[1] The Editor of the present work in a memoir, the first part of which is published in the *Quarterly Journal of Mathematics*, June, 1889, has dealt with the case of a uniform body-force and continuous surface-load.

Clebsch's remark on S. 142 as to a failure of the ordinary theory does not seem fully justified. The theory had in respect to the non-coincidence of loading and bending planes been corrected by Persy in 1834 (Art. 811*), and he had been followed by both Saint-Venant and Bresse with full consideration of this very point. Clebsch while reproducing results exactly equivalent to theirs makes no reference to their writings: see our Arts. 1581*, 14, 171, 177, and 515.

The section concludes with a very severe criticism of that modification of the torsion theory of Coulomb, which supposes the stretch in the longitudinal 'fibres' of a prism under torsion can affect sensibly its torsional moment. I can only suppose the *gewisse Kreise* Clebsch spends his satire on are composed of the authors, whose papers on torsion are referred to in our Arts. 481, 581 and 803. The criticism is severe, but perhaps not unjustified.

[1350.] S. 148–99 (*F. E.* pp. 295–374) of Clebsch's work deal with the subject of thick plates, the edges (not the faces) of which are subjected to load. I believe the method here, as well as several of the results, are original. In the French edition these pages appear as a separate chapter entitled: *Plaques d'épaisseur quelconque.* Clebsch in this portion of his work applies the *semi-inverse* method of Saint-Venant (Arts. 3, 6, 9, 71, etc.) to the problem of thick plates. Suppose the normal to the plane faces of the plate to be taken as the direction of the axis of z, and the plane of x, y to be the mid-plane of the plate. Then Clebsch *assumes*:

$$\widehat{zx} = \widehat{yz} = \widehat{zz} = 0 \ldots\ldots\ldots\ldots\ldots\ldots(1),$$

i.e. he causes the other three stresses to vanish, not those assumed by Saint-Venant for his rod problem: see our Art. 1334. Besides no load on the faces of the plate Clebsch supposes no body-forces, and he then inquires what solutions of the equations of elasticity are possible under these conditions and what system of load they connote on the cylindrical boundary of the plate.

With regard to (1) Clebsch merely writes:

Diese Gleichungen gelten zunächst nur für die Werthe von z, welche den Grenzflächen der Platte entsprechen. Ich werde aber nur diejenigen Zustände untersuchen, für welche diese Gleichungen für jeden Punkt der Platte erfüllt sind. Man sieht, dass dann jedenfalls die auf die cylindrischen Seitenflächen wirkenden Kräfte keine der z-Axe parallele, also zu der Platte normale Componente liefern dürfen, weil sonst wenigstens am Rande jene Spannungen nicht verschwinden würden (S. 149, *F. E.* p. 296).

[1351.] The body stress-equations are now obviously:

$$\frac{d\widehat{xx}}{dx} + \frac{d\widehat{xy}}{dy} = 0, \quad \frac{d\widehat{xy}}{dx} + \frac{d\widehat{yy}}{dy} = 0 \quad \dots\dots(2).$$

Following Saint-Venant's modification of Clebsch and supposing the plate to possess a planar distribution of isotropy, we have to use the stress-strain relations:

$$\left.\begin{aligned}
\widehat{xx} &= a\,\frac{du}{dx} + f'\frac{dv}{dy} + d'\frac{dw}{dz}, & \widehat{yz} &= d\left(\frac{dv}{dz} + \frac{dw}{dy}\right), \\
\widehat{yy} &= f'\frac{du}{dx} + a\,\frac{dv}{dy} + d'\frac{dw}{dz}, & \widehat{zx} &= d\left(\frac{dw}{dx} + \frac{du}{dz}\right), \\
\widehat{zz} &= d'\left(\frac{du}{dx} + \frac{dv}{dy}\right) + c\,\frac{dw}{dz}, & \widehat{xy} &= f\left(\frac{du}{dy} + \frac{dv}{dx}\right)
\end{aligned}\right\} \dots\dots(3),$$

where $a = 2f + f'$: see our Art. 114 and 117, (*b*).

Hence we have by (1):

$$\frac{dv}{dz} + \frac{dw}{dy} = 0, \quad \frac{dw}{dx} + \frac{du}{dz} = 0, \quad \frac{du}{dx} + \frac{dv}{dy} = -\frac{c}{d'}\frac{dw}{dz} \quad \dots\dots(4),$$

and by (2):

$$\left.\begin{aligned}
f\left(\frac{d^2u}{dx^2} + \frac{d^2u}{dy^2}\right) + (f + f')\frac{d}{dx}\left(\frac{du}{dx} + \frac{dv}{dy}\right) + d'\frac{d^2w}{dxdz} = 0, \\
f\left(\frac{d^2v}{dx^2} + \frac{d^2v}{dy^2}\right) + (f + f')\frac{d}{dy}\left(\frac{du}{dx} + \frac{dv}{dy}\right) + d'\frac{d^2w}{dydz} = 0
\end{aligned}\right\} \dots\dots(5).$$

From (4) and (5) by eliminating u, v we find:

$$\frac{d}{dx}\left(\frac{d^2w}{dz^2}\right) = 0, \quad \frac{d}{dy}\left(\frac{d^2w}{dz^2}\right) = 0, \quad \frac{d}{dz}\left(\frac{d^2w}{dz^2}\right) = 0 \dots\dots(6).$$

Thus w must be of the form:

$$w = -\frac{Cz^2}{2} - zF + \mathfrak{f} - C\frac{c}{d'}\frac{x^2 + y^2}{4} \quad \dots\dots(7),$$

where F and $\mathfrak{f}$ are arbitrary functions of x and y, and the last term is taken out of $\mathfrak{f}$ for the convenience of analysis.

Differentiating equations (5) with regard to x and y respectively, adding, and replacing $du/dx + dv/dy$ in it by its value from the third equation of (4), we find by aid of the first two of (6):

$$\frac{d^2}{dx^2}\left(\frac{dw}{dz}\right) + \frac{d^2}{dy^2}\left(\frac{dw}{dz}\right) = 0,$$

or from (7):

$$\frac{d^2F}{dx^2} + \frac{d^2F}{dy^2} = 0 \dots\dots\dots\dots\dots(8).$$

But differentiating the third of (4) with regard to z, and using the first two of (4), we have:

$$\frac{d^2w}{dx^2} + \frac{d^2w}{dy^2} = \frac{c}{d'} \frac{d^2w}{dz^2},$$

whence from (7)

$$\frac{d^2\mathfrak{f}}{dx^2} + \frac{d^2\mathfrak{f}}{dy^2} = 0 \dots\dots\dots\dots\dots(9).$$

Now determine u and v from the first two of (4) and we have:

$$u = \frac{z^2}{2}\frac{dF}{dx} - z\left(\frac{d\mathfrak{f}}{dx} - C\frac{c}{d'}\frac{x}{2}\right) + \phi,$$
$$v = \frac{z^2}{2}\frac{dF}{dy} - z\left(\frac{d\mathfrak{f}}{dy} - C\frac{c}{d'}\frac{y}{2}\right) + \psi \quad \Big\} \dots\dots(10),$$

where ϕ and ψ are arbitrary functions of x, y.

Equations (7) and (10) satisfy identically the first two equations of (4); the third equation of (4), however, introduces a relation between ϕ and ψ. Substitute (10) in that equation, and substitute u, v and w in (5) after using the third of (4), then remembering (8) and (9) we find:

$$\frac{d\phi}{dx} + \frac{d\psi}{dy} = \frac{c}{d'} F,$$
$$\frac{d^2\phi}{dx^2} + \frac{d^2\phi}{dy^2} + \frac{c}{d'}\left(\frac{H}{f} - 1\right)\frac{dF}{dx} = 0, \quad \Big\} \dots\dots\dots(11),$$
$$\frac{d^2\psi}{dx^2} + \frac{d^2\psi}{dy^2} + \frac{c}{d'}\left(\frac{H}{f} - 1\right)\frac{dF}{dy} = 0$$

where H is the plate-modulus $= 2f + f' - d'^2/c$: see our Art. 323.

The last two of these equations evidently contain (8).

Eliminating F by the first of (11) we find from the last two equations:

$$\frac{H}{f}\frac{d^2\phi}{dx^2} + \frac{d^2\phi}{dy^2} + \left(\frac{H}{f} - 1\right)\frac{d^2\psi}{dxdy} = 0,$$
$$\frac{d^2\psi}{dx^2} + \frac{H}{f}\frac{d^2\psi}{dy^2} + \left(\frac{H}{f} - 1\right)\frac{d^2\phi}{dxdy} = 0 \quad \Big\} \dots\dots\dots(12).$$

Equations (12) and (9) suffice to determine ϕ, ψ and $\mathfrak{f}$ and so solve the body-shift equations.

Any values therefore of $\mathfrak{f}$, ϕ and ψ satisfying (9) and (12) give the following expressions for the shifts:

$$
\left.
\begin{aligned}
u &= \frac{d'}{c}\frac{z^2}{2}\left(\frac{d^2\phi}{dx^2} + \frac{d^2\psi}{dx\,dy}\right) + \phi - z\left(\frac{d\mathfrak{f}}{dx} - \frac{c}{d'}\frac{Cx}{2}\right), \\
v &= \frac{d'}{c}\frac{z^2}{2}\left(\frac{d^2\phi}{dx\,dy} + \frac{d^2\psi}{dy^2}\right) + \psi - z\left(\frac{d\mathfrak{f}}{dy} - \frac{c}{d'}\frac{Cy}{2}\right), \\
w &= -\frac{Cz^2}{2} - \frac{d'}{c}z\left(\frac{d\phi}{dx} + \frac{d\psi}{dy}\right) + \mathfrak{f} - \frac{c}{d'}C\frac{x^2+y^2}{4}
\end{aligned}
\right\}
\dots\dots(13),
$$

provided the stresses on the cylindrical face of the plate be given by:

$$
\left.
\begin{aligned}
\widehat{xx} &= \frac{fd'}{c}z^2\left(\frac{d^3\phi}{dx^3} + \frac{d^3\psi}{dx^2\,dy}\right) + H\frac{d\phi}{dx} + (H-2f)\frac{d\psi}{dy} - 2fz\left(\frac{d^2\mathfrak{f}}{dx^2} - C'\right), \\
\widehat{yy} &= \frac{fd'}{c}z^2\left(\frac{d^3\phi}{dx\,dy^2} + \frac{d^3\psi}{dy^3}\right) + H\frac{d\psi}{dy} + (H-2f)\frac{d\phi}{dx} - 2fz\left(\frac{d^2\mathfrak{f}}{dy^2} - C'\right), \\
\widehat{xy} &= \frac{fd'}{c}z^2\left(\frac{d^3\phi}{dx^2\,dy} + \frac{d^3\psi}{dx\,dy^2}\right) + f\left(\frac{d\phi}{dy} + \frac{d\psi}{dx}\right) - 2fz\frac{d^2\mathfrak{f}}{dx\,dy},
\end{aligned}
\right\}
\dots(14)
$$

where
$$
C' = \frac{c}{fd'}(H - f)\frac{C}{2}
$$

Saint-Venant (p. 302) puts $\eta''/(1-\eta')$ for d'/c, and C/η'' he replaces by another constant C'' (C' in his notation); $2f$ he puts $= E'/(1+\eta')$, whence he finds, since $2f + f' - d'^2/c = E'/(1 - \eta'^2)$:

$$
H = 2f/(1-\eta'), \quad H - 2f = 2f\eta'/(1-\eta'),
$$

and
$$
C' = (1+\eta')\frac{C''}{2}
$$

His results then agree with Clebsch's if the accents be removed from E and η, and we note that Clebsch's $\mu =$ Saint-Venant's and our η.

See Clebsch, S. 148–52 and F. E. pp. 295–302.

[1352.] Clebsch next turns to the boundary conditions. Suppose forces X, Y to act at the point x, y, z of the cylindrical boundary and let the outwardly measured normal make an angle p with the x axis, then we must have:

$$
\left.
\begin{aligned}
X &= \widehat{xx}\cos p + \widehat{xy}\sin p, \\
Y &= \widehat{xy}\cos p + \widehat{yy}\sin p
\end{aligned}
\right\}
\dots\dots\dots\dots(15)
$$

Now the stresses contain only powers of z up to the second.
Hence X, Y must be of the form:

$$
\left.
\begin{aligned}
X &= X_0 + X_1 z + X_2 z^2, \\
Y &= Y_0 + Y_1 z + Y_2 z^2
\end{aligned}
\right\}
\dots\dots\dots\dots(16)
$$

Whence we obtain as the boundary-conditions by substituting (14) in (15):

$$X_0 = \left\{ H \frac{d\phi}{dx} + (H - 2f) \frac{d\psi}{dy} \right\} \cos p + f \left(\frac{d\phi}{dy} + \frac{d\psi}{dx} \right) \sin p,$$

$$Y_0 = f \left(\frac{d\phi}{dy} + \frac{d\psi}{dx} \right) \cos p + \left\{ H \frac{d\psi}{dy} + (H - 2f) \frac{d\phi}{dx} \right\} \sin p,$$

$$X_1 = -2f \left\{ \left(\frac{d^2\mathfrak{f}}{dx^2} - C' \right) \cos p + \frac{d^2\mathfrak{f}}{dx\,dy} \sin p \right\},$$

$$Y_1 = -2f \left\{ \frac{d^2\mathfrak{f}}{dx\,dy} \cos p + \left(\frac{d^2\mathfrak{f}}{dy^2} - C' \right) \sin p \right\}, \qquad \Bigg\} \ldots(17).$$

$$X_2 = \frac{fd'}{c} \left\{ \left(\frac{d^3\phi}{dx^3} + \frac{d^3\psi}{dx^2 dy} \right) \cos p + \left(\frac{d^3\phi}{dx^2 dy} + \frac{d^3\psi}{dx\,dy^2} \right) \sin p \right\},$$

$$Y_2 = \frac{fd'}{c} \left\{ \left(\frac{d^3\phi}{dx^2 dy} + \frac{d^3\psi}{dx\,dy^2} \right) \cos p + \left(\frac{d^3\phi}{dx\,dy^2} + \frac{d^3\psi}{dy^3} \right) \sin p \right\}$$

Here are six equations with only three functions ϕ, ψ, $\mathfrak{f}$, hence the six quantities X_0, Y_0, X_1, Y_1, X_2, Y_2 cannot in general be independent.

[1353.] Clebsch now proceeds to an analysis of these separate results. He considers first the terms:

$$X = X_0 + X_2 z^2, \quad Y = Y_0 + Y_2 z^2 \ldots\ldots\ldots\ldots(18).$$

These do not change when z is changed to $-z$, so that the boundary forces are symmetrical about the mid plane of the plate. The condition of the plate is thus stretch without flexure. The shifts will then take the forms:

$$u = u_0 + u_2 z^2, \quad w = -w_1 z \atop v = v_0 + v_2 z^2, \qquad \Big\} \ldots\ldots\ldots\ldots\ldots(19).$$

Consider first the terms u_0 and v_0 only, or let $u_2 = v_2 = 0$; these lead us from (13) by aid of (11) almost at once to:

$$\frac{d\phi}{dx} + \frac{d\psi}{dy} = \kappa, \quad \frac{d\phi}{dy} - \frac{d\psi}{dx} = \kappa',$$

where κ and κ' are constants.

Hence
$$u_0 = \phi = \frac{\kappa x + \kappa' y}{2} + U, \atop v_0 = \psi = \frac{\kappa y - \kappa' x}{2} + V \qquad \Bigg\} \ldots\ldots\ldots\ldots(20),$$

where U, V are solutions of

$$\frac{dU}{dx} + \frac{dV}{dy} = 0, \quad \frac{dU}{dy} - \frac{dV}{dx} = 0 \ldots\ldots\ldots\ldots(21).$$

(a) Neglecting U and V for a time we note that in u_0, v_0, the terms in κ' correspond only to a slight rotation, and those in κ to a *uniform* stretch $= \kappa/2$ in all directions parallel to the mid-plane, and since $\widehat{zz} = 0$, to a uniform squeeze perpendicular to it equal to $\kappa d'/c$. The load necessary to produce this is $(H - f)\kappa$ along the normal at each point of the cylindrical boundary.

(b) Neglecting this uniform strain and turning to that depending on U and V we find from (21) that

$$\left.\begin{array}{l} U + V\sqrt{-1} = \chi_1 (x - y\sqrt{-1}), \\ U - V\sqrt{-1} = \chi_2 (x + y\sqrt{-1}) \end{array}\right\} \dots\dots\dots\dots(22).$$

Hence χ_1 and χ_2 with the assistance of the κ terms can be so determined as to solve the following problem :

A plate is to be so stretched by forces acting on its cylindrical boundary that the squeeze normal to its faces shall be everywhere uniform,

i.e. $dw/dz = - (d\phi/dx + d\psi/dy) d'/c = - \kappa d'/c$,

but all the generators of the cylindrical surface receive arbitrary shifts perpendicular to their length.

Clebsch discusses this problem on S. 158–60, and investigates the required values of X_0, Y_0 for arbitrary shifts when the cylindrical boundary is right circular in § 42, S. 160–4 (*F. E.* pp. 312–16). The problem is of more analytic than practical interest, as it would be extremely difficult to pull out the edges of an actual plate to any chosen change of form.

[1354.] The following section is of more practical value. It is entitled : *Anwendung auf angenäherte Lösung allgemeiner Aufgaben* (S. 164–6 ; *F. E.* pp. 316–19). Clebsch notes that the general problem :—Given the load on the cylindrical boundary to find the shifts and stresses in the plate—is not solvable under the conditions (1) of Art. 1350, for the reason we have given immediately after equation (17) in our Art. 1352. Let us, however, suppose the plate to be of *small* thickness h, and let us apply the *principle of the elastic equivalence of statically equipollent loads* (see our Arts. 8, 9, 21, 100).

Let A and B be the components of the load on a strip hds of the cylindrical boundary, where ds is an element of the contour of the mid-plane. Then by the above principle we have from (16) :

$$A = \int_{-h/2}^{+h/2} X dz = X_0 h + X_2 \frac{h^3}{12}, \quad B = \int_{-h/2}^{+h/2} Y dz = Y_0 h + Y_2 \frac{h^3}{12} \dots (23).$$

Hence we see that X_1 and Y_1 do not occur, and further that the six equations (17) reduce really to two, i.e. we are thrown back on (15). We have indeed

$$
\left.\begin{aligned}
A = &\left\{\left(H\frac{d\phi}{dx} + (H-2f)\frac{d\psi}{dy}\right)h + \frac{fd'}{c}\left(\frac{d^3\phi}{dx^3} + \frac{d^3\psi}{dx^2 dy}\right)\frac{h^3}{12}\right\}\cos p \\
&+ \left\{f\left(\frac{d\phi}{dy} + \frac{d\psi}{dx}\right)h + \frac{fd'}{c}\left(\frac{d^3\phi}{dx^2 dy} + \frac{d^3\psi}{dx dy^2}\right)\frac{h^3}{12}\right\}\sin p, \\
B = &\left\{f\left(\frac{d\phi}{dy} + \frac{d\psi}{dx}\right)h + \frac{fd'}{c}\left(\frac{d^3\phi}{dx^2 dy} + \frac{d^3\psi}{dx dy^2}\right)\frac{h^3}{12}\right\}\cos p \\
&+ \left\{\left(H\frac{d\psi}{dy} + (H-2f)\frac{d\phi}{dx}\right)h + \frac{fd'}{c}\left(\frac{d^3\phi}{dx dy^2} + \frac{d^3\psi}{dy^3}\right)\frac{h^3}{12}\right\}\sin p
\end{aligned}\right\}\dots(24).
$$

These with equation (12) fully determine ϕ and ψ. u, v, w will then be found by retaining only the terms in (13) involving ϕ and ψ.

[1355.] In the following section (S. 167–81; *F. E.* pp. 319–33) Clebsch solves the equations of the previous article for the case of a circular plate. That is to say he supposes the circular plate stretched by any system of load parallel to the mid-plane imposed on its cylindrical boundary. The solution is only approximate, as it proceeds on the assumption of the elastic equivalence of statically equipollent load systems, but it would be more and more nearly true as the thickness of the plate became small as compared with its radius. The investigation is a very fine piece of analysis, but the complexity of its results renders it of little physical value, except perhaps in some one or two special cases, when the results might probably be reached by other and simpler processes. Clebsch concludes with the remark:

Es ist ohne Zweifel möglich, das entsprechende Problem auch für andre Formen der Platte zu lösen, als für die hier angenommene. Indess wird es genügen, in einem Fall Weg und Auflösung vollständig dargestellt zu haben, zumal schon dieser einfachste Fall nicht ohne Verwickelung erscheint (S. 181).

[1356.] § 45 of the treatise (S. 181–4; *F. E.* pp. 334–7) is concerned with the terms corresponding to X_1 and Y_1 in equations (16).

We have[1], retaining only these terms, for the shifts :

$$u = -z\left(\frac{d\mathfrak{f}}{dx} - \frac{c}{d'}\frac{Cx}{2}\right), \quad v = -z\left(\frac{d\mathfrak{f}}{dy} - \frac{c}{d'}\frac{Cy}{2}\right),$$

$$w = -\frac{Cz^2}{2} + \mathfrak{f} - \frac{c}{d'}C\frac{x^2+y^2}{4}, \quad \Bigg\} \quad \ldots\ldots(25),$$

where

$$\frac{d^2\mathfrak{f}}{dx^2} + \frac{d^2\mathfrak{f}}{dy^2} = 0$$

and for the stresses

$$\widehat{xx} = -2fz\left(\frac{d^2\mathfrak{f}}{dx^2} - C'\right), \quad \widehat{yy} = -2fz\left(\frac{d^2\mathfrak{f}}{dy^2} - C'\right), \quad \Bigg\} \quad \ldots\ldots(26).$$

$$\widehat{xy} = -2fz\frac{d^2\mathfrak{f}}{dx\,dy}, \quad \text{where } C' = \frac{c}{fd'}(H-f)\frac{C}{2}$$

Here the stresses all change sign with z, hence the forces which act on the edge of the plate are equal and opposite on either side the mid-plane. Thus the character of the above solution is *one of flexure by couples*. In the mid-plane itself there are no stresses.

Clebsch treats (S. 183) a special case of this, namely when $\mathfrak{f} = 0$. This corresponds to the case we have dealt with in our Art. 323 where the plane faces become paraboloids of revolution. Saint-Venant's *Note* (*F. E.* pp. 337–68) which we have analysed in our Arts. 323–37 treats the whole subject much more fully and satisfactorily.

[1357.] The values of X_1 and Y_1 given by the second pair of equations (17) are not perfectly arbitrary, as there is only one function $\mathfrak{f}$ at our choice. Hence it follows that Clebsch's investigation leads to no solution of the problem of flexure for an arbitrary system of load couples round the boundary. Clebsch in § 46 (S. 184–9; *F. E.* pp. 368–74) mentions the following however as one of the problems which can be solved by the aid of (25) and (26) :

Durch passende, in der angegebenen Weise wirkende Kräftepaare soll die Platte so gebogen werden, dass die Peripherie der Mittelfläche nach der Biegung auf einer beliebig vorgeschriebenen, der ursprünglichen Peripherie sehr nahe kommenden Oberfläche liegt (S. 185).

Let $\chi(x, y, z) = 0$ be the given surface, then we must have in the mid plane at the contour $w \times d\chi/dz = -\chi$. *So soon as the constant* C *is chosen*, the value of $\mathfrak{f}$ becomes determinate. Clebsch works out the particular case of a circular plate (S. 186–8, *F. E.* pp. 371–3).

These results differ from Clebsch's. The errors of the latter are corrected by Saint-Venant (*F. E.* p. 334, *footnote*).

He notes in conclusion that the value of the dilatation deduced from (25) for any form of plate is

$$\theta = Cz\left(\frac{c}{d'} - 1\right),$$

and thus is independent of **f**. Thus there is *only one* way of solving the above problem, when we attach to it the condition that there shall be no dilatation (i.e. $C = 0$).

The problem suggested by Clebsch does not seem one capable of practical realisation in any but a few special cases, which are more easily dealt with by other processes.

With this section Clebsch's treatment of thick plates closes.

[1358.] S. 190–355 of the *Treatise* are entitled : *Theorie elastischer Körper, deren Dimensionen zum Theil sehr klein (unendlich klein) sind* (*F. E.* pp. 407–806). The first separate portion of this deals with *thin rods*, and occupies S. 192–263 (*F. E.* pp. 409–631). Of this S. 242–261 deal with the vibrations of such rods. The second separate portion, S. 264–355, deals with the theory of *thin plates*, S. 331–55 being especially occupied with a discussion of their vibrations.

Clebsch attributes the first exact theory of bodies having one or two dimensions very small to Kirchhoff (see our Art. 1253), and proposes to follow his methods with certain modifications. In particular he deals only with homogeneous isotropic material. This restriction is removed in the *Annotated Clebsch* of Saint-Venant, where isotropy is assumed in the plane of the cross-section only. Clebsch begins his general investigations with the statement of Kirchhoff's principle which we have cited in our Art. 1253, and which does not seem to me so obvious as both Clebsch and Kirchhoff appear to consider it.

[1359.] § 48 (S. 192–7) of the *Treatise* commences the discussion of the problem of the thin rod of uniform cross-section, initially straight and acted upon solely by terminal loads. Clebsch supposes such a rod built up of small elementary cylinders placed end to end, and only acted upon at their terminals by the elastic stresses of the adjacent elements. To each such cylinder he applies the formulae obtained in his solution of Saint-Venant's problem and in justification of this he remarks :

Zwar waren jene Formeln nur bei einer gewissen Vertheilung der

Kräfte streng richtig, aus welchen jene Componenten und Drehungsmomente sich zusammensetzen. Aber die dabei eintretende Ungenauigkeit wird offenbar um so grösser, je grösser der Querschnitt ist, und wird verschwindend klein, wenn der Querschnitt selbst verschwindend klein ist, wie in dem vorliegenden Fall. Wie also auch dann in Wirklichkeit die eintretenden Spannungen über den Querschnitt vertheilt seien, immer wird man sie sich bis auf Grössen höherer Ordnung so vertheilt denken können, wie die oben in dem de Saint-Venant'schen Problem erhaltenen Formeln sei ergeben. Man kann also jene Formeln sofort auf die kleinen Verschiebungen anwenden, welche im Innern eines der gedachten Elemente auftreten (S. 193).

Now the principle of the elastic equivalence of statically equipollent load systems here appealed to depends for its accuracy on the smallness of the loaded surface as compared with the other dimensions of the body; i.e., if l be the length and ϵ a linear dimension of the cross-section of a cylinder, ϵ/l must be small in all practical applications of Saint-Venant's results. Hence when Clebsch applies these results to an elementary cylinder of length δs, we must have $\epsilon/\delta s$ small in order that the application may be legitimate. Now Clebsch takes x, y, z to represent the coordinates of any point in an elementary cylinder referred to axes attached to this element, thus it is obvious that the x and y can be of the order ϵ, and z, being taken in the direction of the axis of the cylinder, can be of the order δs. Thus z must be capable of taking values which are great as compared with ϵ, but on S. 195 Clebsch writes:

Es ist vor allem wichtig, sich über die Ordnung der in diesen Formeln auftretenden Grössen zu orientiren. Bezeichnen wir durch eine Zahl, welche von der Ordnung der Querdimensionen des Stabes ist, so sind x, y, z von der Ordnung ϵ.

This seems to me a grave fault in Clebsch's method of approaching *Kirchhoff's Problem*. He assumes x, y, z all of the same order and this order to be that of ϵ, but if he is to apply the results of Saint-Venant's problem, z^2, yz and xz can be of a far higher order than x^2, xy and y^2. The terms retained on Clebsch's S. 197 do not thus seem necessarily of the same order.

It will be remembered that Kirchhoff himself adopts a different mode of procedure. He obtains equations (see our Art. 1257) for the internal shifts of an element u, v, w, which are true independently of the hypothesis

$$\widehat{xx} = \widehat{yy} = \widehat{xy} = 0\,{}^{1}$$

(such equations are in part given by Clebsch on S. 202). He then states that his equations agree with Saint-Venant's if this hypothesis holds, and in his special examples assumes it to hold: see his *Gesammelte Abhandlungen*, S. 301 and 311, and the *Vorlesungen*, S. 415, 416 and 423. Thus in Kirchhoff's investigations we come at the values of u, v, w last, and on a clearly stated hypothesis, but in Clebsch's we have apparently perfectly general values given for u, v, w, deduced by making a cylinder of finite cross-section dwindle to one of infinitely small cross-section; these values, however, are in reality only *particular* cases of the equations afterwards given on S. 202, and they are obtained by diminishing indefinitely the length of the cylinder, so that their application without further investigation seems to me illegitimate.

In order the better to exhibit Clebsch's procedure and the manner in which he deduces and expands Kirchhoff's results, I cite in the following article Clebsch's expressions for the shifts in an element deduced from the values he has obtained in his treatment of *Saint-Venant's Problem*. I give them, however, in my own notation and with the modifications introduced in the French Edition for a planar distribution of isotropy.

[1360.] Let axes x, y, z be chosen in an element, so that if the rod returns to its unstrained condition the z axis coincides with the axis of the rod, and those of x and y with the principal axes of the cross-section. Suppose for simplicity that the cross-section is symmetrical about these axes. Then let u, v, w be the shifts referred to these axes of coordinates of a point x, y, z, in the immediate neighbourhood of their origin. Let P, Q, R, be the components of the total statical load applied parallel to these axes on a terminal cross-section, and P', Q', R' the moments of this load about the same axes; let ω be the cross-section and κ_1, κ_2 its swing-radii about the axes of x and y respectively; let E be the *longitudinal* stretch-modulus of the rod, η the stretch-squeeze ratio for a longitudinal stretch, and μ the slide-modulus parallel to the cross-section, then Clebsch finds (S. 197):

---

[1] x and z in our notation are interchanged in that of Kirchhoff's *Abhandlungen*.

$$
E\omega u = \left(-\eta P' \frac{xy}{\kappa_1^2} + Q' \frac{\eta (x^2 - y^2) + z^2}{2\kappa_2^2} - R' \frac{zy}{\chi^2} \right)
$$

$$
- \{\eta Rx\} - \left[\eta z \left(P \frac{x^2 - y^2}{2\kappa_2^2} + Q \frac{xy}{\kappa_1^2} \right) + P \frac{z^3}{6\kappa_2^2} \right],
$$

$$
E\omega v = \left(\eta Q' \frac{xy}{\kappa_2^2} + P' \frac{\eta (x^2 - y^2) - z^2}{2\kappa_1^2} + R' \frac{xz}{\chi^2} \right)
$$

$$
- \{\eta Ry\} - \left[\eta z \left(P \frac{xy}{\kappa_2^2} + Q \frac{y^2 - x^2}{2\kappa_1^2} \right) + Q \frac{z^3}{6\kappa_1^2} \right],
$$

......(i).

$$
E\omega w = \left(- Q' \frac{xz}{\kappa_2^2} + P' \frac{yz}{\kappa_1^2} - R' \frac{B_0}{\chi^2} \right)
$$

$$
+ \{Rz\} + \left[\frac{z^2}{2} \left(\frac{Px}{\kappa_2^2} + \frac{Qy}{\kappa_1^2} \right) + \frac{PB_1}{\kappa_2^2} + \frac{QB_2}{\kappa_1^2} - P \frac{xy^2}{\kappa_2^2} - Q \frac{x^2 y}{\kappa_1^2} \right],
$$

where

$$
\chi^2 = \frac{\mu}{E} \left\{ \kappa_1^2 + \kappa_2^2 - \frac{1}{\omega} \iint \left(x \frac{dB_0}{dy} - y \frac{dB_0}{dx} \right) d\omega \right\},
$$

and B_0, B_1 and B_2 are to be determined by the equations of our Art. 1338.

Now Clebsch notes that κ_1, κ_2 and χ are all of the order ϵ, and he says that x, y, z are of the same order, hence in the first place he neglects in u, v, w the expressions in the curled and in the square brackets, i.e. *he retains only the first line of each.* He remarks that if R is much greater in magnitude than the other forces, then the terms in curled brackets must be retained; if P and Q on the other hand are extremely great then the terms in square brackets must be retained (S. 197; *F. E.* p. 415). It seems to me that equations like (i) are better deduced as special solutions of the equations for u, v, w obtained in the following section on the express assumption that

$$
\widehat{xx} = \widehat{yy} = \widehat{xy} = 0.
$$

[1361.] In the following section Clebsch deduces Kirchhoff's equations (ix) of our Art. 1258. Changing the x on the left-hand of these equations to z, and on the right-hand side the ϵ to σ, after transporting it to the third equation; further changing q to r_2 and p to r_1, we have Clebsch's equations of S. 202 (*F. E.* p. 421) in his own notation:

$$
\frac{du}{dz} = ry - r_2 z, \qquad \frac{dv}{dz} = r_1 z - rx,
$$

$$
\frac{dw}{dz} = r_2 x - r_1 y + \sigma
$$

..............(ii).

Now substitute the values of u, v, w from (i) in (ii), we find:

$$ry - r_2 z = -\frac{R'y}{E\omega\chi^2} + \frac{Q'z}{E\omega\kappa_2{}^2} - \left[\eta\left(P\,\frac{x^2-y^2}{2E\omega\kappa_2{}^2} + \frac{Qxy}{E\omega\kappa_1{}^2}\right) + \frac{Pz^2}{2E\omega\kappa_2{}^2}\right],$$

$$r_1 z - rx = -\frac{P'z}{E\omega\kappa_1{}^2} + \frac{R'x}{E\omega\chi^2} - \left[\eta\left(\frac{Pxy}{E\omega\kappa_2{}^2} + \frac{Q\,(x^2-y^2)}{2E\omega\kappa_1{}^2}\right) + \frac{Qz^2}{2E\omega\kappa_1{}^2}\right], \quad \Big\}\ \dots\text{(iii)}.$$

$$r_2 x - r_1 y + \sigma = -\frac{Q'x}{E\omega\kappa_2{}^2} + \frac{P'y}{E\omega\kappa_1{}^2} + \left\{\frac{R}{E\omega}\right\} + \left[z\left(\frac{Px}{E\omega\kappa_2{}^2} + \frac{Qy}{E\omega\kappa_1{}^2}\right)\right]$$

If we neglect the terms in square brackets, we have with Clebsch:

$$r_1 = -P'/(E\omega\kappa_1{}^2), \quad r_2 = -Q'/(E\omega\kappa_2{}^2), \quad r = -R'/(E\omega\chi^2),\ \Big\}$$
$$\sigma = R/(E\omega) \qquad\qquad\qquad \dots\dots\text{(iv)}.$$

Here σ is by previous assumptions small as compared with $r_2 x - r_1 y$. Further we suppose P and Q not to be so great that the terms in the square brackets need be retained.

But suppose P, Q are so great that these terms must be retained; then since r, r_1, r_2 are not functions of x and y, it is obvious that the equations (ii) can no longer hold. But to obtain (ii), Kirchhoff (see our Art. 1258) and Clebsch (S. 202) not only neglect u, v, w on the right as compared with x, y, z but also du/ds, dv/ds, dw/ds. The reason given for this neglect is not very clearly stated. Saint-Venant however, in a footnote (*F. E.* pp. 420–2), endeavours to put the reason for the neglect of the s fluxions of the shifts in a clearer light. He says that since the changes in u are continuous and never very rapid, the order of du/ds for example is $(u_l - u_0)/l$ where l is the length of the rod and u_l, u_0 the terminal shifts measured from axes near these terminals; similarly du/dz is of the order $(u'_{l'} - u'_0)/l'$, where l' is the length of the little elementary cylinder (i.e. ds), and $u'_{l'}$, u'_0 the terminal shifts. Now the numerators, he says, are of the same order of magnitude, but l' is infinitely small as compared with l, whence we may neglect the s fluxions as compared with the z fluxions of the shifts. But neither Clebsch nor Saint-Venant fully explains why, when the terms in P and Q are great, the above reasoning no longer holds and why we must then retain du/ds, dv/ds, dw/ds in the equations (ii). Clebsch merely says that they must be retained if P and Q are great, and that then the terms dr_1/ds, dr_2/ds in them will be very great as compared with r_1, r_2, whence he says it follows that:

$$\frac{dr_2}{ds} = \frac{P}{E\omega\kappa_2{}^2}, \quad \frac{dr_1}{ds} = -\frac{Q}{E\omega\kappa_1{}^2} \dots\dots\dots\text{(v)}.$$

I imagine that these equations are supposed to be deduced in somewhat the following fashion: Substitute the values of P', Q', R' as a first approximation from (iv) in the last equation but one of (i) and we have:

$$w = (r_2 x - r_1 y)\,z + rB_0 + \sigma z$$
$$+ \left[\frac{z^2}{2E\omega}\left(\frac{Px}{\kappa_2{}^2} + \frac{Qy}{\kappa_1{}^2}\right) + \frac{PB_1}{E\omega\kappa_2{}^2} + \frac{QB_2}{E\omega\kappa_1{}^2} - P\,\frac{xy^2}{E\omega\kappa_2{}^2} - Q\,\frac{x^2 y}{E\omega\kappa_1{}^2}\right].$$

Substitute this, remembering that r_2, r_1 and r are functions of s, in

$$\frac{dw}{dz} = \frac{dw}{ds} + r_2 x - r_1 y + \sigma,$$

and we have :

$$\frac{z}{E\omega}\left(\frac{Px}{\kappa_2^2} + \frac{Qy}{\kappa_1^2}\right) = \left(\frac{dr_2}{ds}x - \frac{dr_1}{ds}y\right)z + \frac{dr}{ds}B_0,$$

and therefore :

$$\frac{dr_2}{ds} = \frac{P}{E\omega\kappa_2^2}, \qquad \frac{dr_1}{ds} = -\frac{Q}{E\omega\kappa_1^2}, \qquad \frac{dr}{ds} = 0 \dots\dots\dots(\text{v}).$$

These equations do not seem to me based on very satisfactory reasoning, *supposing the above to be really the method of finding them which Clebsch had in view*. We have not yet shown them to be consistent with the first two equations of (iii) when we introduce the terms du/ds and dv/ds into those equations, but on substitution they will be found to be so. Kirchhoff in his investigations does not appear to touch upon this point, although the equations (v) are of real importance[1]

[1362.] Clebsch gives the following interpretation of the first two results in (v) (he does not refer to the third one, which obviously denotes the constancy of the torsion along the length of the rod) :

Diese Erscheinung hat eine einfache Bedeutung. Man sieht daraus, dass der Stab im Allgemeinen bestrebt sein wird, eine Gestalt anzunehmen, in welcher für keinen seiner Querschnitte die seitlichen Gesammtcomponenten unverhältnissmässig gross werden. Ist es ihm nicht möglich eine derartige Gestalt in allen seinen Theilen anzunehmen, so werden gewisse ausgezeichnete Punkte auftreten, in denen die gegen die Axe des Elements senkrechten Kräfte P, Q vorwiegend werden, und in denen dann zugleich eine der Grössen dr_1/ds, dr_2/ds oder beide sehr grosse Werthe erhalten. Um die geometrische Bedeutung hiervon einzusehen, bemerke man nun, dass für $z=0$, also in der Axe des Stabes [rather for all values of x, y, z] nach (i) :

$$d^2u/dz^2 = Q/(E\omega\kappa_2^2) = -r_2, \qquad d^2v/dz^2 = -P'/(E\omega\kappa_1^2) = r_1 \dots\dots(\text{vi}).$$

Nun ist bereits früher darauf hingewiesen, dass die Grössen links bis auf sehr kleine Grössen die reciproken Krümmungshalbmesser derjenigen Curven bedeuten, welche man aus der Projection der Schwerpunktslinie auf die durch die Axe des Elements und je eine Hauptaxe des Querschnitts gelegten Ebenen erhält. Eben diese Bedeutung haben also, abgesehen vom Zeichen, r_1 und r_2. Und in der Nähe jener ausgezeichneten Punkte muss also wenigstens einer dieser Krümmungshalbmesser sich sehr schnell ändern, da einer wenigstens von den Differentialquotienten dr_1/ds, dr_2/ds verhältnissmässig gross wird (S. 203).

---

[1] Differentiating the first two of (iv) with regard to s and using (v), we have $\frac{dP'}{ds} = Q$, and $\frac{dQ'}{ds} = -P$. These express that for the thin rod in this case, the total shear is the fluxion of the bending-moment, a well-known theorem for small shifts which forms the basis of a good deal of the graphical treatment of such rods : see our Arts. 319 and the third equation of Art. 534.

It seems to me that perhaps as simple a meaning of the equations when P, Q are large is that given in the footnote to our last article.

[1363.] In the fiftieth section of his book Clebsch obtains equations for the equilibrium of the rod, similar to those of Kirchhoff cited in our Art. 1265 but rather more general. Let the total stress across any cross-section at distance s from a terminal be given in relation to the axes x, y, z chosen as in our Art. 1360 by the components P, Q, R at the centroid of the cross-section and the couples P', Q', R'; let X', Y', Z' be the body-forces acting per unit of volume on the element $dx\,dy\,ds$, parallel to axes x', y', z' fixed in space; further let

$$\begin{Bmatrix} U \\ V \\ W \end{Bmatrix} = \iint \begin{Bmatrix} X' \\ Y' \\ Z' \end{Bmatrix} dx\,dy, \quad \begin{Bmatrix} U_1 \\ V_1 \\ W_1 \end{Bmatrix} = \iint \begin{Bmatrix} X' \\ Y' \\ Z' \end{Bmatrix} x\,dx\,dy, \quad \begin{Bmatrix} U_2 \\ V_2 \\ W_2 \end{Bmatrix} = \iint \begin{Bmatrix} X' \\ Y' \\ Z' \end{Bmatrix} y\,dx\,dy \dots \text{(vii)},$$

and let the direction-cosine system of x, y, z with regard to x', y', z' be:

	x'	y'	z'
x	a_1	β_1	γ_1
y	a_2	β_2	γ_2
z	a	β	γ

Then Clebsch obtains the following system of equations[1] from purely statical considerations:

$$\left.\begin{aligned} \frac{dP}{ds} + (rQ - r_2 R) + a_1 U + \beta_1 V + \gamma_1 W &= 0, \\[4pt] \frac{dQ}{ds} + (r_1 R - rP) + a_2 U + \beta_2 V + \gamma_2 W &= 0, \\[4pt] \frac{dR}{ds} + (r_2 P - r_1 Q) + a U + \beta V + \gamma W &= 0, \\[4pt] \frac{dP'}{ds} + (rQ' - r_2 R') - Q + a U_2 + \beta V_2 + \gamma W_2 &= 0, \\[4pt] \frac{dQ'}{ds} + (r_1 R' - rP') + P - a U_1 - \beta V_1 - \gamma W_1 &= 0, \\[4pt] \frac{dR'}{ds} + (r_2 P' - r_1 Q') + a_2 U_1 + \beta_2 V_1 + \gamma_2 W_1 & \\[2pt] - a_1 U_2 - \beta_1 V_2 - \gamma_1 W_2 &= 0 \end{aligned}\right\} \dots\dots\text{(viii)}.$$

[1] We use here as in Art. 1360 P, Q, R for Clebsch's A, B, C to distinguish from the A, B, C of our Art. 1265.

Here P, Q, R, P', Q', R' must be determined by equations (iv) and (v) of our Art. 1361 and r_1, r_2, r, are given by

$$r_1 = a_2 \frac{da}{ds} + \beta_2 \frac{d\beta}{ds} + \gamma_2 \frac{d\gamma}{ds},$$

$$r_2 = a \frac{da_1}{ds} + \beta \frac{d\beta_1}{ds} + \gamma \frac{d\gamma_1}{ds}, \qquad \Big\} \quad \dots\dots\dots\dots\text{(ix)}.$$

$$r = a_1 \frac{da_2}{ds} + \beta_1 \frac{d\beta_2}{ds} + \gamma_1 \frac{d\gamma_2}{ds}$$

Compare our Arts. 1257 and 1265.

We see that the last three of equations (viii) agree with Kirchhoff's equations (xxiv) of our Art. 1265, if we put U, V, W, U_1, V_1, W_1, U_2, V_2, W_2 zero, or suppose no body-forces; further, Kirchhoff's equations (xxiii) are easily deducible from Clebsch's first three, when we put U, V, W zero. We note that Kirchhoff's

$$A = Pa_1 + Qa_2 + Ra, \qquad B = P\beta_1 + Q\beta_2 + R\beta,$$
$$C = P\gamma_1 + Q\gamma_2 + R\gamma \qquad\qquad\qquad \Big\} \quad \dots\dots\text{(x)}.$$

Kirchhoff's equations are indeed Clebsch's (13) on S. 205.

(This section in the *F. E.* occupies pp. 424–30.)

[1364.] The values of the shifts u, v, w, and those of r, r_1, r_2 obtained in our Arts. 1360 and 1361 cannot be applied to the equations of the previous article, since they were obtained on the assumption of no body-forces. Clebsch accordingly in the following section deals with the case of no body-forces; here obviously A, B, C of (x) are constants, hence by aid of (iv) Clebsch puts the last three equations into the form:

$$\kappa_1{}^2 \frac{dr_1}{ds} + \left(\kappa_2{}^2 - \chi^2\right) r_2 r + \frac{1}{E\omega} \left(Aa_2 + B\beta_2 + C\gamma_2\right) = 0,$$

$$\kappa_2{}^2 \frac{dr_2}{ds} + \left(\chi^2 - \kappa_1{}^2\right) r r_1 - \frac{1}{E\omega} \left(Aa_1 + B\beta_1 + C\gamma_1\right) = 0, \Big\} \quad \dots\dots\text{(xi)}.$$

$$\chi^2 \frac{dr}{ds} + \left(\kappa_1{}^2 - \kappa_2{}^2\right) r_1 r_2 = 0$$

These agree in form with Euler's equations for the motion of a heavy body about a fixed point, and thus demonstrate Kirchhoff's *elastico-kinetic analogy*: see our Arts. 1267, 1270 and 1283, (*b*)

Clebsch then solves in general terms equations (xi) on the assumption that A, B, C are zero, or that the terminals of the rod are only acted upon by couples. The general type of the solution is well-known to students of dynamics: see Routh's *Treatise on Rigid Dynamics*, 1877, p. 404 or Schell's *Theorie der Bewegung*, Bd. II., 1880, S. 437–42. The additional matter in the case of the elastic problem is the determination of the direction cosines a, β, γ, &c. in terms of the r's and hence the total shifts in terms of s. See Kirchhoff's discussion of the like problem in our Art. 1267. This section occupies S. 209–15 (*F. E.* pp. 430–7).

[1365] § 52 (S. 215-218; *F. E.* pp. 437-40) deals with the case
in which the cross-section possesses inertial isotropy (i.e. $\kappa_1 = \kappa_2$) and the
terminals of the rod are acted upon solely by couples. Clebsch obtains by
a simpler process than that in the original memoir of Kirchhoff (see our
Art. 1268) the equation to the helix due to a given system of couples.
Compare the results of Wantzel, Binet and Saint-Venant referred to in
our Arts. 1240*, 175*, 1583*, 1593*-5* and 1606*-8*.

[1366.] § 53 (S. 218-222; *F. E.* pp. 440-6) deals with the
practically interesting case of flexure in a plane which contains a
principal axis of each cross-section. Clebsch easily deduces the equation
of the ordinary Bernoulli-Eulerian theory for the *special case of a strut
with one end built-in and the other, or loaded end, free:*

$$E\omega\kappa_2{}^2 \frac{d^2\phi}{ds^2} = C \sin\phi \quad\dots\dots\dots\dots\dots\dots(\text{xii}),$$

where ϕ is the angle between the strained and unstrained positions of
the tangent at s, and C is the longitudinal load. He takes a first
integral of this, and determines by an ingenious bit of analysis on S.
221 (*F. E.* p. 444) that if l be the length of the strut, we must have
always:

$$l\sqrt{\frac{C}{E\omega\kappa_2{}^2}} > \frac{\pi}{2},$$

if there is to be flexure.

The section concludes with a determination of the shift at any point
of the axis of the rod, by formulae drawn from elliptic functions.

[1367.] § 54 (S. 223-9; *F. E.* pp. 446-54) is entitled:
*Zusammenhang mit der gewöhnlichen Theorie. Kleine Verschiebun-
gen.* Clebsch shows very clearly how Kirchhoff's theory leads to
results agreeing with the Bernoulli-Eulerian theory when the
total shifts are small. In particular we may draw attention to the
method in which a term involving strut-action, due to a considerable longitudinal load (C) on the rod, is introduced into the
flexure-equations. We do not cite Clebsch's results here but
refer to our Art. 1373 for the more complete equations, involving
accelerational terms.

[1368.] Clebsch devotes S. 229-42 (*F. E.* pp. 454-468) to the
discussion of rods, which in their unstrained condition are *curved.* He
deduces the requisite equations exactly like Kirchhoff (see our Art.
1264), attributing this extension of the theory to him.

His first three equations (S. 231; *F. E.* p. 457) are the same as

Kirchhoff's (xxiii) of our Art. 1265 with the introduction of the body-force terms U, V, W from (vii), i.e.:

$$\frac{dA}{ds} + U = 0, \quad \frac{dB}{ds} + V = 0, \quad \frac{dC}{ds} + W = 0 \quad \ldots\ldots\ldots\text{(xiii)},$$

where for A, B, C he substitutes their values from (x). These equations hold for all rods whether initially curved or not.

The next three equations on the same page are obtained from the last three of (viii) by substituting

$$E\omega\kappa_1^2 (\bar{r}_1 - r_1), \quad E\omega\kappa_2^2 (\bar{r}_2 - r_2), \quad E\omega\chi^2 (\bar{r} - r)$$

for P', Q', R' (see equations (iv) of our Art. 1361), where $\bar{r}_1$, $\bar{r}_2$ and $\bar{r}$ are the values of r_1, r_2 and r as defined by (ix) *before strain*. The last three equations of (viii) with these substitutions[1] we will term for purposes of reference:

$$\text{equations} \ldots\ldots\ldots\ldots\ldots\ldots\ldots\ldots\ldots\ldots\ldots\ldots\ldots\ldots\text{(xiv)}.$$

[1369.] In § 56 (S. 232–3; *F. E.* pp. 457–9) Clebsch deals with the case of a rod, which when unstrained has a curved axis lying in the plane $x'z'$; this plane passes through a principal axis of each cross-section. The rod is supposed to be without body-forces and bent solely by terminal couples whose axes are perpendicular to its plane. ϕ being the angle between the tangent to the axis of the rod and the axis of z' after strain, $\bar{\phi}$ the value before strain, and the terminal $s = 0$ being fixed, Clebsch easily deduces for the strained form of the axis of the rod:

$$\phi - \bar{\phi} = \frac{Q'}{E\omega\kappa_2^2} s,$$

$$x' = \int_0^s \sin\left(\bar{\phi} + \frac{Q's}{E\omega\kappa_2^2}\right) ds,$$

$$z' = \int_0^s \cos\left(\bar{\phi} + \frac{Q's}{E\omega\kappa_2^2}\right) ds,$$

Q' being the resultant terminal couple.

[1370.] In §§ 57 and 58 Clebsch discusses the small shifts of originally bent rods and shows how the equations are to be integrated (S. 233–242; *F. E.* pp. 459–68). This portion of Clebsch's work seems wholly original and very valuable. He himself remarks:

Von der grössten Wichtigkeit für die Anwendung aber ist die Ausstellung von Formeln, welche sehr kleine Gestaltsveränderungen ursprünglich krummer Stäbe darstellen (p. 233).

Clebsch's theory depends upon the recognition that

$$a' = a - \bar{a}, \quad \beta' = \beta - \bar{\beta}, \quad \gamma' = \gamma - \bar{\gamma}, \quad \text{etc.},$$

---

[1] In making the substitutions, it must be borne in mind that the r's which appear in (viii) are not themselves to be replaced by $(\bar{r} - r)$'s.

are all very small quantities, bars over the symbols denoting quantities before strain. Hence he easily shows that the nine quantities a', β', γ' ...can be expressed, owing to the six relations which hold between the a, β, γ..., in terms of three variables p_1, p_2 and p. He obtains the following formulae:

$$\left.\begin{array}{lll} a_1{}' = p_2\bar{a} - p\bar{a}_2, & a_2{}' = p\bar{a}_1 - p_1\bar{a}, & a' = p_1\bar{a}_2 - p_2\bar{a}_1, \\ \beta_1{}' = p_2\bar{\beta} - p\bar{\beta}_2, & \beta_2{}' = p\bar{\beta}_1 - p_1\bar{\beta}, & \beta' = p_1\bar{\beta}_2 - p_2\bar{\beta}_1, \\ \gamma_1{}' = p_2\bar{\gamma} - p\bar{\gamma}_2, & \gamma_2{}' = p\bar{\gamma}_1 - p_1\bar{\gamma}, & \gamma' = p_1\bar{\gamma}_2 - p_2\bar{\gamma}_1 \end{array}\right\} \ \ldots\ldots(\text{xv}).$$

Replacing r by $1/\rho$ and substituting $a' + \bar{a}$, $\beta + \bar{\beta}$, $\gamma + \bar{\gamma}$, etc. for a, β, γ,... in equations (ix), we find by neglecting products of the p's:

$$\left.\begin{array}{l} \dfrac{1}{\rho_1} - \dfrac{1}{\bar{\rho}_1} = \dfrac{dp_1}{ds} + \dfrac{p_2}{\bar{\rho}} - \dfrac{p}{\bar{\rho}_2}, \\[2mm] \dfrac{1}{\rho_2} - \dfrac{1}{\bar{\rho}_2} = \dfrac{dp_2}{ds} + \dfrac{p}{\bar{\rho}_1} - \dfrac{p_1}{\bar{\rho}}, \\[2mm] \dfrac{1}{\rho} - \dfrac{1}{\bar{\rho}} = \dfrac{dp}{ds} + \dfrac{p_1}{\bar{\rho}_2} - \dfrac{p_2}{\bar{\rho}_1} \end{array}\right\} \ \ldots\ldots\ldots\ldots\ldots(\text{xvi}).$$

Here ρ_1, ρ_2 are the radii of curvature of the rod's central line and ρ is its radius of torsion after strain, $\bar{\rho}_1$, $\bar{\rho}_2$, $\bar{\rho}$ the corresponding quantities before strain.

Their differences are very small and we may neglect their squares when substituting in (xiv).

Let A_l, B_l, C_l be the components parallel to the axes x', y', z' (Art. 1363) of the forces acting on the terminal $s = l$ of the rod, then remembering that the shifts are very small, we can easily see that P, Q, R may be calculated from the values of the forces in the unstrained position of the rod, or they are given by three equations of the type:

$$P = \bar{a}_1 \int_s^l U ds + \bar{\beta}_1 \int_s^l V ds + \bar{\gamma}_1 \int_s^l W ds + A_l\bar{a}_1 + B_l\bar{\beta}_1 + C_l\bar{\gamma}_1\ldots\ldots(\text{xvii}),$$

where U, V, W have the values given in (vii).

Substituting (xvi) in (xiv) Clebsch obtains the following linear differential equations for the p's, the coefficients being of course functions of s:

$$\left.\begin{array}{l} \dfrac{d\nu_1}{ds} + \dfrac{\nu_2}{\bar{\rho}} - \dfrac{\nu}{\bar{\rho}_2} = \tau_1, \\[2mm] \dfrac{d\nu_2}{ds} + \dfrac{\nu}{\bar{\rho}_1} - \dfrac{\nu_1}{\bar{\rho}} = \tau_2, \\[2mm] \dfrac{d\nu}{ds} + \dfrac{\nu_1}{\bar{\rho}_2} - \dfrac{\nu_2}{\bar{\rho}_1} = \tau \end{array}\right\} \ \ldots\ldots\ldots\ldots\ \ldots\ldots(\text{xviii}),$$

where ν_1/κ_1^2, ν_2/κ_2^2, ν/χ^2 are the three differences on the left-hand side of (xvi), and

$$\tau_1 = \frac{1}{E\omega}\left(-Q + \bar{a}U_2 + \bar{\beta}V_2 + \bar{\gamma}W_2\right),$$

$$\tau_2 = \frac{1}{E\omega}\left(P - \bar{a}U_1 - \bar{\beta}V_1 - \bar{\gamma}W_1\right), \qquad \Bigg\} \ \dots\dots(\text{xix}).$$

$$\tau = \frac{1}{E\omega}\left(\bar{a}_2 U_1 + \bar{\beta}_2 V_1 + \bar{\gamma}_2 W_1 - \bar{a}_1 U_2 - \bar{\beta}_1 V_2 - \bar{\gamma}_1 W_2\right)$$

To obtain the shifts u', v', w' parallel to the axes x', y', z' of a point on the central axis of the rod, we have :

$$u' = \int_0^s a\,(1+\sigma)\,ds - \int_0^s \bar{a}\,ds = \int_0^s (a' + \sigma\bar{a})\,ds$$
$$= \int_0^s (p_1\bar{a}_2 - p_2 a_1 + \sigma\bar{a})\,ds, \Bigg\} \ \dots\dots(\text{xx}),$$

with similar expressions for v' and w'.

In (xx) we cannot neglect σ, as a', β', γ' are themselves small, but its value is at once known from $\sigma = R/(E\omega)$.

[1371.] To integrate the above system of equations, we have evidently to deal with two similar groups of the types

$$\frac{d\nu_1}{ds} + \frac{\nu_2}{\bar{\rho}} - \frac{\nu}{\bar{\rho}_2} = \tau_1,$$

$$\frac{dp_1}{ds} + \frac{p_2}{\bar{\rho}} - \frac{p}{\bar{\rho}_2} = \frac{\nu_1}{\kappa_1^2}.$$

Clebsch solves these by a remarkably graceful analytical process. If the end $s=0$ of the rod be built-in, and M_1, M_2, M be the moments of the external load at the other end $s=l$ round the axis system x, y, z attached to that end, then he finds formulae of the following type :

$$p = f\bar{a} + g\bar{\beta} + h\bar{\gamma},$$

where $$f = \int_0^s \left(\frac{\bar{a}_1\nu_1}{\kappa_1^2} + \frac{\bar{a}_2\nu_2}{\kappa_2^2} + \frac{\bar{a}\nu}{\chi^2}\right)ds,$$

and g or h is to be found from f by changing a to β or γ respectively ; p_1 or p_2 is to be found from p by attaching the subscripts 1 or 2 to $\bar{a}$, $\bar{\beta}$, $\bar{\gamma}$. Further

$$\nu = a\bar{a} + b\bar{\beta} + c\bar{\gamma},$$

where $$a = -\int_s^l \left(\bar{a}_1\tau_1 + \bar{a}_2\tau_2 + \bar{a}\tau\right)ds - \frac{1}{E\omega}\left(\frac{\bar{a}_1 M_1}{\kappa_1^2} + \frac{\bar{a}_2 M_2}{\kappa_2^2} + \frac{\bar{a}M}{\chi^2}\right),$$

and b or c is to be found from a by changing a to β or γ respectively ; ν_1 or ν_2 is to be found from ν by attaching the subscripts 1 or 2 to $\bar{a}$, $\bar{\beta}$, $\bar{\gamma}$.

[1372.]   S. 242-261 of Clebsch's *Treatise* are devoted to the equations of motion of thin rods, especially to the cases of vibration of straight rods (*F. E.* pp. 468-628).

In § 59 Clebsch demonstrates that by D'Alembert's principle the general equations of motion are obtained from (viii) by replacing

$$U, \ V, \ W, \ U_1, \ V_1, \ W_1, \ U_2, \ V_2, \ W_2$$

respectively, by

$$
\left.
\begin{aligned}
&U - \Delta\omega\frac{d^2\xi}{dt^2}, \quad V - \Delta\omega\frac{d^2\eta}{dt^2}, \quad W - \Delta\omega\frac{d^2\zeta}{dt^2}, \\[2mm]
&U_1 - \Delta\omega\kappa_2^2\frac{d^2a_1}{dt^2}, \quad V_1 - \Delta\omega\kappa_2^2\frac{d^2\beta_1}{dt^2}, \quad W_1 - \Delta\omega\kappa_2^2\frac{d^2\gamma_1}{dt^2}, \\[2mm]
&U_2 - \Delta\omega\kappa_1^2\frac{d^2a_2}{dt^2}, \quad V_2 - \Delta\omega\kappa_1^2\frac{d^2\beta_2}{dt^2}, \quad W_2 - \Delta\omega\kappa_1^2\frac{d^2\gamma_2}{dt^2}
\end{aligned}
\right\} \quad \ldots\ldots(\text{xxi}),
$$

where Δ is the density of the rod, and ξ, η, ζ the coordinates of a point on its axis referred to the axes x', y', z' fixed in space.

[1373.]   On S. 246 (*F. E.* p. 473) Clebsch gives the complete equations for the motion and equilibrium of an originally straight rod.   The notation is that of our previous articles: see in particular equations (vii), (viii) and (x).

(*a*)   For transverse vibrations:

$$E\omega\kappa_2^2\frac{d^4u}{dz^4} = C\frac{d^2u}{dz^2} + U + \frac{dW_1}{dz} - \Delta\omega\frac{d^2u}{dt^2} + \Delta\kappa_2^2\omega\frac{d^4u}{dz^2dt^2},$$

$$E\omega\kappa_1^2\frac{d^4v}{dz^4} = C\frac{d^2v}{dz^2} + V + \frac{dW_2}{dz} - \Delta\omega\frac{d^2v}{dt^2} + \Delta\kappa_1^2\omega\frac{d^4v}{dz^2dt^2},$$

with the terminal conditions for $z = l$:

$$E\omega\kappa_2^2\left(\frac{d^3u}{dz^3}\right)_{z=l} = -A + C\left(\frac{du}{dz}\right)_{z=l} + (W_1)_{z=l} + \Delta\kappa_2^2\omega\left(\frac{d^3u}{dzdt^2}\right)_{z=l},$$

$$E\omega\kappa_1^2\left(\frac{d^3v}{dz^3}\right)_{z=l} = -B + C\left(\frac{dv}{dz}\right)_{z=l} + (W_2)_{z=l} + \Delta\kappa_1^2\omega\left(\frac{d^3v}{dzdt^2}\right)_{z=l},$$

and

$$E\omega\kappa_2^2\left(\frac{d^2u}{dz^2}\right)_{z=l} = (P')_{z=l},$$

$$E\omega\kappa_1^2\left(\frac{d^2v}{dz^2}\right)_{z=l} = -(Q')_{z=l}.$$

(*b*)   For longitudinal vibrations:

$$E\omega\frac{d^2w}{dz^2} = -W + \Delta\omega\frac{d^2w}{dt^2},$$

with the terminal condition for $z = l$:

$$E\omega\left(\frac{dw}{dz}\right)_{z=l} = C.$$

(c)   For torsional vibrations $(1/\rho = da_2/ds = -d\phi/dz)$:

$$E\omega\chi^2\frac{d^2\phi}{dz^2} = U_2 - V_1 + \Delta\omega(\kappa_1{}^2 + \kappa_2{}^2)\frac{d^2\phi}{dt^2},$$

with the terminal condition for $z = l$:

$$E\omega\chi^2\left(\frac{d\phi}{dz}\right)_{z=l} = R'.$$

The last terms in the four equations of (a), depending on the rotatory inertia of the elements, are really of the same order, namely $(\epsilon/l)^2$, as the terms which Clebsch has neglected in deducing his equations for thin rods ; hence the method by which he brings them into these vibrational equations is not wholly satisfactory.

Of the terms in the longitudinal force C in the same equations Saint-Venant remarks $(F.\ E.$ footnote p. 475) :

Ces termes sont la seule partie réellement influente que Clebsch ait ajoutée aux équations connues de vibration transversale.

It is, so far as I know, true that equations so complete in form as (a) were first given by Clebsch, but Seebeck in his memoir of 1849 introduced the term due to longitudinal traction and used it to deduce N. Savart's theorem : see our Art. 471.

[1374.]   § 60 (S. 247–52 ; *F. E.* pp. 475–80) dealing with longitudinal vibrations contains nothing of novelty.

§ 61 (S. 252–60, *F. E.* pp. 480*gg*–490) treats of the transverse vibrations of straight rods. Clebsch takes the general equations (a) and starts by supposing the rod pivoted at either end. He discusses the equation which gives the form of the z functions in the solution, particularly the case when it has equal roots (S. 256 ; *F. E.* p. 485). Having retained the term in M, he is able to deal with the cases of a rod, a stiff string and a flexible tight string under the same general analysis : see our remarks on Seebeck in Arts. 471–2.

§ 62 (S. 260–1 ; *F. E.* p. 628) briefly refers to torsional vibrations.

§ 63 (S. 261–3 ; *F. E.* pp. 629–31) gives the values of the stresses for the case of a very thin rod initially straight. They are to a first approximation :

$$\widehat{zz} = \frac{1}{\omega}\left(R + \frac{P'y}{\kappa_1{}^2} - \frac{Q'x}{\kappa_2{}^2}\right),$$

$$\widehat{zx} = -\frac{\mu}{E\omega}\frac{R'}{\chi^2}\left(y + \frac{dB_0}{dx}\right),$$

$$\widehat{yz} = \frac{\mu}{E\omega}\frac{R'}{\chi^2}\left(x - \frac{dB_0}{dy}\right).$$

The first stress agrees for a special case with the value given by the old theory : see our Arts. 815* and 71 ; the second two stresses are due only to the torsion, and coincide with the values first given by Saint-Venant : see our Arts. 17 and 1344.

[1375.]   The next portion of Clebsch's *Treatise* (S. 264–355; *F. E.* pp. 632–806) deals with *thin plates*.   On S. 264 is the footnote referring to the services of Kirchhoff and Gehring, which we have mentioned in Art. 1293.   On S. 271 (*F. E.* p. 640) Clebsch gives equations (64) for the shifts which are identical with those given by Kirchhoff numbered (ix) in our Art. 1294.

Now take the expressions (x) found by Kirchhoff for the strains and substitute them in the values of the stresses given in our Art. 1203 (*b*); thus allowing the plate to have three axes of elastic symmetry, we find :

$$\left.\begin{aligned}
\widehat{xx} &= a\,(q_1 z + \sigma_1) + f'\,(-p_2 z + \sigma_2) + e'\,dw_0/dz, \quad \widehat{yz} = d\,dv_0/dz, \\
\widehat{yy} &= f'\,(q_1 z + \sigma_1) + b\,(-p_2 z + \sigma_2) + d'\,dw_0/dz, \quad \widehat{zx} = e\,du_0/dz, \\
\widehat{zz} &= e'\,(q_1 z + \sigma_1) + d'\,(-p_2 z + \sigma_2) + c\,dw_0/dz, \quad \widehat{xy} = f(-2p_1 z + \tau)
\end{aligned}\right\} \dots \text{(i)},$$

where $\qquad a = 2f + f', \quad b = 2d + d', \quad c = 2e + e'.$

Now substitute these in the body-stress equations, the body-forces being by Kirchhoff's principle negligible (see our Art. 1253), and we find :

$$d^2 u_0/dz^2 = 0, \quad d^2 v_0/dz^2 = 0, \quad \frac{d\widehat{zz}}{dz} = 0.$$

Now u_0, v_0, w_0 are independent of x and y, hence $\widehat{zz}$ is independent of x and y; it follows therefore that :

du_0/dz, dv_0/dz and $\widehat{zz}$ are constants, or that $\widehat{yz}$, $\widehat{zx}$ and $\widehat{zz}$ are constants. But these stresses are supposed to vanish at the surface, hence :

$$\widehat{yz} = \widehat{zx} = \widehat{zz} = 0 \dots\dots\dots\dots\dots\dots\dots\text{(ii)}.$$

The method by which this conclusion is reached should be compared with Kirchhoff's reasoning : see our Arts. 1294–5.

Further since $du_0/dz = dv_0/dz = 0$, u_0 and v_0 are constants, but they are to vanish for $z = 0$, therefore we have them both zero.   Next integrating $\widehat{zz} = 0$, to find w_0, we have,

$$w_0 = -\frac{1}{c}\left\{ (e'\sigma_1 + d'\sigma_2)\, z + (e'q_1 - d'p_2)\frac{z^2}{2} \right\},$$

no constant being added as w_0 is to vanish with z.

We can now write down the shifts and stresses completely, i.e.:

$$\left.\begin{aligned}
u &= -p_1 yz + q_1 zx + \sigma_1 x + \tau y, \\
v &= -p_2 yz - p_1 zx + \sigma_2 y, \\
w &= -\frac{q_1}{2}x^2 + p_1 xy + \frac{p_2}{2}y^2 - \frac{1}{c}\left\{ (e'\sigma_1 + d'\sigma_2)\,z + (e'q_1 - d'p_2)\frac{z^2}{2} \right\}, \\
\widehat{xx} &= \left(a - \frac{e'^2}{c}\right)(\sigma_1 + q_1 z) + \left(f' - \frac{d'e'}{c}\right)(\sigma_2 - p_2 z), \\
\widehat{yy} &= \left(f' - \frac{d'e'}{c}\right)(\sigma_1 + q_1 z) + \left(b - \frac{d'^2}{c}\right)(\sigma_2 - p_2 z), \\
\widehat{xy} &= f(-2p_1 z + \tau), \\
\widehat{zx} &= \widehat{yz} = \widehat{zz} = 0
\end{aligned}\right\} \dots \text{(iii)}.$$

These equations agree with the values given in the French *Clebsch* (p. 643), if we put therein

$$\partial_a = \sigma_1, \ \partial_b = \sigma_2, \ g = \tau, \ r_2 = -q_1, \ s_1 = -p_2, \ r_1 = -p_1.$$

They are identical with Clebsch's own values (S. 273), if we put his $r_2 = -q_1, \ s_1 = -p_2, \ r_1 = -p_1$ and suppose uniconstant isotropy, so that our $f = $ his $E/\{2\,(1+\mu)\}$, our $a - e'^2/c = b - d'^2/c = $ his $E/(1-\mu^2)$, our $f' - d'e'/c = $ his $\mu E/(1-\mu^2)$. (Note Clebsch's $\mu = $ the stretch-squeeze modulus, our η.)

Clebsch makes the interesting remark that the values of u, v, w, in (iii) for a thin plate are only special cases of those we have found in equations (13) of our Art. 1351 for the thick plate, if we take in the latter

$$\phi = \sigma_1 x + \tau y, \qquad \psi = \sigma_2 y,$$

$$\mathfrak{f} = -\frac{p_2 + q_1}{4}\,(x^2 - y^2) + p_1 xy, \qquad C = \frac{d'}{c}\,(q_1 - p_2),$$

and put for the case of isotropy in the plane of the plate

$$e' = d'.$$

With regard to the range within which equations (iii) are applicable Clebsch remarks:

Das Element ist nur durch Spannungskräfte ergriffen, welche der Ebene seiner Mittelfläche parallel sind, und durch Kräftepaare, deren Axen in jener Ebene liegen. Man darf deswegen nicht sagen, dass die Spannungen oder die auf den Rand wirkenden Kräfte, welche eine andere Richtung hätten, absolut verschwinden müssen ; aber sie nehmen Werthe an, vermöge deren sie nur Verschiebungen hervorbringen, welche gegenüber den andern Verschiebungen von einer höheren Ordnung sind. Es ist wichtig dies zu bemerken in Bezug auf die Tragweite der hier zu entwickelnden Formeln. Denn betrachten wir den Rand der Platte, so können die auf denselben wirkenden Kräfte entweder im Stande sein, denselben so zu biegen, dass die äussern Kräfte wirklich tangential zur Platte wirken ; und in diesem Fall ist kein Widerspruch vorhanden. Ist aber dies nicht der Fall, so müssen entweder die Kräfte, welche auf den Rand wirken und gegen denselben senkrecht sind, selbst äusserst klein (Grössen höherer Ordnung) werden, oder es müssen sich Ausnahmspunkte der Art ergeben, wie sie hier nicht behandelt werden sollen, und in welchen eigenthümliche grosse Krümmungen eintreten (S. 273-4).

[1376.] § 67 of the treatise (S. 274–282 ; *F. E.* pp. 645–656) shows that, when the shifts are *finite*, the approximate form of the mid-plane is a developable surface : see our Art. 1297. Clebsch discusses this developable surface at considerable length and shows that its introduction leads us to three arbitrary functions ξ_0, η_0, ζ_0, in terms of which all the other elements of the problem (α_1, β_1, γ_1, α_2, β_2, γ_2, α_3, β_3, γ_3, ξ, η, ζ in Kirchhoff's notation: see our Art. 1294) can be expressed. The problem then

156   CLEBSCH.   [1377—1378

reduces itself to the discovery of these three arbitrary functions;
we require, however, in order to ascertain them the equations
of equilibrium of an element of the plate.

[1377.] These Clebsch investigates by some rather lengthy analysis
in his §§ 68 and 69 (S. 282–94; *F. E.* pp. 656–71). In the first of
these sections he applies the principle of virtual work to determine the
relations between the stresses and the load system at every point of the
mid-plane and at every point of its contour (equations (88) S. 289). In
the latter section he substitutes the values obtained in (iii) for the
stresses in these equations. For every point of the mid-plane we have
three equations to be satisfied. These in Clebsch's investigations contain
two functions Q_1, Q_2 defined by two additional equations: see his equation (92), S. 292. These equations are given[1] in the French edition
for three axes of elastic symmetry as (267) on p. 668, and for isotropy
in the plane of the plate as (267 *a*) on the same page. They involve first
fluxions of σ_1, σ_2, τ and first and second fluxions of p_1, p_2, q_1, with regard
to s_1 and s_2 in the notation of our Art. 1294. The contour conditions, five
in number, involve an arbitrary function Δ (in Clebsch's notation) as well
as the above Q_1 and Q_2, so that they are really only equivalent to *four*.
They are given by Clebsch as (93) on S. 294, or with a more general distribution of elasticity in the French edition as (268) on pp. 670–1.
These equations for the finite shifts of thin plates are too complex for
reproduction here, and we must content ourselves with merely referring
to them. So far as we are aware there has been hitherto no practical
application of them.

[1378.] In § 70 (S. 295–9, *F. E.* pp. 671–6) Clebsch indicates the
general stages by which the problem of the finite shifts of thin plates
might be solved. He writes of it:

Dieselbe sondert sich in drei verschiedene Theile. Der erste Theil hat zum
Zweck die Bestimmung der abwickelbaren Fläche, von welcher die Mittelfläche der Platte im gebogenen Zustande nur sehr wenig abweicht. Der
zweite Theil beschäftigt sich sodann mit der Aufsuchung der Dilatationen
σ_1, σ_2, τ_1, welche durch die äussern Kräfte hervorgerufen werden; der dritte
endlich mit den Bestimmungen der kleinen Abweichungen der wirklichen
Gestalt der Mittelfläche von der gefundenen abwickelbaren Fläche (S. 295).

The first part of this problem, the determination of the developable
surface as the approximate form of the strained mid-plane, is considered
in § 70. In § 71 the method of dealing with this part of the problem,
when there are no body-forces and the load on any element of the

---

[1] There are some misprints. In (267), read in third equation γ for γ_1, and in
fourth and fifth equations read $\left(f' - \dfrac{d'e'}{c}\right)$ for $\left(f - \dfrac{d'e'}{c}\right)$. In (267 *a*) a factor has got
inverted and then transposed; the term in curved brackets should be
$$\partial (s_1 - \eta' r_2)/\partial b + (1 - \eta') \partial r_1/\partial a.$$

contour depends only on the magnitude and direction of the element, is especially developed. In this section Clebsch makes a slight reference to how the problem of the finite shifts of a plate, whose mid-surface in the unstrained condition has the form of a developable surface, might be approached (S. 302).

In § 72 Clebsch indicates in the briefest manner how the second and third parts of the problem might be dealt with. He concludes with the remark :

> Ich habe hier eine kurze Skizze von der Reihenfolge der Probleme entwickelt, auf welche man bei der Behandlung des Problems endlicher Biegungen sehr dünner Platten geführt wird. Nur in einem einzigen Falle kann man ohne Weiteres vorschreiten, um die vorgeführten Probleme selbst zu untersuchen ; dann nämlich, wenn die endlichen Biegungen aufhören, und nur die an die ursprüngliche Gestalt der Mittelfläche anzubringenden Correctionen aufzufinden sind. Dieser Fall, in welchem alle Theile der Platte von ihrer ursprünglichen Lage nur sehr wenig abweichen, soll jetzt eingehender behandelt werden (S. 305).

In concluding our remarks on this problem of Clebsch's, it is sufficient to note that the elastic principles involved are simple and the earlier part of the investigation involves no great difficulties, but that the analysis required to solve even simple cases promises to be far too complex for us to hope by aid of it for any results of physical or technical value.

[1379.] § 72 is entitled : *Kleine Verschiebungen* and occupies S. 305–8 (*F. E.* pp. 684–9, where a more general distribution of elasticity is dealt with). Clebsch deduces from his general equations the two equations for the shifts of the mid-plane in its own plane and the corresponding contour-conditions as we have given them in our Arts. 389 and 391, except that he neglects the surface load on the plane faces, i.e. the terms

$$(\widehat{zx})_{+\epsilon}, \ (\widehat{zx})_{-\epsilon}, \ (\widehat{zy})_{+\epsilon} \text{ and } (\widehat{zy})_{-\epsilon}$$

Further he gives the equation for the transverse shift of a point on the mid-plane and the two contour-conditions such as we have given them in our Arts. 384–5, 390, 392–4 except that he again disregards the surface load. His equations are thus more general than those of Kirchhoff, but not so general as those of Saint-Venant. His method is certainly better than Kirchhoff's first method ; it is not so concise but it is more general than Kirchhoff's second method. As depending upon the theory of the finite shifts of thin plates, it is more cumbersome than the method by which Saint-Venant and Boussinesq have deduced still more general results.

Clebsch compares these equations for the small shifts of thin plates with those he has obtained for the case of thick plates (see our Arts. 1351-2), and remarks with regard to the first system of equations, or those for the shifts in the mid-plane:

...dass es genau mit den Systemen (11), (14), (15) [of our Art. 1351-2] übereinstimmt, nur dass hier noch Glieder auftreten, welche von den auf das Innere wirkenden Kräften abhängen, dass hingegen diejenigen Glieder fehlen, welche dort mit höheren Potenzen von *h* multiplicirt erschienen (S. 307).

[1380.]   In § 74 (S. 309-19; *F. E.* pp. 753-763) we have the case of the small shifts of a thin circular plate in its own plane completely solved.   The contour of the plate is supposed either to be subjected to a given system of forces or to be simply fixed.   Clebsch includes body-forces acting parallel to the plane of the plate.   Several serious errors in Clebsch's equations (19) are corrected in the French edition.

In § 75 (S. 319-27, *F. E.* pp. 763-72) the small transverse shifts of a circular plate subjected to any system of body-force are dealt with.

Clebsch's work here amounts to the following process.   Suppose the plate to possess elastic isotropy *in its plane*, and let it be stretched to a traction T uniform in all directions; let H be the plate modulus and w the transverse shift of the point in the mid-plane of the plate defined by r, ϕ, the polar coordinates with respect to the centre.   The body shift-equation for w is the following:

$$\frac{d^4w}{dr^4} + \frac{2}{r}\frac{d^3w}{dr^3} - \frac{1}{r^2}\frac{d^2w}{dr^2} + \frac{1}{r^3}\frac{dw}{dr} - \frac{12T}{Hh^2}\left(\frac{d^2w}{dr^2} + \frac{1}{r}\frac{dw}{dr}\right)$$

$$+ \frac{d^2}{d\phi^2}\left[\frac{2}{r^2}\frac{d^2w}{dr^2} - \frac{2}{r^3}\frac{dw}{dr} + \frac{4}{r^4}w - \frac{12T}{Hh^2}\frac{w}{r^2}\right] + \frac{1}{r^4}\frac{d^4w}{d\phi^4}$$

$$= \frac{12}{Hh^3}\left(R' + \frac{dP''}{dx} + \frac{dQ''}{dy}\right) \quad \ldots\ldots\ldots\ldots(i).$$

Here, if P, Q, R be the components of body-force acting on the element $dx\,dy\,dz$ of the plate of which the thickness is h, and if x and y are $r\cos\phi$, $r\sin\phi$ respectively, then:

$$R' = \int_{-h/2}^{+h/2} R\,dz, \quad P'' = \int_{-h/2}^{+h/2} Pz\,dz, \quad Q'' = \int_{-h/2}^{+h/2} Qz\,dz.$$

(See S. 320; *F. E.* p. 765, and compare our Arts. 384-5, 390, where $\epsilon = h/2$, and Art. 1300 (c)).

Clebsch now supposes $R' + \dfrac{dP''}{dx} + \dfrac{dQ''}{dy}$ to be known in sines and cosines of multiple angles of ϕ, and then expresses w in like form. This gives him a differential equation for a coefficient of one of the terms in w as a function of r alone.   To simplify this he assumes $T = 0$, and it then takes the form :

$$\left(\frac{d^2}{dr^2} + \frac{1}{r}\frac{d}{dr} - \frac{i^2}{r^2}\right)^2 w_i = \frac{12}{Hh^3}\gamma_i \quad\ldots\ldots\ldots\ldots\ldots(ii),$$

where w_i is the coefficient of $\cos i\phi$ or $\sin i\phi$ in w, and γ_i the coefficient of the like terms in $R' + \dfrac{dP''}{dx} + \dfrac{dQ''}{dy}$.

Clebsch then uses his unrivalled powers of analysis to find the most general solution of (ii) for a *complete* circular plate (S. 326 ; *F. E.* p. 771).   He determines the constants of this solution for the particular case in which the edge of the plate is built-in.

[1381.]   § 76 is entitled : *Biegung einer am Rande eingespannten kreisförmigen Platte durch ein einzelnes Gewicht* (S. 327–31 ; *F. E.* pp. 772–8).   The word *eingespannt* is here to be taken in the sense of *eingeklemmt*, "built-in," as Clebsch puts $T = 0$ in using equation (i).   A horizontal circular plate with built-in edge is supposed to be loaded at any single point.   Clebsch remarks :

Diese Aufgabe kann man mit grosser Annäherung so behandeln, als wäre an Stelle des Gewichts eine auf das Innere des betreffenden Elementes wirkende Kraft gegeben, deren Wirkungskreis nur auf einen sehr kleinen Raum beschränkt ist.   Uebrigens ist der Fall, wo mehrere Gewichte an verschiedenen Punkten angebracht sind, ganz ebenso zu behandeln ; die dann entstehenden Verschiebungen sind nichts anderes, als die Summe derjenigen, welche von den einzelnen Gewichten herrühren würden (S. 327).

The solution is obtained as a special case of the results referred to in the previous article.   It is somewhat lengthy, but as being the theoretical answer to a definite practical problem it is given here.   Let w be the deflection at the point r, ϕ of the plate of radius b, loaded at r_0, ϕ_0 with the weight P, then :

$$w = w_0 + \sum_1^\infty w_i \cos i\,(\phi - \phi_0),$$

where for points for which $r < r_0$:

$$w_0 = \frac{3P}{2\pi Hh^3}\left[r^2\left(1 + \log r_0\right) - r_0^2\left(1 - \log r_0\right) + \frac{b^2 - r^2}{2b}\left(2b\log b + \frac{b^2 + r_0^2}{b}\right)\right.$$
$$\left. - \left(b^2 + r_0^2\right)\log b \right],$$

$$w_1 = \frac{3P}{\pi Hh^3}\left[-rr_0\log r_0 - \frac{r^3}{4r_0} + \frac{b^2 - r^2}{2b^2}r\left(\frac{r_0^3}{4b^2} + r_0\log b + r_0\right)\right.$$
$$\left. + \frac{3b^2 - r^2}{2b^3}r\left(\frac{r_0^3}{4b} + br_0\log b\right)\right],$$

$$w_i = \frac{3P}{2i\pi Hh^3}\left[\frac{r^i}{(i-1)r_0^{\,i-2}} - \frac{r^{i+2}}{(i+1)r_0^{\,i}} + \frac{b^2-r^2}{2b^{i+1}}\,r^i\left(\frac{ir_0^{\,i+2}}{(i+1)b^{i+1}} - \frac{(i-2)r_0^{\,i}}{(i-1)b^{i-1}}\right)\right.$$
$$\left. + \frac{(i+2)b^2-ir^2}{2b^{i+2}}\,r^i\left(\frac{r_0^{\,i+2}}{(i+1)b^i} - \frac{r_0^{\,i}}{(i-1)b^{i-2}}\right)\right],$$

and for those for which $r > r_0$:

$$w_0 = \frac{3P}{2\pi Hh^3}\left[(r^2+r_0^{\,2})\log r + \frac{b^2-r^2}{2b}\left(2b\log b + \frac{b^2+r_0^{\,2}}{b}\right) - (b^2+r_0^{\,2})\log b\right],$$

$$w_1 = \frac{3P}{\pi Hh^3}\left[-\frac{r_0^{\,3}}{4r} - rr_0\log r + \frac{b^2-r^2}{2b^2}\,r\left(\frac{r_0^{\,3}}{4b^2}+r_0\log b + r_0\right)\right.$$
$$\left. + \frac{3b^2-r^2}{2b^3}\,r\left(\frac{r_0^{\,3}}{4b}+br_0\log b\right)\right],$$

$$w_i = \frac{3P}{2i\pi Hh^3}\left[\frac{r_0^{\,i}}{(i-1)r^{i-2}} - \frac{r_0^{\,i+2}}{(i+1)r^i} + \frac{b^2-r^2}{2b^{i+1}}\,r^i\left(\frac{ir_0^{\,i+2}}{(i+1)b^{i+1}} - \frac{(i-2)r_0^{\,i}}{(i-1)b^{i-1}}\right)\right.$$
$$\left. + \frac{(i+2)b^2-ir^2}{2b^{i+2}}\,r^i\left(\frac{r_0^{\,i+2}}{(i+1)b^i} - \frac{r_0^{\,i}}{(i-1)b^{i-2}}\right)\right].$$

Clebsch remarks of these results[1] :

Auf so verwickelte Formen führt ein Problem, dessen analoges, bei Stäben, mit Recht unter die elementarsten gezählt wird. Und selbst dann nur gelang es, wenn die Peripherie der Scheibe kreisförmig vorausgesetzt wurde. Inzwischen muss auf die Wichtigkeit des Problems auch für Anwendungen hingewiesen werden, so wie auf die Methode, mit deren Hülfe es vielleicht auch für andere Formen gelingt, die Lösung des Problems herzustellen (S. 330).

[1382.] We may note that the central deflection f_{r_0} for a load at r_0 is given by

$$f_{r_0} = \frac{3P}{2\pi Hh^3}\left(\frac{b^2-r_0^{\,2}}{2} + r_0^{\,2}\log\frac{r_0}{b}\right),$$

which becomes for a central load

$$f_0 = \frac{3Pb^2}{4\pi Hh^3}.$$

These results may be compared with those of our Art. 334.
They agree with the values that Clebsch gives on S. 331 for the deflection at $r = r_0$ due to a central load P. This follows from the general principle that deflection at $r = r_0'$ for loading at $r = r_0$ is equal to deflection at $r = r_0$ for the same loading at $r = r_0'$.

[1383.] Clebsch in § 77 (S. 331-3; *F. E.* pp. 778–81) next passes to the motion of plates, and finds by D'Alembert's principle the terms which must in this case be introduced into the general equations for

[1] There are several misprints in the reproduction of these formulae in the French edition : see p. 776.

finite shifts. For certain quantities P', Q', R', P'', Q'', R'' (the X, Y, Z, X'', Y'', Z'' of our Art. 384) we must substitute :

$$P' - \Delta h \frac{d^2\xi}{dt^2}, \qquad Q' - \Delta h \frac{d^2\eta}{dt^2}, \qquad R' - \Delta h \frac{d^2\zeta}{dt^2},$$

$$P'' - \frac{\Delta h^3}{12} \frac{d^2\alpha}{dt^2}, \qquad Q'' - \frac{\Delta h^3}{12} \frac{d^2\beta}{dt^2}, \qquad R'' - \frac{\Delta h^3}{12} \frac{d^2\gamma}{dt^2},$$

where Δ is the density of the plate, h its thickness, ξ, η, ζ the co-ordinates of a point on the mid-surface referred to fixed axes in space, and α, β, γ the direction-cosines of the angles the normal to the mid-surface makes with the axes of ξ, η, ζ.

Clebsch confines himself, however, to the case of thin plates with small shifts, and deals only with the *transverse* vibrations of these in two special sub-cases, which form the topic of the following two sections of his treatise.

If ζ be the transverse finite shift, we may in this case clearly replace it by the w of our usual notation for small shifts.

[1384.] § 78 is entitled : *Klangfiguren einer kreisförmigen freien Platte* (S. 334–43 ; *F. E.* pp. 781–93). Clebsch shows that, when no forces act on the plate, we must (owing to the conclusions of the previous section) replace :

$$R' + \frac{dP''}{dx} + \frac{dQ''}{dy} \text{ by } - \Delta h \frac{d^2}{dt^2} \left\{ w - \frac{h^2}{12} \left(\frac{d^2w}{dx^2} + \frac{d^2w}{dy^2} \right) \right\},$$

but he at once neglects the term multiplied by h^2 as very small. The remainder of his investigation adds, I think, nothing to that of Kirchhoff : see our Arts. 1241–3. On S. 333 he had acknowledged the latter's services in this matter.

[1385.] § 79 is entitled : *Schwingungen einer kreisförmigen ge-spannten Membrane* (S. 344–7 ; *F. E.* pp. 793–7), but Clebsch's treatment differs from the ordinary one, in that he does not suppose the membrane *perfectly flexible*. His work thus corresponds for membranes to that of Seebeck on strings : see our Art. 472. We must start from equation (i) of our Art. 1380 and retain now the terms in T, while the term in brackets on the right-hand side must be replaced as in our Art. 1384 by

$$- \Delta h \frac{d^2}{dt^2} \left\{ w - \frac{h^2}{12} \left(\frac{d^2w}{dr^2} + \frac{1}{r} \frac{dw}{dr} + \frac{1}{r^2} \frac{d^2w}{d\phi^2} \right) \right\}.$$

Writing $12T/(Hh^2) = 1/b^2$ and $12\Delta/(Hh^2) = 1/a^4$ we have Clebsch's equation of S. 344 (*F. E.* p. 794[1]).

---

[1] Equation (310) p. 793 should give $1/b^2$ and not b^2.

The boundary conditions become

$$\frac{d^2w}{dr^2} + \left(1 - \frac{2f}{H}\right)\left(\frac{1}{r}\frac{dw}{dr} + \frac{1}{r^2}\frac{d^2w}{d\phi^2}\right) = 0, \left.\vphantom{\frac{d^2w}{dr^2}}\right\} \text{ when } r = b \text{ the radius.}$$
$$w = 0$$

The first condition signifies that the couple round the tangent to the contour vanishes at each point. See our Art. 391.

Substituting $w = \Sigma\Sigma R_{mn}\gamma_{mn}\cos(m\phi + a_m)\cos(\kappa_{mn}t + \beta_{mn})$, where γ_{mn}, a_m, β_{mn} are constants, we have a differential equation to determine R_{mn} as a function solely of r. This can be solved by expanding R in a series of ascending powers of r in the usual mode. The solution is really found to be in terms of the same functions, i.e. Bessel's functions, as in the previous case, but the coefficients of the arguments of these functions are different. The transcendental equation for the frequencies of the notes follows in the usual manner from the contour conditions. Clebsch makes no attempt either to solve this equation numerically, or even to calculate the effect on the pitch of the notes of a *slight stiffness* in the membrane.

He concludes this portion of his work by showing that the transcendental equations which occur in the problems of both plate and membrane have real roots, or that the motion is really periodic (S. 347–55; *F. E.* pp. 797–806). The method adopted resembles that of the general problem of § 20 of the *Treatise* (see our Arts. 1328–30), and leads to expressions for the arbitrary constants of the solutions in terms of the initial shifts and velocities.

[1386.] The remainder of Clebsch's treatise is entitled: *Anwendungen*, and occupies S. 356–424 (*F. E.* pp. 807–880). This portion of the work is less exact[1] and more elementary. I content myself with noticing anything that seems of special interest or originality in the problems dealt with, not giving an abstract of the whole.

(a) In § 82 Clebsch investigates in an approximate manner what the cross-section of a rod under longitudinal body-force and load must be, in order that the stress may be everywhere uniform. Let Z be the load per unit volume, and ω the cross-section at a distance z from one end, and let P be the total load at $z = l$ and T the uniform tractive stress; then approximately

$$T\omega = P + \int_z^l Z\omega dz,$$

---

[1] Clebsch rightly insists on the importance of recognising the approximate character of most of the ordinary practical formulae:

Es wird in den betreffenden Fällen immer auf den Mangel an Strenge hingewiesen werden; eine Gewissenhaftigkeit, welche ebenso natürlich als nothwendig erscheint, und welche dennoch leider in Schriften über Anwendungen dieser Art nur zu häufig vermisst wird (S. 356).

or, by differentiation since T is a constant,

$$d\omega/\omega = -Zdz/T,$$

whence

$$\omega = \frac{P}{T} e^{\frac{1}{T}\int_z^l Zdz}.$$

Suppose, for example, we require the cross-section when Z is due to a centrifugal force measured by a^2z per unit mass. Let Δ be the density of the rod, then we have:

$$\omega = \frac{P}{T} e^{\frac{\Delta a^2}{2T}(l^2-z^2)}.$$

(S. 360–3; F. E. pp. 812–5.)

(b) In § 85 Clebsch investigates the fail-point in beams variously supported. He adopts a *stress-limit*, which in the simple cases dealt with does not lead him to results differing from those which would have arisen in considering relative strength from the standpoint of a *strain-limit*: see our Arts. 4 (γ), 5, 169 (c), 320–1, and 1327.

(c) On S. 392–6 we have the case of a uniformly loaded continuous beam freely supported on $(n+1)$ equally distant points of support. This problem had already been dealt with by Lamarle and others: see our Arts. 576, 947, 949, etc. The analysis is here clear and the results concisely given.

(d) In § 88 Clebsch deals with the problem of 'solids of equal resistance' by what is really only the approximate Bernoulli-Eulerian theory: see our Arts. 4*, 5*, 16*, 915*, etc., and compare them with our Art. 56 Case (4). Clebsch supposes all the cross-sections of the rod to be *similar* figures and that the maximum stress in all the cross-sections is the same. Further he supposes the plane of flexure to contain a principal axis of each cross-section. Suppose h to be the distance of the most strained 'fibre' from the 'neutral axis' and κ the swing radius about its neutral axis of the cross-section distant z from one end of the rod; let h_1 and κ_1 be the corresponding values when $z = l$, and let T be the given maximum stress. Then Clebsch finds the following differential equation to determine the linear dimensions of successive cross-sections:

$$\frac{d}{dz}\left(h^2 \frac{dh}{dz}\right) = \frac{h_1^2 p}{3T\kappa_1^2} h^2,$$

where p is the body-force per unit volume of the rod at z.

Clebsch solves this in two cases: (i) when only gravity, i.e. a constant p, acts on the rod; (ii) when no body-force but a terminal load P at $z = l$ produces flexure.

In the first case we have:

$$h = \frac{1}{30} \frac{h_1^2 p}{T\kappa_1^2} (l-z)^2 \ldots\ldots \text{ a parabola.}$$

In the second case we have, correcting a misprint :

$$h^3 = \frac{P h_1{}^4}{T \omega_1 \kappa_1{}^2} (l - z) \ldots \ldots \text{ a cubical parabola.}$$

(S. 396–402 ; *F. E*. pp. 852–8.)

(e) § 89 is entitled *Biegung bei sehr grossem Zug oder Druck in der Richtung der Längsaxe. Säulenfestigkeit* (S. 402–8 ; *F. E*. pp. 859–65). The equation for flexure now takes the form

$$E \omega \kappa^2 \frac{d^2 u}{dz^2} = M \pm R (u_l - u),$$

where M is the bending moment of all the forces perpendicular to the axis of the rod, R the component of the longitudinal force and u_l the deflection for $z = l$. Clebsch integrates this for R positive or negative. In the latter case, which corresponds to the case of a strut, Clebsch notes the inconsistency of the ordinary theory, in which the strut only bends for certain definite values of the load and for no intervening ones. He attributes this failure of the theory to the fact that for long rods the shifts become finite and we must fall back on the results of our Art. 1366. This correction had of course already been made by Lagrange : see our Art. 110*. But the full theory of finite shifts, we have seen, also leads to the Bernoulli-Eulerian value of the buckling-load and with this Clebsch seems to be content, although that theory has been by no means verified experimentally. He indicates in brief terms the point as to the relative magnitudes of the buckling and compressive loads, which had been previously worked out by Lamarle : see our Arts. 1258*–9*.

[1387.] In § 90 (S. 409–13 ; *F. E*. pp. 866–71), Clebsch discusses an important practical problem, namely : the discovery of the strains and stresses in a system of bars pinned together at their terminals, or in a framework. In this section the body forces are supposed negligible and the terminal loads in no case sufficiently great to produce buckling. Thus the system will be without flexure. If there be no *supernumerary* bars, we know that in this case the stresses in the various members can be easily ascertained by the method of reciprocal figures ; if supernumerary bars exist, however, we are compelled to use *ab initio* the elastic properties of the bars.

Suppose x_i, y_i, z_i to be the coordinates of the ith node of such a frame before strain and u_i, v_i, w_i its shifts after strain, these latter quantities being very small ; further let E_{ik} be the stretch-modulus of the bar joining the nodes i and k, r_{ik} its unstrained, $r_{ik} + \rho_{ik}$ its strained length, ω_{ik} its cross-section ; let X_i, Y_i, Z_i be the components of the load at the ith node. Then we must have for the equilibrium of that node

$$X_i + \Sigma_k \frac{E_{ik}\, \omega_{ik}\, \rho_{ik}\, (x_k - x_i)}{r_{ik}^2} = 0,$$

$$Y_i + \Sigma_k \frac{E_{ik}\, \omega_{ik}\, \rho_{ik}\, (y_k - y_i)}{r_{ik}^2} = 0,$$

$$Z_i + \Sigma_k \frac{E_{ik}\, \omega_{ik}\, \rho_{ik}\, (z_k - z_i)}{r_{ik}^2} = 0,$$

where the summation is to extend for all nodes k united by bars to the node i.

If there be n nodes we have thus $3n$ equations; there are also $3n$ unknowns, namely the u, v, w for the n nodes. $\Big($Obviously ρ_{ik} is given by

$$\rho_{ik} = \frac{(x_i - x_k)\,(u_i - u_k) + (y_i - y_k)\,(v_i - v_k) + (z_i - z_k)\,(w_i - w_k)}{r_{ik}}.\Big)$$

But of the $3n$ equations 6 must be the equations of statical equilibrium between the external loads X, Y, Z. Hence there are 6 of the shifts undetermined by the above equations. This is what we should expect, as we might give the system any displacement of translation or rotation as a whole without affecting the elastic equations, and such displacement involves six arbitrary constants. As a rule certain nodes and directions of rods will be fixed, and these will give the additional conditions sufficient to solve the problem, even when there are also unknown reactions at some of the nodes to be determined. Clebsch works out this solution by solving for u, v, w, in the special case when a loaded node is supported by any number of bars attached to fixed points.

[1388.] The problem of the preceding article becomes more complex when there are joints or nodes attached to points in the bars themselves as well as at their terminals, so that flexure takes place. Clebsch's § 91 (S. 413–20; *F. E.* pp. 872–79), entitled: *Stabsysteme mit Biegung*, deals with this problem. He supposes the bars to be pin-jointed, the cross-section of each to be uniform and their weight to be negligible. His method is indicated in the following sentence:

Das allgemeine Princip wird auch hier darin bestehen, dass man die Verschiebungen der Knotenpunkte zunächst wie bekannte Grössen behandelt, aus ihnen die eintretenden elastischen Kräfte bestimmt, mit welchen die Stäbe in ihren Knotenpunkten reagiren, und endlich die Gleichgewichtsbedingungen für die in den Knotenpunkten wirkenden äussern und elastischen Kräfte aufstellen ; Gleichungen die schliesslich genau hinreichen um aus ihnen die eingeführten Verschiebungen zu bestimmen (S. 413).

Clebsch describes clearly the various stages in the problem, writing down the general types of the equations involved. He afterwards applies his principle to an isosceles truss with a vertical strut, the truss being supported at the terminals of the base and loaded at the vertex. Both the general investigation and the special example are of interest, but we must refer our reader to the *Treatise* itself for the details.

[1389.] Clebsch's *Treatise* concludes with a section on *Torsion*. He introduces Coulomb's theory for a prism of circular cross-section and discusses the maximum stress. The method adopted is unsatisfactory and Clebsch himself expressed his discontent with it. In a letter to Saint-Venant he authorised its withdrawal, and there is thus no need to criticise it here.

[1390.] The work we have been considering was published when the author was twenty-nine years of age and was based upon his lectures at Karlsruhe delivered in 1861. However unsuitable it may seem as an introduction to the study of elasticity for technical students, we cannot but be surprised by the wealth and ingenuity of analysis which the young mathematician was able to bring to the service of the more theoretical parts of elasticity. It is this fertility of analytical method which makes his work so suggestive and has given it a place among the classics of our subject, notwithstanding its incomplete and somewhat one-sided character.

Die physikalische, überhaupt die naturwissenschaftliche Auffassung lag ihm, bei allen Kenntnissen, die er im Einzelnen besass, verhältnissmässig fern und interessirte ihn nicht besonders. Nur in seiner ersten Arbeit, seiner Inauguraldissertation: *Ueber die Bewegung eines Ellipsoids in einer Flüssigkeit*, so wie später noch einmal in einer Untersuchung über Circularpolarisation, findet sich bei ihm eine Vergleichung der abgeleiteten Resultate mit dem Experimente. Ihn interessirte vielmehr die mathematische Fragestellung, insofern sie geschickte analytische Behandlung verlangte: er gehörte, auch hierin Jacobi ähnlich, der rein mathematischen Richtung an, welche den abstracten Gedanken um seiner selbst willen verfolgt.

This excellent characterisation of Clebsch I take from S. 5–6 of a pamphlet the joint labour of several of his friends: *Alfred Clebsch. Versuch einer Darlegung und Würdigung seiner wis-*

senschaftlichen Leistungen (Leipzig, 1873), to which I must refer
the reader for details of Clebsch's life and work.

[1391.] In the preface to his *Treatise* on elasticity Clebsch
especially excludes from the scope of his work all applications
of elasticity to the theory of light but suggests that he may later
deal with this wide field (S. v). All that has been preserved to
us of Clebsch's labours in this field are the memoirs in Bd. 57
(1860) and Bd. 61 (1863) of *Crelles Journal* (see our Arts. 1324
and 1392) and a posthumous publication entitled: *Prinzipien der
mathematischen Optik* edited by A. Kurz and separately published
in 1887 (Augsburg) as an offprint from the *Blätter für das
bayerische Realschulwesen*. This journal being probably more
inaccessible to our readers than the offprint, we shall refer only
to the pages of the latter.

This short treatise on optics containing 53 pages is based upon the
supposition that the ethereal molecules of masses m, m_1, act upon each
other with a force of the form

$$mm_1 f(r),$$

where r is their central distance (S. 1). Thus Clebsch really starts with
a hypothesis that must lead him to rari-constant elastic equations the
moment he neglects the terms in differential coefficients higher than the
second, which he at first retains in order to explain dispersion in
Cauchy's manner (S. 15). When however he does neglect these
differential coefficients (S. 17–19), we find he has equations which
apparently involve nine independent constants for a crystal with three
rectangular axes of elastic symmetry and two independent constants in
the case of isotropy. But this independence is only apparent; take for
example the equations of isotropy in S. 19 which are of the type

$$\frac{d^2 u}{dt^2} = (a+b)\nabla^2 u + 2b\,\frac{d\theta}{dx}.$$

It will be found by substituting the values on p. 18 for $u_1 - u$ etc.
in the fundamental equations (3) of S. 3 that with the notation of our
Art. 143, we have

$$b = \tfrac{1}{2}\Sigma m\chi\,(r_0)\,x_0{}^2 y_0{}^2,$$

and

$$a + b = \tfrac{1}{2}\Sigma m\chi\,(r_0)\,x_0{}^4 - \Sigma m\chi\,(r_0)\,x_0{}^2 y_0{}^2.$$

Consequently

$$a = \tfrac{1}{2}\left\{\Sigma m\chi\,(r_0)\,x_0{}^4 - 3\Sigma m\chi\,(r_0)\,x_0{}^2 y_0{}^2\right\}$$

$$= 0,$$

by equation (xxxiv) of the same article.

Thus we see that Clebsch's equations are really rari-constant, and the fact that there are relations among his elastic constants would, I think, considerably modify the rest of his investigations.

For double-refraction Clebsch obtains conditions identical with those of Cauchy and Saint-Venant. Thus in other notation his relations (44) and (46) (S. 21–2) agree with (xxxix) of our Art. 148, if we put $d = d'$, $e = e'$ and $f = f'$ therein.

Much of the pamphlet contains extremely interesting analysis to which we may draw the attention of those interested in optical theories. We have referred above to all that directly concerns us when we are dealing with elasticity.

[1392.] *Ueber die Reflexion an einer Kugelfläche. Journal für die reine und angewandte Mathematik*, Bd. 61, S. 195–262. Berlin, 1863. This memoir is dated October 30, 1861. It opens with the following paragraph explaining its object:

Obgleich das Problem der Reflexion von Lichtstralen an einer gegebenen Fläche längst die Aufmerksamkeit der Geometer in vielfacher Hinsicht auf sich gezogen hat, so ist doch niemals der Versuch gemacht worden, aus den Bewegungsgleichungen selbst die Gesetze dieser Erscheinungen zu deduciren, und so theoretisch eine sichere Basis für Untersuchungen dieser Art zu gewinnen. Der einzige Fall, den man betrachtete, war der einer unendlich ausgedehnten brechenden Ebene, auf welche ebene Wellen fallen ; und so kam es, dass die geometrischen Sätze der Dioptrik und Katoptrik mit dem was man heute eigentlich Optik zu nennen gewohnt ist, nur durch einige Betrachtungen der Enveloppentheorie lose und gewaltsam verknüpft sind.

[1393.] Clebsch attempts in this memoir to investigate the motion of an isotropic elastic medium, when the disturbances are totally reflected from the surface of a sphere contained in it. He deals in fact with a special case of an extension of *Lamé's Problem* (see our Art. 1111*) to media in vibratory motion. In order to simplify the surface conditions, which he remarks are still both theoretically and physically somewhat obscure, he takes the simplest hypothesis possible:

dass nämlich die Kugel *vollständig* reflectire, und dass, bei Abwesenheit gebrochener Wellen, die einfallenden und die reflectirenden Bewegungen an der Oberfläche der Kugel sich in ihrer Summe genau aufheben (S. 195).

The last words are somewhat obscure, but what Clebsch really does is to put the total shift (due to incident and reflected disturbances) at the surface of his spherical reflector zero. In other words, *he fixes his ether at the surface of a totally reflecting body*[1]. With this simplification Clebsch remarks that the problem becomes a purely mathematical one, namely: to develop the differential coefficient of a function of x, y, z, with regard to one of these variables in a series of spherical harmonics, if the development of the function itself in spherical harmonics be given. The solution is obtained simply and symmetrically by replacing the spherical harmonic by the corresponding homogeneous function of the nth degree.

The results of the investigation are exceedingly complex for the optically important case of a wave-length very small as compared with the radius of the reflecting surface, but they are capable of being easily dealt with approximately, if the wave-length be large as compared with this radius.

[1394.] § 1 of the memoir is entitled: *Zurückführung der Gleich-ungen der Elasticität auf getrennte partielle Differentialgleichungen* and occupies S. 196–9. Clebsch adopts as his equations for an isotropic medium the type:

$$\frac{d^2u}{dt^2} = (a^2 - b^2)\frac{d\theta}{dx} + b^2\nabla^2 u \quad\text{...................} \quad \text{(i)},$$

so that his $b^2 = \mu/\rho$, and his $a^2 = (\lambda + 2\mu)/\rho$ of this *History*. He then gives a demonstration that the most general solution for the shifts is of the form:

$$\left.\begin{aligned} u &= \frac{dP}{dx} + \frac{dW}{dy} - \frac{dV}{dz}, \\ v &= \frac{dP}{dy} + \frac{dU}{dz} - \frac{dW}{dx}, \\ w &= \frac{dP}{dz} + \frac{dV}{dx} - \frac{dU}{dy} \end{aligned}\right\} \quad\text{....................} \quad \text{(ii)},$$

where P is a solution of:

$$\frac{d^2P}{dt^2} = a^2\nabla^2 P \quad\text{..........................} \quad \text{(iii)},$$

---

[1] This fixing of the boundary as a condition for total reflection is interesting in the light of Sir William Thomson's hypothesis that the ether may be treated as an elastic solid fixed at infinity (and I suppose at totally reflecting surfaces placed in it). See the *Philosophical Magazine*, November, 1888. Is it possible in this case to absolutely neglect the waves 'reflected' from the infinite boundary as Helmholtz asserts? Must not the steady motion be considered to have existed from an 'infinite' time? If we cannot neglect the reflected waves what becomes of them and how do they affect the problem of cosmic temperature?

and U, V, W are solutions of:

$$\frac{d^2\phi}{dt^2} = b^2 \nabla^2 \phi \quad \dots\dots\dots\dots\dots\dots \text{(iv)}.$$

Compare these results with those of Lamé and Kirchhoff referred to in our Arts. 1078* and 1309.

[1395.] § 2 entitled: *Entwickelung von P, U, V, W nach Kugelfunctionen*, occupies S. 200–2. In this section Clebsch introduces a function M_n which may be considered as consisting of two parts. M_n is a *solid* spherical harmonic of the nth degree, i.e. a homogeneous function of the nth degree in x, y, z which satisfies $\nabla^2 M_n = 0$, but its arbitrary coefficients are functions of r and t, where r is to be put equal to r, i.e. $\sqrt{x^2 + y^2 + z^2}$, after all differentiations have taken place. Thus if ∇^2 be taken in its most general sense to operate on $M_n = f(r, x, y, z)$ we have to take into consideration the differentiation of M_n with regard to the r which occurs in the coefficients of M_n treated as a solid harmonic in x, y, z. Thus we find, if Δ^2 be used instead of ∇^2 to denote the special sense of an operator on x, y, z only :

$$\nabla^2 M_n = \frac{1}{r^2}\frac{d}{dr}\left(r^2 \frac{dM_n}{dr}\right) + \frac{2}{r}\frac{d}{dr}\left(x\frac{dM_n}{dx} + y\frac{dM_n}{dy} + z\frac{dM_n}{dz}\right) + \Delta^2 M_n$$

$$= \frac{d^2 M_n}{dr^2} + \frac{2n+2}{r}\frac{dM_n}{dr}.$$

Hence we shall obtain a solution of an equation of the type :

$$\frac{d^2\phi}{dt^2} = c^2 \nabla^2 \phi,$$

if we take

$$\phi = \sum_0^\infty M_n,$$

and determine the coefficients of M_n as functions of r and t from the equation :

$$\frac{d^2 M_n}{dt^2} = c^2 \left(\frac{d^2 M_n}{dr^2} + \frac{2n+2}{r}\frac{dM_n}{dr}\right).$$

[1396.] A method based upon this property of the function M_n for solving the body-shift equations (i) and (ii) is developed by Clebsch in the following sections (§§ 3 and 4, S. 202–8). Many of the fundamental properties of *solid* spherical harmonics were here published, I believe, for the first time.

On S. 208 Clebsch gives the following expression for the type of shift :

$$u = \sum^\infty u_n \dots\dots\dots\dots\dots\dots\dots\dots\dots\text{(v)},$$

where :

$$u_n = \frac{dP_{n+1}}{dx} + r^{2n+1}\frac{d}{dx}\frac{Q_{n-1}}{r^{2n-1}} + z\frac{dT_n}{dy} - y\frac{dT_n}{dz}$$

and

$$Q_{n-1} = -\frac{1}{r}\frac{d}{dr}\left\{\frac{M_{n-1}}{2n-1} - \frac{M'_{n-1}}{n}\right\},$$

$$P_{n+1} = \frac{1}{r^{2n+2}}\frac{d}{dr}\left\{r^{2n+3}\left(\frac{M_{n+1}}{2n+3} + \frac{M'_{n+1}}{n+1}\right)\right\}$$

$$\Bigg\} \quad \ldots\ldots\ldots\text{(vi)},$$

and where M_n, M_n', T_n are arbitrary solid spherical harmonics of the nth degree. Their coefficients are, however, functions of r, t. Those of M_n satisfy the equation :

$$\frac{d^2\phi}{dt^2} = a^2\left\{\frac{d^2\phi}{dr^2} + \frac{2n+2}{r}\frac{d\phi}{dr}\right\} \quad\ldots\ldots\ldots\ldots\text{(vii)},$$

and those of M_n' and T_n satisfy the equation :

$$\frac{d^2\phi}{dt^2} = b^2\left\{\frac{d^2\phi}{dr^2} + \frac{2n+2}{r}\frac{d\phi}{dr}\right\} \quad\ldots\ldots\ldots\ldots\text{(viii)},$$

with the condition that M_0' is to be zero.

[1397.]　In the particular case when the motion is not vibratory we must put $d^2\phi/dt^2 = 0$, we then obtain a solution in solid spherical harmonics of the equations of elasticity for a medium subjected only to surface forces. Results for this special case of an elastic solid in equilibrium were given by Sir William Thomson in a paper published in the same year as Clebsch's, but read on November 27, 1862, a year later than Clebsch's was written. Sir William Thomson's conclusions will be dealt with in our Chapter XIV. They differ considerably in form from Clebsch's as cited above. When ϕ is independent of t, we have

$$\phi = \Sigma\,(H_n + H_n'/r^{2n+1}),$$

where H_n and H_n' are solid spherical harmonics of order n. Substituting we ultimately obtain the shifts in a form which is explicitly free from the somewhat complex coefficients involving the elastic constants a^2 and b^2 [i.e. $(m+n)/\rho$ and n/ρ] which occur in Sir W. Thomson's form of the solution.

[1398.]　In the fifth section (S. 209-11) Clebsch integrates the equations of types (vii) and (viii), and shows that ϕ in (vii) is of the form :

$$\phi = \frac{f^n\,(\mathbf{r}-at)+F^n\,(\mathbf{r}+at)}{\mathbf{r}^{n+1}} - \frac{n\,(n+1)}{2}\frac{f^{n-1}\,(\mathbf{r}-at)+F^{n-1}\,(\mathbf{r}+at)}{\mathbf{r}^{n+2}}$$

$$+ \frac{(n-1)\,n\,(n+1)\,(n+2)}{2\,.\,4}\frac{f^{n-2}\,(\mathbf{r}-at)+F^{n-2}\,(\mathbf{r}+at)}{\mathbf{r}^{n+3}}$$

$$-\,\ldots+\,\ldots\pm\frac{1\,.\,2\ldots 2n}{2\,.\,4\ldots 2n}\frac{f\,(\mathbf{r}-at)+F\,(\mathbf{r}+at)}{\mathbf{r}^{2n+1}}\ldots\ldots\ldots\ldots\text{(ix),}$$

where a must be changed to b in the case of (viii), and the indices attached to f and F denote derivatives of those functions.

[1399.] To complete the solution it is necessary to determine from the surface conditions the arbitrary functions f and F, which it must be remembered are to be solid spherical harmonics in x, y, z of the order n. This Clebsch achieves very ingeniously. He demonstrates on S. 206 that:

$$n\,(2n+1)\,Q_{n-1} = -\left\{\frac{du_n}{dx}+\frac{dv_n}{dy}+\frac{dw_n}{dz}\right\},$$

$$(n+1)\,(2n+1)\,P_{n+1} = -r^{2n+3}\left\{\frac{d}{dx}\frac{u_n}{r^{2n+1}}+\frac{d}{dy}\frac{v_n}{r^{2n+1}}+\frac{d}{dz}\frac{w_n}{r^{2n+1}}\right\},$$

$$n\,(n+1)\,T_n = x\left(\frac{dw_n}{dy}-\frac{dv_n}{dz}\right)+y\left(\frac{du_n}{dz}-\frac{dw_n}{dx}\right)$$

$$+\,z\left(\frac{dv_n}{dx}-\frac{du_n}{dy}\right)\ldots\ldots\text{(x).}$$

Now on the right there are no differentiations with regard to r. Hence if u, v, w, the shifts over two given concentric surfaces, say of radii r_1 and r_2, be given as functions of two position-angles, we can express them by surface spherical harmonics and by dividing or multiplying at the same time by the proper power of r_1 and r_2 replace these surface harmonics by solid harmonics. Now u_n, v_n, w_n are known as the nth solid harmonics in u, v, w respectively, and these results may therefore be directly substituted in (x), as there is no differentiation with regard to r. Thus by aid of (vi) and (x) we can determine the six arbitrary functions which occur in M_n, M'_n and T_n as exhibited in (ix). It might seem from this that the motion was fully determined when the arbitrary shifts over two concentric spherical surfaces are given at each instant of time, but Clebsch guards himself against this assumption by the remark that though the surfaces of a spherical shell were fixed, its material could still move.

[1400.] § 6 is entitled: *Vollständige Behandlung des Falles, wo eine, auf einer bestimmten Kugelfläche gegebene Bewegung sich ins Unendliche ausbreiten kann*, and occupies S. 211–15. Suppose a number of centres of disturbance of given character to be at finite distances apart in an infinite medium; then the shifts U, V, W which they would produce at any points of the medium are known. Now introduce into the medium a spherical surface of no disturbance and let the additional shifts be represented by u, v, w; then the conditions to be satisfied over the surface are:

$$u + U = 0, \quad v + V = 0, \quad w + W = 0.$$

These will suffice to determine u, v, w in the manner suggested in the previous article. The shifts at any point of space will then be

$$u + U, \quad v + V, \quad w + W.$$

It must be noted, however, that there is an additional fact to be taken into consideration. One of the rigid spherical boundaries has really been taken at an infinite distance, hence, Clebsch asserts, there can be no inward bound wave or the terms in equation (ix) involving the function $F(r + at)$ and its derivatives cannot exist. This conclusion Clebsch attributes on S. 211 to Hèlmholtz.

Now U, V, W may be expressed over the surface of the given sphere in terms of spherical harmonics and hence quantities like

$$\overline{Q}_{n-1}, \quad \overline{P}_{n+1}, \quad \overline{T}_{n},$$

corresponding to them found from equations similar to (x). But these lead us to the values of Q_n, P_n, T_n, for we must clearly have :

$$Q_n = -\overline{Q}_n, \quad P_n = -\overline{P}_n, \quad T_n = -\overline{T}_n,$$

at the spherical surface. In this manner Clebsch finds linear differential equations with constant coefficients for the arbitrary functions which occur in the values of P_n, Q_n, T_n. Clebsch notes the following facts with regard to these equations :

(i) If the incident disturbance be periodic, the reflected motion (i.e. one corresponding to the particular integral) will also be periodic, i.e. the shifts will not contain the time outside the arguments of sine or cosine.

(ii) A free motion of the system (i.e. one corresponding to the complementary function) is possible, even when there are no external centres of disturbance, and a spherical surface is rigidly fixed in the medium. But this motion cannot contain any periodic terms except for special values of a/b, and these values of a/b appear, to judge from the equations in the case of the smallest values of n, to be negative and therefore impossible. A general proof of this Clebsch has, however, been unable to find (S. 215).

These two results are of considerable interest; in particular the free motion of a mass of elastic material fixed to two rigid enclosing surfaces is deserving of closer investigation. It may possibly be found that some such masses are incapable of isochronous vibrations. The impossibility of free vibrations in Clebsch's case, however, is solely the result of his neglecting the inward bound wave.

Since the free vibration in the case (ii) of a fixed spherical surface is, according to Clebsch, non-periodic, so every periodic disturbance will be duly reflected, for the only possible case of failure, that of equality between the periods of forced and of free vibrations, cannot occur.

[1401.] In § 7, entitled: *Einfachste Bewegungen. Einführung der Functionen f, ϕ* (S. 216–21), Clebsch deals with the special case in which the incident disturbance is given by a single term of the period $2\pi/k$, or where $\bar{P}_n, \bar{Q}_n, \bar{T}_n$ are of the form[1]:

$$\bar{P}_n = \bar{P}'_n e^{kt\sqrt{-1}}, \quad \bar{Q}_n = \bar{Q}'_n e^{kt\sqrt{-1}}, \quad \bar{T}_n = \bar{T}''_n e^{kt\sqrt{-1}} \ldots\ldots\ldots (\text{xi}).$$

But these lead at once to

$$\frac{d^2 M_n}{dt^2} = -k^2 M_n,$$

and thus to determine M_n and M'_n we have from (vii) and (viii) the equations:

$$\left.\begin{array}{l} a^2 \left(\dfrac{d^2 M_n}{dr^2} + \dfrac{2n+2}{r} \dfrac{dM_n}{dr} \right) + k^2 M_n = 0, \\[2ex] b^2 \left(\dfrac{d^2 M_n'}{dr^2} + \dfrac{2n+2}{r} \dfrac{dM_n'}{dr} \right) + k^2 M_n' = 0 \end{array}\right\} \ldots\ldots\ldots\ldots(\text{xii}).$$

Clebsch then shows that:

$$M_n = \left\{ H_n' F_n \left(\frac{rk}{a} \right) + H_n'' \Phi_n \left(\frac{rk}{a} \right) \right\} e^{kt\sqrt{-1}},$$

$$M_n' = \left\{ H_n''' F_n \left(\frac{rk}{b} \right) + H_n^{\text{IV}} \Phi_n \left(\frac{rk}{b} \right) \right\} e^{kt\sqrt{-1}},$$

$$T_n = \left\{ H_n^{\text{V}} F_n \left(\frac{rk}{b} \right) + H_n^{\text{VI}} \Phi_n \left(\frac{rk}{b} \right) \right\} e^{kt\sqrt{-1}},$$

where H_n, H_n' etc. are solid harmonics of the nth order, and $F_n(s), \Phi_n(s)$ are solutions of the equation:

$$\frac{d^2 \Omega}{ds^2} + \frac{2n+2}{s} \frac{d\Omega}{ds} + \Omega = 0.$$

These solutions are as follows:

$$F_n(s) = \{ f_n(s) + \phi_n(s) \sqrt{-1} \} e^{s\sqrt{-1}},$$

$$\Phi_n(s) = \{ f_n(s) - \phi_n(s) \sqrt{-1} \} e^{-s\sqrt{-1}},$$

where:

$$\left.\begin{array}{l} f_n(s) = \dfrac{1}{s^{n+1}} - \dfrac{(n-1)\,n\,(n+1)\,(n+2)}{2\,.\,4\,.\,s^{n+3}} + \dfrac{(n-3)\ldots\ldots(n+4)}{2\,.\,4\,.\,6\,.\,8\,.\,s^{n+5}} - \ldots, \\[2ex] \phi_n(s) = \dfrac{n\,(n+1)}{2\,.\,s^{n+2}} - \dfrac{(n-2)\ldots(n+3)}{2\,.\,4\,.\,6\,.\,s^{n+4}} + \dfrac{(n-4)\ldots\ldots(n+5)}{2\,.\,4\ldots10\,.\,s^{n+6}} - \ldots \end{array}\right\} \ldots(\text{xiii}),$$

or,

$$-s\phi_0, \quad s^2 f_1, \quad -s^3 \phi_2, \quad s^4 f_3, \ldots\ldots$$

$$sf_0, \quad s^2 \phi_1, \quad s^3 f_2, \quad s^4 \phi_3, \ldots\ldots$$

---

[1] To avoid the confusion of the symbol t being used with two different meanings, I have slightly changed Clebsch's notation throughout.

are the consecutive numerators and denominators respectively of the convergents to the continued fraction

$$\tan s = -\frac{0}{1} + \cfrac{1}{\dfrac{1}{s}+} \cfrac{1}{\dfrac{3}{s}+} \cfrac{1}{\dfrac{5}{s}+} \cfrac{1}{\dfrac{7}{s}+\text{etc.}}.$$

[1402.] From the values of M_n, M_n' and T_n, Clebsch finds those of Q_{n-1} and P_{n+1}, and ultimately u_n, v_n and w_n (S. 220). He thus reduces the problem to the determination of $H_n'......H_n^{\text{VI}}$, which will follow at once from (xi), since Q_n, P_n and T_n are known in terms of $\bar{Q}_n$, $\bar{P}_n$, $\bar{T}_n$ by Art. 1400.

In the case of a single spherical surface, over which there is no shift, no function of r + at can occur (Art. 1400), so that the terms in F_n disappear, and Clebsch easily finds H_n'', H_n^{IV} and H_n^{VI} as functions of the given quantities $\bar{P}_n$, $\bar{Q}_n$ and $\bar{T}_n$.

[1403.] §§ 8–18 (S. 222–62) deal with the case of a single centre of disturbance, and like previous portions of the memoir are very characteristic of Clebsch's remarkable power of analysis.

In this case the solution of the equation

$$\frac{d^2\phi}{dt^2} = a^2 \nabla^2 \phi$$

for a disturbance symmetrical about the source x_0, y_0, z_0 consists of terms of the form:

$$\frac{\sin\frac{k}{a}(R-at)}{R} \text{ and } \frac{\cos\frac{k}{a}(R-at)}{R},$$

where $R = \sqrt{(x-x_0)^2 + (y-y_0)^2 + (z-z_0)^2}$.

These may be replaced by the exponential term

$$R^{-1} e^{-\frac{k}{a}(R-at)\sqrt{-1}},$$

and the shifts due to a disturbance of the kind considered will be of the type:

$$U = C\frac{x-x_0}{R}\frac{d}{dR}\left(\frac{e^{-\frac{k}{a}R\sqrt{-1}}}{R}\right)e^{kt\sqrt{-1}}$$

$$+ \frac{C'''(y-y_0)-C''(z-z_0)}{R}\frac{d}{dR}\left(\frac{e^{-\frac{k}{b}R\sqrt{-1}}}{R}\right)e^{kt\sqrt{-1}}......(\text{xiv}),$$

by our Art. 1394, the C's being constants.

If U_1 be the first, U_2 the second term of U, we have :

$$U_1 : V_1 : W_1 :: x - x_0 : y - y_0 : z - z_0,$$

and $U_1{}^2 + V_1{}^2 + W_1{}^2$ is a function only of R and t. Thus the total shift is in the direction of motion of the wave, and is a function only of the distance from the centre of disturbance and of the time. We are therefore dealing with a spherical wave of longitudinal vibrations.

For the second terms we have :

$$U_2 (x - x_0) + V_2 (y - y_0) + W_2 (z - z_0) = 0,$$
$$U_2 C' + V_2 C'' + W_2 C''' = 0,$$

while $U_2{}^2 + V_2{}^2 + W_2{}^2$ is not independent of

$$\frac{x - x_0}{R}, \quad \frac{y - y_0}{R}, \quad \frac{z - z_0}{R},$$

or the direction-cosines of the line from the centre of disturbance to the point disturbed. We see that the shifts U_2, V_2, W_2 are perpendicular to this line, and are also parallel to the plane

$$C'x + C''y + C'''z = 0.$$

They correspond therefore to a wave of transverse vibrations.

[1404.] The next stage in the problem is to find the value of

$$\frac{1}{R} e^{-mR\sqrt{-1}}, \text{ or of } \frac{\cos mR}{R} \text{ and } \frac{\sin mR}{R},$$

in solid spherical harmonics, m being a constant. This is accomplished by Clebsch in § 9 (S. 224—7). Although it is impossible to reproduce the whole of the analysis of this special case, these expansions may be given here as they seem likely to be serviceable in a number of problems quite independent of the present one. I may cite them as follows:

$$\frac{1}{R} \begin{Bmatrix} \cos mR \\ \sin mR \end{Bmatrix} = m\, Y_0 \left[f_0\,(mr_0) \begin{Bmatrix} \cos (mr_0) \\ \sin (mr_0) \end{Bmatrix} \mp \phi_0\,(mr_0) \begin{Bmatrix} \sin (mr_0) \\ \cos (mr_0) \end{Bmatrix} \right]$$
$$\times \left[f_0\,(mr) \sin (mr) + \phi_0\,(mr) \cos (mr) \right]$$
$$\mp 3m^3 Y_1 \left[f_1\,(mr_0) \begin{Bmatrix} \sin (mr_0) \\ \cos (mr_0) \end{Bmatrix} \pm \phi_1\,(mr_0) \begin{Bmatrix} \cos (mr_0) \\ \sin (mr_0) \end{Bmatrix} \right]$$
$$\times \left[f_1\,(mr) \cos (mr) - \phi_1\,(mr) \sin (mr) \right]$$
$$+ 5m^5 Y_2 \left[f_2\,(mr_0) \begin{Bmatrix} \cos (mr_0) \\ \sin (mr_0) \end{Bmatrix} \mp \phi_2\,(mr_0) \begin{Bmatrix} \sin (mr_0) \\ \cos (mr_0) \end{Bmatrix} \right]$$
$$\times \left[f_2\,(mr) \sin (mr) + \phi_2\,(mr) \cos (mr) \right]$$
$$\mp 7m^7 Y_3 \left[f_3\,(mr_0) \begin{Bmatrix} \sin (mr_0) \\ \cos (mr_0) \end{Bmatrix} \pm \phi_3\,(mr_0) \begin{Bmatrix} \cos (mr_0) \\ \sin (mr_0) \end{Bmatrix} \right]$$
$$\times \left[f_3\,(mr) \cos (mr) - \phi_3\,(mr) \sin (mr) \right]$$

$+$ etc. .... (xv).

Here f_n and ϕ_n are given by our (xiii), and Y_n is the solid spherical harmonic :

$$Y_n = (rr_0)^n P_n (\cos \phi), \text{ where } \cos \phi = \frac{xx_0 + yy_0 + zz_0}{rr_0},$$

and
$$R = \sqrt{r^2 + r_0^2 - 2rr_0 \cos \phi}.$$

As an identity we may of course put on the right-handside of (xv) $r = r$ and $r_0 = r_0$, but the form in which we have left it shows us at once how to apply the operators Δ^2 and ∇^2 of our Art. 1395 to it. This application is required in the further course of Clebsch's analysis.

[1405.] In § 10 (S. 228-9) Clebsch expands U, V, W in solid spherical harmonics by the aid of (xv). He thus obtains the disturbance, due to the source at x_0, y_0, z_0, at any point x, y, z on a spherical surface of radius r with centre at the origin. In § 11 (S. 229-33) he deals with the case of an incident wave of purely longitudinal vibrations (U_1, V_1, W_1) and he shows that such a wave always produces reflected waves of both longitudinal and transverse vibrations. In § 12 (S. 233-6) we have the case of an incident wave of purely transverse vibrations (U_2, V_2, W_2); it is shown that with one exception, the reflected wave consists partly of longitudinal and partly of transverse vibrations. § 13 (S. 236-40) deals with the exceptional case of no reflected longitudinal vibrations. This case occurs when the resultant of the shifts U_2, V_2, W_2 is parallel to a plane which is perpendicular to the line joining the centre of disturbance to the centre of the reflecting spherical surface, i.e. in the notation of our Art. 1403, we have

$$C' : C'' : C''' :: x_0 : y_0 : z_0.$$

[1406.] So far Clebsch has confined himself to a single centre of disturbance. In § 14 (S. 240-6) and § 15 (S. 247-50) he deals with the special problem of determining the system of centres of disturbance which if distributed over a spherical surface inside the reflecting sphere would produce the reflected motion (see our Art. 1400 and compare Art. 1312). A different system is necessary for the two types of vibration, and what is more the distribution of systems of disturbance is quite different for waves of different periods. The whole investigation, although the results are very complex, is of interest, especially when we compare it with similar investigations dealing with fluid motion in and about spheres by the method of images.

[1407.] § 16 (S. 250–4) is entitled : *Untersuchung des Falles, wo bei massig grosser Wellenlänge der Radius der reflectirenden Kugel sehr klein ist.* Clebsch commences by remarking that the case in which, ϵ being the radius of the reflecting sphere, $k\epsilon/a$ and $k\epsilon/b$ are very large,—a case of great importance in the application of elastical theory to optics,— does not admit of any great simplification in the fórmulae. On the other hand the case in which these quantities are small, or the radius of the reflecting sphere is small as compared with the wave-length, admits of great simplification. We can in this case for the incident motion replace for points in the neighbourhood of the reflecting sphere :

$$\frac{e^{-\frac{k}{a}R\sqrt{-1}}}{R} \text{ and } \frac{e^{-\frac{k}{b}R\sqrt{-1}}}{R} \text{ by } \frac{e^{-\frac{k}{a}r_0\sqrt{-1}}}{r_0} \text{ and } \frac{e^{-\frac{k}{b}r_0\sqrt{-1}}}{r_0} \text{ respectively.}$$

Starting from these Clebsch determines the principal terms in the various functions on which the values of u, v, w depend. He shows that it is only in the immediate neighbourhood of the reflecting sphere that its influence is of large magnitude, but that at greater distances, it is of the order of the radius of the sphere (ϵ): see his S. 254. He divides his investigation into two parts. The first occupies § 17 (S. 254–9) and deals with the reflected disturbance at points remote from the reflecting sphere but not necessarily from the disturbing centre. The approximate results are given on S. 256, but they are too long to be cited here. On the other hand they take simpler forms, when the disturbed point is at a great distance alike from the centre of disturbance and from the reflecting sphere. In this case the type of shift, *due to the reflected motion only,* is given by :

$$u = \frac{3k\epsilon}{(b^2 + 2a^2)\,rr_0} \left\{ \frac{b^2}{a} Cl \cos\phi \sin k\left(t - \frac{r}{a} - \frac{r_0}{a}\right) \right.$$

$$+ aC\left(l_0 - l\cos\phi\right)\sin k\left(t - \frac{r}{b} - \frac{r_0}{a}\right)$$

$$+ bGl \sin\phi \cos\chi \sin k\left(t - \frac{r}{a} - \frac{r_0}{b}\right)$$

$$+ \frac{a^2}{b} G\left(l_0' \sin\psi_0 - l\sin\phi\cos\chi\right)\sin k\left(t - \frac{r}{b} - \frac{r_0}{b}\right)\Big\} \ldots\ldots\ldots(\text{xvi}),$$

where :

$2\pi/k$ is the period of the disturbance,
C, C', C'', C''' determine its amplitude as in our Art. 1403,
$G = \sqrt{C'^2 + C''^2 + C'''^2}$,
$r =$ distance of disturbed point from centre of reflecting sphere,
$r_0 =$ distance of centre of disturbance from the centre of sphere,
l, m, n are the direction cosines of r, and l_0, m_0, n_0 of r_0,
ϕ is the angle between r and r_0,
χ is the angle between the planes rr_0 and $C'x + C''y + C'''z = 0$,
ψ_0 is the angle between the perpendicular p_0 to the latter plane and

r_0, while l_0', m_0', n_0' are the direction cosines of the perpendicular to p_0 and r_0.

Obviously it is only r, l, m, n, ϕ and χ which change with the position of the disturbed point.

Clebsch's results on S. 257 do not all agree with the above. He gives for his constants n and q the values $\dfrac{Kk\epsilon}{arr_0^2}$ and $\dfrac{k\epsilon}{brr_0^2}$ where I find in his notation the same values multiplied by the factor $\dfrac{3a^2}{b^2 + 2a^2}$.

On S. 257–8 Clebsch draws a number of conclusions as to the general character of the vibrations. These follow at once from the trigonometrical form in which we have displayed his results in (xvi). Indeed in that form, they are obvious on inspection.

[1408.] The second part of Clebsch's investigations deals with the disturbance at points very close to the reflecting sphere, when the centre of disturbance is supposed to be at a considerable distance. This occupies the final section § 18 (S. 259–62). In the notation of the preceding article Clebsch finds shifts *due to the reflected motion only*, of the type:

$$u = \frac{Ck}{ar_0} \left\{ \frac{\epsilon}{r} l_0 + \frac{a^2 - b^2}{2\,(b^2 + 2a^2)} \left(\frac{r^2}{\epsilon^2} - 1 \right) (l_0 - 3l \cos \phi) \right\} \sin k \left(t - \frac{r_0}{a} \right)$$

$$+ \frac{Gk}{br_0} \left\{ \frac{\epsilon}{r} l_0' \sin \psi_0 + \frac{a^2 - b^2}{2\,(b^2 + 2a^2)} \left(\frac{r^2}{\epsilon^2} - 1 \right) (l_0' \sin \psi_0 - 3l \sin \phi \cos \chi) \right\}$$

$$\times \sin k \left(t - \frac{r_0}{b} \right) \quad\ldots\ldots\ldots\ldots\text{(xvii)}.$$

Thus in the neighbourhood of the reflecting sphere we have only to deal with two waves, one of longitudinal and one of transverse vibrations.

Clebsch instead of discussing the motion as given by (xvii) adds to it the shift due to the direct action of the disturbance, i.e. the real part of

$$u = e^{kt\sqrt{-1}} \left\{ - Cl_0 \frac{d}{dr_0} \left(\frac{e^{-\frac{k}{a}r_0\sqrt{-1}}}{r_0} \right) + (C''n_0 - C'''m_0) \frac{d}{dr_0} \left(\frac{e^{-\frac{k}{b}r_0\sqrt{-1}}}{r_0} \right) \right\}$$

$$\ldots\ldots\ldots\text{(xviii)},$$

so far as terms of the order $1/r_0$ are concerned.

The only difference this makes is that we must read $\dfrac{\epsilon}{r} - 1$ for $\dfrac{\epsilon}{r}$ in the first term within each pair of curled brackets in equations of the type (xvii), so that u, v, w now vanish for $r - \epsilon$, as of course they ought to do.

[1409.] From the values of the shifts as expressed in the above manner Clebsch forms expressions for the amplitude of the longitudinal and transverse vibrations in the immediate neighbourhood of the reflecting sphere. He concludes :

(i) That the amplitude of the longitudinal vibrations will be greatest for points whose directions from the centre of the reflecting sphere are nearly perpendicular to the line joining that centre to the centre of disturbance.

(ii) That the amplitude of the transverse vibrations will be greatest for points whose directions from the centre of the reflecting sphere lie near the plane which passes through the line joining the centre of the sphere to the centre of disturbance (r_0) and the line p_0 perpendicular to the plane $C'x + C''y + C'''z = 0$ (termed by Clebsch the *Axe der Bewegung, Axe der einfallenden Schwingungen* and also *Axe der Schwingungen*).

(iii) That as the values of u, v, w, do not alter when x, y, z are changed to $-x, -y, -z$, respectively, the formulae to this approximation give no trace of a *shadow*.

It would be interesting to know, if this result be true for other than elastic media. We might easily place in the electro-magnetic field a non-conducting sphere the radius of which would be small as compared with the wave-length of a possible disturbance. Would such a sphere have a shadow ?

[1410.] Although Clebsch, as usual, seems more interested in his analytical processes than in their physical applications, and makes no attempt to deduce numerical results, there is still so much of physical suggestion in his memoir, that quite apart from the analytical merits, it will repay close study. Special applications to several simple physical problems appear to be placed by it within reach of ordinary calculation, while the contributions it offers to the theory of solid spherical harmonics are of wider physical value than is suggested by the title of the memoir.

[1411.] It seems well to consider in this chapter the memoir by Gehring to which we have had occasion to refer in our Arts. 1292–3 and 1375 and which is closely related to the researches of Kirchhoff and Clebsch. It was published at Berlin in 1860 as a dissertation for the doctorate and is entitled : *De aequationibus differentialibus quibus aequilibrium et motus laminae crystallinae definiuntur.* It contains 30 quarto pages, and is dedicated to Kirchhoff.

[1412.] The object of the memoir is stated in the following introductory paragraph :

In problemate aequilibrii et motus laminae elasticae tractando cl. Sophie Germain, Lagrange, Poisson conjecturas fecerunt, quas falsas esse cl. Kirchhoff in diarii Crelliani tomo xl demonstravit. Qui novam theoriam deduxit ex principiis, quae quidem non minus sunt hypothetica, quae tamen similia sunt iis, quibus cl. Jac. Bernoulli usus est ad baculi elastici aequilibrium et oscillationes definienda et quae theoriam praebuerunt satis experimentis congruentem. In recentiore commentatione (Diar. Crell. lvi) cl. Kirchhoff theorema proposuit generale, cujus ope fere omnia elasticitatis problemata accuratissime solvi possunt. Unde mihi liceat, nulla facta conjectura physica, solum mathematicis deliberationibus utens, aequationes deducere, quarum integratione aequilibrium et motus laminae elasticae crystallinae vel non crystallinae definiuntur sub ea conditione, ut tantum infinite paullum (*sic!*) ex aequilibrii statu lamina progrediatur. Quae theoria a me constituetur, in casu laminae non crystallinae easdem praebet aequationes ac inventas a cl. Kirchhoff et confirmat igitur conjecturas, quibus usus est. Sed non minus facile casus laminae crystallinae ea continetur (p. 5).

One or two remarks may be made on this. The historical reference is evidently based on the statement in Kirchhoff's memoir on plates : see our Art. 1234. But it is very inexact. Lagrange so far as he went made no false conjecture, and Poisson's work ought not to be placed on the same footing with that of Sophie Germain. The criticism of Kirchhoff's first hypothesis as hypothetical is just (see our Art. 1236), but the author can hardly mean that he has really deduced the plate equations *nulla facta conjectura physica, solum mathematicis delibe-rationibus utens !* That would indeed be a feat equally brilliant with the discovery of the whole theory of elasticity in Taylor's Theorem : see our Arts. 928*, 299–300. The general theorem of Kirchhoff's which is referred to is that of our Art. 1253. Finally by a "crystalline body" the author means one having 21 independent elastic constants ; the two things are, however, by no means necessarily identical.

[1413.] Pages 5–12 determine in a general manner the value of the shifts for a plate of isotropic material and correspond to our Arts. 1293–4. Gehring does not define what he means by a *lamina*, but his method shows that so far his results are only true for a plate of in-definitely small thickness. The reason for neglecting certain terms and retaining others is rather vaguely based (p. 10 of the memoir) on a reference to a similar neglect in Kirchhoff's memoir on thin rods: see our Art. 1258. So far Gehring's results would appear to be true for *finite* shifts, and they agree with those given by Clebsch on S. 270–1 of his *Treatise* or by Kirchhoff in his *Vorlesungen* : see our Art. 1294.

[1414.] Gehring next proceeds to find an expression for the elastic potential of the plate. This is still I think true, admitting Kirchhoff's assumptions, up to p. 14 for finite shifts, and the result on that page agrees with Kirchhoff's *Vorlesungen* S. 455 (see our Art. 1294). We have then (pp. 15–18) the equations for the longitudinal and transverse shifts of the plate supposing it to be only infinitely little displaced from the position of equilibrium. The final results ought to agree with equations (13) and (14) of Kirchhoff's S. 459 (see our Art. 1296), but they do not. To begin with, the value of f_2 given in equation 24 (b) is quite wrong, and the value for k given on p. 17 is likewise wrong. To bring Gehring's results into unison with Kirchhoff's, it is necessary to replace the coefficients $1 + \Theta$ in equations (28) and (29) by $\frac{1}{2}(1 + \Theta)$, but far more considerable modifications would have to be made in the steps by which these equations are reached. It is to be noted that Gehring works out fully only the equations of the shifts in the plane of the plate; for the transverse shift he cites Kirchhoff's results: see our Arts. 1298–9. Gehring's $d\xi/da$ is the $1 + du/dx$ of Kirchhoff, and his $d\eta/db$ the $1 + dv/dy$ of Kirchhoff.

[1415.] The second part of Gehring's paper occupies pp. 18–30 and deals with the equations for the longitudinal and transverse shifts of a thin plate whose elastic material has twenty-one constants. The results ought to be of importance, for few plates possess elastic isotropy, and for testing various physical theories it is often desirable to deal mathematically with material possessing considerable elastic complexity.

We have here to determine the value of the elastic potential subject to the relations (see our Art. 1294)

$$\widehat{zx} = \widehat{yz} = \widehat{zz} = 0,$$

z being the direction of the normal to the plate.

These relations enable us to express σ_{yz}, σ_{zx} and s_z as functions of the three remaining strains σ_{xy}, s_x and s_y and thus to express the elastic potential per unit volume as a function of σ_{xy}, s_x, s_y and the 21 constants. This is done by Gehring on pp. 18–21. His results thus far appear to be correct, but I think might be somewhat simplified. The next stage is to substitute the values of these strains (e.g. those of our Art. 1294) in the elastic potential and integrate it through the thickness of the plate. But Gehring makes errors in the value of all three of the expressions:

$$\int_{-c}^{+c} x_x^2 dz, \qquad \int_{-c}^{+c} x_y^2 dz, \qquad \int_{-c}^{+c} y_y^2 dz,$$

which he gives towards the bottom of p. 21 (2c is here the thickness of the plate, and x_x, y_y, x_y correspond in our notation to s_x, s_y and σ_{xy}). The denominator 3 in the last terms of the values of these expressions ought not to be there. The terms thus wrong involve only the *first power of the thickness*, and therefore their error ought only to affect the equations for the shifts in the plane of the plate.

Gehring gives the equations for the transverse vibrations (one equation for the shift at any point of the mid-plane and the boundary-conditions) as (43) and (44) on p. 28, and the equations for the co-planar vibrations of the plate (two for the shifts at any point of the mid-plane in the plane of the plate and the two boundary-conditions) as (46), (47) and (48) on p. 29. He then remarks that these equations agree with those for an isotropic plate if we make the proper assumptions as to the relations of the 21 elastic constants which he gives. *But it will be found that this is not the fact for the last set of equations* (46–48), or Gehring's results for the longitudinal vibrations are erroneous. The first set of equations (43–44) do give the correct results for the case of isotropy, or *pro tanto* we have confirmation of the correctness of Gehring's equations for the transverse vibrations of a 21-constant plate.

In conclusion Gehring remarks of these equations :

Integratio aequationum (43) et sequentium tam difficilis videtur esse, ut in hodierno scientiae analyticae statu fieri non possit (p. 30).

Nor is the difficulty confined only to the 21-constants, even the equations for a thin plate of isotropic material had up to that time only been solved for the special case of a circular boundary.

[1416.] *Summary.* The three German elasticians with whose researches we have dealt in this chapter mark a very great advance in the mathematical treatment of elastic problems. Franz Neumann stands, however, on a somewhat different footing from Kirchhoff and Clebsch. His style is clearer and he keeps more in mind the physical bearings of his analysis. He possesses much originality and in his investigations on photo-elasticity and the elasticity of crystals he breaks almost untrodden ground, which both physicists and mathematicians have hardly yet exhausted. Clebsch, while by far the greatest analyst of the three, puts physics (and the technologists for whom he is professedly writing) in the background. Stimulated by Saint-Venant's work, he has not Saint-Venant's practical experience, and in simplifying the latter's results for prisms and in extending his processes to plates, he is guided rather by love of the analytical processes involved than by their practical applications. No mathematician, however, can read Clebsch's *Treatise* without recognising the suggestive character of its analysis, and appreciating the mental power of its author. Kirchhoff's researches in the field of elasticity, like Lamé's, suffer to some extent from being out of touch with physical experience,

This is markedly the case in those contributions to our subject, which border on electro-magnetism and optics. But Kirchhoff's treatment of both the rod and plate problems, if it cannot be said to have been final, still advanced those problems a long stage. Future discussions will probably serve to define better the limits within which Kirchhoff's assumptions are legitimate, and will possibly add further terms, of minor importance except in special cases, to his expressions for the strain-energy; they will hardly, however, displace Kirchhoff's investigations as the latter have done Poisson's and Cauchy's. In our chapter on Boussinesq we shall endeavour to give some general comparison of the French and German methods of dealing with rod and plate problems.

CHAPTER XIII.

BOUSSINESQ.

SECTION I.

Memoirs dealing directly with Elasticity and Molecular Action.

[1417.] ONE of the most distinguished of the pupils of Saint-Venant is M. J. Boussinesq, member of the *Institut,* and at present Professor of the Faculty of Science, Paris. A *Notice sur les travaux scientifiques de M. J. Boussinesq (Notice* I.) was published at Lille in 1880, when Boussinesq was a candidate for membership of the *Institut,* and an *Extrait de la Notice sur les titres et travaux scientifiques de M. J. Boussinesq...et supplément à cette Notice pour les travaux publiés depuis cette époque (Notice* II.) was published in 1885, also at Lille, when Boussinesq was again a candidate. In 1880 Saint-Venant made an *Analyse succincte des travaux de M. Boussinesq, professeur à la Faculté des sciences de Lille,* which appeared in a lithographed form. The *Notices* I. and II. as well as the *Analyse succincte* form a very useful bibliographical guide to Boussinesq's researches prior to 1885, but my *résumé* and criticism of his work in the present chapter are based on the perusal of the memoirs themselves. Boussinesq's investigations extend far beyond elasticity, dealing in particular with light, heat, hydro-dynamics and the philosophical basis of the fundamental principles of dynamical science.

[1418.] *Étude nouvelle sur l'équilibre et le mouvement des corps solides élastiques dont certaines dimensions sont très-petites par rapport à d'autres.* Journal de mathématiques, 2ᵉ Série, T. XVI., pp. 125–274. Paris, 1871. This, the *Premier Mémoire* with the above title and with the sub-title: *Des tiges*, was presented to the *Académie* April 3, 1871, and analysed in the *Comptes rendus*, T. LXXII., pp. 407–10.

Let x be the direction of the tangent at any point to the central line of a bar, beam or rod, and let y, z, two lines at right-angles, be taken in the plane of the cross-section. Then Saint-Venant in 1853–6 (see our Arts. 1 and 69) had obtained two solutions of the general equations of elasticity on the assumptions that $\widehat{yy} = \widehat{zz} = \widehat{yz} = 0$, and that the central line is initially straight. These solutions were shewn on the principle of the elastic equipollence of statically equivalent load systems (see our Arts. 8, 21, 100) to correspond to the torsion of a prism about its axis and to the flexure of a prism either under an isolated central load or as a terminally loaded cantilever. In obtaining these solutions Saint-Venant had not supposed elastic isotropy, but merely that the elasticity was the same in all planes perpendicular to the central axis. He applied his results to a great variety of cross-sections, and shewed that they did not justify the earlier hypotheses of Cauchy and Poisson: see our Arts. 29 and 75. With regard to Saint-Venant's solutions there is an important distinction between that for the case of torsion and that for the case of flexure. In the former case the shears $\widehat{xy}$ and $\widehat{xz}$ are fundamental, and their values must be ascertained in order to calculate the torsional resistance of a rod, however small the dimensions of its cross-section as compared with its length. In the latter case the shears $\widehat{xy}$ and $\widehat{xz}$ are shewn to be practically negligible whenever the dimensions of the cross-section are small, i.e. in the case of what is really a *rod*, and the discovery of their values is only needed as a step towards shewing that they are negligible, and so justifying the Bernoulli-Eulerian theory.

Clebsch in his *Treatise* (see our Art. 1332) had sought the most general solution of the equations of elasticity subject to the conditions $\widehat{yy} = \widehat{zz} = \widehat{yz} = 0$, and had thus reached a solution of those equations embracing both the flexure and torsion problems of Saint-Venant. But as in all Clebsch's work this result was only obtained for the case of bi-constant isotropy. He dealt also with the case of rods of double curvature. Kirchhoff in a memoir of 1858 (see our Art. 1251) had endeavoured to give a complete theory of strain in thin rods with an initially curved central line. But a defect of Kirchhoff's theory has been pointed out by Saint-Venant (see our Art. 316), and the objections against it are again raised by Boussinesq in the present memoir (pp. 127–9 and § VII. pp. 176–81). In the equations (vii) of our Art. 1257, Kirchhoff neglects the terms du/ds, dv/ds and dw/ds. Now Boussinesq points out (p. 179) that this assumption has no à priori justification, but that the terms neglected appear to be of the same order as those retained. He cites cases in which the assumption would not be true, but remarks that it is satisfied in general when Saint-Venant's hypothesis, $\widehat{yy} = \widehat{zz} = \widehat{yz} = 0$, is fulfilled. Thus it is satisfied in Saint-Venant's cases of flexure and torsion but not when there is an appreciable longitudinal or buckling load (see our Arts. 911* and 1361):

En résumé, la théorie de M. Kirchhoff conduit, dans le cas de tiges dont la contexture est symétrique par rapport à leurs sections normales, aux vraies formules approchées de la flexion et de la torsion; mais elle me paraît reposer sur une hypothèse douteuse à priori, consistant à admettre que les sections normales, primitivement égales entre elles, sont encore, sur une longueur finie, égales après les déplacements. Elle a aussi l'inconvénient de laisser parmi les quantités qu'elle néglige comme trop petites, les actions tangentielles exercées, dans le cas de la flexion inégale, à travers les divers éléments plans d'une de ces sections, forces qu'il est cependant intéressant d'étudier, puisque leur résultante est égale et contraire à celle des actions extérieures qui produisent la flexion (pp. 128–9).

[1419.] Boussinesq in the present memoir endeavours to amplify the labours of previous investigators by a discussion of the following topics:

(a) He seeks to demonstrate that Saint-Venant's assumption ($\widehat{yy} = \widehat{zz} = \widehat{yz} = 0$) is legitimate and necessary for thin rods. This assumption amounts to saying that the mutual action of the fibres at a finite distance from their extremities is invariably directed along their tangents. The demonstration (and the resulting equations for a thin rod) Boussinesq considers the fundamental part of his memoir:

Je les expose pour le cas général où des actions quelconques seraient appliquées, non-seulement près des extrémités, mais encore sur la masse entière de la tige, et où celle-ci serait hétérogène, mais de contexture symétrique par rapport à ses sections normales, et formée de fibres qui, isolées, subiraient les mêmes déformations latérales si on les soumettait à de simples tensions, produisant sur toutes la même dilatation longitudinale (p. 127).

(b) The general equations for the strain of a thin rod are given. These correspond closely to Clebsch's results for Saint-Venant's problem dealt with, however, on the supposition that the elasticity is not isotropic but the same for each cross-section: see our Arts. 1332 and 1360.

(c) Boussinesq points out a certain analogy between hydrodynamics and the torsion of prisms. Another hydrodynamic analogy had been previously noticed by Thomson and Tait in their *Treatise on Natural Philosophy*, Art. 705. Oxford, 1867.

(d) He discusses (from the general elastic equations however) a problem already dealt with more fully by Seebeck, namely, the influence of rigidity on the transverse vibrations of a string: see our Arts. 471–2.

We will now consider these points in some detail.

[1420.] After the introduction, which deals with the historical aspect of the problem, Boussinesq passes in §§ I. and II. (pp. 130–44) to a general discussion of the equations of elasticity and the expressions of the stresses in terms of the strains for various types of elastic media.

In the first section (pp. 132–5) Boussinesq gives a proof of the relations of compatibility of the types:

$$\left.\begin{aligned} \frac{d^2 s_x}{dy\,dz} &= \frac{1}{2}\frac{d}{dx}\left(\frac{d\sigma_{zx}}{dy}+\frac{d\sigma_{xy}}{dz}-\frac{d\sigma_{yz}}{dx}\right), \\ \frac{d^2\sigma_{yz}}{dy\,dz} &= \frac{d^2 s_y}{dz^2}+\frac{d^2 s_z}{dy^2} \end{aligned}\right\} \quad \dots\dots\dots (i),$$

first stated by Saint-Venant: see our Arts. 112 and 190.

In the second section two special cases of distribution of elastic homogeneity are considered, (a) when the medium is symmetrical about the plane yz: see our Art. 78; (b) when the medium is isotropic round the axis of x. In the latter case we may write:

$$\left.\begin{aligned} \widehat{xx} &= \lambda'\theta + 2\mu' s_x, & \widehat{yz} &= \mu\sigma_{yz} \\ \widehat{yy} &= \lambda\theta + \nu s_x + 2\mu s_y, & \widehat{zx} &= \mu''\sigma_{zx}, \\ \widehat{zz} &= \lambda\theta + \nu s_x + 2\mu s_z, & \widehat{xy} &= \mu''\sigma_{xy} \end{aligned}\right\} \quad \dots\dots\dots\dots(ii).$$

Boussinesq by aid of these equations expresses the strains in terms of the stresses (p. 140). He further shews that if W be the strain energy per unit volume:

$$\left.\begin{aligned} \widehat{xx} &= \frac{dW}{ds_x}, & \widehat{yz} &= \frac{dW}{d\sigma_{yz}}, \\ s_x &= \frac{dW}{d\widehat{xx}}, & \sigma_{yz} &= \frac{dW}{d\widehat{yz}} \end{aligned}\right\} \quad \dots\dots\dots\dots\dots (iii).$$

[1421.] § III. (pp. 144–50) is entitled : *Étude d'une tige de très-petite section. Considérations préliminaires.* Boussinesq takes for his elastic body one which is sensibly cylindrical for a length comparable with its transverse dimensions, the total length being much greater than these latter dimensions. He supposes the total shifts to be as consider-able as one pleases, but the strains at each point to be small. *u, v, w* represent the *small* shifts relative to some chosen point of any point of the small element of the *rod* bounded by two adjacent cross-sections. The plane *yz* is taken parallel to the unstrained position *relative to the element* of some cross-section of the element. All this is in practical agreement with Kirchhoff's treatment : see our Art. 1257. Boussinesq considers the cross-section (ω) to have any number of cavities and that the contour of the cross-section (*s*) may thus consist of several closed curves. Further the constitution of the material of the rod is supposed to vary from one point to another, very gradually along the axis of *x*, but rapidly and even abruptly if desired along certain lines in the plane of the cross-section. No load is applied to the curved surface of the rod, but only to the terminal cross-sections.

Boussinesq then proves the following identity, *U, V, W* being any functions of *x, y, z*, which are continuous over ω, and the same statement holding for their first derivatives, except along lines at which the material abruptly changes its constitution :

$$\int^\omega \left\{ \widehat{xy}\frac{dU}{dy} + \widehat{zx}\frac{dU}{dz} + \widehat{yy}\frac{dV}{dy} + \widehat{zz}\frac{dW}{dz} + \widehat{yz}\left(\frac{dV}{dz} + \frac{dW}{dy}\right) \right\} d\omega$$
$$= \int^\omega \left\{ U\left(\frac{d\widehat{xx}}{dx} + \rho X\right) + V\left(\frac{d\widehat{xy}}{dx} + \rho Y\right) + W\left(\frac{d\widehat{zx}}{dx} + \rho Z\right) \right\} d\omega \dots \text{(iv)},$$

the integrations extending all over the cross-section ω, and *X, Y, Z* being the body-forces. This result easily flows from multiplying the body stress-equations by *U, V, W* and integrating by parts over the area of the cross-section the sum of the results so obtained.

[1422.] We now come to the fundamental part of Boussinesq's argument (pp. 148–53). I must confess that it by no means carries conviction to my mind. Boussinesq aims at demonstrating that Saint-Venant's assumption :

$$\widehat{yy} = \widehat{zz} = \widehat{yz} = 0 \dots\dots\dots\dots\dots\dots (\alpha)$$

is practically true, or that these stresses are negligible as compared with the remaining three when no load is applied to the surface of the rod except near its extremities. The assumption (α) may possibly be incorrect for the parts of the rod very near the extremities.

Boussinesq's argument seems to be of the following kind :

Considering only portions of the cross-section where the elastic constitution of the material of the rod is continuous, it is natural to suppose the stresses here are also continuous. But where the axis of y meets the surface $\widehat{yy} = \widehat{zz} = \widehat{yz} = 0$, hence by Maclaurin's Theorem (Boussinesq does not appeal to this theorem, but I think there is an implicit assumption of it) we must have results of the type:

$$\widehat{yy} = (y - y') \left(\frac{d\widehat{yy}}{dy}\right)_0 + z \left(\frac{d\widehat{yy}}{dz}\right)_0 + \text{ terms involving the square of the}$$
linear dimensions of the cross-section.

Here y' is the distance of the origin of coordinates from the point at which the axis of y cuts the contour of the cross-section. Similar results will hold for all the other stresses $\widehat{zz}, \widehat{yz}, \widehat{xy}, \widehat{zx}$, which vanish at certain points of the contour. But to quote Boussinesq's words:

Donc, la section ω ayant toutes ses dimensions très-petites, les forces $\widehat{yy}, \widehat{zz}, \widehat{yz}, \widehat{zx}, \widehat{xy}$ ne peuvent qu'être fort petites dans toute son étendue par rapport aux valeurs absolues moyennes de leurs dérivées premières en y et z (p. 148).

It seems to me that this argument fails because it does not state what are the quantities relative to which the y and z of the cross-section are small; y and z cannot be *absolutely small*. In other words exactly the same objections apply to Boussinesq's theory as to Cauchy's, Poisson's and Neumann's expansions of the stresses in terms of the coordinates of a point in the plane of the cross-section: see our Arts. 466*, 618*, 29, 75 and 1225–6. Boussinesq continues:

D'ailleurs la continuité, sur une longueur finie de la tige, des $\widehat{xx}, \ldots \widehat{yz}, \ldots$ et de leurs dérivées en y, z, exige que les dérivées en x de toutes ces quantités ne soient pas d'un ordre de grandeur plus élevé que l'ordre de ces quantités elles-mêmes, si ce n'est toutefois aux points voisins des extrémités de la tige, ou plus généralement, de ceux où la constitution de la matière et les conditions dans lesquelles elle se trouve varieraient brusquement dans le sens des x. Si l'on fait abstraction de ces points tout particuliers, les deux dernières équations (ϵ)[1] pourront être réduites à

$$\frac{d\widehat{yy}}{dy} + \frac{d\widehat{yz}}{dz} = 0, \quad \frac{d\widehat{yz}}{dy} + \frac{d\widehat{zz}}{dz} = 0 \quad \ldots\ldots\ldots\ldots\ldots(\text{v}).$$

[1] This symbol refers to the body-stress equations: see for example our Art. 1517*.

En effet, si nous considérons, par exemple, la seconde des équations (ϵ), les termes $d\widehat{yy}/dy$, $d\widehat{yz}/dz$ pourront être, soit de l'ordre de $d\widehat{xy}/dx$, soit incomparablement plus petits, soit incomparablement plus grands. Dans les deux premiers cas, $\widehat{yy}$ et $\widehat{yz}$ seront, d'après ce qui précède, négligeables par rapport à $\widehat{xy}$, et l'on pourra poser en comparaison $\widehat{yy} = 0$, $\widehat{yz} = 0$; dans le troisième cas, la seconde équation (ϵ) se réduira sensiblement aux deux termes $d\widehat{yy}/dy$, $d\widehat{yz}/dz$, car le dernier ρY n'est jamais que de l'ordre de $d\widehat{xy}/dx$. Donc on pourra poser toujours $d\widehat{yy}/dy + d\widehat{yz}/dz = 0$ (pp. 148–9).

The argument here is that $\widehat{yy}$ and $d\widehat{yy}/dy$ are of very different orders of small quantities, while $\widehat{xy}$ and $d\widehat{xy}/dx$ are of the same order. I do not see what step in the reasoning hinders $\widehat{yy}$ from being a function of the form $c \sin y$, say, which vanishes for $y = \pm \pi$, the units of the linear dimension of the cross-section being taken as small as we please. In this case $\widehat{yy}$ and $d\widehat{yy}/dy$ do not seem to be of a totally different order, and it would therefore appear that Boussinesq's argument is not sufficient.

[1423.] Assuming Boussinesq's conclusions as to the order of quantities, it follows that when the elastic distribution is symmetrical with regard to the plane of yz, the second fluxions with regard to x of the slides σ_{yz}, σ_{zx}, σ_{xy} and the stretches s_y, s_z will be negligible as compared with their second fluxions with regard to y and z. This follows at once from the expressions for the strains in terms of the stretches, if we remember the above relations between the order of the fluxions of the stresses. Hence from the relations of type (i) we have:

$$\frac{d^2 s_x}{dy\,dz} = \frac{d^2 s_x}{dy^2} = \frac{d^2 s_x}{dz^2} = 0 \quad\dots\dots\dots\dots\dots\text{(vi)},$$

or, if χ_1, χ_2, χ_3 be arbitrary functions of x:

$$s_x = \chi_1 + \chi_2 z + \chi_3 y \quad\dots\dots\dots\dots\dots\text{(vii)}.$$

Putting $U = 0$, $V = v$, $W = w$ in equation (iv), Boussinesq obtains (p. 149) by aid of (v):

$$\int^\omega (\widehat{yy}\, s_y + \widehat{zz}\, s_z + \widehat{yz}\, \sigma_{yz})\, d\omega = 0 \quad\dots\dots\dots\dots\text{(viii)}.$$

Further by putting:

$$U = 0, \qquad V = C_3\left(\chi_1 z + \tfrac{1}{2}\chi_2 z^2\right) + C_1\left(\chi_1 y + \chi_2 zy + \tfrac{1}{2}\chi_3 y^2\right) - \tfrac{1}{2} C_2 \chi_3 z^2,$$
$$W = \tfrac{1}{2} C_3 \chi_3 y^2 + C_2\left(\chi_1 z + \tfrac{1}{2}\chi_2 z^2 + \chi_3 zy\right) - \tfrac{1}{2} C_1 \chi_2 y^2,$$

we find:
$$\int^\omega (C_1 \widehat{yy} + C_2 \widehat{zz} + C_3 \widehat{yz})\, s_x d\omega = 0 \quad\dots\dots\dots\text{(ix)},$$

where C_1, C_2, C_3 are any constants whatever.

By substituting for the strains in (viii), expressing the integral as the sum with positive coefficients of the integrals of the squares of four expressions linear in the stresses $\widehat{yy}$, $\widehat{zz}$ and $\widehat{yz}$, a result obtained by aid of (ix), Boussinesq deduces that for a thin rod loaded only at the terminals:

$$\widehat{yy} = \widehat{zz} = \widehat{yz} = 0 \dots\dots\dots\dots (x),$$

or, Saint-Venant's assumption: see our Art. 1422.

[1424.] Boussinesq in a later memoir has again returned to this point as to the relative magnitude of the fluxions of the stresses: see our Art. 1433. The supposition he makes in that memoir, namely: that the variation of the stresses parallel to the axis of a rod or to the mid-plane of a plate is very small as compared with their variation in the plane of the cross-section of the rod or perpendicular to the mid-plane of the plate, does not seem to me established by the arguments used:

Si donc on fait abstraction de ces régions restreintes, l'équilibre d'un tronçon[1] quelconque à fort peu près prismatique présentera cette circonstance, que les composantes...des pressions et les déformations...y seront sensiblement les mêmes, soit tout le long d'une même *fibre longitudinale* perpendiculaire aux bases du prisme, s'il s'agit d'une tige, soit sur toute l'étendue d'une *couche* quelconque parallèle aux bases du prisme, s'il s'agit d'une plaque. Au contraire, les mêmes pressions et déformations varieront en général d'une manière très-notable dans les sens des dimensions transversales d'une tige ou dans celui de l'épaisseur d'une plaque. Il est d'ailleurs évident que les actions extérieures directement appliquées à la masse du tronçon (y compris l'inertie dans le cas d'un équilibre dynamique), et celles qui le sont à la portion de la superficie du corps qui fait partie de la surface du tronçon, n'ont qu'une influence minime sur les forces $\widehat{xx}$, ... $\widehat{yz}$, ... toutes ces actions n'étant presque rien en comparaison de celles qui agissent sur le reste du corps et dont l'ensemble donne lieu aux réactions intérieures $\widehat{xx}$, ... $\widehat{yz}$, ... (p. 164 of the memoir cited in our Art. 1433).

For the case of a rod this supposition leads to

$$\frac{d}{dx}(\widehat{xx}, \widehat{yy}, \widehat{zz}, \widehat{yz}, \widehat{zx}, \widehat{xy}) = 0 \dots\dots\dots\dots(A).$$

But Boussinesq shews that the narrower assumptions

$$\frac{d}{dx}(\widehat{xy}, \widehat{zx}) = 0, \quad \frac{d^2}{dx^2}(\widehat{yy}, \widehat{zz}, \widehat{yz}) = 0 \dots\dots\dots(B)$$

are sufficient to lead to the same solution as that which we are

[1] A small prismatic element of the rod bounded by two adjacent cross-sections, or of the plate bounded by the faces and two pairs of planes at right angles perpendicular to the plane face of the plate.

discussing from the first memoir. In fact they lead us to the results (*b*) of our Art. 317, and these last results[1] give us with some easy analysis (pp. 166–172 of the second memoir) the results (vii) and (x) of the last article.

Saint-Venant has reproduced Boussinesq's argument, and in our Art. 318 we have already cited his version of it, expressing at the same time our doubts as to its sufficiency.

[1425.] Changing the notation of Art. 317 to that of our present discussion the last two conditions of (*b*) become :

$$\frac{d\sigma_{zz}}{dx} = 0, \quad \frac{d\sigma_{xy}}{dx} = 0,$$

or,
$$\frac{ds_x}{dz} = -\frac{d^2w}{dx^2}, \quad \frac{ds_x}{dy} = -\frac{d^2v}{dx^2}.$$

Hence by (vii) :

$$\chi_2 = -\frac{d^2w}{dx^2}, \quad \chi_3 = -\frac{d^2v}{dx^2} \quad \dots\dots\dots\dots(xi),$$

or since χ_2 and χ_3 are independent of y and z, the second fluxions with regard to x of w and v may be supposed to be taken at the point $y = z = 0$. In the case where the curvature is small, we see that χ_2 and $-\chi_3$ represent the changes in curvature of the central line in the planes zx and xy respectively. Thus :

$$\chi_2 = \frac{1}{R_y} - \frac{1}{R_y^0}, \quad -\chi_3 = \frac{1}{R_z} - \frac{1}{R_z^0} \quad \dots\dots\dots\dots (xi)'$$

(see p. 185 of the memoir of 1879), where R_y, R_z are the radii of curvature in the planes zx, xy, after strain, and R_y^0, R_z^0 those before strain.

Further χ_1 is evidently the stretch of the central line of the rod, or s_x^0, say. Hence we have obtained a physical interpretation of the as yet undetermined functions in (vii). Considering the portion of the rod on one side of any cross-section ω, let the moments of the applied load and the body-forces on this portion round the axes[2] of x, y and z be respectively M_x, M_y, M_z, then since (x) holds we have :

$$\left. \begin{array}{l} \displaystyle\int^\omega \widehat{xx}z\,d\omega = M_y, \\[2ex] \displaystyle-\int^\omega \widehat{xx}y\,d\omega = M_z, \\[2ex] \displaystyle\int^\omega (y\widehat{xz} - z\widehat{xy})\,d\omega = M_x \end{array} \right\} \quad \dots\dots\dots\dots (xii).$$

[1] In the present investigation it is x, in the investigation of Art. 317 it is z, which is the prismatic axis.

[2] The axes are supposed to be taken so that a right-handed screw-motion in the positive direction of x turns y towards z, and so with cyclic interchange for each axis.

The conditions (x) lead us as in our Art. 78 at once to

$$\widehat{xx} = E s_x, \qquad s_y = -\eta_1 s_x, \qquad s_z = -\eta_2 s_x,$$

whence from (vii) we find :

$$M_y = \mathfrak{E}\omega\kappa^2{}_y \chi_2, \qquad M_z = -\mathfrak{E}\omega\kappa^2{}_z \chi_3,$$

and if F be the component parallel to the axis of x of the whole system of load and body-forces acting on one side of ω :

$$F = \int^{\omega} \widehat{xx}\, d\omega = \mathfrak{E}\omega\chi_1.$$

Hence : $\chi_1 = F/\mathfrak{E}\omega, \quad \chi_2 = M_y/\mathfrak{E}\omega\kappa^2{}_y, \quad \chi_3 = -M_z/\mathfrak{E}\omega\kappa^2{}_z \ldots\ldots$ (xiii).

These determine the value of s_x and give in fact the elements of the solution of the problem of the *thin* rod so far as they are due to extension and flexure. Here Boussinesq has followed Bresse (see our Art. 515) and treated the cross-section as having a density equal to the variable stretch-modulus. Thus the centroid is found from the conditions :

$$\int^{\omega} Ey\, d\omega = \int^{\omega} Ez\, d\omega = 0,$$

while $\mathfrak{E}\omega = \int^{\omega} E\, d\omega, \quad \mathfrak{E}\omega\kappa_y{}^2 = \int^{\omega} Ez^2 d\omega, \quad \mathfrak{E}\omega\kappa_z{}^2 = \int^{\omega} Ey^2 d\omega,$

define $\mathfrak{E}$, κ_y and κ_z.

[1426.] If we seek v and w from equations (8) of our Art. 78, we determine them to be of the following form :

$$\left.\begin{aligned}
v &= \chi_5 - \chi_4 z + \tfrac{1}{2}\{\epsilon\chi_1 z + (\eta_2\chi_3 + \epsilon\chi_2)\, z^2\} - \eta_1\left(\chi_1 y + \chi_2 yz + \tfrac{1}{2}\chi_3 y^2\right), \\
w &= \chi_6 + \chi_4 y + \tfrac{1}{2}\{\epsilon\chi_1 y + (\eta_1\chi_2 + \epsilon\chi_3)\, y^2\} - \eta_2\left(\chi_1 z + \chi_3 yz + \tfrac{1}{2}\chi_2 z^2\right)
\end{aligned}\right\} \ldots \text{(xiv)},$$

where χ_4, χ_5, χ_6 are undetermined functions of x only.

Now the equations which still remain to be satisfied are the first body-stress equation, or

$$\left.\begin{aligned}
&\frac{d\widehat{xy}}{dy} + \frac{d\widehat{zx}}{dz} + \rho X + \frac{d}{dx}\{E\,(\chi_1 + \chi_2 z + \chi_3 y)\} = 0, \\
&\text{and the equation} \\
&\widehat{xy}\, dz - \widehat{zx}\, dy = 0, \text{ over the contour of the section}
\end{aligned}\right\} \ldots\ldots \text{(xv)}.$$

If the values of σ_{zx}, σ_{xy} be calculated in terms of the fluxions of u and of the values of v and w given in (xiv), and then $\widehat{xy}$, $\widehat{zx}$ be determined from their values in (7) of our Art. 78, we find a partial differential equation for u involving only u and χ_4, χ_5, χ_6, simple functions of x, as unknowns, together with a surface condition involving the same quantities. Now these equations will not suffice to determine the four unknowns, but Boussinesq on pp. 160–1 shews that they completely determine the values of $\widehat{xy}$ and $\widehat{zx}$.

[1427.] In the special case where the elasticity and density are everywhere uniform and the body-force X is constant over the cross-section, Boussinesq works out completely the equations to determine $\widehat{xy}$ and $\widehat{zx}$.

The load being applied only at definite points of the rod, F of equation (xiii) is only a function of x in so far as it involves the body-force X, and thus

$$dF/dx = -\rho X\omega,$$

or

$$d\chi_1/dx = -\rho X/E.$$

Further if S_y and S_z be the total shearing loads on ω parallel to the axes of y and z:

$$S_y = -dM_z/dx \text{ and } S_z = dM_y/dx \text{ (see our Art. 1361, ftn.)}.$$

Hence: $d\chi_2/dx = S_z/E\omega\kappa_y^2,$ $d\chi_3/dx = S_y/E\omega\kappa_z^2 \ldots\ldots$ (xvi).

Thus equation (xv) becomes

$$\frac{d\widehat{xy}}{dy} + \frac{d\widehat{zx}}{dz} + \frac{S_z z}{\omega\kappa_y^2} + \frac{S_y y}{\omega\kappa_z^2} = 0,$$

or

$$\frac{d}{dy}\left(\widehat{xy} + \frac{S_y y^2}{2\omega\kappa_z^2}\right) + \frac{d}{dz}\left(\widehat{zx} + \frac{S_z z^2}{2\omega\kappa_y^2}\right) = 0,$$

whence we can take, if ϕ be an arbitrary function of y and z:

$$\widehat{xy} = \frac{d\phi}{dz} - \frac{S_y y^2}{2\omega\kappa_z^2}, \qquad \widehat{zx} = -\frac{d\phi}{dy} - \frac{S_z z^2}{2\omega\kappa_y^2}\ldots\ldots\ldots\ldots (xvii).$$

Turning to equation (7) of our Art. 78, we find

$$\sigma_{xy} = \frac{e\widehat{xy} - h'''\widehat{zx}}{ef - h''h'''},$$

$$\sigma_{zx} = \frac{f\widehat{zx} - h''\widehat{xy}}{ef - h''h'''}.$$

But $$\frac{d}{dz}\left(\sigma_{xy} - \frac{dv}{dx}\right) = \frac{d}{dy}\left(\sigma_{zx} - \frac{dw}{dx}\right).$$

Hence by aid of (xvi) and (xvii) we find:

$$f\frac{d^2\phi}{dy^2} + (h'' + h''')\frac{d^2\phi}{dy\,dz} + e\frac{d^2\phi}{dz^2} + 2\left(ef - h''h'''\right)\frac{d\chi_4}{dx}$$

$$+ z\left\{\frac{h'''S_z}{\omega\kappa_y^2} - (ef - h''h''')\frac{2\eta_2}{E}\frac{S_y}{\omega\kappa_z^2} - (ef - h''h''')\frac{\epsilon}{E}\frac{S_z}{\omega\kappa_y^2}\right\}$$

$$- y\left\{\frac{h''S_y}{\omega\kappa_z^2} - (ef - h''h''')\frac{2\eta_1}{E}\frac{S_z}{\omega\kappa_y^2} - (ef - h''h''')\frac{\epsilon}{E}\frac{S_y}{\omega\kappa_z^2}\right\} = 0. \quad (xviii).$$

This result is in agreement with Boussinesq's (44), p. 162, except that he uses thlipsinomic while we are using tasinomic constants: see Art. 445.

The second equation of (xv) gives us for the contour condition :

$$d\phi + \frac{S_z z^2}{2\omega\kappa_y^2} dy - \frac{S_y y^2}{2\omega\kappa_z^2} dz = 0 \quad \dots\dots \dots\dots \text{(xix)}.$$

Finally from (xii) we have

$$M_x = -\int^\omega \left(\frac{d\phi}{dy} y + \frac{d\phi}{dz} z + \frac{S_z yz^2}{2\omega\kappa_y^2} - \frac{S_y zy^2}{2\omega\kappa_z^2} \right) d\omega \dots\dots \text{(xx)}.$$

On pp. 162–5 Boussinesq considers the case of a section containing one or more holes. The problem here involves the usual modifications due to cyclosis in dealing with the function ϕ.

It will be noted that (xviii) and (xix) above are more general than Saint-Venant's results given in our Art. 82 as equation (19).

It will be remembered that Saint-Venant finds for his flexural moment M, $d^2M/dx^2 = 0$: see our Art. 80. This follows at once from the second and third body-stress equations which give, since Saint-Venant supposes no body-forces : $d\widehat{xy}/dx = 0$, $d\widehat{zx}/dx = 0$: see our Art. 79, equation (11). Boussinesq neglects the terms

$$d\widehat{xy}/dx + \rho Y \text{ and } d\widehat{zx}/dx + \rho Z,$$

in his second and third body-stress equations, for he says ρY, ρZ are of the same order at most as $d\widehat{xy}/dx$ and $d\widehat{zx}/dx$, and these he holds to be negligible as compared with terms like $d\widehat{yy}/dy + d\widehat{yz}/dz$: see his p. 149 and our Art. 1422. Now his analysis leads to $\widehat{yy} = \widehat{zz} = \widehat{yz} = 0$. Hence I find it difficult to understand how $d\widehat{xy}/dx + \rho Y$ can be small as compared with $d\widehat{yy}/dy + d\widehat{yz}/dz$ unless we have absolutely :

$$\frac{d\widehat{xy}}{dx} + \rho Y = 0, \qquad \frac{d\widehat{zx}}{dx} + \rho Z = 0 \dots\dots\dots\dots \text{(xxi)}.$$

If we take the exact assumptions of the second memoir (Art. 1424)

$$d\widehat{xy}/dx = d\widehat{zx}/dx = 0,$$

then $Y = Z = 0$, and Boussinesq's apparently more general solution leads us again to Saint-Venant's, involving $d^2M/dx^2 = 0$.

But if Y and Z be not zero then it is impossible to put

$$\widehat{yy} = \widehat{zz} = \widehat{yz} = 0,$$

for these quantities can (for example at certain points of a heavy beam other than those of external loading) be infinitely greater than $\widehat{xy}$ or $\widehat{zx}$: see a paper by the Editor : *On the Flexure of Heavy Beams, Quarterly Journal of Mathematics*, Vol. XXIV., p. 106. Thus so far as Boussinesq's theory is more general than Saint-Venant's, in that it appears to allow of body-forces, I doubt its accuracy. Let us make the additional assumption of the second memoir that such body-forces have only a vanishingly small influence on the stresses (p. 164 of the second

memoir). We easily find by differentiating the first body-stress equation
with regard to x and using (xxi):

$$\rho\left(\frac{dY}{dy} + \frac{dZ}{dz} - \frac{dX}{dx}\right) = \frac{d^2\widehat{xx}}{dx^2} \quad\dots\dots\dots\dots\text{(xxii)}.$$

Now if the body-forces are to be wholly neglected we have
$d^2\widehat{xx}/dx^2 = 0$, which leads to Saint-Venant's results, or χ_1, χ_2, χ_3 must be
all linear functions of x, whence by (xi) and (xiii) the axial shifts can
only be algebraic functions of the third degree in x, and either the load-
system, or the original form of the rod must be extremely limited. If
on the other hand the body-forces are not zero, it appears that a certain
relation must be satisfied between them and the surface load. For the
surface load and the body-forces fully determine χ_1, χ_2, χ_3, and hence
(xxii) and (vii) give a relation between them, which as a rule will
not be satisfied.

If we do not take $d^2\widehat{xx}/dx^2 = 0$, but extremely small, then it seems
necessary that χ_1, χ_2 and χ_3 should be extremely small, or the total
longitudinal load and the changes of curvature very small; but it must
still be remembered that in this case, even at points distant from the
points of application of the external load, $\widehat{yy}$, $\widehat{zz}$ and $\widehat{yz}$, although absolutely
small, are not at every point necessarily small *relatively* to $\widehat{xy}$ and $\widehat{xx}$.

To sum up this part of Boussinesq's investigation : It does not seem
to sufficiently justify the ordinary assumption of the Bernoulli-Eulerian
hypothesis ($\widehat{zz} = \widehat{yy} = \widehat{yz} = 0$) for the cases either of a sensible continuous
loading or of body-forces, while in the cases in which continuous loading
and body-forces produce insensible effects, it does not bring out clearly
that the stresses neglected can at certain points be of the same order
as some of those retained ; further it does not fully solve the difficulties
involved in the result $d^2\widehat{xx}/dx^2 = 0$, or what really amounts to the same
thing

$$d^2M_y/dx^2 = d^2M_z/dx^2 = d^2F/dx^2 = 0.$$

[1428.] Pp. 165–76 of the memoir are occupied with a discussion of
the shape of the distorted rod after the strain. This is obtained by
combining the shifts of short prismatic elements and should be compared
with the similar investigation due to Kirchhoff : see our Arts. 1257
et seq.

Pp. 176–81 contain the criticism of Kirchhoff's treatment of rods, to
which we have already referred : see our Art. 1418.

[1429.] § VIII., which occupies pp. 181–94, is entitled : *Dé-
composition de l'action totale exercée sur un tronçon de la tige en
six actions élémentaires, qui produisent respectivement une extension ou une contraction, deux flexions égales, deux flexions inégales
et une torsion.* This is an analysis into its component parts of the
solution we have sketched in the above pages, and it resembles

Clebsch's treatment of Saint-Venant's problem in S. 85–94 of his *Theorie der Elasticität* (see our Arts. 1333–9) except that Clebsch dealt only with the equations for bi-constant isotropy and with the simple case of an initial straight central line.

Boussinesq points out that in the case of the flexure problem the $d\chi_4/dx$ of our equation (xviii), Art. 1427, is zero, when either (1) the cross-section has a centre of figure, or (2) the axis of z (or y) is an axis of symmetry and the elastic structure is symmetrical about the plane of zx (or xy) (p. 188).

Further since $d\widehat{xy}/dx$ and $d\widehat{zx}/dx$ are either zero or negligible, it follows from (xvii), that $d\phi/dz$ and $d\phi/dy$ are sensibly independent of x, if S_y and S_z the total shears are constant; and hence from (xviii) that the like holds for $d\chi_4/dx$, which is therefore essentially a constant. Thus for flexure in the cases of symmetry mentioned above $d\chi_4/dx$ is zero, and for torsion since S_y and S_z are then zero $d\chi_4/dx$ may be treated as practically a constant.

Boussinesq remarks that the case of torsion is the only one which requires us to integrate (xviii), for in the case of flexure the slides σ_{xy} and σ_{zx} are negligible (pp. 174, 186 and 194).

[1430.] The next section of the memoir (pp. 194–204) deals more especially with the general laws of torsion. In this case σ_{xy} and σ_{zx} have always to be found by the integration of a differential equation. Putting the total shears S_y and S_z zero, and $d\chi_4/dx =$ a constant $= \tau$, we have from (xvii) and (xviii):

$$\widehat{xy} = \frac{d\phi}{dz}, \qquad \widehat{zx} = -\frac{d\phi}{dy}, \left.\begin{array}{c} \\ \\ \end{array}\right\} \dots\text{(xxiii)},$$

$$f\frac{d^2\phi}{dy^2} + (h'' + h''')\frac{d^2\phi}{dy\,dz} + e\frac{d^2\phi}{dz^2} + 2\,(ef - h''h''')\,\tau = 0$$

while from (xx) $M_x = -\int^\omega \left(\frac{d\phi}{dy}\,y + \frac{d\phi}{dz}\,z\right) d\omega$

$$= 2\int^\omega \phi\,d\omega \dots \dots \dots \dots \dots \dots \text{(xxiv)},$$

as is easily seen by integrating by parts and using (xix), which now gives $\phi =$ a constant for the contour; but this constant may be supposed included in the value of ϕ so that $\phi = 0$ over the contour. Boussinesq shews that in the special case where $h'' + h''' = 0$ equations (xxiii) and (xxiv) are related to those for the steady motion of a viscous fluid in a tube, the cross-section of the tube being an orthogonal projection of that of the rod, at least for the case when the cross-section consists of an area without cyclosis (pp. 195–9). The steady velocity of the viscous fluid corresponds to the ϕ of the torsional problem.

Boussinesq proves the following proposition (p. 199) :

Les forces exercées aux divers points d'une section sont partout dirigées suivant les courbes $\phi = const.$, qui seraient celles d'égale vitesse dans des tubes, et elles sont égales en chaque point, par unité de surface, à la dérivée de ϕ suivant la normale menée en ce point à la courbe $\phi = const.$, qui y passe ; elles ont la même expression que le glissement relatif, dans un tube, de deux couches liquides adjacentes.

If then the curves $\phi = $ const. are constructed for equal increments of the constant, this family will in a manner reproduce the peculiarities of the contour, but members of the family must in general be closer together along a short than a long diameter. Hence the stress which varies as the constant increment of ϕ divided by the perpendicular distance (dn) between two adjacent members of the family will in general be a maximum upon the shorter diameters of the cross-section. Further ϕ is in general a maximum at the central parts of the section (hence $d\phi/dy = 0$, $d\phi/dz = 0$ there), and thus at these parts the stress is a minimum, so that we should expect $d\phi/dn$ to reach its maximum value at points on the contour, but by what precedes these will be the points on it nearest to the centre. Boussinesq goes further and demonstrates on pp. 200–2, that the components $d\phi/dy$ and $d\phi/dz$ of $d\phi/dn$ cannot be maxima or minima in the interior of the cross-section.

Boussinesq terminates this portion of his memoir by a discussion of the modifications introduced into the torsion moment when there is cyclosis of the cross-section, i.e. when the rod contains a hollow. This case is of special interest from its application to the theory of flaws in bars : see our Art. 1348, (e).

On pp. 204–9 he records the cases in which solutions of the torsion or flexure equations have been obtained, citing the results of Saint-Venant : see our Arts. 18–42 and 83–97, and referring to that of Clebsch for a section bounded by confocal ellipses in a footnote on pp. 209–10 : see our Art. 1348, (e).

[1431.] § xi. of the memoir (pp. 210–26) is entitled : *Exemples divers d'équilibre et de mouvement d'une tige rectiligne dont les déformations totales sont très-petites.* In this section Boussinesq deduces from the general equations of the earlier part of his memoir the special equations for the longitudinal, transverse and torsional vibrations of rods.

He deals also with cases in which a mass or masses are attached to a vibrating rod. He does not integrate these equations, but refers on this point to the special investigations of Navier, Poisson, Poncelet, Saint-Venant and Phillips: see our Arts. 272*–3*, 466*–71*, 577*–81*, 988*–92*, 104, 203–23, and 680.

[1432.] § XII. of the memoir (pp. 226–40) is entitled : *Étude d'une tige rectiligne soumise à une traction antérieure aux déplacements. Vibrations des cordes en tenant compte de la rigidité.* Boussinesq cites from his memoir on liquid waves (see our Art. 1442) the results he has obtained for the body-stress equations when there exists a considerable initial stress. He works out the particular case of a single initial traction $\widehat{xx}_0$, and develops at considerable length the form taken by the equations of the earlier part of the memoir, when this initial stress $\widehat{xx}_0$ exists in the direction of the axis of a rod. He applies his general results to obtain the equation due to Seebeck for the vibrations of a slightly stiff string (see our Art. 471), and he deduces the result (ii) of our Art. 472 for the case $i = 1$ with a slightly different form of statement, *viz.* the effect of the stiffness of a string upon its fundamental note is the same as if its total tension P were increased from P to $P + \dfrac{\pi^2}{l^2} E\omega\kappa^2$, or the stiffness produces a constant increase in the apparent tension. Since E is not sensibly changed by large tensions approaching even the rupture strength, we see that this law of increase holds for all variations of P which do not produce great changes in ω.

[1433.] The above memoir by Boussinesq is by no means easy reading and it does not appear to me to possess the clearness and conclusiveness of parts of his later work. It seems well to take in conjunction with it a supplement written in 1876, but first published in 1879. It is entitled: *Complément à une étude de 1871 sur la théorie de l'équilibre et du mouvement des solides élastiques dont certaines dimensions sont très-petites par rapport à d'autres.* The first section of this paper containing some general remarks on the negligible terms in the equilibrium-equations for plates and rods, and the second and third sections dealing with rods only were published in the *Journal de mathématiques*, T. V. pp. 163–94. Paris, 1879.

[1434.] The first two sections (pp. 163–81) we have practically dealt with in our consideration of the earlier memoir: see our Art 1424. We may note, however, two or three additional points which occur on pp. 179–81.

(a) To a first approximation, or on the supposition (A) of our Art.

1424, $\widehat{dxx}/dx = 0$, and thus the slides σ_{xy}, σ_{zx} depend entirely on the torsion, or on the existence of the couple M_x. For in this case the functions χ_1, χ_2, χ_3 become absolute constants and therefore by (xvi), S_z and S_y are zero. This should be compared with the rather vaguer statements referred to in our Arts. 1427 and 1429.

(b) The strains s_y, s_z and σ_{yz} are entirely independent of M_x, or the torsion while altering the form of the cross-section does not alter the form of the projection of the cross-section on a plane perpendicular to the central line.

(c) From a slight extension of (b) Boussinesq proves geometrically the theorem demonstrated in our Art. 181 (d), namely : that the same amount of torsion is produced when the same couple twists the rod or prism round any axis whatever parallel to its central axis.

[1435.] Section III. of the memoir (pp. 181–94) is entitled : *Application à la théorie des tiges.* Boussinesq remarks that the theory in which the relations $\widehat{yy} = \widehat{zz} = \widehat{yz} = 0$ hold, applies in *absolute rigour* only to prismatic rods of length infinitely greater than the linear dimensions of their cross-sections. It may, in practice however, be applied with considerable exactness even to rods the central line of which is a curve of double curvature, and this application Boussinesq proceeds to make in the following manner.

Let y and z be the principal axes of any cross-section, and x the tangent to the central line at this cross-section ; let s measure an arc of the central line from some fixed point up to this cross-section, and $s + \delta s$ to an adjacent cross-section ; let $a_0 ds$ be the angle between the principal axis y in the cross-section at s and the projection upon this cross-section of the principal axis y in the cross-section at $s + \delta s$; let $R_y{}^0$ and $R_z{}^0$ be as before (see our Art. 1425) the radii of curvature in the planes of zx and xy, all before strain. Let the corresponding quantities after strain be a, R_y and R_z, and let $s_x{}^0$ be the stretch of the central line at s. Then $a - a_0$ is very nearly equal to τ the angle of torsion at s, and if Q be the total thrust on the cross-section at s, M_x be the torsional couple, M_y and M_z be the bending moments in the planes zx and xy respectively we have (see our Art. 1425) :

$$Q = -\mathbb{E}\omega s_x{}^0, \qquad\qquad M_x = \nu\omega^2 (a - a_0),$$
$$M_y = \mathbb{E}\omega\kappa_y{}^2 \left(\frac{1}{R_y} - \frac{1}{R_y{}^0}\right), \quad M_z = \mathbb{E}\omega\kappa_z{}^2 \left(\frac{1}{R_z} - \frac{1}{R_z{}^0}\right) \left.\right\} \ldots \text{(xxv)},$$

where ν is a constant to be determined from the solution of equations (xxiii) and (xxiv). To describe the whole system of force upon a particular cross-section we require besides these quantities to know the total shears S_y and S_z parallel respectively to the axes of y and z, (p. 185). Boussinesq further subjects the elementary prism of length ds to certain external forces:

J'appellerai ρ la densité moyenne primitive du tronçon, dont la masse vaudra par suite $\rho\omega ds$, et je désignerai par $\rho X\omega ds$, $\rho Y\omega ds$, $\rho Z\omega ds$ les composantes totales des actions extérieures dont il s'agit. Quant à leurs moments par rapport à Oy, Oz, les deux forces $\rho Y\omega ds$, $\rho Z\omega ds$, dont les bras de levier seront comparables à ds, n'en donneront que de négligeables, et ceux de $\rho X\omega ds$ seront en général insensibles, surtout si les composantes longitudinales de l'action extérieure ne sont pas distribuées trop inégalement de part et d'autre du centre de gravité des sections. L'autre axe Ox étant parallèle à la force $\rho X\omega ds$, il y aura seulement à compter le moment des actions extérieures transversales par rapport à l'axe Ox ou à l'élément ds de fibre moyenne: j'appellerai $\beta\rho\omega^{\frac{3}{2}}ds$ ce moment, dont $\beta\sqrt{\omega}$ sera en quelque sorte la valeur par unité de masse, valeur comparable à la force qui le produit multipliée par un bras de levier de l'ordre des dimensions transversales de la tige ou de l'ordre de $\sqrt{\omega}$ (p. 187).

The general equations of Statics will then give us relations between the values of M_x, M_y, M_z, Q, S_y, and S_z corresponding to the cross-section at s and those corresponding to that at $s + \delta s$. Boussinesq confines his attention to the following cases:

(a) Slightly strained rod, originally without tortuosity and straight, i.e. $a_0 = 0$, $R_y{}^0 = R_z{}^0 = 0$. Boussinesq finds, pp. 189—90:

$$\left.\begin{array}{ll} \dfrac{dQ}{dx} = \rho\omega X, & \dfrac{dM_x}{dx} = -\beta\rho\omega^{\frac{3}{2}}, \\[3mm] S_y = -\dfrac{dM_z}{dx}, & S_z = \dfrac{dM_y}{dx}, \\[3mm] \dfrac{dS_y}{dx} = \dfrac{Q}{R_z} - \rho\omega Y, & \dfrac{dS_z}{dx} = \dfrac{Q}{R_y} - \rho\omega Z \end{array}\right\} \quad \dots\dots (\text{xxvi}).$$

These lead to

$$\frac{d^2 M_z}{dx^2} + \frac{Q}{R_z} - \rho\omega Y = 0, \qquad \frac{d^2 M_y}{dx^2} - \frac{Q}{R_y} + \rho\omega Z = 0 \dots (\text{xxvii}).$$

In the case of a negligible total thrust Q, the last four results of (xxvi) give us the well-known results of Graphical Statics, that the shear-curve is the sum-curve of the load-curve, and the bending-moment-curve the sum-curve of the shear-curve. The terms Q/R_z and Q/R_y will not, however, as Boussinesq remarks, be in general negligible as compared with dS_y/dx and dS_z/dx. They cannot, for example, be neglected in cases of longitudinal tension, or again in those of buckling action.

(b) The rod is symmetrical with regard to a plane and symmetrically strained with regard to this plane.

Let the plane be that of xy, then $a_0 = 0$, $R_y{}^0 = 0$, $a = 0$, $M_x = 0$, $R_y = 0$, $S_z = 0$, $Z = 0$, and we find :

$$\frac{dQ}{ds} = \rho\omega X - \frac{S_y}{R_z}, \qquad S_y = -\frac{dM_z}{ds},$$

$$\frac{dS_y}{ds} = \frac{Q}{R_z} - \rho\omega Y \qquad\qquad\Bigg\} \quad \ldots\ldots\ldots (\text{xxviii}).$$

These give

$$\frac{d^2M_z}{ds^2} + \frac{Q}{R_z} - \rho\omega Y = 0,$$

$$\frac{dQ}{ds} - \frac{1}{R_y}\frac{dM_z}{ds} - \rho\omega X = 0 \quad\Bigg\} \quad \ldots\ldots\ldots\ldots (\text{xxix}).$$

The results (xxv) substituted in either (xxvii) or (xxix) determine for cases (a) or (b) the form of the strained central line, i.e. the so-called elastic line.

Boussinesq remarks (p. 192) that the thrust Q and the bending moment M_z enter into both the equations (xxix), and in such fashion that one cannot be made zero without the other being in general compelled to satisfy two incompatible equations. It is usually impossible to set up longitudinal without transverse vibrations or *vice versâ* in a curved rod. This point had already been noticed by Resal for the case of a rod with a circular central line in his *Traité de Mécanique générale*, T. II. p. 153.

[1436.] The above investigations only determine the total shears S_y and S_z. If it be required to determine the stresses $\widehat{xy}$ and $\widehat{zx}$, then, for a rod only moderately bent, the formulae and equations of our Arts. 1425-7 may be safely applied to a second approximation,— the first approximation being considered as that in which these stresses are neglected altogether. As a case in which the flexure-slides σ_{xy}, σ_{zx} could be worked out Boussinesq suggests the problem of a small torsion applied to a rod under considerable flexure (p. 194).

Boussinesq's results for rods of double curvature should be compared with those of Saint-Venant and of Bresse discussed in our Arts. 1584*-1592*, 1597*-1608* and 534.

In a footnote at the conclusion of his memoir Boussinesq refers to Thomson and Tait's *Treatise on Natural Philosophy*, Arts. 702-3 where they deal with the case of a constrained torsion, which they term *simple torsion*.

[1437.] (i) *Étude nouvelle sur l'équilibre et le mouvement des corps solides élastiques dont certaines dimensions sont très-petites par rapport à d'autres. Second Mémoire. Des plaques planes. Journal de mathématiques*, T. XVI. pp. 241-274 (see also *Comptes rendus*, T. LXXII. pp. 449-52). Paris, 1871.

14—2

(ii) *Complément à une étude de* 1871 *sur la théorie de l'équilibre et du mouvement des solides élastiques dont certaines dimensions sont très-petites par rapport à d'autres. Suite* IV. *Équations
d'équilibre d'une plaque. Journal de mathématiques,* T. V. pp.
329–44. Paris, 1879.

These, the second parts of the memoirs of 1871 and 1879
respectively, deal with thin plates; the results of the first are
apparently supposed to hold only for *plane* plates, but those of
the second are considered to be true also for curved plates or
shells. The two papers are best dealt with together.

[1438.] If the axes of x, y be in the tangent plane to the mid-
plane of the plate at any point, then Boussinesq takes in his first
memoir (p. 246):

$$\widehat{zz} = \widehat{yz} = \widehat{zx} = 0 \quad \dots\dots\dots\dots\dots\dots \text{ (i)},$$

and so obtains the remaining stresses as linear functions of s_x, s_y, σ_{xy}.
He takes:

$$\widehat{xx} = K\left(\beta s_x + \beta' s_y + \beta'' \sigma_{xy}\right),$$

$$\widehat{yy} = K\left(\beta_1 s_x + \beta_1' s_y + \beta_1'' \sigma_{xy}\right),$$

$$\widehat{xy} = K\left(\gamma s_x + \gamma' s_y + \gamma'' \sigma_{xy}\right),$$

where β, β', β'', β_1, β_1', β_1'', γ, γ', γ'' are independent of z but can
vary with x and y, while K is a function, continuous or otherwise, of z
and may vary very slightly with x and y.

The general investigation is similar to that adopted by Saint-Venant
(see our Arts. 384–9). In the case, however, of elastic isotropy
parallel to the mid-plane of the plate the H of equation (vi) of our Art.
385 is equal to $\dfrac{3\beta}{2\epsilon^3}\displaystyle\int_{-\epsilon'}^{+\epsilon''} Kz^2 dz$ in Boussinesq's notation, where $-\epsilon'$, ϵ'' are
the values of z at the surfaces of the plate, and are supposed to be
slightly variable with x and y.

The contour-conditions at the edge of the plate are reduced to two
(pp. 250–1, 257–8) in the same manner as had been previously adopted
by Thomson and Tait, although Boussinesq independently discovered
the method: see our Arts. 488*, 394, 1440–1 and 1522-4.

[1439.] Boussinesq on pp. 268–74 of the first memoir considers
the effect of great initial stresses parallel to the mid-plane of a plane
plate. He deals especially with the case of a tightly and uniformly
stretched membrane, the notes of which are influenced by its stiffness.
His results may be easily deduced from our Arts. 384–5 and 390. In
Art. 390 put $\widehat{xx}_0 = \widehat{yy}_0 = Q/(2\epsilon)$ and $\widehat{xy}_0 = 0$, then (vi) of Art. 385, having

regard to (iii) of Art. 384, may be written for the case of a vibrating plate of density ρ :

$$2\epsilon\rho \frac{d^2 w_0}{dt^2} = Q\left(\frac{d^2 w_0}{dx^2} + \frac{d^2 w_0}{dy^2}\right) - \frac{2H\epsilon^3}{3}\left(\frac{d^2}{dx^2} + \frac{d^2}{dy^2}\right)^2 w_0 \ \dots\dots \ \text{(ii)}.$$

Assuming the last term on the right in the case of a slightly stiff membrane to be small as compared with the first, we may suppose the solution still to be of the membrane type

$$w_0 = \Sigma W_i (A_i \cos m_i at + B_i \sin m_i at),$$

where $a^2 = Q/(2\epsilon\rho)$ and

$$\frac{d^2 W_i}{dx^2} + \frac{d^2 W_i}{dy^2} + m^2{}_i W_i = 0.$$

Substituting in small terms we find (ii) may be written so far as terms in m_i are concerned :

$$2\epsilon\rho \frac{d^2 w_0}{dt^2} = \left(Q + \frac{2H\epsilon^3 m_i{}^2}{3}\right)\left(\frac{d^2 w_0}{dx^2} + \frac{d^2 w_0}{dy^2}\right),$$

or, the effect of a slight stiffness in the membrane is to increase the apparent tension in the case of a note of period $2\pi/m_i a$ by the amount $\frac{2}{3}H\epsilon^3 m_i{}^2$: see our Art. 1300, (c).

[1440.] The most unsatisfactory part of the investigation undoubtedly lies in the assumption (i) of our Art. 1438 :

$$\widehat{zz} = \widehat{zx} = \widehat{yz} = 0,$$

and this point is discussed more at length in the second memoir. The investigation of the second memoir has been reproduced by Saint-Venant in a somewhat modified and simplified form : see our Arts. 385–8. Neither the arguments of the original memoir nor of Saint-Venant's modification seem to me convincing, especially for the case of curved plates or shells : see in particular our Art. 1296 *bis*. Boussinesq in the course of his memoirs refers to the researches of Navier, Poisson, Kirchhoff and Gehring : see our Arts. 258*, 474*, 1233, 1292 and 1411. In a footnote at the end (p. 344) of his second memoir Boussinesq acknowledges that Thomson and Tait had preceded him in giving a true explanation of the difficulty as to the contour-conditions in the case of a plate (see our Arts. 488* and 394). He further refers to his controversy with Lévy (see our Art. 397), which would hardly have arisen had Thomson and Tait's *Treatise* been better known in France : see our Arts. 1441, 1522–4 and Chapter XIV.

On the last pages (pp. 342–4) of the second memoir are some interesting remarks upon Saint-Venant's principle of the elastic equivalence of statically equipollent systems of loading: see our Arts. 8, 9, 21 and 100.

[1441.] In the *Journal de mathématiques* (T. III. pp. 219–306, Paris, 1877) will be found a long memoir by Maurice Lévy entitled: *Mémoire sur la théorie des plaques élastiques planes,* in which the author questions what I have termed the Thomson-Tait reconciliation of Poisson's and Kirchhoff's contour-conditions for a thin plate, attributing that reconciliation, however, to Boussinesq: see our Arts. 488* and 397. The author works out with considerable fulness of analysis a solution for plates of finite thickness, and endeavours to shew by means of his solution that *three* contour-conditions are in general necessary for every elementary strip, and that the terms neglected by Poisson involving cubes and higher powers of the thickness (see our Arts. 477*–9*) cannot in general be neglected. What Lévy does is practically to introduce terms into the stresses which in certain cases may be made to allow for the *local perturbations* produced by the replacing one statical system of contour-load by an equipollent one. This replacement is essential to the Thomson-Tait reconciliation and is legitimate for thin plates owing to Saint-Venant's general principle of the elastic equivalence of statically equipollent load systems. But it is certainly of importance to measure the amount of the local perturbation due to the replacement. This had been practically done by Thomson and Tait in their *Treatise* in 1867 (see our Art. 488*), and therefore a rediscussion of the Kirchhoff-Poisson boundaries conditions in 1877 was somewhat late.

Lévy's memoir, however, led to a controversy with Boussinesq, which will be found in a series of articles in the *Comptes rendus,* as follows:

(I.) J. Boussinesq: *Sur les conditions aux limites dans le problème des plaques élastiques,* T. 85, pp. 1157–9. Paris, 1877. (Points out that Lévy's terms give only certain *local perturbations,* i.e. are not sensible far from the contour.)

(II.) M. Lévy: *Quelques observations au sujet d'une Note de M. Boussinesq. Ibid.* pp. 1277–80. (Asserts that the contour-load might produce rupture in one case, though it might not when it was replaced by an equipollent statical system, and that therefore the replacement cannot be elastically legitimate.)

(III.) J. Boussinesq: *Sur la question des conditions spéciales au contour des plaques élastiques*, T. 86, pp. 108–10. Paris, 1878. (Points out very forcibly that both Lévy and Poisson have already reduced their contour-conditions to *three* for each generator of the edge instead of three for each point of the generator, and so have already applied that very principle of the elastic equivalence of equipollent loads the truth of which Lévy is disputing.)

(IV.) M. Lévy: *Quelques observations sur une nouvelle Note de M. Boussinesq... Ibid.* pp. 304–7. (Accuses Boussinesq of "obscuring by empirical considerations an extremely clear question" and asserts that "his 'incontestable principles' cannot prevail against the fundamental principles of mechanics." The statement is repeated that the so-called perturbations are not local to the edge, but occur throughout the plate.)

(V.) J. Boussinesq: *Sur les conditions spéciales au contour des plaques. Ibid.* pp. 461–3. (A temperate reply to IV. pointing out that the terms introduced by Lévy are of the order $e^{-\frac{\pi n}{2\epsilon}}$, where 2ϵ is the small thickness of the plate, and n an element of normal to the contour. Hence they vanish at a small distance from the contour. Further these terms would vary with every distribution of the load along a generator of the bounding cylinder of the plate. Thus there would be an infinite number of solutions satisfying Poisson's three conditions and yet differing from each other as much as they differed from Kirchhoff's solution. Thus Poisson's conditions do not really suffice to determine Lévy's terms.)

The whole controversy might have been avoided by an early investigation of the order of Lévy's terms, such an investigation had been given ten years previously by Thomson and Tait: see our Arts. 1522–4, and Chapter XIV.

[1442.] *Théorie des ondes liquides périodiques. Mémoires présentés...à l'Académie des Sciences. Sciences mathématiques et physiques*, T. XX. pp. 509–615. Paris, 1872. This memoir was presented to the *Académie*, April 19, 1869, with additions of November 29, 1869 and September 5, 1870. Portions only concern our present inquiry and we will refer to them briefly here.

[1443.] § 1 (pp. 513–7) is entitled: *Équations des mouvements continus d'un milieu quelconque.* Here Boussinesq considers the type of body-stress equation which arises when the squares and products of the shift-fluxions cannot be neglected, see our Arts. 1617* and 234.

He shews that if θ be given by

$$1 + \theta = (1 + u_x)(1 + v_y)(1 + w_z) - v_z w_y (1 + u_x) - w_x u_z (1 + v_y)$$
$$- u_y v_x (1 + w_z) + u_y v_z w_x + u_z v_x w_y,$$

then Π being any element of volume

$$\frac{d}{dt}\left(\frac{\Pi}{1+\theta}\right) = 0, \quad \text{or} \quad \Pi = \Pi_0 (1 + \theta),$$

i.e. θ is the dilatation.

The body-stress equations are then shewn to be of the type:

$$\left. \begin{array}{l} \dfrac{d\widehat{xx}}{dx}\dfrac{d\theta}{du_x} + \dfrac{d\widehat{xx}}{dy}\dfrac{d\theta}{du_y} + \dfrac{d\widehat{xx}}{dz}\dfrac{d\theta}{du_z} \\[2mm] + \dfrac{d\widehat{xy}}{dx}\dfrac{d\theta}{dv_x} + \dfrac{d\widehat{xy}}{dy}\dfrac{d\theta}{dv_y} + \dfrac{d\widehat{xy}}{dz}\dfrac{d\theta}{dv_z} \\[2mm] + \dfrac{d\widehat{xz}}{dx}\dfrac{d\theta}{dw_x} + \dfrac{d\widehat{xz}}{dy}\dfrac{d\theta}{dw_y} + \dfrac{d\widehat{xz}}{dz}\dfrac{d\theta}{dw_z} \end{array} \right\} = \rho \left\{ \frac{d^2 u}{dt^2} - X \right\},$$

ρ being the primitive density.

If the squares and products of the shift-fluxions can be neglected, this becomes

$$\left. \begin{array}{l} (1 + \theta)\left(\dfrac{d\widehat{xx}}{dx} + \dfrac{d\widehat{xy}}{dy} + \dfrac{d\widehat{xz}}{dz}\right) \\[3mm] -\left(\dfrac{d\widehat{xx}}{dx} u_x + \dfrac{d\widehat{xx}}{dy} v_x + \dfrac{d\widehat{xx}}{dz} w_x\right) \\[3mm] -\left(\dfrac{d\widehat{xy}}{dx} u_y + \dfrac{d\widehat{xy}}{dy} v_y + \dfrac{d\widehat{xy}}{dz} w_y\right) \\[3mm] -\left(\dfrac{d\widehat{xz}}{dx} u_z + \dfrac{d\widehat{xz}}{dy} v_z + \dfrac{d\widehat{xz}}{dz} w_z\right) \end{array} \right\} = \rho \left\{ \frac{d^2 u}{dt^2} - X \right\},$$

where $\theta = u_x + v_y + w_z$.

If the fluxions of the stresses are themselves so small that their products with the shift-fluxions may be neglected, we obtain the usual body-stress equations of elasticity.

[1444.] *Note* 3 (pp. 584–604), *où sont établies des relations générales et nouvelles entre l'énergie interne d'un corps, fluide ou solide, et ses pressions ou forces élastiques.* This *Note* gives the general relations between the strain-energy and the stresses of a medium. In a footnote (pp. 585–6) Boussinesq refers to Rankine's introduction of the term *potential energy* and discusses the *internal potential energy of a medium.* The object of the *Note* is recited in the following words:

La méthode employée au paragraphe 1 ne donne pas seulement les équations exactes des mouvements continus des corps élastiques, isotropes ou hétérotropes, solides ou fluides ; elle permet encore, lorsque la température de ces corps est supposée assez voisine du zéro absolu pour qu'on puisse, dans le calcul des actions mutuelles de leurs molécules, faire abstraction des mouvements vibratoires d'amplitude insensible, ou calorifiques, et aussi, dans un autre cas très-général dont nous allons parler, d'exprimer complètement leurs forces élastiques en fonction des dérivées partielles des déplacements u, v, w par rapport aux coordonnées primitives x, y, z, et de celles de leur énergie interne par rapport à six variables dont cette énergie dépend. En supposant très-petites les dérivées partielles de u, v, w en x, y, z, les résultats ainsi obtenus sont d'accord avec ceux que fournit une méthode basée sur le calcul des variations, et que M. de Saint-Venant a employée (p. 584).

See our Arts. 127 and 237.

The other very general case referred to above is that in which the elements of volume into which the medium may be divided, have primitively any temperatures whatever, are rendered afterwards impermeable to heat, and have their temperature a function at each instant only of the actual form and dimensions of the element at that instant.

[1445.] Boussinesq represents the internal potential energy, i.e. strain-energy, by Φ and obtains nine relations typified by the following three :

$$\widehat{xx}\,\frac{d\theta}{du_x} + \widehat{xy}\,\frac{d\theta}{dv_x} + \widehat{xz}\,\frac{d\theta}{dw_x} = \frac{d\Phi}{du_x},$$

$$\widehat{xx}\,\frac{d\theta}{du_y} + \widehat{xy}\,\frac{d\theta}{dv_y} + \widehat{xz}\,\frac{d\theta}{dw_y} = \frac{d\Phi}{du_y},$$

$$\widehat{xx}\,\frac{d\theta}{du_z} + \widehat{xy}\,\frac{d\theta}{dv_z} + \widehat{xz}\,\frac{d\theta}{dw_z} = \frac{d\Phi}{du_z}.$$

Solving these equations for $\widehat{xx}$, $\widehat{xy}$, $\widehat{xz}$, we have :

$$\widehat{xx} = \frac{1}{1+\theta}\left\{(1+u_x)\,\frac{d\Phi}{du_x} + u_y\,\frac{d\Phi}{du_y} + u_z\,\frac{d\Phi}{du_z}\right\},$$

$$\widehat{xy} = \frac{1}{1+\theta}\left\{v_x\,\frac{d\Phi}{du_x} + (1+v_y)\,\frac{d\Phi}{du_y} + v_z\,\frac{d\Phi}{du_z}\right\},$$

$$\widehat{xz} = \frac{1}{1+\theta}\left\{w_x\,\frac{d\Phi}{du_x} + w_y\,\frac{d\Phi}{du_y} + (1+w_z)\,\frac{d\Phi}{du_z}\right\}.$$

Boussinesq now remarks that Φ does not in reality depend upon the *nine* shift-fluxions but on the three stretches and three slide-*cosines*. See our Art. 1621*.

These are given by the types :

$$s_x = -1 + \sqrt{(1+u_x)^2 + v_x{}^2 + w_x{}^2},$$

$$c_{yz} = \frac{u_y u_z + (1+v_y)\,v_z + (1+w_z)\,w_y}{(1+s_y)\,(1+s_z)}.$$

Boussinesq now introduces a new set of variables connected with the stretches and slides by relations of the type (see our Art. 1622*):

$$s_x = -1 + \sqrt{1 + 2\epsilon_x}, \qquad c_{yz} = \frac{\eta_{yz}}{\sqrt{(1 + 2\epsilon_y)(1 + 2\epsilon_z)}}.$$

He then finds stresses of the types:

$$\widehat{xx} = \frac{1}{1 + \theta}\left[\frac{d\Phi}{d\epsilon_x}(1 + u_x)^2 + \frac{d\Phi}{d\epsilon_y}u_y^2 + \frac{d\Phi}{d\epsilon_z}u_z^2\right.$$
$$\left. + 2\frac{d\Phi}{d\eta_{yz}}u_y u_z + 2\frac{d\Phi}{d\eta_{zx}}u_z(1 + u_x) + 2\frac{d\Phi}{d\eta_{xy}}u_y(1 + u_x)\right],$$

$$\widehat{yz} = \frac{1}{1 + \theta}\left[\frac{d\Phi}{d\epsilon_x}v_x w_x + \frac{d\Phi}{d\epsilon_y}w_y(1 + v_y) + \frac{d\Phi}{d\epsilon_z}v_z(1 + w_z)\right.$$
$$+ \frac{d\Phi}{d\eta_{yz}}\{(1 + v_y)(1 + w_z) + v_z w_y\} + \frac{d\Phi}{d\eta_{zx}}\{v_z w_x + v_x(1 + w_z)\}$$
$$\left. + \frac{d\Phi}{d\eta_{xy}}\{v_x w_y + (1 + v_y)w_x\}\right],$$

where
$$\frac{d\Phi}{d\epsilon_x} = \frac{1}{1 + s_x}\frac{d\Phi}{ds_x} - \frac{1}{(1 + s_x)^2}\left(c_{zx}\frac{d\Phi}{dc_{zx}} + c_{xy}\frac{d\Phi}{dc_{xy}}\right),$$
$$\frac{d\Phi}{d\eta_{yz}} = \frac{1}{(1 + s_y)(1 + s_z)}\frac{d\Phi}{dc_{yz}}.$$

On the substitution of these latter results in the former we have expressions for the stresses in terms of the differentials of Φ with regard to the six strains.

[1446.] Suppose the shift-fluxions are so small that their products may be neglected, then the slide-cosines c become the slides σ and the equations reduce to

$$\widehat{xx} = (1 - s_y - s_z)\frac{d\Phi}{ds_x} + (2u_z - \sigma_{zx})\frac{d\Phi}{d\sigma_{zx}} + (2u_y - \sigma_{xy})\frac{d\Phi}{d\sigma_{xy}},$$
$$\widehat{yz} = (1 - s_x - s_y - s_z)\frac{d\Phi}{d\sigma_{yz}} + w_y\frac{d\Phi}{ds_y} + v_z\frac{d\Phi}{ds_z} + v_x\frac{d\Phi}{d\sigma_{zx}} + w_x\frac{d\Phi}{d\sigma_{xy}}.$$

Boussinesq next assumes Φ to be of the following form:

$$\Phi = \text{const.} + A_1 s_x + A_2 s_y + A_3 s_z + B_1 \sigma_{yz} + B_2 \sigma_{zx} + B_3 \sigma_{xy} + \Phi_1,$$

where Φ_1 is a homogeneous function of the second degree in the strain-components, and A_1, A_2, A_3, B_1, B_2, B_3 are the primitive differentials of Φ with respect to $s_x, \ldots, \sigma_{yz}, \ldots$, i.e. its differentials when there is zero strain. We find

$$\widehat{xx} = A_1(1 - v_y - w_z) + B_2(u_z - w_x) - B_3(v_x - u_y) + \frac{d\Phi_1}{ds_x},$$
$$\widehat{yz} = B_1(1 - u_x - v_y - w_z) + A_2 w_y + A_3 v_z + B_2 v_x + B_3 w_x + \frac{d\Phi_1}{d\sigma_{yz}}.$$

These results agree with those obtained by Saint-Venant, if it be noted that he takes

$$\Phi = \text{const.} + A_1(s_x + \tfrac{1}{2}s_x^2) + A_2(s_y + \tfrac{1}{2}s_y^2) + A_3(s_z + \tfrac{1}{2}s_z^2) + B_1\sigma_{yz}(1 + s_y + s_z)$$
$$+ B_2\sigma_{zx}(1 + s_z + s_x) + B_3\sigma_{xy}(1 + s_x + s_y) + \Phi_1',$$

which gives

$$\Phi_1 = \tfrac{1}{2}(A_1 s_x^2 + A_2 s_y^2 + A_3 s_z^2) + B_1\sigma_{yz}(s_y + s_z) + B_2\sigma_{zx}(s_z + s_x) + B_3\sigma_{xy}(s_x + s_y) + \Phi_1',$$

and leads to his formulæ : see our Arts. 237–9.

In § 6 (pp. 594–7) Boussinesq gives a geometrical interpretation of the derivatives of Φ, which he considers renders his mode of dealing with the problem more satisfactory than those of Saint-Venant and Cauchy (p. 599).

A somewhat different mode of investigating the same problem is given on pp. 599–604. Boussinesq assumes that the stresses are linear functions of the nine strain-fluxions, and then investigates what form they can possibly take so that the motion of the body as a whole shall not produce stress across any plane within it.

[1447.] *Recherches sur les principes de la Mécanique, sur la constitution moléculaire des corps et sur une nouvelle théorie des gaz parfaits. Journal de mathématiques,* T. XVIII., pp. 305–60. Paris, 1873. This memoir was presented to the *Académie des Sciences et des Lettres de Montpellier* on July 8, 1872, and published in the *Mémoires* for the same year, T. VIII., pp. 109–56. See *Notice* I. pp. 62–3[1].

There is much in this memoir which is suggestive with regard to the molecular and atomic constitutions of bodies and the relations of these to thermal and cohesive properties. The particular molecular hypothesis adopted by Boussinesq embodies the assumption of *modified action* (p. 307 : see our Arts. 276, 305), but it supposes that the accelerations of the various material points of an isolated system are solely functions of their *actual* mutual distances. Boussinesq's arguments in favour of this do not seem to me at all conclusive (p. 313). It does not appear how far he intends to take the ether into account in his isolated system of material points, but I have indicated elsewhere that at least one molecular hypothesis leads to intermolecular action being a function of the *velocity* of the molecules relative to the ether, and

---

[1] In the *Analyse succincte*, Saint-Venant writes of this memoir:

C'est une synthèse que M. Boussinesq a entreprise comme ont fait d'autres esprits élevés. Il en tire une foule d'explications, de judicieuses distinctions, et une théorie des gaz parfaits. Mais la nécessité où il est de faire quelques hypothèses nous détermine à nous abstenir d'ajouter ce vaste essai à ses nombreux titres (p. 18).

thus the accelerations of the material points being functions of the velocities or indirectly of *past* relative distances (see *Lond. Math. Soc. Proceedings*, Vol. XX., p. 297, 1888). Further Boussinesq supposes (p. 327) that the action between two atoms of the same molecule does not depend in an appreciable degree on the distances between atoms belonging to other molecules, but this again seems to me doubtful in the case of 'kin' atoms in different molecules, the equality of the free periods of which renders it very probable that they largely influence each other's action (see *American Journal of Mathematics*, Vol. XIII., p. 361, 1890).

With suppositions such as the above, Boussinesq, starting from the principle of energy, deduces various principles of thermodynamics, elasticity, fluidity and melting. Thus laws attributed to Gay-Lussac, Mariotte, Joule, Regnault, Delaroche and Bérard are deduced without appeal to the kinetic theory of gases as propounded by D. Bernoulli and developed by Clausius:

J'espère que la théorie nouvelle paraîtra étayée sur des suppositions en moindre nombre et plus vraisemblables (p. 310).

It does not appear that Boussinesq's theory would admit of that interchange of atoms between the molecules of a solid which has been supposed by Maxwell and other physicists to be continually taking place.

Je supposerai l'état chimique du corps assez stable pour que les positions relatives moyennes des atomes qui composent une même molécule restent à peu près les mêmes durant tous les phénomènes étudiés...(p. 327).

Thus in this theory the energy of atomic movement is independent of intermolecular distances, while the energy of molecular movement depends solely upon intermolecular distances (pp. 328–9).

[1448.] The part of the memoir most closely connected with our subject is § VIII. (pp. 350–5) entitled: *Action moléculaire dans un corps isotrope; solidité et fluidité.* This matter is also discussed in a paper entitled: *Note sur l'action réciproque de deux molécules. Comptes rendus*, T. LXV., pp. 44–6. Paris, 1867.

Boussinesq starts with the axioms that intermolecular force must depend: (i) on the initial distance between two molecules and its direction, (ii) on the manner in which relative molecular displacements vary throughout a small region enclosing the two molecules. The latter condition is that which we have called the *hypothesis of modified action* (see our Arts. 276 and 305) and leads to biconstant formulae in the case of elastic isotropy. Boussinesq obtains a type of intermolecular force from his axioms which would lead to biconstant formulae and to constant initial tractions "qui représente chez les fluides la pression dans l'état primitif" (*C. R.* p. 46). He does not discuss their meaning in the case of an ordinary elastic body.

The type of intermolecular action found for two molecules whose distance r has been increased by a small distance δr is of the form :

$$\phi = A - B\frac{\delta\rho}{\rho} + C\frac{\delta r}{r},$$

where, ρ being the density, $-\delta\rho/\rho$ is the dilatation; A, B and C are functions of r. This he considers can be thrown into the form :

$$\phi = F(r + \delta r, \ \rho + \delta\rho) + F_1(r)\frac{\delta r}{r},$$

where F and F_1 are certain functions. Of this result he writes :

Ainsi, dans un milieu isotrope peu écarté de son état primitif d'équilibre, l'action moléculaire se compose de deux forces : l'une, que j'appellerai de première espèce, ne varie qu'avec la distance actuelle des deux molécules considérées et la densité actuelle du milieu ; la seconde, que j'appellerai de deuxième espèce, dépend de la distance primitive des deux molécules et du petit écartement qu'elles ont subi à l'époque actuelle (p. 352).

The 'actions of the first kind' Boussinesq considers build up the elasticity of fluids. The 'actions of the second kind' are what constitute solidity. Boussinesq appeals to experience (p. 353) to shew that the actions of the second kind vanish for ratios of δr to r exceeding certain very small positive values. The disappearance of the second term constitutes the transition from the solid to the fluid state. Boussinesq attributes the fact that Navier, Lamé and Clapeyron arrived in their early investigations at uniconstant isotropy to their neglect of the first term in the above value for ϕ. He seems to indicate that the addition of the fluid term will lead to biconstant formulae :

On trouverait en effet celles-ci en ajoutant aux expressions anciennes et incomplètes des actions normales N la pression constante, fonction de la densité *actuelle*, que donnent les actions de première espèce, et qui introduirait, outre une partie principale, antérieure aux déplacements observés, un terme proportionnel à la petite dilatation θ (p. 353).

This appears to be the same idea as had occurred to Rankine, but the truth of which we have seen reason to call in question : see our Arts. 424, 429 and 431.

[1449.] *Note complémentaire au Mémoire précédent.—Sur les principes de la théorie des ondes lumineuses qui résulte des idées exposées au § VI. Journal de mathématiques*, T. XVIII., pp. 361–90.

Paris, 1873. This also appears in the *Annales de chimie*, T. xxx., pp. 539–65. Paris, 1873. It is a general explanation and a reply to certain criticisms of the principles involved in Boussinesq's elastic theory of waves of light. The author puts extremely clearly the arguments in favour of his hypotheses and shews that his theory is really based on physical conceptions, i.e. does more than substitute

à l'analyse mécanique des phénomènes une sorte de symbole analytique d'une généralité telle, qu'ils y soient tous compris (p. 361).

I have made use of this *Note* in explaining the hypotheses of the memoir of 1868: see our Art. 1478. It would carry us too far into the subject of light to even briefly analyse its contents here. It concludes with two supplements to the memoir of 1868 dealing with more approximate formulae than those there given for the aberration of light (pp. 383–90). Compare the *Comptes rendus*, T. 74, pp. 1573–6, 1872, and T. 76, pp. 1293–6, 1873.

[1450.] *Sur deux lois simples de la résistance vive des solides.* *Comptes rendus*, T. 79, pp. 1324–8 and 1407–11. Paris, 1874. This memoir contains a general proof of the hypothesis first adopted by Homersham Cox in 1849, when dealing with the transverse re-silience of bars (see our Art. 1435*) and afterwards shewn by Saint-Venant to hold for a considerable number of special cases (see our Arts. 368–9). By means of this hypothesis we are able to determine very approximately the maximum *shift* (or deflection) and the period of the principal vibration for a considerable range of problems, but, as we have pointed out earlier in this *History*, the expression for the maximum *strain* obtained in this manner is, as a rule, not sufficiently approximate to be of practical value : see our Art. 371, (iii). Boussinesq attributes the first statement of the hypothesis to Saint-Venant, but this is incorrect (see our Art. 201), although the deduction of the period of the principal vibration and the legitimate use of the hypothesis (i.e. the demonstration of its applicability in a considerable range of special cases) is certainly due to the French scientist.

Suppose a mass P of elastic material to have certain portions of its external surface free and others rigidly fixed, and let a mass Q of very small volume, but possessed of a considerable velocity, strike the mass P in a definite point and become fixed to it without however modifying its elasticity; then, we require some hypothesis by which we can easily approximate to the motion after the impact of the system consisting of the concentrated mass Q and the extended mass P.

When the mass P is so small as compared with Q that the effect of its inertia may be neglected, the problem reduces to a simple statical one, but when the masses P and Q are comparable the problem becomes more complex, for the total motion of the system must then be considered as the resultant of an infinite series of simple harmonic motions and it is necessary to calculate the amplitudes and periods of these motions. Saint-Venant had been led to the following approximate laws (which are practically an extension of Cox's hypothesis) by the exact calculation of a number of special cases:

If in any problem the expressions for the shifts are reduced to their principal term (or term of longest period), and if the ratio of P to Q does not exceed a certain limit (which can be as great as 2, 3 or sometimes even 4) then the square of the reciprocal of the period of vibration and that of the amplitude of the oscillations of the concentrated mass Q are both inversely proportional to the sum of this mass Q and of the products obtained by multiplying each element dP of the extended mass by the square of the ratio of its statical shift to the analogous shift of the concentrated mass.

These are the simple approximate laws which Boussinesq proposes to demonstrate in the present memoir.

[1451.] Let u, v, w be the shifts of any point of the elastic body P after impact, then with the usual assumptions the stresses will be linear functions of the first space-fluxions of the shifts. Hence, if the shifts be represented by the expressions

$$u = \Sigma \phi_1 \left(\frac{A}{n} \sin nt + B \cos nt \right),$$
$$v = \Sigma \phi_2 \left(\frac{A}{n} \sin nt + B \cos nt \right), \quad \Bigg\} \quad \dots\dots\dots\dots\dots(i),$$
$$w = \Sigma \phi_3 \left(\frac{A}{n} \sin nt + B \cos nt \right)$$

where ϕ_1, ϕ_2, ϕ_3 are functions of x, y, z the space-coordinates only, then the stresses will be given by equations of the type

$$\widehat{xx} = \Sigma \widehat{xx_0} \left(\frac{A}{n} \sin nt + B \cos nt \right),$$
$$\widehat{yz} = \Sigma \widehat{yz_0} \left(\frac{A}{n} \sin nt + B \cos nt \right) \quad \Bigg\} \quad \dots\dots \dots\dots (ii),$$

where $\widehat{xx_0}$, ... $\widehat{yz_0}$, ... are functions of x, y, z only. Hence, if we suppose no body-forces to act on P, the body-shift equations become of the type:

$$\frac{d\widehat{xx_0}}{dx} + \frac{d\widehat{xy_0}}{dy} + \frac{d\widehat{zx_0}}{dz} + \rho n^2 \phi_1 = 0 \dots\dots\dots\dots\dots(iii),$$

for each particular value of n.

Equations (iii) shew that any individual set of the functions ϕ_1, ϕ_2, ϕ_3 are the shifts that would be produced by applying to the mass P body-forces $n^2\phi_1$, $n^2\phi_2$, $n^2\phi_3$ parallel to the three axes of x, y, z respectively. For each such set we must have at points of the external surface which are rigidly-fixed:

$$\phi_1 = \phi_2 = \phi_3 = 0,$$

and at points which are free:

$$\widehat{xx}_0 \cos \alpha + \widehat{xy}_0 \cos \beta + \widehat{zx}_0 \cos \gamma = 0,$$

$$\left.\begin{array}{c} \\ \\ \end{array}\right\} \quad \ldots\ldots\ldots\ldots \text{(iv)},$$

with two similar equations, where α, β, γ are the direction-angles of the normal to the surface-element at the free point. These results are sufficient to give the ϕ's and it remains to be indicated how the A's and B's would be determined from the initial conditions of the system.

[1452.] Let ϕ_1', ϕ_2', ϕ_3' and n' be a second system of values of ϕ_1, ϕ_2, ϕ_3 and n, satisfying equations like (iii). Multiply the three equations of type (iii) by ϕ_1', ϕ_2', ϕ_3' respectively, add and integrate over the volume U of the whole system, P and Q; we find integrating by parts and using the surface conditions (iv):

$$n^2 \iiint (\phi_1\phi_1' + \phi_2\phi_2' + \phi_3\phi_3') \rho \, dU$$

$$= \iiint \left\{ \widehat{xx}_0 \frac{d\phi_1'}{dx} + \ldots + \widehat{yz}_0 \left(\frac{d\phi_2'}{dz} + \frac{d\phi_3'}{dy} \right) + \ldots \right\} dU \ \ \ldots\ldots \text{(v)}.$$

Now we have seen that $\widehat{xx}_0, \ldots \widehat{yz}_0, \ldots$ and ϕ_1, ϕ_2, ϕ_3 are the stresses and shifts due to a certain elastic system in equilibrium, hence these stresses will be linear functions of the space-fluxions of the shifts involving the usual 21 coefficients, i.e. they will be differentials of a quadratic function of the space-fluxions of the shifts. It follows then that the expression on the right-hand side of (v) under the sign of integration is *symmetrical* with regard to ϕ_1, ϕ_2, ϕ_3 and ϕ_1', ϕ_2', ϕ_3', or, we must have:

$$n^2 \iiint (\phi_1\phi_1' + \phi_2\phi_2' + \phi_3\phi_3') \rho \, dU = n'^2 \iiint (\phi_1\phi_1' + \phi_2\phi_2' + \phi_3\phi_3') \rho \, dU,$$

whence, if n be not equal to n':

$$\iiint (\phi_1\phi_1' + \phi_2\phi_2' + \phi_3\phi_3') \rho \, dU = 0 \ \ \ldots\ldots\ldots\ldots \text{(vi)}.$$

Equation (vi) enables us to determine the values of A and B from the initial conditions at time $t = 0$. Thus if u_0, v_0, w_0 be the initial shifts, and $\dot{u}_0$, $\dot{v}_0$, $\dot{w}_0$ be the initial speeds, we have from (i) and (vi):

$$A = \frac{\iiint (\dot{u}_0\phi_1 + \dot{v}_0\phi_2 + \dot{w}_0\phi_3) \, \rho \, dU}{\iiint (\phi_1{}^2 + \phi_2{}^2 + \phi_3{}^2) \, \rho \, dU},$$

$$B = \frac{\iiint (u_0\phi_1 + v_0\phi_2 + w_0\phi_3) \, \rho \, dU}{\iiint (\phi_1{}^2 + \phi_2{}^2 + \phi_3{}^2) \, \rho \, dU}$$

$$\left.\begin{array}{c} \\ \\ \\ \\ \end{array}\right\} \quad \ldots\ldots\ldots\ldots \text{(vii)},$$

the integrals being extended throughout the whole system U. Returning to equation (v), let $n' = n$ and therefore $\phi' = \phi$, we then have :

$$n^2\iiint(\phi_1{}^2 + \phi_2{}^2 + \phi_3{}^2)\,\rho dU = 2\iiint W dU \ldots\ldots\ldots\ldots(\text{viii}),$$

where W is the quadratic function of the space-fluxions of the ϕ's which would be the strain-energy for the shifts ϕ_1, ϕ_2, ϕ_3.

So far Boussinesq's investigation is practically identical with that of Clebsch given in our Arts. 1329–30, but the form of his results renders them immediately applicable to the problem of resilience.

[1453.] In the problem of resilience at the instant of the blow u_0, v_0, w_0 are zero, and so also are the speeds $\dot{u}_0$, $\dot{v}_0$, $\dot{w}_0$ except at the elementary volume immediately surrounding the point x, y, z at which the impact of Q takes place. Now the values of ϕ_1, ϕ_2, ϕ_3 are clearly such that they leave undetermined an arbitrary constant factor, and we can so choose that factor that $\phi_1{}^2 + \phi_2{}^2 + \phi_3{}^2 = 1$ at the point x, y, z. But ϕ_1, ϕ_2, ϕ_3 will then represent the direction-cosines of the shift of the point x, y, z for a simple component vibration. Thus the numerator in the value of A is the momentum of the impinging body Q resolved in the direction in which Q makes the oscillation of period $2\pi/n$. If this momentum be represented by QV we have :

$$B = 0, \qquad A = \frac{QV}{\iiint(\phi_1{}^2 + \phi_2{}^2 + \phi_3{}^2)\,\rho dU} \ldots\ldots\ldots\ldots(\text{ix}).$$

Further, if f be the amplitude of the vibrations corresponding to n, we have $f = A/n$, or by (viii) :

$$f = \frac{QVn}{2\iiint W dU} \ldots\ldots\ldots\ldots\ldots\ldots\ldots(\text{x}).$$

[1454.] When Q is very great as compared with P, we can suppose $\rho = 0$, except at the point x, y, z of the total system. In this case equations of the type (iii) shew us that only one mode of vibration is possible which is that corresponding to a statical system $\phi_1{}^0$, $\phi_2{}^0$, $\phi_3{}^0$ in which there is no suppositious body force $\rho n^2\phi_1{}^0$, $\rho n^2\phi_2{}^0$, $\rho n^2\phi_3{}^0$ on any element dU of the system except on the concentrated mass Q at x, y, z, where there is a force Qn^2, the direction of which is given by $\phi_1{}^0$, $\phi_2{}^0$, $\phi_3{}^0$.

Cox and Saint-Venant's hypothesis would thus be exactly true, if we might neglect the inertia of P. Supposing we cannot neglect this inertia, there will then be several systems of values for ϕ_1, ϕ_2, ϕ_3. But we shall now shew that the expressions for u, v, w may still be reduced with a certain degree of approximation to their principal terms, that is, to those which correspond to values of ϕ_1, ϕ_2, ϕ_3 close to $\phi_1{}^0$, $\phi_2{}^0$, $\phi_3{}^0$. Let

$$\Delta\phi_1{}^0 = \phi_1 - \phi_1{}^0, \quad \Delta\phi_2{}^0 = \phi_2 - \phi_2{}^0, \quad \Delta\phi_3{}^0 = \phi_3 - \phi_3{}^0,$$

and let us calculate the value of $\iiint W dU$. We find, since W is a quadratic function of ϕ_1, ϕ_2, ϕ_3:

$$\iiint W dU = \iiint W_0 dU + \iiint W_\Delta dU$$
$$+ \iiint \left\{ \widehat{xx_0}^0 \frac{d\Delta\phi_1^0}{dx} + \dots + \widehat{yz_0}^0 \left(\frac{d\Delta\phi_2^0}{dz} + \frac{d\Delta\phi_3^0}{dy} \right) + \dots \right\} dU \dots \text{(xi)},$$

where W_0 and W_Δ are the same functions of ϕ_1^0, ϕ_2^0, ϕ_3^0 and $\Delta\phi_1^0$, $\Delta\phi_2^0$, $\Delta\phi_3^0$ respectively as W is of ϕ_1, ϕ_2, ϕ_3.

Now we have three equations of the type:

$$\frac{d\widehat{xx_0}^0}{dx} + \frac{d\widehat{xy_0}^0}{dy} + \frac{d\widehat{xz_0}^0}{dz} = 0.$$

If these be multiplied respectively by $\Delta\phi_1^0$, $\Delta\phi_2^0$, $\Delta\phi_3^0$, added and integrated by parts over the volume U, we find that the last integral of equation (xi) is zero, because over the surface of the system either (a) the surface stresses are zero, or (b) at fixed points ϕ and ϕ^0 vanish, or (c) at the element round the point x, y, z, $\Delta\phi_1^0$, $\Delta\phi_2^0$, $\Delta\phi_3^0$ are zero since the direction of the statical displacement is taken to agree with that of the dynamical and these have ϕ_1, ϕ_2, ϕ_3 and ϕ_1^0, ϕ_2^0, ϕ_3^0 respectively for direction-cosines. Hence (xi) reduces to

$$\iiint W dU = \iiint W_0 dU + \iiint W_\Delta dU \dots\dots\dots\dots\text{(xii)}.$$

Now $\Delta\phi_1^0$, $\Delta\phi_2^0$, $\Delta\phi_3^0$ are clearly of the order P/Q as compared with ϕ^0, for they vanish with P and the stresses must be linear in terms of the applied load. Thus it follows, since W_Δ is of the order $(\Delta\phi^0)^2$, that it is of the order $(P/Q)^2$ as compared with W_0. Hence if $(P/Q)^2$ is negligible we may neglect the second term in $\iiint W dU$, and we accordingly find:

$$\left. \begin{aligned} n^2 &= \frac{2\iiint W_0 dU}{Q + \iiint\{(\phi_1^0)^2 + (\phi_2^0)^2 + (\phi_3^0)^2\}\, dP}, \\ f &= \frac{QVn}{2\iiint W_0 dU} \end{aligned} \right\} \quad \dots\dots\dots\text{(xiii)},$$

since

$$\iiint\{(\phi_1^0)^2 + (\phi_2^0)^2 + (\phi_3^0)^2\}\, \rho dU = Q + \iiint\{(\phi_1^0)^2 + (\phi_2^0)^2 + (\phi_3^0)^2\}\, dP,$$

remembering the value of $(\phi_1^0)^2 + (\phi_2^0)^2 + (\phi_3^0)^2$ at (x, y, z). These are the analytical expressions of the laws stated above.

[1455.] Boussinesq shews in the penultimate paragraph of his memoir how the above results are easily extended to the case when the blow of the impinging body is not concentrated on a very small region, but there are several concentrated masses producing impacts at the same instant (pp. 1410–1).

He concludes the memoir with the following words:

Remarquons enfin que, dans les problèmes les plus usuels, le mouvement vibratoire étudié est de même sens pour tous les points du système : alors les inerties des diverses parties dP de la masse disséminée agissent à chaque instant de manière à accroître leurs déplacements dus aux inerties des masses heurtantes ou concentrées, et la valeur $\sqrt{\phi_1^2 + \phi_2^2 + \phi_3^2}$ de l'écart proportionnel de chacune de ces parties est plus grande qu'elle ne serait sans cela, c'est à dire pour $P = 0$. Ainsi le dénominateur de l'expression (xiii) de n^2 est approché par défaut. Mais, vu la formule (xii), l'intégrale $\iiint W dU$ y est aussi évaluée par défaut dans le numérateur. Ces erreurs se compensent par suite en partie, et l'on conçoit que la formule (xiii) de n^2 soit encore assez approchée, comme l'a reconnu M. de Saint-Venant, même pour des valeurs assez grandes du rapport de P à Q (p. 1411). See our Arts. 366–69.

[1456.] *Sur la construction géométrique des pressions que supportent les divers éléments plans se croisant en un même point d'un corps, et sur celle des déformations qui se produisent autour d'un tel point. Journal de mathématiques, T. III., pp. 147–152. Paris, 1877.*

This paper contains an elegant and simple method of proving the fundamental theorems in stress and strain without using any of the properties of surfaces of the second degree. It might advantageously be followed by elementary text-books on Elasticity and Geology.

Let T_1, T_2, T_3 be the principal tractions and s_1, s_2, s_3 the principal stretches, each set in descending order of magnitude. Instead of considering these two systems as they stand, Boussinesq first subtracts from the members of either half the sum of the greatest and least tractions, or of the greatest and least stretches respectively. He thus obtains, if $R = \frac{1}{2}(T_1 - T_3)$ and $S = \frac{1}{2}(s_1 - s_3)$, the systems:

$$R, \; T, \; -R \quad \text{and} \quad S, \; s, \; -S,$$

where T and s are what the mean principal traction and stretch become ; clearly T and s have values lying respectively between R, $-R$ and S, $-S$. These second systems evidently only differ from the first by the superposition of either a uniform pressure or a uniform stretch respectively in all directions, and consequently the maximum and minimum values of stress and strain obtained from these two reduced systems will have the same direction as those of the two primitive systems.

[1457.] Let the direction-cosines of any plane over which the stress is F be $\cos\alpha$, $\cos\beta$, $\cos\gamma$, then for the reduced system

$$F = \sqrt{R^2(\cos^2\alpha + \cos^2\gamma) + T'^2\cos^2\beta} = \sqrt{R^2 - (R^2 - T'^2)\cos^2\beta},$$

and F will have direction angles α', β', γ' such that:

$$\cos \alpha' = \frac{R \cos \alpha}{F}, \quad \cos \beta' = \frac{T \cos \beta}{F}, \quad \cos \gamma' = -\frac{R \cos \gamma}{F}$$

From these results Boussinesq easily deduces the following construction:

A partir de l'origine et dans le plan des deux plus grandes forces principales R, T, on mènera, d'un même côté de la force principale moyenne T, deux droites inclinées, sur cette force moyenne, l'une de l'angle donné β que fait avec elle la normale à l'élément superficiel proposé, l'autre de l'angle β' dont le cosinus vaut $(T/F) \cos \beta$, en donnant à celle-ci la longueur

$$F = \sqrt{R^2 - (R^2 - T^2) \cos^2 \beta} \; ;$$

puis on imprimera à ces deux droites deux rotations égales et contraires autour de la force principale moyenne T: à l'instant où la première droite viendra coïncider avec la normale à l'élément plan, la seconde représentera la pression qui lui est appliquée (pp. 148–9).

[1458.] Clearly the maximum value of F is reached in the plane of xz (or that of R, $-R$), and it then has the value R. The angle χ between F and the normal to the plane across which it acts is given by

$$\cos \chi = \frac{R (\cos^2 \alpha - \cos^2 \gamma) + T \cos^2 \beta}{F},$$

and therefore when $\beta = \pi/2$, $\cos \chi = \cos 2\alpha$. Thus the traction and shear components of F for the plane xz are respectively

$$F \cos 2\alpha \quad \text{and} \quad F \sin 2\alpha.$$

We see then that (for the primitive as well as the reduced system) the maximum shear is across a plane the normal to which bisects the angle between the greatest and least tractions and its magnitude $= R = \frac{1}{2} (T_1 - T_3)$. This is Hopkins' Theorem: see our Art. 1368*.

Thus the greatest and least total stresses, the greatest and least tractive stresses and the greatest shear all lie in one plane, i.e. that of the greatest and least principal tractions (p. 150).

[1459.] If the stretches are small,—so that their squares, as is usually the case, may be neglected,—then precisely similar results follow for the distribution of strain. In the reduced system the shift of one terminal of a line of unit length relative to the other terminal gives, if it be measured perpendicular to the line itself, the change in direction of the given line. This change of angle is numerically greatest for the bisectors of the directions of greatest and least stretch and is then equal to S and $-S$ respectively. Hence the change of angle between these two bisectors will be the maximum slide and has for its value $2S = s_1 - s_3$, or the difference between the greatest and least stretches.

[1460.] *Sur les problèmes des températures stationnaires, de la torsion et de l'écoulement bien continu, dans les cylindres ou les tuyaux dont la section normale est un rectangle à côtés courbes ou est comprise entre deux lignes fermées. Journal de mathématiques*, T. VI., pp. 177–186. Paris, 1880. This memoir is really a discussion of the solution of the equation

$$\frac{d^2u}{dx^2} + \frac{d^2u}{dy^2} = 0,$$

by conjugate functions. It refers to Thomson and Tait's solution of the torsion problem in terms of such functions and to the hydrodynamic analogies of those authors and of Boussinesq himself: see our Art. 1430 and Chapter XIV.

[1461.] *Calcul des dilatations linéaires éprouvées par les éléments matériels rectilignes appartenant à une portion infiniment petite d'une membrane élastique courbe, que l'on déforme, et démonstration très simple du théorème de Gauss sur la déformation des surfaces inextensibles. Recueil de la Société des sciences de Lille.* T. VIII., pp. 381–90. Lille, 1880. See also the *Comptes rendus*, T. LXXXVI., pp. 816–8. Paris, 1878. This is a geometrical investigation of the stretch of an element at the origin on the surface

$$2z = rx^2 + 2sxy + ty^2,$$

when this surface is strained into

$$2z = r'x^2 + 2s'xy + t'y^2.$$

Boussinesq obtains a general expression for the stretch, which he then supposes to be zero,—or applies to the case of an inextensible membrane. In this case Gauss's theorem as to curvature follows at once, and expressions for the shifts of any point in the neighbourhood of the origin are obtained.

The analysis is easy and the results are not very complex in form.

[1462.] *Formules de la dissémination du mouvement transversal dans une plaque plane indéfinie. Comptes rendus*, T. CVIII., pp. 639–45. Paris, 1889. Fourier in his *Théorie analytique de la chaleur* (§§ 411–2) gives an integral of the equation for the transverse vibrations of an infinite elastic plate: see also our Art. 207*. Suppose the mid-plane of this plate to coincide with the plane of xy, then the equation to be solved is of the form

$$\frac{d^2w}{d(bt)^2} + \left(\frac{d^2}{dx^2} + \frac{d^2}{dy^2}\right)^2 w = 0 \quad\ldots\ldots\ldots\ldots\ldots(i),$$

where b^2 is a constant for the plate and may be taken as a factor of t^2 : see our Art. 385.

Fourier's solution of 1818 applied only to initial shifts. Boussinesq proposes in the first place to generalise it by considering also initial velocities. He does not seem to have noticed that this more general case had also been dealt with by Fourier in his *Théorie analytique* of 1822. Thus he gives a solution of (i) subject to the conditions for $t = 0$, that:

$$w = f(\xi, \eta), \quad dw/dt = f_1(\xi, \eta)$$

at the point ξ, η of the plane x, y. Here f and f_1 are two functions of ξ, η which vary gradually from point to point of the plane and vanish at an infinite distance.

Boussinesq further discusses what he holds to be the delicate and rather obscure point as to the real value taken by Fourier's solution when $t = 0$.

The solution obtained by Boussinesq is given by

$$w = \frac{1}{\pi} \iint f(x + 2a\sqrt{bt},\ y + 2\beta\sqrt{bt}) \sin \rho^2 \, da d\beta$$

$$+ \frac{1}{\pi} \int_0^t dt \iint f_1(x + 2a\sqrt{bt},\ y + 2\beta\sqrt{bt}) \sin \rho^2 \, da d\beta,$$

where $\rho^2 = a^2 + \beta^2$, and the limits for the integrations with regard to a and β are determined by

$$\xi = x + 2a\sqrt{bt}, \quad \eta = y + 2\beta\sqrt{bt},$$

ξ and η being taken over the whole area of the initial disturbance.

This should be compared with Poisson's solution given in our Art. 425.

[1463.] *Leçons synthétiques de Mécanique générale servant d'introduction au cours de Mécanique physique de la Faculté des Sciences de Paris.* Paris, 1889.

This work of 132 pages discusses the general mechanical principles which may be supposed to govern the systems of molecules by aid of which the physicist conceptualises the action of physical bodies. As in the note of 1891 (see our Art. 1464) Boussinesq draws the important distinction between the actual velocities and accelerations of individual molecules and the mean *local* velocity and *local* acceleration, which are those of a particle conceived to consist of an infinitely great number of individual molecules (pp. 72–7). The seventh lecture deals with general notions as to stress and leads up to the topics of the eighth which

1464] BOUSSINESQ. 223

contains some judicious remarks on the physiological and psychological aspects of force. Boussinesq holds that the only intelligible conception of force is the mass product of acceleration :

Et gardons-nous de confondre cette quantité précise, constituant le seul sens positif ou démontré des forces mécaniques, avec la signification relativement vague *d'effort musculaire* mais surtout avec celle, encore moins définie, de *cause physique*, que le mot *force* rappelle également ; notions qu'il faut laisser à d'autres champs d'étude, où notre esprit ne peut malheureusement prétendre qu'à un degré de clarté médiocre (p. 89).

Boussinesq divides the total internal energy of a body into two parts, namely, an elastic and a thermal energy (pp. 105-6). He demonstrates that the former or strain-energy (*l'énergie de ressort, l'énergie potentielle d'élasticité* etc.) depends only on the initial and final configurations of the system, provided the system be in space of uniform temperature, and the changes of configuration be made so slowly that the equilibrium of temperature is infinitely little destroyed at each instant : see our Chapter XIV.

[1464.] In a note in the *Comptes rendus* (T. CXII., pp. 1054-6, Paris, 1891) entitled : *Théorie élastique de la plasticité et de la fragilité des corps solides*, M. Brillouin starts from the hypothesis,—that to any definite homogeneous strain of a body corresponds always an absolutely unique system of elastic stresses, but to a definite system of elastic stresses there does not necessarily correspond a definite strain. This leads him up to some remarks on the Poisson-Navier hypothesis of intermolecular force. He holds that this hypothesis is not fundamentally erroneous, but requires modification owing to the fact that individual molecules are in motion. This motion may be oscillatory and of small amplitude in the case of a true solid, but it may still be sufficient to modify intermolecular action. Great pressures convert the movement of oscillation into one of translation, and this forms the explanation of set, rupture, flow, etc.

This note of Brillouin led to the publication by Boussinesq of another entitled : *Sur l'explication physique de la fluidité (Comptes rendus,* T. CXII., pp. 1099-1102, Paris, 1891) referring to similar opinions expressed by himself in his *Leçons synthétiques...*(see our Art. 1463) and in still unpublished lectures delivered at the Sorbonne in 1887-9. Boussinesq cites from the manuscript of his lectures a general description of how he has applied ideas similar to those of Brillouin to throw light on the elasticity, viscosity and internal friction of fluids.

Section II.

Memoirs on Wave Motion and the Elastic Theory of the Ether.

[1465.] *Essai sur la théorie de la lumière.* Comptes rendus, Tome LXI., pp. 19–21. Paris, 1865. This is an abstract of a memoir by the author. It commences with the following words :

Le Mémoire que j'ai l'honneur de soumettre à l'Académie des Sciences est relatif à la théorie de l'élasticité (p. 19).

It does not appear, however, to have been ever published in its entirety, although doubtless portions of it were incorporated in the memoirs of 1867 and 1868 : see our Arts. 1467 and 1478.

The first part of the memoir appears to have contained the deduction of the equations of elasticity for an isotropic medium when terms of the second order in the shift-fluxions are retained. The second part of the memoir applied the theory developed in the first to the vibrations which constitute light. The theories of double refraction of Fresnel, MacCullagh and Neumann were deduced as special cases, but the author appears to have met with the difficulty that the explanation of the dispersive power in this manner involves a *pouvoir considérable d'extinction*. Boussinesq merely suggests that in transparent bodies there may be :

une action spéciale, destinée à contre-balancer ce pouvoir d'extinction, et par suite à diminuer l'opacité (p. 21).

[1466.] *Équations des petits mouvements des milieux isotropes comprimés.* Comptes rendus, T. LXV., pp. 167–70. Paris, 1867. This is an abstract of a memoir afterwards published at length in Liouville's *Journal*: see our Arts. 1467–71. The equations are here obtained by a process slightly different from that of the memoir in the *Journal*.

[1467.] *Mémoire sur les ondes dans les milieux isotropes déformés.* Journal de mathématiques, T. XIII., pp. 209–41. Paris, 1868.

Boussinesq studies in this memoir the vibratory motion of an isotropic medium which has been subjected to "initial stress"

(see our Arts. 616*, 1210* and 129). This initial stress may be
of two kinds: (i) a traction uniform in all directions which does
not change the isotropy of the medium to aeolotropy and which
may be represented by K, and (ii) a system of initial principal
tractions A, B, C, producing aeolotropy symmetrical with respect
to three planes at right angles in an element of the elastic solid.
Of these latter tractions Boussinesq writes:

> Nous admettrons que, dans la portion considérée du corps, les
> éléments plans normalement pressés ou tirés par les actions déformatrices
> gardent la même direction à tous les instants consécutifs, et que ces
> forces A, B, C varient avec le temps de manière à conserver entre elles
> les mêmes rapports B/A, C/A. Si nous désignons par a, b, c trois
> nombres constants, proportionnels à A, B, C, et du même ordre de
> petitesse que les dilatations linéaires éprouvées par le corps pendant sa
> déformation, les rapports A/a, B/b, C/c seront égaux entre eux, et à une
> même fonction F du temps t. La fonction $F(t)$ peut d'ailleurs être
> quelconque: elle se réduit à une constante, si les actions déformatrices
> restent les mêmes toujours; elle sera nulle ou constante à partir d'une
> certaine valeur de t, si A, B, C deviennent elles-mêmes nulles ou
> constantes au bout d'un certain temps. Quoi qu'il en soit, cette fonction
> étant supposée connue, la constitution du corps, à chaque instant, ne
> dépendra plus que de a, b, c (pp. 211-2).

This quotation indicates Boussinesq's assumptions.

[1468.] Referring back to our Art. 231, we see that the ϵ, ϵ', ϵ''
of that article might have been taken equal to the a, b, c of Boussinesq.
He assumes the stresses to be linear functions of the shift-fluxions and
the coefficients of these functions to be linear functions of a, b, c.
Then by considerations (i) of symmetry with regard to the planes of
the initial principal tractions, (ii) of the initial isotropy which must
continue to exist in whole or part according as $a = b = c$, or two of them
are equal, and (iii) of the invariability of the stresses when the body
is rotated as a whole, expressions are deduced for the stresses as
functions of a, b, c, K and *eight* constants p, l, l', l'', m, m', m'' and n
(pp. 212-8). The method adopted involves no appeal either to the
rari-constant molecular theory or to the principle of work. If the latter
be adopted an additional relation is found between the constants of
the form:

$$l'' = n - p \ldots\ldots\ldots\ldots \ldots\ldots\ldots\ldots\ldots(i).$$

Here p is the ratio $A/a = B/b = C/c$. Thus independently of a, b, c
there will be K and *seven* other constants.

[1469.] We can easily deduce Boussinesq's results from those
of the memoir of Saint-Venant discussed in our Arts. 230-2. This

memoir immediately follows Boussinesq's in the same volume of the *Journal*. But the functions of the initial stresses which occur in the expressions for the elastic stresses are, it must be remembered, quoted by Saint-Venant from a memoir which proceeds on rari-constant lines (see our Art. 232). Boussinesq reaches less general expressions, but they are not open to a criticism of the same kind. His method is however too long for reproduction here.

Referring to Art. 232 let us take

$$\widehat{xx}_0 = T + P\epsilon, \quad \widehat{yy}_0 = T + P\epsilon', \quad \widehat{zz}_0 = T + P\epsilon'',$$
$$\widehat{yz}_0 = 0, \quad \widehat{zx}_0 = 0, \quad \widehat{xy}_0 = 0.$$

These agree in form with Boussinesq's values for the initial stresses. We then have the following types of traction and shear:

$$\begin{aligned} \widehat{xx} &= (T + P\epsilon)(1 + s_x - s_y - s_z) + \{a + l\epsilon + m(\epsilon' + \epsilon'')\} s_x \\ &\quad + \{\delta' + p\epsilon'' + q(\epsilon' + \epsilon)\} s_y + \{\delta' + p\epsilon' + q(\epsilon + \epsilon'')\} s_z, \\ \widehat{yz} &= T\sigma_{yz} + P\left(\epsilon'' \frac{dv}{dz} + \epsilon' \frac{dw}{dy}\right) + \{\delta + r\epsilon + s(\epsilon' + \epsilon'')\}\sigma_{yz} \end{aligned} \right\} \quad ....(ii),$$

where, as is shewn in Art. 231:

$$a = 2\delta + \delta', \quad m = 2r + p, \quad l + m = 4s + 2q \ldots\ldots\ldots\ldots(iii).$$

Here there are *fourteen* constants in the expressions for the stresses, but the three relations (iii) reduce them to *eleven*, which exactly agrees with Boussinesq's number (*twelve*) when the relation (i), due to the principle of work (and tacitly assumed by Saint-Venant in the form of equations (ii) of our Art. 231) is adopted. The equations (ii) are exactly Boussinesq's in form, although he uses different constants[1].

[1470.] Substituting in the body stress-equations we find as a type of the body shift-equations:

$$\begin{aligned} \rho \frac{d^2u}{dt^2} &= \{\delta + \delta' + (p + r)(\epsilon + \epsilon' + \epsilon'') + (q + s - p - r)\epsilon\} \frac{d\theta}{dx} \\ &\quad + \{\delta + T + r(\epsilon + \epsilon' + \epsilon'') - (r - s)\epsilon\} \nabla^2 u \\ &\quad + (P - r + s)\left(\epsilon \frac{d^2u}{dx^2} + \epsilon' \frac{d^2u}{dy^2} + \epsilon'' \frac{d^2u}{dz^2}\right) \\ &\quad + (q + s - p - r)\frac{d}{dx}\left(\epsilon \frac{du}{dx} + \epsilon' \frac{dv}{dy} + \epsilon'' \frac{dw}{dz}\right) \ldots\ldots\ldots\ldots (iv). \end{aligned}$$

[1] The following give the change in notation from our results (ii) to Boussinesq's on his p. 218; our constants being on the left of the equalities: $T = K$, $P = p$, $\epsilon = a$, $\epsilon' = b$, $\epsilon'' = c$, $a = l + 2m - K$, $\delta = m - K$, $\delta' = l + K$, $l = l' + 2(n + m' - m'')$, $m = l' + n + 2(m' + m'')$, $p = l'$, $q = l' + l'' + p$, $r = m' + m''$, $s = m'$. Further among the constants of our equation (v) compared with Boussinesq's equation (8) p. 221: $\rho = \delta$, $\lambda_1 = \lambda\delta$, $\lambda_2 = \lambda'\delta$, $\mu_1 = \mu\delta$, $\mu_2 = \rho\delta$, $\sigma_1 = \sigma\delta$, $\lambda_3 = \nu\delta$.

Put $\delta + \delta' + (p + r)(\epsilon + \epsilon' + \epsilon'') = \lambda_1,$ $q + s - p - r = \lambda_2 = \lambda_3,$

$\delta + T' + r(\epsilon + \epsilon' + \epsilon'') = \mu_1,$ $-(r - s) = \mu_2,$

$P - r + s = \sigma_1,$

then we have for the type of shift-equation :

$$\rho \frac{d^2 u}{dt^2} = (\lambda_1 + \lambda_2 \epsilon) \frac{d\theta}{dx} + (\mu_1 + \mu_2 \epsilon) \nabla^2 u + \sigma_1 \left(\epsilon \frac{d^2 u}{dx^2} + \epsilon' \frac{d^2 u}{dy^2} + \epsilon'' \frac{d^2 u}{dz^2} \right)$$

$$+ \lambda_3 \frac{d}{dx} \left(\epsilon \frac{du}{dx} + \epsilon' \frac{dv}{dy} + \epsilon'' \frac{dw}{dz} \right) \ldots\ldots\ldots\ldots\ldots \ldots\ldots (v).$$

Boussinesq's equations (p. 221) agree with this, excepting that λ_3 is not necessarily equal to λ_2, unless appeal be made to the principle of work.

The constants $\lambda_1, \lambda_2, (\lambda_3), \mu_1, \mu_2, \sigma_1$ are independent, but if the initial tractions A, B, C as defined in our Art. 1467 are zero, then $P = 0$, and we have $\sigma_1 = \mu_2$.

[1471.] Boussinesq now supposes the quantities A, B, C (or, $P\epsilon$, $P\epsilon'$, $P\epsilon''$) to become after a given epoch constant, and investigates the motion of a plane wave in the medium whose vibrational shifts satisfy equations of the type (v), the quantities $\epsilon, \epsilon', \epsilon''$ being very small. He shews (pp. 222–3) that there will be waves of vibrations (a) almost in and (b) almost normal to the wave front (*quasi-transverse* and *quasi-longitudinal* waves), the divergence depending on terms of the same order as $\epsilon, \epsilon', \epsilon''$. There will be *exactly* transverse waves if $\mu_2 + \lambda_3 = 0$, and *exactly* longitudinal if :

$$(\mu_2 + \lambda_2) \epsilon = (\mu_2 + \lambda_2) \epsilon' = (\mu_2 + \lambda_2) \epsilon'',$$

or, in general if $\mu_2 + \lambda_2 = 0$. Thus if the principle of work hold the conditions reduce to the single one $\mu_2 + \lambda_2 = 0$, which will be found by (iii) to reduce to $l = m$. In the case of rari-constant isotropy we ought to have in equations (ii) of our Art. 231, $d = d'$, $e = e'$, $f = f'$, or $p = r$, $q = s$; whence it follows that $l = m$ involves $p = q$ and $r = s$, or a perfectly isotropic medium. Hence no *exactly* transverse waves can be propagated in a rari-constant isotropic medium, however initially strained, unless the medium remain isotropic.

[1472.] Boussinesq next proceeds to discuss the quasi-transverse and quasi-longitudinal waves, but to do more than indicate his results would lead us too far into the theory of light. On pp. 223–37 he deals with the directions of vibration, plane of polarisation, wave-surface, etc. of quasi-transverse waves.

In the case of $\sigma_1 = 0$ Boussinesq obtains exactly Fresnel's wave-surface; when σ_1 is not zero he shews (p. 229) how in this case to deduce the wave-surface from Fresnel's. In the former case the plane of polarisation is, as in Fresnel's theory, perpendicular to the direction of vibration. It is only possible for σ_1 to be zero, without at the same time μ_2 being zero, if P be not zero, or since double refraction then depends on the finiteness of P, it is only possible in an isotropic medium in which the initial strains are different in different senses, i.e. A, B, C must not in this case be zero. If $\sigma_1 = \mu_2$, but differs from zero, which arises when $P = 0$, or the initial stresses reduce to a uniform traction T in all directions, then the wave-surface is exactly that of Fresnel, but the direction of vibration lies in the plane of polarisation, or Boussinesq's theory agrees with that of Neumann and MacCullagh. Thus the multi-constant equations of an isotropic medium subjected to initial tractions can be made to cover the theories of both Fresnel and Neumann as special cases.

[1473.] In the discussion of the *quasi-longitudinal* vibrations (pp. 237–8) Boussinesq shews that if the medium be such that it can propagate exactly transverse and exactly longitudinal vibrations, then the velocity of the longitudinal waves will be the same *whatever be their direction*. He considers it very improbable that this manner of propagating longitudinal waves can be characteristic of any but an exactly isotropic medium. Such isotropy, however, he holds to be inconsistent with the physical properties of a doubly refracting medium, or he concludes that such a medium ought not to have the power of propagating waves of *exactly* transverse or *exactly* longitudinal vibrations in all directions. This argument Boussinesq suggests may be taken in conjunction with the others raised by Saint-Venant against Green's theory; see our Arts. 147, 229 and 265.

[1474.] The memoir concludes with a *Généralisation* (pp. 238–41), in which Boussinesq supposes the tractions A, B, C to remain for a certain time proportional to one set ϵ, ϵ', ϵ'' of deformations, and after the lapse of this time to become proportional to another set. He shews that the general results of the memoir still hold, and in particular that the distribution of elasticity is still *ellipsoidal:* see our Arts. 139 and 142.

[1475.] *Étude sur les vibrations rectilignes et sur la diffraction dans les milieux isotropes et dans l'éther des cristaux. Journal de mathématiques,* T. XIII., pp. 340–71. Paris, 1868. *Comptes rendus,* T. LXV., pp. 672–3, 1867. This memoir starts in the first place from the equations for the vibrations of an isotropic elastic medium of the type:

$$\frac{d^2u}{dt^2} = (\lambda + \mu)\frac{d\theta}{dx} + \mu\nabla^2 u \ \ldots\ldots\ldots\ldots\ (i),$$

and supposes the vibrations to be (*a*) rectilinear and (*b*) of very short period. It supposes the amplitude and direction of the vibrations to vary from point to point of the medium, and then investigates their laws.

If χ and ϕ be functions of x, y, z, and l, m, n be the direction-cosines of the rectilinear vibration of period τ, Boussinesq seeks solutions of the system of equation (i) of the types:

$$
\left.
\begin{aligned}
u &= l\chi \cos \frac{2\pi}{\tau} (t - \phi), \\
v &= m\chi \cos \frac{2\pi}{\tau} (t - \phi), \\
w &= n\chi \cos \frac{2\pi}{\tau} (t - \phi)
\end{aligned}
\right\} \quad \dots\dots\dots\dots\dots\text{(ii)},
$$

where, ω being the velocity of the wave, the terms in $(\tau\omega)^2$ are supposed negligible in the final equations as a result of (b).

Obviously $\phi =$ a constant is the equation to the wave-front.

In the case of the above isotropic medium, it is found that the rectilinear vibrations are either accurately longitudinal or accurately transversal, i.e. are either perpendicular or parallel to the wave-front (pp. 342–5). Boussinesq finds that for longitudinal vibrations only three surfaces are possible wave-fronts namely parallel planes, coaxial circular cylinders and concentric spheres (pp. 348–9) and the same is true for transverse waves, if the vibrations be supposed limited in direction to the lines of curvature (pp. 351–3). In the case of longitudinal vibrations however the amplitude has the same value for all points of the same wave-front (pp. 348–9), while for transverse waves the amplitude varies from one point to another of the same *curve of vibration* in the inverse ratio of the distance of this curve from the neighbouring curve of vibration. Boussinesq terms a *curve of vibration* the curve the tangent to which is the direction of vibration of the particle at the point of contact. Thus in transverse vibrations the curves of vibration are a family of curves lying in the front of the wave (pp. 350–3).

Pour établir toutes ces lois, nous avons supposé que u, v, w variaient d'une manière continue d'un point aux points voisins; ce n'est qu'à cette condition que l'on peut poser les équations (i), et négliger dans (3) [equations obtained from our (i) by substituting (ii) in them and equating to zero the coefficients of the cosine and sine of $\frac{2\pi}{\tau} (t - \phi)$]

les termes en $\tau^2\omega^2$. Or cette condition n'est pas satisfaite à une trop petite distance du centre de l'ébranlement, dans les ondes sphériques. Donc les lois obtenues ne sont vraies qu'à partir d'une onde sphérique centrale, dont nous appellerons plus loin ζ le rayon, qui est très-petit et presque insensible (p. 353).

[1476.] The memoir in the next place (pp. 353–65) discusses *quasi-transversal* vibrations in a medium of which the elastic equations are of the form :

$$\frac{d^2u}{dt^2} = (1 + a)\left\{(\lambda + \mu)\,\frac{d\theta}{dx} + \mu\nabla^2 u\right\},$$

$$\frac{d^2v}{dt^2} = (1 + b)\left\{(\lambda + \mu)\,\frac{d\theta}{dy} + \mu\nabla^2 v\right\}, \qquad \text{(iii)},$$

$$\frac{d^2w}{dt^2} = (1 + c)\left\{(\lambda + \mu)\,\frac{d\theta}{dz} + \mu\nabla w\right\}$$

a, b, and c being small as compared with unity.

These are of the same type as those Boussinesq deduces from his elastic theory of light (see our Art. 1480) for a doubly refracting medium. They are also practically identical with those of Sarrau and others. Boussinesq considers only the quasi-transverse vibrations corresponding to waves propagated from the origin of coordinates and his conclusions are indicated in the following words :

Ces ondes sont celles de Fresnel, et les vibrations sont dirigées sensiblement, en chacun de leurs points, suivant la projection, sur le plan tangent à l'onde en ce point, du rayon qui y aboutit. Les lignes de vibration sont à très-peu près des ellipses sphériques, ayant leurs foyers sur les axes optiques ; leurs trajectoires orthogonales sont des courbes sphériques de même nature. L'amplitude est soumise à trois lois ; elle varie : 1° suivant un même rayon, en raison inverse de la distance à l'origine ; et de plus, sur une même onde : 2° suivant une même ligne de vibration, en raison inverse de la distance de cette ligne à la ligne de vibration voisine ; 3° suivant une trajectoire orthogonale aux lignes de vibration, en raison inverse de la distance de cette trajectoire à la trajectoire voisine. En appelant r le rayon mené de l'origine à un point quelconque, U, U' les angles qu'il fait avec les deux axes optiques, ces trois lois reviennent à dire que le carré de l'amplitude est égal à une constante divisée par le produit $r^2\sin U\sin U'$. C'est la formule qu'obtient M. Lamé, dans ses *Leçons sur l'élasticité*, § 126, par une tout autre voie et pour des milieux biréfringents d'une autre espèce (p. 341).

[1477.] The above results hold generally for elastic media of the type (iii). On pp. 365–71 Boussinesq applies the laws he has deduced to the special optical problems of diffraction and of the definition (*la délimitation*) of rays of light. To discuss his results would, however, lead us beyond our proper field.

[1478.] *Théorie nouvelle des ondes lumineuses. Journal de mathématiques*, T. XIII., pp. 313–39. Paris, 1868. This memoir was presented to the *Académie* on August 5, 1867, and a résumé of the theory appeared in the *Comptes rendus*, T. LXV., pp. 235–9, 1867. Various additions to the memoir containing expansions of the theory will be referred to in the sequel.

[1479.] Boussinesq's theory is an elastical one and therefore must be referred to in this *History*, but the details of its application to the phenomena of light lie outside our field and we must refer the reader to the original memoirs for them. Boussinesq makes the following assumptions :

(*a*) The free ether may be regarded as an isotropic elastic solid.

(*b*) Its density and elasticity inside and outside transparent bodies is sensibly the same.

(*c*) The velocity of wave motion in the ether of space is so different from the velocity of sound in solid bodies, that it is reasonable to suppose that it is the ether in transparent bodies and not the material medium of those bodies which transmits light.

(*d*) The vibrations of the ether produce vibrations of the same period in the molecules of the transparent body, but the amplitudes of these are so small that they do not produce any sensible *elastic* action between the particles of the body.

(*e*) The displacements of the molecules of the body are functions of the shifts of the ether in the immediate neighbourhood of those molecules and in certain cases of the shift-speeds. Boussinesq expresses this analytically by saying that to a first approximation we may assume the shifts of a particle of the body to be linear functions of the shifts and shift-fluxions with regard to space (and in some cases also with regard to time) of the ether. If u, v, w be the shifts of the point x, y, z of the ether, he writes (*Note* of 1872, pp. 364–5 : see our Art. 1449) :

Les déplacements u_1', v_1', w_1' d'une molécule pondérable deviennent donc des fonctions de u, v, w et de leurs dérivées partielles des divers ordres, fonctions qu'on peut supposer linéaires (vu la petitesse excessive des variables) et sans termes constants, comme on le fait toujours en cas pareil, par l'emploi de la série de Taylor, quand on étudie une fonction aux environs d'un point pour lequel elle s'annule, et qu'on n'aperçoit aucune raison de supposer ses dérivées premières nulles ou discontinues en ce point. Les déplacements moyens u_1, v_1, w_1, suivant les axes, de la matière pondérable contenue à l'intérieur d'un petit volume quelconque s'obtiendront en multipliant la masse de chacune des molécules qui en font partie par son déplacement parallèle à l'axe considéré, et en divisant la somme des produits pareils par la masse totale des molécules ; ces déplacements moyens seront donc aussi, à fort peu près, des fonctions linéaires, sans termes constants, des déplacements u, v, w de l'éther, en un point pris à l'intérieur du volume considéré ou tout près, et de leurs dérivées par rapport à x, y, z.

(f) The product of the density (ρ) of the ether into the component of its shift parallel to any axis is of the same order as the
product of the density (ρ_1) of the body into the component of its shift
in the same direction. Boussinesq in reality bases this upon (d); he
considers the vibrations in the ether only to produce discordant actions
between the ponderable molecules, actions incapable of continuing in
and for themselves, i.e. giving rise to no elastic forces in the material
body. In this case:

les quantités de mouvement prises, suivant trois axes rectangulaires
de coordonnées, par la matière pondérable, ne peuvent grandir d'un instant
à l'autre qu'autant que celles de l'éther grandissent elles-mêmes : l'hypothèse
la plus naturelle qu'on puisse faire, sur les rapports qu'ont entre elles les
premières et les secondes de ces quantités de mouvement, consiste à admettre
qu'elles sont du même ordre de grandeur (*Note* of 1873, p. 363 : see our Art.
1449).

[1480.] The resultant per unit volume of the elastic forces parallel
to the axis of x due to the ether is of the form

$$(\lambda + \mu)\frac{d\theta}{dx} + \mu\nabla^2 u,$$

or is of the order, $$\frac{\mu}{\rho} \times \rho\,\frac{d^2 u}{dx^2},$$

i.e.   (velocity of light)$^2 \times \rho\,\dfrac{d^2 u}{dx^2}$.

Similarly the elastic forces parallel to the axis of x due to the elasticity
of the material body will be of the order

$$(\text{velocity of sound})^2 \times \rho_1\,\frac{d^2 u_1}{dx^2}.$$

Hence it follows from (f) that the ratio of these forces is of the same
order as the ratio of the square of the velocity of light to that of the
velocity of sound, and accordingly the latter force may be neglected as
compared with the former in dealing with the motion.

If the transparent body has no velocity of translation as a whole
comparable with the velocity of light-vibrations, the mean accelerations
of its ponderable particles will be

$$\frac{d^2 u_1}{dt^2}, \qquad \frac{d^2 v_1}{dt^2}, \qquad \frac{d^2 w_1}{dt^2},$$

and consequently its components of inertia per unit volume due to the
vibratory motion will be the products of these quantities and ρ_1, or,
what is the same thing, the total reactions, per unit volume of the
ether, exerted on the ether by the ponderable particles of the body will be

$$-\rho_1\frac{d^2 u_1}{dt^2}, \qquad -\rho_1\frac{d^2 v_1}{dt^2}, \qquad -\rho_1\frac{d^2 w_1}{dt^2},$$

because the sole force acting on the ponderable particles is due to the impulse of the ether upon them.

Hence the equations for the motion of the ether in a transparent body are of the type :

$$(\lambda + \mu)\frac{d\theta}{dx} + \mu\nabla^2 u - \rho_1\frac{d^2 u_1}{dt^2} = \rho\frac{d^2 u}{dt^2} \quad \dots\dots\dots\dots \text{(i)},$$

(p. 318 of the Memoir of 1868).

[1481.] Boussinesq now makes u_1 a function of u, v, w and their fluxions and retains only the first powers of these quantities: see (e) of our Art. 1479. The nature of this function will depend on the aeolotropic or isotropic character of the medium (pp. 319–21). From the equation which results by substituting u_1 in (i) Boussinesq deduces the laws of dispersion and of rotatory polarisation (pp. 321–7), of double refraction (pp. 328–31), and of 'elliptic' double refraction with application to the case of quartz (pp. 331–38). He makes on pp. 338–9 a few remarks on the conditions at the interface of two media. He points out that if the shifts in the two media along the interface be made equal and the stresses across the interface, then the elasticity of the ether in both media being the same, all the first space-fluxions of the shifts will be equal at the interface. The latter condition of continuity, which does not hold in the case of unequal elasticities, seems needful in order to obtain expressions for the intensities of the reflected and refracted waves, which will agree with experiment.

C'est pourquoi nous avons cru devoir admettre la constance d'élasticité de l'éther dans deux milieux adjacents. Quant à la constance de sa densité, elle n'est pas nécessaire à notre théorie ; mais elle nous paraît une condition naturelle de la constance d'élasticité, et nous la regardons comme vraisemblable (p. 339).

[1482.] *Addition au mémoire intitulé : Théorie nouvelle des ondes lumineuses. Journal de mathématiques*, T. XIII., pp. 425–438. Paris, 1868. This paper extends Boussinesq's theory to the explanation of the laws of the following phenomena : § I, the refractive and rotatory powers of a mixture of various transparent substances,—the theory easily leads to the usual approximately correct laws,—§ II, the magnetic rotation of the plane of polarised light, and § III, the aberration of light when the transmitting body is in motion,—Fresnel's formula for the velocity of the wave of light in terms of the velocity of the medium and its refractive index being deduced. A more complete consideration of the problem of aberration will be found in the *Note Complémentaire* of 1873: see our Art. 1449.

[1483.] *Sur les lois qui régissent, à une première approximation, les ondes lumineuses propagées dans un milieu homogène et transparent d'une contexture quelconque.* Journal de mathématiques, T. XVII., pp. 167–76. Paris, 1872.

Taking the equations of his memoir of 1868 (see our Art. 1480) of the type:

$$\rho \frac{d^2u}{dt^2} + \rho_1 \frac{d^2u_1}{dt^2} = (\lambda + \mu) \frac{d\theta}{dx} + \mu \nabla^2 u,$$

Boussinesq neglects the rotatory and dispersive powers of the medium and supposes u_1, v_1, and w_1 to be linear functions of u, v, w. He shews that by a proper choice of axes, they can be given by

$$u_1 = \alpha u - \zeta v + \epsilon w,$$
$$v_1 = \beta v - \delta w + \zeta u,$$
$$w_1 = \gamma w - \epsilon u + \delta v,$$

where $\alpha, \beta, \gamma, \delta, \epsilon, \zeta$ are constants depending on the nature of the medium. All known transparent bodies are but slightly aeolotropic from the optical standpoint and hence $\alpha - \beta, \beta - \gamma, \gamma - \alpha, \delta, \epsilon, \zeta$ are very small quantities. Boussinesq on this assumption investigates the motion of plane-waves in such a medium. If δ, ϵ, ζ be zero, he deduces formulae agreeing with those of Fresnel for double refraction. If they be not zero, there are two waves whose vibrations are very nearly but not accurately transverse. Boussinesq studies the laws of these quasi-transverse waves, and shews that they may be enunciated, like those for the accurately transverse waves of Fresnel, by the use of the 'optic ellipsoid of elasticity' with the aid of a certain right-line of given length or 'optical axis of asymmetry', which passes in a given direction through the centre of the ellipsoid. For crystals which have one of their mineralogical axes perpendicular to the plane of the others, Boussinesq's results agree with those of Fresnel. For crystals, where this perpendicularity does not hold, this agreement is not a necessity of Boussinesq's analysis, and special experiments, Boussinesq considers, would have to be made to determine whether these crystals also may be considered as having three rectangular planes of optical symmetry (Fresnel's hypothesis) or whether they obey the more general laws possible according to the present memoir.

[1484.] There can be little doubt, having regard both to the comprehensiveness of its results and to the clearness of its hypotheses, that Boussinesq's elastic theory of light is in most respects superior to the elastic theories proposed by MacCullagh, Neumann, Green, Lamé and others. Should elastic theories be destined, as seems probable, to be replaced by an electro-magnetic theory, there are yet, probably, many points of Boussinesq's investigations which might be usefully transferred from the one theory to the other [1].

SECTION III.

The Application of Potentials to the Theory of Elasticity.

[1485.] *Les déplacements qu'entraînent de petites dilatations ou condensations quelconques produites, dans tout milieu homogène et isotrope indéfini, sont calculables à la manière d'une attraction newtonienne. Comptes rendus*, T. XCIV., pp. 1648–50. Paris, 1882.

The body-shift equations for small vibrations may be written in the form :

$$\left(\frac{d^2}{dt^2} - b^2\nabla^2\right)(u, v, w) = (a^2 - b^2)\frac{d\theta}{d(x, y, z)} \ldots\ldots\ldots (i),$$

where $b^2 = \mu/\rho$ and $a^2 = (\lambda + 2\mu)\rho$: see our Art. 1394.

We have at once :

$$d^2\theta/dt^2 = a^2\nabla^2\theta \ldots\ldots\ldots\ldots\ldots\ldots (ii).$$

Let θ at the point x_1, y_1, z_1 be given by $\theta(x_1, y_1, z_1, t)$ and let us consider matter distributed throughout space of which the density at time t equals $\theta/4\pi$; thus, this density will vary with

---

[1] That the investigations of the *elastic* theory of light are not rendered utterly useless by the adoption of other constitutions for the ether has been pointed out by Sir William Thomson in a paper published for the first time in Vol. III. of his *Mathematical and Physical Papers*, pp. 436—65. Cambridge, 1890.

the time. Its potential at a point x, y, z distant r from x_1, y_1, z_1 will be

$$\Phi = \frac{1}{4\pi} \iiint \frac{\theta\,(x_1,\,y_1,\,z_1,\,t)}{r}\, d\varpi \ldots\ldots\ldots\ldots\text{(iii)},$$

where $d\varpi$ is an element of volume at x_1, y_1, z_1.

Then $\qquad\qquad \nabla^2\Phi = -\,\theta\,(x,\,y,\,z,\,t)\ldots\ldots\ldots\ldots\ldots\text{(iv)},$

by Poisson's Theorem.

Boussinesq now puts:

$$\nabla^2\Phi = \frac{1}{4\pi} \iiint \nabla^2\theta\,\frac{d\varpi}{r} = \frac{1}{4\pi a^2} \iiint \frac{d^2\theta}{dt^2}\,\frac{d\varpi}{r} = \frac{1}{a^2}\,\frac{d^2\Phi}{dt^2},$$

referring to equations (i) and (ii). His first equality seems to me legitimate only if we suppose θ and $d\theta/dx$, $d\theta/dy$, $d\theta/dz$ to vanish at an infinite distance. For, I presume, it is obtained by putting $\dfrac{d}{dx}\left(\dfrac{1}{r}\right) = -\dfrac{d}{dx_1}\left(\dfrac{1}{r}\right)$ and integrating by parts. Whence if $d\sigma$ be an element of an infinite bounding surface we must have $\iint \dfrac{\theta\,(x_1,\,y_1,\,z_1,\,t)}{r}\,d\sigma$ and $\iint \dfrac{1}{r}\,\dfrac{d\theta\,(x_1,\,y_1,\,z_1,\,t)}{d\,(x_1,\,y_1,\,z_1)}\,d\sigma$ zero.

Thus we find that θ and Φ are to be found from the equations:

$$d^2\Phi/dt^2 = a^2\nabla^2\Phi, \quad \text{and} \quad \nabla^2\Phi = -\,\theta,$$

while (i) gives us for the corresponding parts of the shifts:

$$(u,\,v,\,w) = -\,d\Phi/d\,(x,\,y,\,z).$$

Thus the shifts are equal to the Newtonian attractions due to a distribution of matter of density varying with the time and proportional to the corresponding dilatation.

In addition to the above there may be parts of u, v, w for which $\theta = 0$; they are evidently given from (i) by solving the equations:

$$\frac{d^2}{dt^2}\,(u,\,v,\,w) = b^2\nabla^2\,(u,\,v,\,w).$$

If we integrate both sides of equation (ii) over the volume of a surface so large that $d\theta/dx$, $d\theta/dy$, $d\theta/dz$ vanish at its boundary, we have

$$\frac{d^2}{dt^2} \iiint \theta\,d\varpi = 0,$$

or $\iiint \theta\,d\varpi$ is constant through the motion. Thus the total mass

of which Φ is the potential is constant and finite if θ initially
does not differ from zero except in limited regions. From this
result Boussinesq (p. 1650) easily deduces that the potential
Φ due to θ tends to zero as the time increases.

[1486.] *Application des potentiels à l'étude de l'équilibre et du
mouvement des solides élastiques, principalement au calcul des
déformations et des pressions que produisent, dans ces solides, des
efforts quelconques exercés sur une petite partie de leur surface ou
de leur intérieur ; Mémoire suivi de notes étendues sur divers points
de physique mathématique et d'analyse.* Paris, 1885. This work
of 722 pages is the most considerable contribution of Boussinesq
to the theory of Elasticity.

This volume also forms T. XIII. of the *Recueil de la Société des Sciences
de Lille.* Lille, 1885. Portions of its contents appeared in separate memoirs
contributed to the *Comptes rendus* for the years 1878–83 : see T. LXXXVI.,
pp. 1260–3; T. LXXXVII., pp. 402–5, 519–22, 687–9, 978–9, 1077–8;
T. LXXXVIII., pp. 277–9, 331–3, 375–8, 701–4, 741–3 ; T. XCIII., pp.
703–6, 783–5 ; T. XCV., pp. 1052–4 ; 1149–52 ; T. XCVI., pp. 245–8. In
addition certain results of the theory were published by Saint-Venant
on pp. 374–407 *a* and pp. 881–8 of the French edition of *Clebsch :* see
our Art. 338.

The object of the work is summed up on pp. 15–19 of the
Introduction : But et résumé de ce travail. It is to discuss the
solution of the following three problems :

(*a*) A small portion of the surface of a large mass of elastic
material is subjected to local stress, it is required to find the strain
at other points.

(*b*) A small portion of the surface of a large mass of elastic
material is subjected to a given deformation, to find the stresses
due to this deformation.

(*c*) A small portion of the interior of a large mass of elastic
material is subjected to a given body-force (e.g. magnetic action),
to find the strains produced by this body-force.

In solving these problems Boussinesq considers the shifts of
the material to be zero at an infinite distance from the small
portion subjected to local load, but he remarks that this condition
is only introduced to fix our ideas.

En réalité, les phénomènes produits dans cette région resteraient, sans doute, à peu près les mêmes, si le corps était entièrement libre dans l'espace ; car ses parties éloignées, vu la grandeur relative de leur étendue et, par suite, de leur masse totale, seraient maintenues sensiblement immobiles par leur inertie, durant bien plus de temps qu'il n'en doit falloir à la petite partie, de masse insignifiante, qui est contiguë à la région d'application des forces exercées, pour suivre les impulsions qu'on lui imprime et se mettre dans un état quasi-permanent de déformation et de tension, c'est-à-dire à fort peu près, dans l'état d'équilibre correspondant à la fixation de l'ensemble du corps et à l'intensité actuelle effective des actions extérieures qu'il supporte à l'endroit que l'on considère (p. 19).

The work will thus be seen to deal with the influence of local stress or strain in producing stress or strain at other parts of an elastic solid.

[1487.] We need not stay long over pp. 15–49 of the work, which give a brief *résumé* of the conclusions reached in the remainder of the volume. We may, however, just refer to one or two points in these pages :

(*a*) Lamé and Clapeyron's solution of the problem of an infinite elastic solid bounded by a plane subjected to an arbitrary distribution of tractive load (see our Art. 1018*) is noticed on pp 19–23. Boussinesq remarks on the extreme difficulty of obtaining physical results from a solution in quadruple integrals.

Further with regard to Lamé, Boussinesq points out that if $U = \int dm/r$ be the ordinary potential of a mass m, then just as

$$\nabla^2 U = -4\pi\rho,$$

ρ being the density, so if $V = \int r\, dm$ be the 'direct potential,' then $\nabla^2\nabla^2 V = -8\pi\rho$ and is not zero as Lamé supposes : see our Art. 1062*. Lamé's solution thus applies only to elastic cases where there is no internal action (represented by ρ), but the 'direct potential' has most value in exactly the reverse case, where there is internal action or ρ cannot be put zero (pp. 31–3).

(*b*) Pp. 23–36 give an account of the action of local stresses which deserves very careful study by those who assert that the mathematicians have failed to perceive the influence of surface loading in producing local strain. The several points of this *résumé* are discussed at greater length in our treatment of the body of the book itself.

(*c*) On pp. 35–6 Boussinesq points out that a "local perturbation", or local strain produced by the local application of a load system in statical equilibrium, decreases less rapidly with the distance in the case of a body with its three dimensions comparable among themselves, than when one or two of its dimensions are small as compared with a third. For example the influence of torsional couples applied round the normal to the contour of a plate has been shewn by Thomson and Tait (see our Art. 1523) to be practically insensible at twice the thickness of the plate from the edge, but the strain produced by a similar local application of forces in equilibrium to the surface of an infinite elastic mass only decreases as we depart from the centre of application inversely as the cube of the distance : see our Art. 1521.

(d) § 8 of the *Introduction* (pp. 36–41) contains some valuable remarks on the difference between solutions by 'potentials' and by Fourier's series, the former tending in many cases to exhibit and the latter to obscure the physical laws of the phenomena investigated. To obtain the physical characteristics of a phenomenon we want to ascertain how a small action at one point influences the condition of affairs at a second, and we shall rarely be able to ascertain this from a Fourier's series when we increase indefinitely the dimensions of a medium in two or more directions. The Fourier's series will then in general be replaced by a multiple integral, and the evaluation of this integral with respect to certain auxiliary variables leads to a simpler solution, which it will generally be needful to further integrate over the region to which the action has been applied. This simpler solution is what is more directly obtained by the method of potentials.

(e) Finally we may note that pp. 42–9 give a *résumé* of some interesting general properties of the potential. One of these is of considerable interest and simplicity, and according to Boussinesq it had not yet been noticed. It consists in the following statement : The Laplacian ∇^2 of any function at a point is equal to one third of the mean value of the second derivative of the function taken for all possible directions round the point (p. 44).

[1488.] Boussinesq next discusses and defines the various types of potential with which he is about to deal.

Let dm be the element of a mass m situated at the point x_1, y_1, z_1, and let the integration be over the whole of the mass m. Then, if $r = \sqrt{(x - x_1)^2 + (y - y_1)^2 + (z - z_1)^2}$, we have the following types of potential at the point x, y, z :

the ordinary or *inverse potential* $U = \int \dfrac{dm}{r}$, ... (iii),

the *direct potential* $V = \int r\, dm$, ... (iv),

(see our Art. 1062*),

the *logarithmic potential with three variables*

$$\psi = \int \log (z - z_1 + r)\, dm, \dots\dots\dots\dots\dots (v).$$

The latter will be called the *first* logarithmic potential; Boussinesq introduces the words 'with three variables' to distinguish it from the *cylindrical potential* or *logarithmic potential with two variables* given by $\int \log r\, dm$, where $r = \sqrt{(x - x_1)^2 + (y - y_1)^2}$

Finally we have
the *second logarithmic potential with three variables*,

$$\Psi = \int [- r + (z - z_1) \log (z - z_1 + r)]\, dm \ldots\ldots\ldots \text{(vi)}.$$

If $z_1 = 0$, or the matter be spread over the plane xy we have $\psi = \int \log (z + r)\, dm$, which is the form in which we shall generally have to deal with it.

We easily find
$$\left.\begin{array}{l} \dfrac{d\psi}{dz} = \int \dfrac{dm}{r} = U, \\[2mm] \dfrac{d\Psi}{dz} = \int \log (z - z_1 + r)\, dm = \psi \end{array}\right\} \ldots\ldots\ldots\ldots\text{(vii)}.$$

Hence if $\nabla^2 U = 0$, we have at once $\nabla^2 \psi = 0$ and $\nabla^2 \Psi = 0$. Further, as in our Art. 1062* $\nabla^2 \nabla^2 V = 2\nabla^2 U = 0$. These relations are deduced on the supposition that none of the matter m lies in the space *within* which we are considering the values of these potentials (pp. 57–61).

[1489.] The most general types of solution by aid of potential functions are given by Boussinesq in a memoir of 1888 entitled:

Équilibre d'élasticité d'un solide sans pesanteur, homogène et isotrope, dont les parties profondes sont maintenues fixes, pendant que sa surface éprouve des pressions ou des déplacements connus, s'annulant hors d'une région restreinte où ils sont arbitraires. Comptes rendus, T. CVI. pp. 1043-8 and 1119–23. Paris, 1888.

I propose to indicate Boussinesq's solutions at this stage and return for the discussion of special cases to the *Application des potentiels*.

Boussinesq's problem is described in the following words:

Comme le corps dont il s'agit n'est à considérer que dans le voisinage de sa région superficielle, sensiblement plane, assujettie aux pressions ou aux déplacements de valeurs autres que zéro, l'on peut, en adoptant pour plan des xy le plan tangent en un point central de cette région et pour axe de z la normale correspondante dirigée vers l'intérieur, le regarder comme limité d'un côté par ce plan et indéfini dans tous les autres sens, ou comme remplissant la moitié de l'espace où les ordonnées z sont positives (p. 1043).

Boussinesq supposes biconstant isotropy. He writes at the surface $z = 0$:

$$p_x = -\frac{1}{2\mu}\,\widehat{zx}, \quad p_y = -\frac{1}{2\mu}\,\widehat{yz}, \quad p_z = -\frac{1}{2\mu}\,\widehat{zz} \quad \ldots\ldots(i),$$

$$u = u_0, \qquad v = v_0, \qquad w = w_0 \ldots\ldots\ldots\ldots (ii),$$

where $p_x, p_y, p_z, u_0, v_0, w_0$ are functions of x and y only. Further he considers u, v, w and u_0, v_0, w_0, to vanish at infinity. He then proposes to solve the body-shift equations of type [1]

$$\frac{d\theta}{dx} + k\nabla^2 u = 0 \quad \ldots\ldots\ldots\ldots\ldots\ldots (iii),$$

(where $k = 1 - 2\eta$, η being the stretch-squeeze ratio), subject to one of the following conditions:

(a) u_0, v_0 and w_0 are known functions of x and y;

(b) p_x, p_y and p_z are known functions of x and y;

(c) u_0, v_0 and p_z are known functions of x and y;

(d) p_x, p_y and w_0 are known functions of x and y.

The sub-case of (b) where p_z is arbitrary, but $p_x = p_y = 0$, had been solved in 1828 by Lamé and Clapeyron by aid of quadruple integrals: see our Art. 1019*. A simplified form of this case had been given by Boussinesq in 1878, and the complete solutions of (a) and (b) before the end of 1882. In (a) and (b) however he had been preceded by V. Cerruti (*Ricerche intorno all' equilibrio de' corpi elastici isotropi. Reale Accademia dei Lincei, Serie 3ª, Memorie della Classe di scienze fisiche...* T. XIII., pp. 81–122. Roma, 1882). In the memoir of 1888 Boussinesq solves for the first time (c) and the most general form of (d).

[1490.] Let U, V, W, ϕ, ϕ_1, Φ be functions of x, y, z which have their Laplacian ∇^2 zero, and their derivatives in x, y, z finite and

---

[1] He states this problem in a slightly different form on pp. 50—3 of his *Application des potentiels...* and shows on pp. 54—6 that the solution is unique for the case (b) referred to below. His method is similar to that of Kirchhoff and Clebsch: see our Arts. 1278, and 1331.

continuous even for $z = 0$, then if $k' = 2k + 1$, we have the following two systems of solutions for (iii):

$$
\left.
\begin{aligned}
u &= \frac{dU}{dz} - \frac{z}{k'} \frac{d}{dx} \left(\frac{dU}{dx} + \frac{dV}{dy} + \frac{dW}{dz} \right), \\[2mm]
v &= \frac{dV}{dz} - \frac{z}{k'} \frac{d}{dy} \left(\frac{dU}{dx} + \frac{dV}{dy} + \frac{dW}{dz} \right), \\[2mm]
w &= \frac{dW}{dz} - \frac{z}{k'} \frac{d}{dz} \left(\frac{dU}{dx} + \frac{dV}{dy} + \frac{dW}{dz} \right)
\end{aligned}
\right\} \quad \ldots\ldots\ldots (iv),
$$

and:

$$
\left.
\begin{aligned}
u &= \frac{d\phi}{dx} - \frac{z}{k'} \frac{d\Phi}{dx} - 2 \frac{d\phi_1}{dy}, \\[2mm]
v &= \frac{d\phi}{dy} - \frac{z}{k'} \frac{d\Phi}{dy} + 2 \frac{d\phi_1}{dx}, \\[2mm]
w &= \frac{d\phi}{dz} - \frac{z}{k'} \frac{d\Phi}{dz} + \Phi
\end{aligned}
\right\} \quad \ldots\ldots\ldots\ldots\ldots (v).
$$

These results may be easily verified by substitution in (iii). The solution (v) leads at once by (i), for $z = 0$, to:

$$
\left.
\begin{aligned}
p_x &= -\frac{d}{dx} \left(\frac{d\phi}{dz} + \frac{k}{k'} \Phi \right) + \frac{d^2\phi_1}{dy\,dz}, \\[2mm]
p_y &= -\frac{d}{dy} \left(\frac{d\phi}{dz} + \frac{k}{k'} \Phi \right) - \frac{d^2\phi_1}{dx\,dz}, \\[2mm]
p_z &= -\frac{d}{dz} \left(\frac{d\phi}{dz} + \frac{1+k}{k'} \Phi \right)
\end{aligned}
\right\} \quad \ldots\ldots\ldots\ldots (vi).
$$

By aid of (iv)—(vi) we can now indicate the method of solving the several cases (a), (b), (c) and (d).

[1491.] *Case* (a). Take U, V, W the potentials due to distributions of matter of densities $-u_0/2\pi$, $-v_0/2\pi$, $-w_0/2\pi$ respectively over the plane xy, then u, v, w will satisfy the body-shift equations, vanish at an infinite distance from the origin of disturbance and be equal to u_0, v_0, w_0 at the plane xy. Thus we have

$$
\left.
\begin{aligned}
U &= -\frac{1}{2\pi} \iint \frac{u_0}{r}\, dx\, dy, \\[2mm]
V &= -\frac{1}{2\pi} \iint \frac{v_0}{r}\, dx\, dy, \\[2mm]
W &= -\frac{1}{2\pi} \iint \frac{w_0}{r}\, dx\, dy
\end{aligned}
\right\} \quad \ldots\ldots\ldots\ldots\ldots (vii),
$$

where r is the distance between the point of the plane $z = 0$ at which the shift is u_0, v_0, w_0 and the point of the mass at which the shift is u, v, w.

[1492.]   *Case* (b).   We may divide this into two sub-cases, which combined will give us the general solution for the case of any surface loading.

Subcase (i). $p_x = p_y = 0$, $p_z =$ any arbitrary function of x and y. This is Lamé and Clapeyron's case.

In equations (vi) put $\phi_1 = 0$ and $k' d\phi/dz + k \Phi = 0$, and we have $p_x = p_y = 0$, $d^2\phi/dz^2 = kp_z$, when $z = 0$.

Assume
$$\phi = -\frac{k}{2\pi} \iint \log (z + r)\, p_z\, dx\, dy,$$

whence
$$\Phi = \frac{k'}{2\pi} \iint \frac{p_z}{r}\, dx\, dy \qquad \Bigg\} \dots\dots\dots\dots (viii).$$

Let p_0 be the true normal pressure $= 2\mu p_z$, then by aid of (v) and (viii) we find the following general expressions for the shifts, etc. :

$$u = -\frac{1}{4\pi\mu} \left\{ \frac{d^2 \iint r p_0 d\omega}{dx\, dz} + (1 - 2\eta)\frac{d}{dx} \iint \log (z + r)\, p_0 d\omega \right\},$$

$$v = -\frac{1}{4\pi\mu} \left\{ \frac{d^2 \iint r p_0 d\omega}{dy\, dz} + (1 - 2\eta)\frac{d}{dy} \iint \log (z + r)\, p_0 d\omega \right\}, \Bigg\} \dots \text{(ix)},$$

$$w = -\frac{1}{4\pi\mu} \left\{ \frac{d^2 \iint r p_0 d\omega}{dz^2} - (3 - 2\eta) \iint \frac{p_0 d\omega}{r} \right\},$$

$$\theta = \frac{1}{2\pi\mu} (1 - 2\eta)\frac{d}{dz} \iint \frac{p_0 d\omega}{r}$$

where $d\omega$ is an element of the plane xy.

[1493.]   Boussinesq in his *Application des potentiels...*(pp. 65–6) obtains, for $z = 0$, the values of $\iint (z\rho d\omega/r^3)$ and $\iint (z^3\rho d\omega/r^5)$, where $\rho d\omega$ is an element at x_1, y_1 of a mass distributed over the surface $z = 0$, and r is the distance of x, y, z from this element.   He writes :

$$x_1 = x + R \cos \chi, \quad y_1 = y + R \sin \chi,$$

and therefore $r^2 = z^2 + R^2$.

Whence, if we put $R = zq$, $dR = zdq$ (z constant), we have, if $\rho (x_1, y_1)$ be the value of ρ at x_1, y_1 :

$$\iint \frac{z\rho d\omega}{r^3} = \int_0^{2\pi} d\chi \int_0^\infty \rho (x + zq \cos \chi,\ y + zq \sin \chi)\frac{q\, dq}{(1 + q^2)^{\frac{3}{2}}}, \text{ whence}$$

$$\iint \frac{z\rho d\omega}{r^3} = 2\pi\rho (x, y).$$

$$\text{Similarly :} \quad \iint \frac{z^3\rho d\omega}{r^5} = \frac{2\pi}{3} \rho (x, y) \qquad \Bigg\} \text{ when } z = 0 \dots\dots\dots\dots\dots(\mathbf{x}).$$

Thus we find for the value of θ in (ix) when $z = 0$, $\theta = -p_0/(\lambda + \mu)$. Further we have $(dw/dz)_{z=0} = -\frac{1}{2}p_0/(\lambda + \mu)$. Thus both the dilatation (i.e. in this case negative, or a rarefaction) and the squeeze at the surface are proportional to the normal pressure. The former of these results had been previously obtained by Lamé and Clapeyron: see our Art. 1019*.

It will be noted that in this subcase:

$$\frac{du}{dy} - \frac{dv}{dx} = 0, \quad \text{and} \quad \frac{dp_x}{dy} - \frac{dp_y}{dx} = 0 \quad \ldots\ldots\ldots\ldots(\text{xi}).$$

The first of these equations expresses that the twist about the axis of z is zero.

[1494.] *Subcase* (ii). Let us take $\Phi = -\dfrac{k'}{1+k}\dfrac{d\phi}{dz}$ in (vi), then we have besides $p_z = 0$:

$$\left.\begin{array}{l} p_x = -\dfrac{1}{1+k}\dfrac{d^2\phi}{dx\,dz} + \dfrac{d^2\phi_1}{dy\,dz}, \\[2mm] p_y = -\dfrac{1}{1+k}\dfrac{d^2\phi}{dy\,dz} - \dfrac{d^2\phi_1}{dx\,dz} \end{array}\right\} \ldots\ldots\ldots\ldots(\text{xii});$$

whence, remembering $\nabla^2\phi$ and $\nabla^2\phi_1$ are both zero, we have for $z = 0$:

$$\frac{d^3\phi}{dz^3} = (1+k)\left(\frac{dp_x}{dx} + \frac{dp_y}{dy}\right), \quad \frac{d^3\phi_1}{dz^3} = -\left(\frac{dp_x}{dy} - \frac{dp_y}{dx}\right) \ldots (\text{xiii}).$$

Boussinesq now takes for ϕ and ϕ_1 second logarithmic potentials (see our Art. 1488), for which we have the second differential in z equal to the ordinary potential.

Let S_1 and S_2 be the shearing loads applied to the surface $z = 0$ parallel to the axes of x and y; then $S_1 = 2\mu p_x$, $S_2' = 2\mu p_y$, and:

$$\frac{d^3\phi}{dz^3} = \frac{1-\eta}{\mu}\left(\frac{dS_1}{dx} + \frac{dS_2}{dy}\right), \quad \frac{d^3\phi_1}{dz^3} = -\frac{1}{2\mu}\left(\frac{dS_1}{dy} - \frac{dS_2}{dx}\right),$$

whence we can satisfy all the conditions by taking

$$\left.\begin{array}{l} \phi = -\dfrac{1-\eta}{2\pi\mu}\left\{\dfrac{d}{dx}\iint\{z\log(z+r) - r\}\,S_1 d\omega + \dfrac{d}{dy}\iint\{z\log(z+r) - r\}\,S_2 d\omega\right\}, \\[3mm] \phi_1 = -\dfrac{1}{4\pi\mu}\left\{\dfrac{d}{dx}\iint\{z\log(z+r) - r\}\,S_2 d\omega - \dfrac{d}{dy}\iint\{z\log(z+r) - r\}\,S_1 d\omega\right\}, \\[3mm] \Phi = \dfrac{3-4\eta}{4\pi\mu}\left\{\dfrac{d}{dx}\iint\log(z+r)\,S_1 d\omega + \dfrac{d}{dy}\iint\log(z+r)\,S_2 d\omega\right\} \end{array}\right\}$$
$$\ldots\ldots\ldots\ldots\ldots\ldots(\text{xiv}).$$

Equations (xiii) were first obtained by Boussinesq, and are given on p. 80 of his *Application des potentiels*. Their solution (xiv) is due to Cerruti; the above method of reaching that solution is, however,

Boussinesq's. We have only to substitute (xiv) in (v) to completely solve this Subcase. Thus we have obtained solutions of *Cases* (a) and (b) in terms of generalised potentials. Similar investigations occupy pp. 62–80 of Boussinesq's *Treatise*. On pp. 182–6 of the *Treatise*, however, Boussinesq puts into a slightly more compact form the results of *Case* (b). Let us re-write our conclusions in the following manner:

Subcase (i).

$$\left. \begin{aligned} u &= \frac{d\phi}{dx} + \frac{z}{k}\frac{d^2\phi}{dx\,dz}, & \frac{\widehat{zx}}{2\mu} &= \frac{z}{k}\frac{d^3\phi}{dx\,dz^2}, \\ v &= \frac{d\phi}{dy} + \frac{z}{k}\frac{d^2\phi}{dy\,dz}, & \frac{\widehat{yz}}{2\mu} &= \frac{z}{k}\frac{d^3\phi}{dy\,dz^2}, \\ w &= -\frac{1+k}{k}\frac{d\phi}{dz} + \frac{z}{k}\frac{d^2\phi}{dz^2}, & \frac{\widehat{zz}}{2\mu} &= \frac{z}{k}\frac{d^3\phi}{dz^3} - \frac{1}{k}\frac{d^2\phi}{dz^2}. \end{aligned} \right\} \dots(\mathrm{xv}).$$

Here, if $\quad \Psi_z = \iint (z \log(z+r) - r)\, p_0 d\omega$, we have

$$\phi = -\frac{k}{4\mu\pi}\frac{d\Psi_z}{dz}$$

Subcase (ii).

$$\left. \begin{aligned} u &= \frac{d\phi}{dx} + \frac{z}{1+k}\frac{d^2\phi}{dx\,dz} - 2\frac{d\phi_1}{dy}, & \frac{\widehat{zx}}{2\mu} &= \frac{z}{1+k}\frac{d^3\phi}{dz^2\,dx} + \frac{1}{1+k}\frac{d^2\phi}{dx\,dz} - \frac{d^2\phi_1}{dy\,dz}, \\ v &= \frac{d\phi}{dy} + \frac{z}{1+k}\frac{d^2\phi}{dy\,dz} + 2\frac{d\phi_1}{dx}, & \frac{\widehat{yz}}{2\mu} &= \frac{z}{1+k}\frac{d^3\phi}{dz^2\,dy} + \frac{1}{1+k}\frac{d^2\phi}{dy\,dz} + \frac{d^2\phi_1}{dx\,dz}, \\ w &= -\frac{k}{1+k}\frac{d\phi}{dz} + \frac{z}{1+k}\frac{d^2\phi}{dz^2}, & \frac{\widehat{zz}}{2\mu} &= \frac{z}{1+k}\frac{d^3\phi}{dz^3}. \end{aligned} \right.$$

Here, if $\qquad \Psi_x = \iint \{z \log(z+r) - r\}\, S_1 d\omega,$

and $\qquad\qquad \Psi_y = \iint \{z \log(z+r) - r\}\, S_2 d\omega,$

$$\phi = -\frac{1+k}{4\pi\mu}\left(\frac{d\Psi_x}{dx} + \frac{d\Psi_y}{dy}\right), \quad \phi_1 = -\frac{1}{4\pi\mu}\left(\frac{d\Psi_y}{dx} - \frac{d\Psi_x}{dy}\right)$$

$$\dots\dots\dots\dots(\mathrm{xvi}).$$

Hence we find for (xv) and (xvi) combined the stresses across any plane parallel to the boundary:

$$(\widehat{zx},\ \widehat{yz},\ \widehat{zz}) = \frac{1}{2\pi}\frac{d^3}{dz^3}(\Psi_x, \Psi_y, \Psi_z) - \frac{z}{2\pi}\frac{d^3}{d(x,y,z)\,dz^2}\left(\frac{d\Psi_x}{dx} + \frac{d\Psi_y}{dy} + \frac{d\Psi_z}{dz}\right)$$

$$\dots\dots\dots\dots(\mathrm{xvii}).$$

For a single element of surface stress $p_0 d\omega$, $S_1 d\omega$, $S_2 d\omega$, we easily deduce, remembering the relations (vii) of Art. 1488:

$$(\widehat{zx},\ \widehat{yz},\ \widehat{zz}) = -\frac{3z\,d\omega}{2\pi r^3}\left(\frac{x}{r}S_1 + \frac{y}{r}S_2 + \frac{z}{r}p_0\right)\frac{(x,y,z)}{r}\ \dots(\mathrm{xviii}).$$

Now the last factor in each case is the direction-cosine of the ray OP joining the elementary area parallel to the surface at a point P with the point Q at which the load is applied ; further, the factor

$$d\omega\left(\frac{x}{r}S_1 + \frac{y}{r}S_2 + \frac{z}{r}p_0\right)$$

is the component of the load parallel to OP ; hence Boussinesq (p. 187) propounds the following law :

Toute action extérieure exercée en un point de la surface d'un solide se transmet à l'intérieur, sur les couches matérielles parallèles à la surface, sous la forme de pressions dirigées exactement à l'opposé de ce point, et qui égalent, pour l'unité d'aire, le produit du coefficient $3/2\pi$ par la composante, suivant leur propre sens, de la force extérieure donnée, par l'inverse du carré de la distance r au même point d'application et par le rapport de la profondeur z de la couche à cette distance r.

[1495.]   *Case* (c). u_0, v_0 and p_z given over $z = 0$.
We can combine (iv) and (v) of Art. 1490 in the following manner. Take $W = 0$, $\phi = \phi_1 = 0$.   We have at the surface[1] $z = 0$:

$$u_0 = \frac{dU}{dz}, \quad v_0 = \frac{dV}{dz}, \quad p_z = \frac{k}{k'}\left(\frac{du_0}{dx} + \frac{dv_0}{dy}\right) - \frac{1+k}{k'}\frac{d\Phi}{dz}.$$

These lead us at once to

$$U = -\frac{1}{2\pi}\iint\frac{u_0 d\omega}{r}, \quad V = -\frac{1}{2\pi}\iint\frac{v_0 d\omega}{r},$$

$$\Phi = \frac{k'}{2\pi(1+k)}\iint\frac{p_z d\omega}{r} - \frac{k}{2\pi(1+k)}\left(\frac{d}{dx}\iint\frac{u_0 d\omega}{r} + \frac{d}{dy}\iint\frac{v_0 d\omega}{r}\right)\bigg\} \quad \text{...(xix)}.$$

Writing
$$\frac{1}{2\pi}\iint\frac{p_z d\omega}{r} = -P_z,$$

we find finally

$$(u, v) = \frac{d(U, V)}{dz} - \frac{z}{1+k}\frac{d}{d(x, y)}\left(\frac{dU}{dx} + \frac{dV}{dy} - P_z\right),$$

$$w = \frac{k}{1+k}\left(\frac{dU}{dx} + \frac{dV}{dy} - \frac{k'}{k}P_z\right) - \frac{z}{1+k}\frac{d}{dz}\left(\frac{dU}{dx} + \frac{dV}{dy} - P_z\right)\bigg\} \quad \text{...(xx)}.$$

[1496.]   *Case* (d). w_0 and p_x, p_y given over $z = 0$.   In this case

[1] We easily deduce the last of these equations from

$$(dw/dz)_{(z=0)} = -\frac{1}{k'}\left(\frac{du_0}{dx} + \frac{dv_0}{dy}\right), \quad \text{and} \quad \theta_{(z=0)} = \frac{k'-1}{k'}\left(\frac{du_0}{dx} + \frac{dv_0}{dy}\right) - \frac{1}{k'}\frac{d\Phi}{dz}.$$

combining (iv) and (v) we take $U = V = 0$ and $d\phi/dz + \Phi = 0$, whence at the surface $z = 0$ we have:

$$w_0 = \frac{dW}{dz}, \quad p_x = -\frac{k+1}{k'}\frac{d^2\phi}{dx\,dz} + \frac{d^2\phi_1}{dy\,dz} - \frac{k}{k'}\frac{d^2W}{dx\,dz},$$
$$\left. p_y = -\frac{k+1}{k'}\frac{d^2\phi}{dy\,dz} - \frac{d^2\phi_1}{dx\,dz} - \frac{k}{k'}\frac{d^2W}{dy\,dz} \right\} \quad \ldots \text{(xxi)}.$$

From the last two equations we find over $z = 0$:

$$\frac{d^3\phi}{dz^3} = \frac{k'}{k+1}\left(\frac{dp_x}{dx} + \frac{dp_y}{dy}\right) + \frac{k}{k+1}\left(\frac{d^2w_0}{dx^2} + \frac{d^2w_0}{dy^2}\right),$$

$$\frac{d^3\phi_1}{dz^3} = -\left(\frac{dp_x}{dy} - \frac{dp_y}{dx}\right).$$

Whence:
$$\phi = -\frac{k}{k+1}W - \frac{k'}{2\pi(k+1)}\left(\frac{dP_x}{dx} + \frac{dP_y}{dy}\right),$$

$$\phi_1 = \frac{1}{2\pi}\left(\frac{dP_x}{dy} - \frac{dP_y}{dx}\right),$$

where
$$P_x = \iint (z\log(z+r) - r)\,p_x\,d\omega,$$
$$P_y = \iint (z\log(z+r) - r)\,p_y\,d\omega.$$

$$\left. \vphantom{\frac{dP_x}{dx}} \right\} \quad \ldots\text{(xxii)}.$$

Further,
$$W = -\frac{1}{2\pi}\iint \frac{w_0\,d\omega}{r},$$

Substituting these values of ϕ, ϕ_1 and W in (iv) and (v), we find after some reductions:

$$u = \frac{1}{\pi}\iint \frac{p_x\,d\omega}{r} - \frac{1}{2\pi(1+k)}\frac{d\chi}{dx},$$

$$v = \frac{1}{\pi}\iint \frac{p_y\,d\omega}{r} - \frac{1}{2\pi(1+k)}\frac{d\chi}{dy},$$

$$w = -\frac{1}{\pi}\frac{d}{dz}\iint \frac{w_0\,d\omega}{r} - \frac{1}{2\pi(1+k)}\frac{d\chi}{dz},$$

where
$$\chi = \frac{d}{dx}\iint rp_x\,d\omega + \frac{d}{dy}\iint rp_y\,d\omega + (1-k)\iint \frac{w_0\,d\omega}{r} - \frac{d^2}{dz^2}\iint rw_0\,d\omega$$

$$\ldots\ldots\ldots\ldots\ldots\ldots\text{(xxiii)}.$$

This completes the set of solutions proposed by Boussinesq as the subject of his memoir. We will now return to his *Application des potentiels*, referring where necessary to the above equations (i)—(xxiii).

[1497.] In the *Application des potentiels* Boussinesq discusses *Case* (b) above by constructing it from three simpler types of integrals (pp. 62-80). For each of these three simpler types he then analyses the nature of the shifts, strains and stresses due to a single element of the potential expressions (pp. 81–98, and 107–8). These investigations, interesting as they undoubtedly are, have still not the same practical importance as that which corresponds to our *Case* (b), *Subcase* (i), for a simple pressure upon a small element of the surface. This occupies pp. 99–107, and we must devote some further space to it.

Referring to our equations (ix) let us consider only the element $p_0 d\omega$ of normal pressure on the surface. Let U represent the shift parallel to the surface at any point distant r from the loaded element in a direction making an angle α with the direction of the pressure, and let w be the shift perpendicular to the surface, then we easily find for points outside $d\omega$:

$$
\left.
\begin{aligned}
U &= \frac{p_0 d\omega}{4\pi\mu r} \left\{ \cos\alpha - \frac{1 - 2\eta}{1 + \cos\alpha} \right\} \sin\alpha, \\
w &= \frac{p_0 d\omega}{4\pi\mu r} \{ \cos^2\alpha + 2(1 - \eta) \}
\end{aligned}
\right\} \quad \dots\dots\dots \text{(xxiv)}.
$$

U is obviously zero for $\alpha = 0$, and again[1] for $\alpha = \cos^{-1}\frac{1}{2}(\sqrt{5 - 8\eta} - 1)$. Between these values U is positive, while between the latter value and $90°$ it is negative. For $\alpha = 90°$, we have

$$
U = -\frac{p_0 d\omega}{4\pi\mu r}(1 - 2\eta) \quad \text{and} \quad W = \frac{p_0 d\omega}{4\pi\mu r}(2 - 2\eta),
$$

or we see that each circle on the surface is depressed more than its radius is diminished.

Although the formulae (xxiv) do not hold when r is vanishingly small, they still do not give infinite values for U and w, for $d\omega$ will be of the order r^2 and therefore $d\omega/r$ vanishes for an infinitely small element. We are obliged in fact for the shifts near $r = 0$ to return to the more complete formula (ix). On the other hand, if p_0 and $d\omega$ be both finite, (xxiv) will hold for all points of the body sufficiently distant from this region (p. 104).

Turning to the stresses, Boussinesq determines what is the total stress across an elementary plane perpendicular to the axis of z, in other words we have to determine $\widehat{zz}$ and $\widehat{zR}$, where $R = \sqrt{r^2 - z^2}$ is the

_____

[1] If η varies from $\frac{1}{2}$ (i.e. for $\mu = 0$) to $\frac{1}{4}$ (i.e. for $\lambda = \mu$), this angle decreases from $90°$ to $68° 32'$ about.

distance from the axis of z to the point at which we are investigating the stress. By the aid of θ as given in (ix) we easily find:

$$\left.\begin{aligned}
\widehat{zz} &= \lambda\theta + 2\mu\frac{dw}{dz} = -\frac{3p_0 d\omega}{2\pi}\frac{\cos^2\alpha}{r^2}\cos\alpha, \\
\widehat{zR} &= \mu\left(\frac{dU}{dz}+\frac{dw}{dR}\right) = -\frac{3p_0 d\omega}{2\pi}\frac{\cos^2\alpha}{r^2}\sin\alpha
\end{aligned}\right\} \quad \ldots\ldots(\text{xxv}).$$

Thus the stress across any elementary plane parallel to the bounding surface $= \dfrac{3p_0 d\omega}{2\pi}\dfrac{\cos^2\alpha}{r^2}$ and is directed from the point at which the elementary pressure $p_0 d\omega$ is applied[1]. From the result Boussinesq draws the following conclusion (p. 105):

Throughout the whole surface of a sphere touching the surface of the given plane at the point of application of the normal force $p_0 d\omega$, the stress, which an elementary plane parallel to the surface of the body is subjected to, is constant. It varies directly as the force $p_0 d\omega$ is directed along the chord from the point of application and for different spheres varies inversely as their surfaces.

This result is independent of the elastic constants of the material; thus we see that the distribution of stress over any plane parallel to the surface is the same for all isotropic bodies (p. 106): see our Art. 1494, *Subcase* (i).

[1498.] Boussinesq next proceeds to discuss the depressions upon the plane surface of an indefinitely great elastic solid due to various distributions of normal pressure (pp. 109–201).

The value of w_0, or the surface depression, is given very easily by (ix), it is:

$$w_0 = \frac{1-\eta}{2\pi\mu}\iint\frac{p_0 d\omega}{r} \quad \ldots\ldots\ldots\ldots\ldots\ldots(\text{xxvi}).$$

Now suppose p_0 taken as a surface density, and let us choose as origin of coordinates the centroid and as axes of x and y the principal axes of this distribution of matter. If P_0 be the total pressure and K_1, K_2 the swing-radii of p_0 round the axes of x and y respectively, further if

$$x = R\cos\chi, \quad y = R\sin\chi,$$

---

[1] Boussinesq has applied this result to an interesting special case bearing on the influence of surface-loading in the problem of beams: *Philosophical Magazine*, Vol. 32, pp. 483–4. London, 1891.

then, by a well-known theorem in potentials due to Poisson (see Minchin's *Treatise on Statics*, Vol. II., p. 307, 1889) we have approximately :

$$w_0 = \frac{1-\eta}{2\pi\mu} \frac{P_0}{R} \left\{ 1 + \frac{K_1^2 + K_2^2}{4R^2} + \frac{3}{4} \frac{K_2^2 - K_1^2}{R^2} \cos 2\chi \right\} \dots \text{(xxvii)}.$$

This may be written

$$w_0 = \frac{1-\eta}{2\pi\mu} \frac{P_0}{R} \left\{ 1 + \frac{K_1^2 + K_2}{4R^2} \left(1 + 3 \frac{K_2^2 - K_1^2}{K_2^2 + K_1^2} \cos 2\chi \right) \right\}.$$

Now the maximum value possible for the term $\dfrac{K_2^2 - K_1^2}{K_2^2 + K_1^2} \cos 2\chi$ is not greater than unity and $K_1^2 + K_2^2$ must be less than the square of the greatest radius-vector of the area to which the pressure is applied, hence in writing

$$w_0 = \frac{1-\eta}{2\pi\mu} \frac{P_0}{R} \dots \dots \text{(xxviii)},$$

we are at most neglecting only $\left(\dfrac{\text{maximum radius-vector}}{\text{distance from centroid}}\right)^2$ of the result. Thus in the case of pressure applied to an area the depression at a distance ten times the maximum radius-vector would be given with less than 1 p.c. error by (xxviii). It is often much less than this, thus for uniform pressure over a circular area, the depression at 4 times the radius is given by (xxviii) with less than 8 p.c. error.

We may note another point with regard to the above result. Suppose the total force P_0 to be zero; then it does not follow that $P_0 (K_1^2 + K_2^2)$ will be zero, but we see that a system of pressures in statical equilibrium, if applied to a small region on the surface of the body will not produce a sensible depression at a small distance from that region, i.e. the depression diminishes as the inverse cube of the distance (pp. 118–9).

[1499.] Returning to (xxvi) we can, either by direct expansion or by an easy application of Legendre's coefficients (see Ferrers' *Spherical Harmonics*, Chapter III.), find w_0 for a distribution of pressure symmetrical round a point of the surface. We should have $p_0 = f(\rho)$, if ρ be the radius vector, and if ρ_1 be the limiting radius of the area to which we apply pressure, R the distance from its centre of the point on the surface at which the depression is w_0 we obtain :

$$w_0 = \frac{1-\eta}{\mu} \int_R^{\rho_1} \overset{\rho > R}{p_0 d\rho} \left\{ 1 + \left(\frac{1}{2}\right)^2 \frac{R^2}{\rho^2} + \left(\frac{1}{2} \cdot \frac{3}{4}\right)^2 \frac{R^4}{\rho^4} + \left(\frac{1}{2} \cdot \frac{3}{4} \cdot \frac{5}{6}\right)^2 \frac{R^6}{\rho^6} + \dots \right\}$$

$$+ \frac{1-\eta}{\mu} \int_\rho^R \overset{\rho < R}{p_0 d\rho} \left\{ \frac{\rho}{R} + \left(\frac{1}{2}\right)^2 \frac{\rho^3}{R^3} + \left(\frac{1}{2} \cdot \frac{3}{4}\right)^2 \frac{\rho^5}{R^5} + \left(\frac{1}{2} \cdot \frac{3}{4} \cdot \frac{5}{6}\right)^2 \frac{\rho^6}{R^6} + \dots \right\}$$

$$\dots \dots \text{(xxix)}.$$

Boussinesq gives a number of interesting cases of this. For example, at a great distance from the loaded area, only the early terms of the second integral are required, and we have

$$w_0 = \frac{1-\eta}{\mu} \int_0^{\rho_1} p_0 d\rho \left(\frac{\rho}{R} + \frac{1}{4} \frac{\rho^3}{R^3} \right),$$

which agrees with (xxvii).

Further the depression w_c at the centre of the loaded area comes from the first term of the first integral and equals

$$w_c = \frac{1-\eta}{\mu} \int_0^{\rho_1} p_0 d\rho \dots\dots\dots\dots\dots (\text{xxix}) \; bis.$$

This shews us that all surface annuli of the same small breadth whatever be their radii, will when subjected to the same stress per unit area produce the same central depression.

Boussinesq finds the value of the central depression for $p_0 \propto \rho^{n-1}$, $p_0 \propto \rho_1^{n-1} - \rho^{n-1}$ and $p_0 \propto (\rho_1^2 - \rho^2)^{-\frac{1}{2}}$ (pp. 119–20). The first two cases correspond to distributions of pressure vanishing at the centre and at the edge of the loaded area respectively. For the same total pressure the depression in the second case is double that in the first; the mean depression in the second case is, however, only $\frac{8}{7}$ that in the first (pp. 125–6). On pp. 121–6 Boussinesq gives expressions for the depression at the edge of the loaded area and for the mean depression over that area.

[1500.] On pp. 126–139 an interesting proposition is proved, and its relation to a corresponding proposition in the case of a circular plate is discussed at some length. Consider two pressures p_0 and p_0' applied to the plane face of an infinite elastic solid uniformly round two circumferences of radii ρ_1 and ρ_1' so that the total loads $2\pi\rho_1 d\rho_1 p_0$ and $2\pi\rho_1' d\rho_1' p_0'$ are equal, then by (xxvi) the depression due to the first load at a point on the second circumference is given by

$$w_0 = \frac{1-\eta}{2\pi\mu} \int_0^{2\pi} \frac{p_0 \rho_1 d\chi d\rho_1}{\sqrt{\rho_1^2 + \rho_1'^2 - 2\rho_1\rho_1' \cos\chi}},$$

and this is exactly equal to

$$w_0' = \frac{1-\eta}{2\pi\mu} \int_0^{2\pi} \frac{p_0' \rho_1' d\chi' d\rho_1'}{\sqrt{\rho_1'^2 + \rho_1^2 - 2\rho_1'\rho_1 \cos\chi'}},$$

or to the depression at the first circumference due to the load on the second, since the total loads are equal. Now suppose the load on one circumference not to be uniformly distributed, then it will produce the same *mean* depression round the second circumference wherever it be placed. Hence the mean depression round the second circumference must be the same for the same load on the first circumference, whatever the law of distribution be. Hence we conclude: *That two equal loads distributed round the perimeters of two concentric circles produce the same mean depressions at each others' circumferences.* A precisely identical law holds for two loaded circumferences concentric with a circular plate the edge of which is either built-in or supported: see our Arts. 336 and 1382.

[1501.] Boussinesq next deals (pp. 139–42) with the case in which the pressure is uniformly distributed over the loaded area. We must first notice a method by which the equation (xxvi) may be easily transformed into an integrable form, when the pressure varies only with the distance from the centre of the area.

Take the point P at which the depression is to be found outside the loaded area and let it be the origin of polar coordinates r and ϕ, ϕ being measured from the line R from the point P to the centre O of the disc. Then by (xxvi)

$$w_0 = \frac{1-\eta}{2\pi\mu} \iint \frac{p_0 r\, dr\, d\phi}{r},$$

$$= \frac{1-\eta}{2\pi\mu} \iint p_0\, dr\, d\phi.$$

Now transforming the variables to ρ and the angle marked ψ in the figure

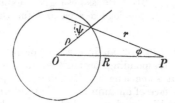

we easily find :

$$w_0 = \frac{1-\eta}{\pi\mu} \int_0^{\rho_1} \int_0^{\pi} \frac{p_0 \rho\, d\rho\, d\psi}{\sqrt{R^2 - \rho^2 \sin^2\psi}}.$$

This may be expressed in the form

$$w_0 = 2\frac{1-\eta}{\pi\mu} \int_0^{\pi/2} \frac{d\psi}{\sin^2\psi} \int_{\rho=\rho_1}^{\rho=0} p_0\, d\sqrt{R^2 - \rho^2 \sin^2\psi}, \quad R > \rho_1 \ldots(\text{xxx}).$$

Now let the point P lie inside the loaded area, the notation remain-

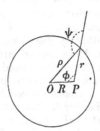

ing the same, then for $\rho = 0$ to R, the integral may be expressed in the form (xxx), but for $\rho = R$ to ρ_1 a slightly different form is needful.

As before, but taking now the annulus outside R:

$$w_0 = \frac{1-\eta}{2\pi\mu} \iint p_0 dr d\phi,$$

$$= -\frac{1-\eta}{\pi\mu} \int_R^{\rho_1} \int_0^\pi \frac{p_0 \rho d\rho d\phi}{\sqrt{\rho^2 - R^2 \sin^2\phi}},$$

or since it involves only $\sin^2\phi$ we have

$$w_0 = 2\frac{1-\eta}{\pi\mu} \int_0^{\pi/2} d\phi \int_{\rho=R}^{\rho=\rho_1} p_0 d\sqrt{\rho^2 - R^2 \sin^2\phi}.$$

Whence we have for a point inside the loaded area:

$$w_0 = 2\frac{(1-\eta)}{\pi\mu} \int_0^{\pi/2} d\psi \left\{ \int_{\rho=R}^{\rho=0} \frac{p_0 d\sqrt{R^2 - \rho^2 \sin^2\psi}}{\sin^2\psi} \right.$$

$$\left. + \int_{\rho=R}^{\rho=\rho_1} p_0 d\sqrt{\rho^2 - R^2 \sin^2\psi} \right\}, \quad R < \rho_1 \;\ldots\ldots(\text{xxxi}).$$

See Boussinesq's pp. 113–7.

[1502.] For the case of p_0 constant we can throw the above into the simple forms:

$$\left. \begin{aligned} w_0 &= 2\frac{1-\eta}{\pi\mu} p_0 \int_0^{\pi/2} \frac{d\psi}{\sin^2\psi} (R - \sqrt{R^2 - \rho_1^2 \sin^2\psi}), \quad R > \rho_1 \\ \text{and} \quad w_0 &= \frac{2(1-\eta)}{\pi\mu} p_0 \int_0^{\pi/2} \sqrt{\rho_1^2 - R^2 \sin^2\psi}\, d\psi, \quad R < \rho_1 \end{aligned} \right\} \ldots(\text{xxxii}).$$

Hence, if w_c be the central depression, w_e that at the edge of the loaded area and $p_0 = P_0/(\pi\rho_1^2)$ we have

$$w_c = \frac{1-\eta}{\pi\mu}\frac{P_0}{\rho_1}, \qquad w_e = \frac{1-\eta}{\pi\mu}\frac{P_0}{\rho_1}\frac{2}{\pi}.$$

Thus: $(w_c - w_e)/w_c = \cdot 363$ about.

Values of w_0 for other points may be obtained directly from (xxxii) or from the series-expression in (xxix). The mean value of w_0 over the loaded area or w_m, is easily found from the second result in (xxxii) by the consideration that:

$$w_m = \frac{2\pi}{\pi\rho_1^2} \int_0^{\rho_1} w_0 R dR,$$

$$= \frac{1-\eta}{\pi\mu}\frac{P_0}{\rho_1}\frac{8}{3\pi} = \tfrac{1}{3}w_e = \frac{8}{3\pi} w_c.$$

See Boussinesq's pp. 125 and 140.

[1503.] It must be noted that a discontinuous change in the normal pressure occurs at the circumference $\rho = \rho_1$, and Boussinesq on pp. 142–9 undertakes an interesting investigation of the nature of the surface-stress along this line of discontinuity. From the results (xxv) we find:

$$\widehat{yz} = -\frac{3}{2\pi} \iint \frac{z^2 (y_1 - y)\, p_0 d\omega}{r^5},$$

where

$$r^2 = (x_1 - x)^2 + (y_1 - y)^2 + z^2,$$

and p_0 is a function of x_1, y_1, or $= f_1(x_1, y_1)$. With the notation of our Art. 1493 we find:

$$\widehat{yz} = -\frac{3}{2\pi} \int_0^{2\pi} \sin \chi\, d\chi \int_0^\infty \frac{f_1 (x + zq \cos \chi,\ y + zq \sin \chi)\, q^2 dq}{(1 + q^2)^{\frac{5}{2}}}.$$

In the limiting case $z = 0$, f_1 takes the value $f_1 (x + \epsilon \cos \chi, y + \epsilon \sin \chi)$ which p_0 has at an infinitely small distance ϵ from x, y. Noting that

$$\int_0^\infty \frac{q^2 dq}{(1 + q^2)^{\frac{5}{2}}} = \tfrac{1}{3},$$

we have still for limiting case

$$\widehat{yz} = -\frac{1}{2\pi} \int_0^{2\pi} f_1 (x + \epsilon \cos \chi,\ y + \epsilon \sin \chi) \sin \chi\, d\chi,$$

and this will in general *not* be zero, if f_1 has sensible variations along the circumference of the circle of infinitely small radius ϵ described round x, y as centre. For example, if x, y be on the boundary of a curve the pressure on one side of which is p_1 and on the other p_2, we have, if the axis of y be parallel to the normal to the curve:

$$\widehat{yz} = -\frac{1}{2\pi} \left[\int_0^\pi p_1 \sin \chi\, d\chi + \int_\pi^{2\pi} p_2 \sin \chi\, d\chi \right]$$

$$= \frac{1}{\pi} (p_2 - p_1).$$

Similarly we can show that under the conditions just stated: $\widehat{zx} = 0$.

Thus all along the circumference of a circular area (on the plane surface bounding an infinite elastic solid) to which a normal pressure p_0 is applied, a radial shearing stress of magnitude p_0/π would have to be applied. In order to avoid the difficulty of this shearing stress at points where the pressure suddenly changes its value, Boussinesq supposes that although the distribution of pressure over the main part of a loaded area may be any whatever, still at the edge of the area the pressure decreases more or less rapidly to zero.

L'effet d'un tel décroissement, supposé purement local, sur les abaissements w_0 produits ailleurs que très près du bord, est d'ailleurs totalement insignifiant ; vu qu'il n'introduit ou ne supprime qu'une pression totale insensible, ne fournissant qu'un potentiel $\iint \dfrac{p_0 d\omega}{r}$ négligeable aux distances finies. En d'autres termes, *ce décroissement supprime les discontinuités qui, sans son existence, se produiraient sur le bord ; mais il n'a aucune influence générale, ou n'oblige à modifier aucune conclusion concernant les déplacements produits à des distances sensibles du contour de la région d'application* (pp. 148–9).

This argument seems quite legitimate.

[1504.] Pp. 149–158 are occupied with some interesting special cases of loaded areas on the plane boundary of an otherwise infinite elastic solid. Let the total load as before be P_0 and its distribution be given by p_0, let w_c be the central, w_e the edge depression of the area, ρ_1 its radius and ω its magnitude.

Then Boussinesq finds :

p_0	w_c	w_e	$(w_c - w_e)/w_c$
$\dfrac{P_0}{\omega}(1+\gamma - 2\gamma\rho^2/\rho_1{}^2)$ (γ being arbitrary)	$\dfrac{3+\gamma}{3}\,\overline{w}$	$\dfrac{18-2\gamma}{9\pi}\,\overline{w}$	$\dfrac{\gamma(3\pi+2)+9(\pi-2)}{3\pi(3+\gamma)}$
$\dfrac{2P_0}{\omega}(1-\rho^2/\rho_1{}^2)$ ($\gamma=1$)	$\dfrac{4}{3}\,\overline{w}$	$\dfrac{16}{9\pi}\,\overline{w}$	·576
$\dfrac{2P_0}{\omega}\rho^2/\rho_1{}^2$ ($\gamma=-1$)	$\dfrac{2}{3}\,\overline{w}$	$\dfrac{20}{9\pi}\,\overline{w}$	$-\,·061$
$\gamma = -\dfrac{9(\pi-2)}{3\pi+2}$ $= -·8993$	$\dfrac{8}{3\pi+2}\,\overline{w}$	$\dfrac{8}{3\pi+2}\,\overline{w}$	0
$\dfrac{P_0}{\omega}$ ($\gamma=0$)	$\overline{w}$	$\dfrac{2}{\pi}\,\overline{w}$	·363
$\dfrac{P_0}{2\omega}\dfrac{1}{\sqrt{1-\rho^2/\rho_1{}^2}}$	$\dfrac{\pi}{4}\,\overline{w}$	$\dfrac{\pi}{4}\,\overline{w}$	0

The value of $\overline{w}$ in the Table is $\dfrac{1-\eta}{\pi\mu}\dfrac{P_0}{\rho_1}$, or the special w_c of our Art. 1502. In the last case the distribution of pressure is so chosen that the whole of the loaded area is *equally depressed*. Since for a uniformly loaded area the mean depression $= \dfrac{8}{3\pi}\,\bar{w}$ (Art. 1502), we see that for the same total load the equally depressed area has a slightly less depression than the mean of the equally loaded area, the ratio of the two depressions being $3\pi^2 : 32$.

On pp. 159–62 Boussinesq extends the case of equal depression to pressure applied over an elliptic loaded area with principal axes $2a$, $2b$. He shews that if the pressure be applied according to the law

$$p_0 = \frac{P_0}{2\omega}\left(1 - \frac{x^2}{a^2} - \frac{y^2}{b^2}\right)^{-\frac{1}{2}},$$

then the elliptic area will be equally depressed[1].

[1505.]   Pp. 167–81 of the work are occupied, partly, with a discussion of some problems on the potential due to Beltrami, and partly with the application of the principle of inversion as in electricity to obtain the depression due to other distributions of pressure. One special case of inversion is worked out on pp. 177–8 and on pp. 180–1. Boussinesq points out how a distribution of pressures may be obtained giving no depression at all outside a given contour, although the pressures themselves extend to regions beyond this contour.

[1506.]   Pp. 182–9 relate to Cerruti's solution for the case of any surface stress whatever over the plane boundary of an otherwise infinite solid. The results are, however, easily obtained by Boussinesq in terms of the second logarithmic potential : see our Art. 1488. Boussinesq on pp. 188–9 discusses an interesting special case which may be just referred to here.

Suppose a single shearing force S_1 applied to the surface element $d\omega$ at the origin.   Then referring to our Equations (xv)–(xvii) we have

$$\Psi_y = \Psi_z = 0, \quad \Psi_x = \{z \log(z + r) - r\}\,S_1 d\omega.$$

Hence we find for the depression w_0 at the surface :

$$w_0 = \frac{S_1 d\omega}{4\pi(\lambda + \mu)}\,\frac{x}{x^2 + y^2}.$$

The contour and the maximum-slope lines of this surface are obviously given by systems of circles lying in the plane $z = 0$, passing through

---

[1] On pp. 162–6 Boussinesq investigates the value of the depression *outside* the uniformly depressed area, i.e. in the last case of the above table.   He finds :

$$w = \frac{\bar{w}}{2} \sin^{-1}\frac{\rho_1}{R}, \; (\rho_1 < R).$$

the origin and having their centres on the axes of x and y respectively. In front and behind the point of application of the shear there will be congruent mound and hollow.

[1507.]   On pp. 190-201 Boussinesq deals with various forms of the potential solution which are embraced in those we have already discussed in our Arts. 1490-6. He points out that as a rule the depression produced in the plane surface of an infinite solid is not proportional to the pressure, i.e. w_0 does not at each point vary as $\widehat{zz}$ $_{(z=0)}$; and he indicates that for pressures applied over limited portions of the surface this proportionality cannot hold.

It is easy, however, to find distributions of normal pressure which will give a depression proportional to that pressure at each point. Looking back at equation (xv) we find for $z = 0$, $\widehat{zz}$ proportional to $d^2\phi/dz^2$, and w_0 to $d\phi/dz$. Hence if we were to take $\phi = e^{-az} \chi(x, y)$, we should have achieved our object provided $\nabla^2\phi = 0$, or

$$\frac{d^2\chi}{dx^2} + \frac{d^2\chi}{dy^2} + a^2\chi = 0.$$

A solution of this is

$$\chi = C \cos(mx + \beta) \cos(m'y + \beta'),$$

provided $m^2 + m'^2 = a^2$.

Thus, $\dfrac{\widehat{zz}}{2\mu} = -\dfrac{a^2}{k}(az + 1)\, e^{-az}\, C \cos(mx + \beta) \cos(m'y + \beta')$,

and we notice that the mean pressure over the plane xy is zero, and that the effect of this pressure, owing to the factor e^{-az}, gets very small at distances from the surface which are only a few times the dimensions of the rectangles within which the surface pressure is alternately positive and negative. A pressure of this kind seems to be, however, of purely theoretical interest.

[1508.]   Boussinesq next turns to the extremely important problem of determining the stresses when a rigid body of known shape is pressed with a given force upon an elastic medium bounded by an infinite plane.   The discussion under its general or special aspects occupies pp. 202-55 and 713-9 of the volume. Boussinesq deals on pp. 202-10 and 713-15 with the general statement of the problem. He supposes that the bodies in contact are smooth, i.e. $\widehat{zx}$ and $\widehat{yz} = 0$ at the surface of contact. Next the total pressure between the bodies is given, and finally if w_c be the central depression $w_c - w_0$ is given over the surface of contact.

Practically we are chiefly concerned with the case in which the surface of contact is a small area only in the plane xy, and it is easy to see in this case, that were the elastic body to have even a slight curvature expressed by $z = \phi(x, y)$, and the rigid body a form given by $z = \psi(x, y)$, we should have

$$w_c - w_0 = \phi(x, y) - \psi(x, y) \ldots\ldots\ldots\ldots(xxxiii),$$

or, the same condition as if we had supposed the elastic body plane and the form of the rigid body given by

$$z = \psi(x, y) - \phi(x, y).$$

Further since the surface of contact is small the part of any convex body in contact with the elastic medium may be taken as an elliptic paraboloid. If R_1 and R_2 be its principal radii of curvature, and the axes of x and y be taken in these directions, we shall have

$$\psi(x, y) = -\tfrac{1}{2}\left(\frac{x^2}{R_1} + \frac{y^2}{R_2}\right),$$

and in the special case of a plane elastic surface

$$w_0 = w_c - \tfrac{1}{2}\left(\frac{x^2}{R_1} + \frac{y^2}{R_2}\right)\ldots\ldots\ldots\ldots(xxxiv).$$

It is obviously necessary for the equilibrium of the system that the resultant pressure should be in the normal to the rigid body at the origin, or that, if the pressure be due to the weight of that body, its centroid should lie on the normal. Boussinesq considers on pp. 204–6 the more general case in which the exact orientation of the rigid body in the position of equilibrium is one of the unknown quantities of the problem. He points out how the problem breaks up into simpler problems of which the solution may be obtained, but he does not solve these problems for any special case.

[1509.] Besides the conditions we have considered in the previous article, there are certain others to be fulfilled at the contour of the surface of contact. This contour will itself have to be determined by the total amount of pressure, and along this at first undetermined contour we must have $\widehat{zz} = 0$, or zero pressure. This condition is discussed at some length by Boussinesq on pp. 208–10, and the reader is referred to our Art. 1503, as an indication to the sort of considerations which arise.

[1510.] Turning to various special cases we may note the following:

Case (a). *The rigid body is a solid of revolution, the end of the axis of which is in contact with the plane boundary of an infinite elastic solid.*

The solution of this case is indicated by Boussinesq in a footnote on pp. 206-7, and it may be obtained by aid of formulae due to Beltrami[1] and discussed at some length by Boussinesq on pp. 167-74. These formulae are the following. Let a circular area be covered with a surface density $\Delta(\rho)$, which is a function only of the distance ρ from the centre, then if $V(\rho)$ be the potential of the area and ρ_1 its radius, we have for the density, $\rho < \rho_1$:

$$\Delta(\rho) = -\frac{1}{\pi^2\rho}\frac{d}{d\rho}\int_\rho^{\rho_1}\left(\frac{\beta d\beta}{\sqrt{\beta^2-\rho^2}}\frac{d}{d\beta}\int_0^\beta\frac{V(\gamma)\gamma d\gamma}{\sqrt{\beta^2-\gamma^2}}\right)\ldots\ldots(\text{xxxv}),$$

and for the potential at a point in the plane of the area, $\rho' > \rho_1$:

$$V(\rho') = \frac{2}{\pi}\int_0^{\rho_1}\left(\frac{d\beta}{\sqrt{\rho'^2-\beta^2}}\frac{d}{d\beta}\int_0^\beta\frac{V(\gamma)\gamma d\gamma}{\sqrt{\beta^2-\gamma^2}}\right)\ldots\ldots(\text{xxxvi}).$$

Now by our equation (xxvi) w_0 is the potential due to a distribution of density $(1-\eta)p_0/(2\pi\mu)$ over the pressed area. Hence we have the following values for the pressure produced by a depression $w_0(\rho)$ inside the circle and for the depression $w_0(\rho')$ outside the circle:

$$\left.\begin{array}{l}p_0 = -\dfrac{2\mu}{1-\eta}\dfrac{1}{\pi\rho}\dfrac{d}{d\rho}\int_\rho^{\rho_1}\left(\dfrac{\beta d\beta}{\sqrt{\beta^2-\rho^2}}\dfrac{d}{d\beta}\int_0^\beta\dfrac{w_0(\gamma)\gamma d\gamma}{\sqrt{\beta^2-\gamma^2}}\right),\\[2ex] w_0(\rho') = \dfrac{2}{\pi}\int_0^{\rho_1}\left(\dfrac{d\beta}{\sqrt{\rho'^2-\beta^2}}\dfrac{d}{d\beta}\int_0^\beta\dfrac{w_0(\gamma)\gamma d\gamma}{\sqrt{\beta^2-\gamma^2}}\right)\end{array}\right\}\ldots(\text{xxxvii}).$$

It must be remembered (see our Art. 1508) that $w_c - w_0(\rho)$ is the quantity which is a given function $-\psi(\rho)$ of ρ, and that w_c will then have to be determined so that $\int_0^{\rho_1}2\pi\rho p_0 d\rho$ equals the total load P_0. This gives us after some changes:

$$\left.\begin{array}{l}P_0 = \dfrac{4\mu}{1-\eta}\int_0^{\rho_1}\dfrac{(w_c+\psi(\gamma))\gamma d\gamma}{\sqrt{\rho_1^2-\gamma^2}}\\[2ex] = \dfrac{4\mu}{1-\eta}\left(w_c\rho_1+\int_0^{\rho_1}\dfrac{\psi(\gamma)\gamma d\gamma}{\sqrt{\rho_1^2-\gamma^2}}\right)\end{array}\right\}\ldots\ldots\ldots(\text{xxxviii}).$$

The last integral can be evaluated if $\psi(\rho)$ be known, and thus w_c may be found; p_0 and $w_0(\rho')$ can then be ascertained by (xxxvii). These integrals have been evaluated by Cerruti for the case of a paraboloid of revolution, or when $\psi(\rho)$ is of the form $C\rho^2$: see pp. 43-4 of the memoir cited in our Art. 1489.

Case (b). *General Case. A rigid solid of any shape is pressed against the plane surface of an infinite elastic solid.* By our equation

[1] These formulae were first given by Beltrami in 1881: see his memoir, *Sulla theoria delle funzioni potenziali simmetriche. Memorie dell' Accad. delle Scienze di Bologna*, Ser. IV., T. II. pp. 462-3. Bologna, 1881.

(xxvi), $w_0 = \dfrac{1-\eta}{2\pi\mu} \displaystyle\iint \dfrac{p_0 d\omega}{r}$, and thus w_0 is the potential due to a surface

distribution $\dfrac{1-\eta}{2\pi\mu} p_0$. Hence we may state the most general problem in the following manner: The potential V at all points of an area in the plane of xy is given $= w_0$, what is the distribution of density over this area which would produce this potential? We have the following equations to solve:

$\nabla^2 V = 0$, for all points in space lying outside the given area;

$\dfrac{dV}{dz} = 0$, outside the given area for $z = 0$;

and $V = w_0$ within the given area.

Boussinesq shews (p. 223) that the solution for V is unique, and that the required pressure is that given by:

$$p_0 = -\frac{\mu}{1-\eta}\left(\frac{dV}{dz}\right)_{z=0}.$$

See his pp. 221–4.

Case (c). *Case of a flat rigid disc pressed on the plane surface of an infinite elastic solid.* In this case w_0 must be constant over the area of the disc, if we suppose the load to be so applied that the face of the disc remains parallel to the initially unstrained surface of the elastic solid. V is therefore constant $(= w_0)$ over the area covered by the disc. Hence the law of distribution of load over the area is precisely the same as that of the electric charge upon the same disc supposed to be a conductor insulated and charged with electricity (p. 225). For the case of a loaded elliptic disc we have seen that:

$$p_0 = \frac{P_0}{2\omega}\left(1 - \frac{x^2}{a^2} - \frac{y^2}{b^2}\right)^{-\frac{1}{2}} \quad \text{(see our Art. 1504).}$$

Boussinesq now shews that the depression is given by

$$w_0 = \frac{1-\eta}{2\pi\mu} P_0 \int_\nu^\infty \frac{d\nu}{\sqrt{(a^2+\nu^2)(b^2+\nu^2)}} \ \ldots\ldots\ldots \text{(xxxix)},$$

where for points inside the area covered by the disc, the lower limit of integration ν equals zero, while for the points (x, y) outside that area ν is determined by

$$\frac{x^2}{a^2+\nu^2} + \frac{y^2}{b^2+\nu^2} = 1. \quad \text{(See pp. 226–9.)}$$

[1511.] The case of rigid discs pressing upon elastic surfaces leads Boussinesq on pp. 213–21 to some discussion of the difficulties arising from discontinuity at the contour of these discs, and then to some general remarks on the nature of such discontinuity in a variety of problems in mathematical physics.

The values we have obtained for the pressure at the edges of circular and elliptic discs shew that, if they remained absolutely rigid, the pressure at their edges would become infinite. Hence either the elastic solid would be ruptured at the edge, or the edge itself would be broken away. Generally of course the varying pressure over the face of the disc will cause the disc itself to bend. These remarks seem to throw considerable light upon the phenomena of *punching*. In particular if we can apply such results for "infinitely thick plates" to the plates dealt with by Tresca[1], we find it intelligible why the portion of an elastic surface under a punch curves itself to avoid an infinite curvature at the edge of the punch, and why if the punch be forcibly pressed upon the surface, it sets into a concavity under the punch. A network of lines across the area covered by the punch remains unchanged after set has been produced; this is explained naturally enough by the concave form taken by the surface beneath the punch :

Il est bon toutefois de remarquer qu'il ne suffirait pas complètement, par lui-même, à la faire admettre ; car, la région périphérique étant incontestablement d'après la loi de répartition obtenue, beaucoup plus pressée que le centre, rien ne dit qu'un écrasement doive se produire, à aucun moment, dans la région centrale supposée même être restée plane, ni, par suite, que les caractères de structure qu'elle présente doivent disparaître, alors que le contour éprouve au contraire, des altérations profondes (p. 215).

In discussing the general occurrence of discontinuity in mathematical physics Boussinesq refers to discontinuous solutions obtained by Thomson and Tait and by St Venant in the case of re-entering angles of prisms under torsion (see our Art. 290), by himself in various hydrodynamical problems, by Rankine and himself in the case of pulverulent masses (see our Arts. 1613–15) and by Tresca in the case of the flow of plastic solids (pp. 217–21).

[1512.] *Case (d). Case of any rigid surface pressing at a point of synclastic curvature upon the plane surface of an infinite elastic solid.*
If, as in our Art. 1508, R_1 and R_2 be the principal radii of curva-

---

[1] *Recueil des Savants étrangers*...T. xx., p. 731. Paris, 1872.

ture, and P the total pressure in the direction of the normal, we have by (xxxiv) to determine w_0 from

$$w_0 = w_c - \tfrac{1}{2}\left(\frac{x^2}{R_1} + \frac{y^2}{R_2}\right) \quad \dots\dots\dots\dots (xl),$$

w_0 being the potential due to some distribution of density $(1-\eta)\,p_0/(2\pi\mu)$ over the area of contact, which we shall assume, pending justification, to be an ellipse of semi-axes a and b. At the contour of this ellipse we must have $p_0 = 0$, to avoid discontinuity. Boussinesq proceeds as follows (p. 231). He divides the ellipse into a number of concentric, similar and similarly situated ellipses of semi-axes ζa, ζb, where ζ varies from 0 to 1. Over each of these ellipses he distributes as a density the total mass $3P_0\zeta^2 d\zeta$ according to the law of electrical distribution on an insulated elliptic conducting disc. After some slight algebraic changes he easily finds from the results in our Art. 1510 *Case* (c), that the density at x, y, due to one of these discs, will be

$$\frac{3P_0\zeta d\zeta}{2\pi ab\sqrt{\zeta^2 - \dfrac{x^2}{a^2} - \dfrac{y^2}{b^2}}}.$$

Integrating from $\zeta = \sqrt{x^2/a^2 + y^2/b^2}$ to 1, we have for the actual density of the entire system of discs

$$p_0 = \frac{3P_0}{2\pi ab}\sqrt{1 - \frac{x^2}{a^2} - \frac{y^2}{b^2}} \quad \dots\dots\dots\dots\dots (xli).$$

Hence we find for the potential of the system

$$w_0 = \frac{3(1-\eta)}{4\pi\mu}\,P_0\int_\nu^\infty \left(1 - \frac{x^2}{a^2 + \nu^2} - \frac{y^2}{b^2 + \nu^2}\right)\frac{d\nu}{\sqrt{(a^2+\nu^2)(b^2+\nu^2)}} \quad \dots (xlii),$$

where the limit of integration ν is zero for points inside the ellipse of contact, and is determined by

$$\frac{x^2}{a^2 + \nu^2} + \frac{y^2}{b^2 + \nu^2} = 1 \quad \dots\dots\dots\dots (xlii)\ \textit{bis}$$

for points outside the ellipse of contact.

Comparing (xlii) with (xl) we see that the latter will be satisfied, if:

$$w_c = \frac{3(1-\eta)}{4\pi\mu}\,P_0\int_0^\infty \frac{d\nu}{D}, \quad \frac{1}{R_1} = \frac{3(1-\eta)}{2\pi\mu}\,P_0\int_0^\infty \frac{d\nu}{(a^2+\nu^2)\,D},$$
$$\left.\frac{1}{R_2} = \frac{3(1-\eta)}{2\pi\mu}\,P_0\int_0^\infty \frac{d\nu}{(b^2+\nu^2)\,D}\right\} \quad \dots\dots (xliii),$$

where $D = \sqrt{(a^2+\nu^2)(b^2+\nu^2)}$.

Boussinesq shews (pp. 234–5) that there are always unique real values of a and b to be found from the two last of these equations if R_1 and R_2 are positive. He easily demonstrates that:

$$w_c = \frac{a^2}{2R_1} + \frac{b^2}{2R_2},$$

whence we find :

$$w_0 = \frac{a^2 - x^2}{2R_1} + \frac{b^2 - y^2}{2R_2} + \frac{3(1-\eta)}{4\pi\mu} P_0 \int_0^\nu \left(\frac{x^2}{a^2 + \nu^2} + \frac{y^2}{b^2 + \nu^2} - 1 \right) \frac{d\nu}{D} \dots \text{(xliv)},$$

the integral vanishing for points inside, and its limit ν being given by (xlii) *bis* for points outside the ellipse.

He further shews that this solution gives continuity in the slope of the tangent plane to the elastic surface along the boundary of the area of contact, i.e. along the curve which projects into the ellipse $\frac{x^2}{a^2} + \frac{y^2}{b^2} = 1$. This might have been expected from the fact that $p_0 = 0$ by (xli) along that curve. Thus all the necessary conditions are satisfied by the solution (xli)–(xliv), and as the solution must be unique (see our Art. 1489, *fin.*) this is the solution sought. Thus we see that the surface of contact is really limited by an ellipse, the principal axes of which are tangents to the principal normal sections of the rigid body. The pressure at any point is proportional to the ordinate through that point of any ellipsoid having this ellipse for its section by a principal plane. Further the mean value of w_0 within the elliptic area $[= \iint w_0 dx dy/(\pi ab)]$ is found to be $\frac{3}{4} w_c$ or $\frac{9}{8}$ of the constant depression (see our Equation xxxix) which would be produced by a flat punch bounded by the ellipse of contact and subjected to the same normal pressure (p. 240).

By adding any arbitrary additional pressure distributed over the ellipse according to the law discussed in Case (c), and therefore giving only a uniform additional depression over the surface of contact, we have a solution of the important case of a cylindrical punch with any elliptic cross-section and a face curved to an elliptic paraboloid, the punch being subjected to any arbitrary pressure along its axis perpendicular to the surface.

[1513.] It only remains to indicate how a and b may be determined. Let $\frac{1}{a} = \int_0^\infty \frac{d\nu}{(a^2 + \nu^2)D}$, $\frac{1}{\beta} = \int_0^\infty \frac{d\nu}{(b^2 + \nu^2)D}$ and $e^2 = 1 - b^2/a^2$, then $\beta/a = R_2/R_1$, and if E and F be the complete elliptic integrals of the first and second orders, we have to find e from :

$$\frac{R_2}{R_1} = \frac{\beta}{a} = \frac{(1-e^2)(1 - E/F)}{e^2 - (1 - E/F)} \quad \dots\dots\dots\dots\dots \text{(xlv)}.$$

Boussinesq expands the right-hand side in powers of e and shews that if ϵ be the usually small quantity, $\dfrac{e^4}{128}$ + higher powers of e, then :

$$\frac{b}{a} = \left(\frac{R_2}{R_1}\right)^{\frac{2}{3+\epsilon}} \qquad\qquad \text{(xlvi)}.$$

For values of $R_2/R_1 < \cdot 1$, we may take very approximately $\epsilon = 0$. For other values numerical tables could easily be prepared from Legendre's Tables by aid of (xlv). For $e = 0$ to 1, ϵ passes from 0 to 1, or b/a varies from the $\frac{2}{3}$ to the $\frac{1}{2}$ power of R_2/R_1. To find a and b, we have only to remark that $\dfrac{1}{a}$, or $\dfrac{2\pi\mu}{3(1-\eta)P_0R_1}$, is a known quantity; but we have $a^3 = a\,(F - E)/e^2$, which accordingly determines a and therefore b, since e and with it F and E are known. Boussinesq shews that very approximately :

$$a = \left[\frac{3\,(1-\eta)}{16\mu} P_0 R_1 \left\{3\left(\frac{R_1}{R_2}\right)^{\frac{1}{3}} - 1\right\}\right]^{\frac{1}{3}} \left(\frac{R_2}{R_1}\right)^{\frac{\epsilon'}{3}} \quad \text{(xlvii)},$$

where ϵ' is of the form $\dfrac{e^4}{128} + \ldots$

Approximately therefore when R_1 and R_2 are not too widely different we may determine a and b from :

$$a = \left[\frac{3\,(1-\eta)}{16\mu} P_0 R_1 \left\{3\left(\frac{R_1}{R_2}\right)^{\frac{1}{3}} - 1\right\}\right]^{\frac{1}{3}} \quad \text{and} \quad \frac{b}{a} = \left(\frac{R_2}{R_1}\right)^{\frac{2}{3}} \quad \text{(xlviii)}.$$

See pp. 241–8.

On pp. 249–55 Boussinesq proves certain properties of the potential having relation to the distribution of electricity on elliptic discs and ellipsoids, but with no special reference to elastic problems.

[1514.] On pp. 715–9 of his volume Boussinesq makes an extension of the above results to the case of two smooth *elastic* bodies pressed normally against each other at any point. He remarks that, when a rigid body of synclastic curvature presses against an elastic body also of synclastic curvature the problem to be solved is the same as when a rigid paraboloid of reduced form (see our Art. 1508) presses upon a plane elastic surface. This auxiliary paraboloid produces in the plane an indent of definite elliptic contour and with a definite pressure at each point given by (xli). If, therefore, when two elastic surfaces of synclastic curvature press against each other, we choose two auxiliary rigid paraboloids which under the same total pressures produce in planes surfaces of

the same materials respectively as the two elastic bodies indents
of the same elliptic contours, there will be the same normal
pressures at corresponding points in the two cases, and these
normal pressures will be the normal pressures for the unreduced
surfaces. Accordingly to solve the problem we have only to
choose two auxiliary rigid paraboloids giving the same elliptic
contours of contact with planes, and, before reduction, the same
surface of synclastic curvature as the surface of contact of the two
elastic bodies. Taking the common normal to these bodies as axis
of z, we should satisfy all conditions by taking the surface of
contact of the form :

$$z = a_1 x^2 + 2a_2 xy + a_3 y^2.$$

The constants a_1, a_2, a_3 must then be determined from the
three conditions involved in the elliptic areas of contact having
the same position and dimensions of principal axes for both
bodies. For in this case, since the pressures as given by (xli) will
be the same for the two solids, and since the shearing stresses
are zero at the surface of contact, all the conditions of the problem
will be fulfilled.

[1515.]   Boussinesq remarks, p. 719, that this important
property of the form of the elementary surface of contact of two
elastic bodies pressed normally against each other was first recognised by Hertz in his memoir: *Ueber die Berührung fester
elastischer Körper, Journal für...Mathematik*, Bd. xcii. S. 156–71.
Berlin, 1882. Hertz recognised that the laws of this contact are
approximately true for the impact of smooth elastic bodies and
applied it especially to the case of the impact of two solid spheres.
Boussinesq discusses Hertz's problem at length, and we shall
consider it here, as it belongs essentially to the theory of the
application of the potential to elastic problems.

[1516.]   Let r_1 and r_2 be the radii of the two spheres, and r' the
radius of their spherical surface of contact considered positive when
it is of the same sign as r_2 and opposite to r_1. Then, if R_1 be the
radius of curvature at the vertex of the first auxiliary rigid paraboloid
which under the same pressure would make the same central depression and same area of contact in a plane boundary as that made in the
first sphere : $1/r_1 = 1/R_1 - 1/r'$.

Making $a = b$ in (xliii) and (xliv) we easily find :

$$w_c = \frac{3}{8} \frac{(1-\eta) \, P_0}{\mu a}, \quad \text{and} \quad w_c = a^2/R_1.$$

Hence
$$a^3/R_1 = \frac{3(1-\eta)}{8\mu} P_0.$$

Similarly for the second sphere, if R_2 correspond to R_1, and η' and μ' to η and μ,

$$w_c' = \frac{3}{8} \frac{(1-\eta') \, P_0}{\mu' a}, \quad \text{and} \quad a^3/R_2 = \frac{3(1-\eta')}{8\mu'} P_0.$$

Thus we find, if

$$\zeta_1 = \frac{1-\eta}{\mu}, \quad \text{and} \quad \zeta_2 = \frac{1-\eta'}{\mu'},$$

that
$$\zeta_1 R_1 = \zeta_2 R_2.$$

But
$$1/r_2 = 1/R_2 + 1/r',$$

so that we have
$$\zeta_2 \left(\frac{1}{r'} + \frac{1}{r_1} \right) = \zeta_1 \left(\frac{1}{r_2} - \frac{1}{r'} \right),$$

or
$$\frac{1}{r'} = \frac{1}{\zeta_1 + \zeta_2} \left(\frac{\zeta_1}{r_2} - \frac{\zeta_2}{r_1} \right),$$

$$= \frac{1}{2} \left(\frac{1}{r_2} - \frac{1}{r_1} \right)$$

for the special case when the spheres are of the same elastic material.

The total approach ξ of the centres of the two spheres and the radius of the circle of contact are given by

$$\left. \begin{aligned} \xi &= a^2 \left(\frac{1}{R_1} + \frac{1}{R_2} \right) = a^2 \left(\frac{1}{r_1} + \frac{1}{r_2} \right), \\ a^3 \left(\frac{1}{r_1} + \frac{1}{r_2} \right) &= \tfrac{3}{8} \left(\zeta_1 + \zeta_2 \right) P_0 \end{aligned} \right\} \quad \ldots\ldots\ldots\ldots\ldots \text{(xlix)}.$$

It follows from these equations that $P_0 \propto \xi^{\frac{3}{2}}$, or the pressure varies as the square root of the cube of the approach of the centres.

The strains will be the greatest at the centre of the area of contact, where we find for the normal squeezes the values $\dfrac{3P_0}{4\,(\lambda + \mu)\,\pi a^2}$ and $\dfrac{3P_0}{4\,(\lambda' + \mu')\,\pi a^2}$, respectively, while the lateral squeezes are just half these values : see Lamé and Clapeyron's result in our Art. 1493 and equation (xli) of our Art. 1512.

[1517.] To justify the application of these formulae to the collision of two spheres, Boussinesq makes (p. 717) the following remarks :

Quand le rapprochement des deux sphères est dû à un choc, les seules déformations perceptibles ont lieu près de la surface de contact, dans des parties dont la masse totale et les inerties sont insignifiantes, eu égard à leurs tensions. Ainsi l'équilibre intérieur régi par les formules (in Art. 1516) y existe sensiblement à tout instant du choc. Le système élastique de deux sphères, ou plus généralement de deux corps contigus à formes massives et arrondies, est donc de ceux où la *force vive* se trouve *séparée* presque entièrement de la *force de ressort*, de manière à n'en troubler que peu les lois ; et la réaction mutuelle P y est, même à l'état de mouvement, simple fonction du rapprochement ξ des parties en présence non encore déformées sensiblement. C'est bien ce que suppose la théorie élémentaire du choc direct des corps élastiques, confirmée par l'expérience dans ce cas de corps massifs.

It must be remarked that this assumption supposes the relative velocity of rebound equal to the relative velocity of impact, or Newton's coefficient of restitution $e = 1$. This certainly does not hold in the case of large masses of metal, where e is more nearly zero. The assumption supposes no energy to be lost in the form of elastic vibrations and of course none in the form of permanent changes of shape : see our Arts. 209–10 and 217.

Following Hertz and Boussinesq, we have if m_1 and m_2 be the masses of the spheres :

$$\frac{d^2\xi}{dt^2} = -\left(\frac{1}{m_1} + \frac{1}{m_2}\right) P_0$$

$$= -\frac{8\,(m_1 + m_2)}{3 m_1 m_2} \frac{\beta^{-\frac{1}{2}}}{(\zeta_1 + \zeta_2)} \xi^{\frac{3}{2}},$$

if

$$\beta = \frac{1}{r_1} + \frac{1}{r_2};$$

whence

$$\left(\frac{d\xi}{dt}\right)^2 = \frac{32}{15} \frac{m_1 + m_2}{m_1 m_2} \frac{\beta^{-\frac{1}{2}}}{(\zeta_1 + \zeta_2)} (\xi_0^{\frac{5}{2}} - \xi^{\frac{5}{2}}),$$

where ξ_0 is the maximum approach.

If v be the velocity of impact we have to determine ξ_0 :

$$v^2 = \frac{32}{15} \frac{m_1 + m_2}{m_1 m_2} \frac{\beta^{-\frac{1}{2}}}{(\zeta_1 + \zeta_2)} \xi_0^{\frac{5}{2}}.$$

Hence the maximum value of the radius of the area of contact, a_0, is given by

$$a_0 = \left(\frac{15}{32} \frac{m_1 m_2}{m_1 + m_2} \frac{\zeta_1 + \zeta_2}{\beta^2} v^2\right)^{\frac{1}{5}} \quad \ldots\ldots\ldots\ldots \quad \ldots\ldots\ldots \quad (1).$$

The semi-duration τ of the impact is easily found from

$$\tau = \frac{1}{v}\int_0^{\xi_0} \frac{d\xi}{\sqrt{1 - \left(\frac{\xi}{\xi_0}\right)^{\frac{5}{2}}}} = \frac{\xi_0}{v}\int_0^1 \frac{d\gamma}{\sqrt{1 - \gamma^{\frac{5}{2}}}}.$$

According to Hertz the definite integral = 1·4716, nearly, whence we have :

$$\tau = 1\cdot4716 \left(\frac{\beta}{v}\right) \left\{\frac{15}{32}\frac{m_1 m_2}{m_1 + m_2}(\zeta_1 + \zeta_2)\right\}^{\frac{2}{5}} \quad \ldots\ldots\ldots \text{(li)}.$$

The results (1) and (li) for the radius of the area of contact and for the semi-duration of the impact ought to be capable of easy experimental investigation.

For the impact of a sphere (radius r) against a *perfectly rigid* plane, we have only to put in the above results $m_1 = m_2$, $\zeta_1 = \zeta_2$, $\beta = 2/r$ and v equal to twice the velocity of impact.

[1518.]　The sixth section of Boussinesq's *Treatise* occupies pp. 256–75, and discusses general properties of the ordinary and logarithmic potentials. Boussinesq first shews that the ordinary potential and the first logarithmic potential with three variables are finite even when the subjects of integration become infinite at a point, or along a part of a line. He then proceeds to prove a general theorem (pp. 260–5) by aid of which we can easily evaluate the differentials of these potentials with regard to x, y, z for a point occupied by matter. This leads him to a demonstration of Poisson's theorem and various allied theorems for the direct and logarithmic potentials. Thus, if V, U and ψ be the direct, ordinary (or inverse) and first logarithmic potentials (as in Art. 1488) we must replace the results of that article for points occupied by "potentiating matter" by :

$$\nabla^2\nabla^2 V = -8\pi\rho, \quad \nabla^2 U = -4\pi\rho, \quad \nabla^2\psi = 4\pi\int_z^\infty \rho\, dz\ldots\ldots \text{(lii)}.$$

[1519.]　The following section is entitled: *Équilibre d'élasticité d'un solide indéfini, sollicité dans une étendue finie par des forces extérieures quelconques* (pp. 276–95).

In a footnote on p. 281 Boussinesq gives the following general solution of the equations of isotropic elasticity when written in the form of type :

$$(\lambda + \mu)\frac{d\theta}{dx} + \mu\nabla^2 u + X = 0,$$

i.e. when X is the body-force per unit volume :

$$u = -\frac{dH}{dx} + \nabla^2 A, \quad v = -\frac{dH}{dy} + \nabla^2 B, \quad w = -\frac{dH}{dz} + \nabla^2 C,$$

where
$$\frac{\lambda + 2\mu}{\lambda + \mu}\, \nabla^2 H = \frac{d\left(\nabla^2 A\right)}{dx} + \frac{d\left(\nabla^2 B\right)}{dy} + \frac{d\left(\nabla^2 C\right)}{dz},$$

and
$$\nabla^2\nabla^2 A = -\frac{1}{\mu} X, \quad \nabla^2\nabla^2 B = -\frac{1}{\mu} Y, \quad \nabla^2\nabla^2 C = -\frac{1}{\mu} Z.$$

Further we have :
$$\theta = \frac{\mu}{\lambda + \mu}\, \nabla^2 H$$

$$\cdots \text{(liii)}.$$

These solutions are easily seen to satisfy the body-shift equations. Boussinesq does not notice that they had been previously given by Sir W. Thomson : see our Chapter XIV. He has in the previous paragraphs of the section discussed various special cases of them. These solutions correspond to the case of an infinite elastic medium fixed at infinity under a system of body-forces applied to a finite volume, or, what is practically the same thing, to the case of a finite elastic body with a fixed boundary when body-force is applied to a very small element of it at considerable distance from the boundary. Since the solution is in terms of potentials, which together with their differentials vanish at an infinite distance from the field of force, the condition as to the fixed boundary is satisfied.

Case (a). Take $X = Y = 0$, but Z not zero.
Then by (liii) if C be the *direct* potential due to a distribution of matter of density $Z/(8\pi\mu)$, $d\varpi$ being an element of volume,

$$C = \int \frac{rZ}{8\pi\mu} d\varpi,$$

and we have
$$H = \frac{\lambda + \mu}{8\pi\mu\,(\lambda + 2\mu)}\, \frac{d\left(\int rZ d\varpi\right)}{dz},$$

with the following type of solution :

$$u = -\frac{\lambda + \mu}{8\pi\mu\,(\lambda + 2\mu)}\, \frac{d^2\left(\int rZ d\varpi\right)}{dx\,dz}, \quad v = -\frac{\lambda + \mu}{8\pi\mu\,(\lambda + 2\mu)}\, \frac{d^2\left(\int rZ d\varpi\right)}{dy\,dz},$$

$$w = -\frac{\lambda + \mu}{8\pi\mu\,(\lambda + 2\mu)}\, \frac{d^2\left(\int rZ d\varpi\right)}{dz^2} + \frac{1}{8\pi\mu}\, \nabla^2\left(\int rZ d\varpi\right)$$

$$\cdots \text{(liv)}.$$

Obviously the addition of three such special solutions gives us the most general solution for all the body-forces finite. Before stating this general solution, we must note that u, v, w are the shifts at the point x, y, z, and that Z in the potential integral is the force at the point x_1, y_1, z_1 where $r = \sqrt{(x - x_1)^2 + (y - y_1)^2 + (z - z_1)^2}$ and $d\varpi = dx_1 dy_1 dz_1$. In order to bear this in mind we shall write for Z, Z_1 when it is the subject of integration. We then find for the general solution shifts of the type :

Case (b).

$$u = \frac{1}{4\pi\mu} . \iint \left\{ \frac{X_1}{r} - \frac{\lambda + \mu}{2(\lambda + 2\mu)} \frac{d}{dx} \left(X_1 \frac{dr}{dx} + Y_1 \frac{dr}{dy} + Z_1 \frac{dr}{dz} \right) \right\} d\varpi,$$

$$\theta = \frac{1}{4\pi(\lambda + 2\mu)} \int \left(X_1 \frac{d\frac{1}{r}}{dx} + Y_1 \frac{d\frac{1}{r}}{dy} + Z_1 \frac{d\frac{1}{r}}{dz} \right) d\varpi \qquad \Bigg\} \dots \text{(lv)}.$$

This solution is really due to Sir W. Thomson: see our Chapter XIV. Boussinesq discusses various modes of reaching it on pp. 284–91 of his *Treatise.*

Case (c). Consider a single force Z_1 applied to a small volume $d\varpi$ which may be taken at the origin. We find that, if U denote the shift perpendicular to the force at a point distant r from its point of application, r making an angle a with the positive direction of the force:

$$U = \frac{1}{32\pi\mu(1 - \eta)} \frac{Z_1 d\varpi}{r} \sin 2a,$$

$$w = \frac{1}{32\pi\mu(1 - \eta)} \frac{Z_1 d\varpi}{r} (7 - 8\eta + \cos 2a), \qquad \Bigg\} \dots\dots\text{(lvi)}.$$

$$\theta = -\frac{Z_1 d\varpi}{4\pi(\lambda + 2\mu)} \frac{\cos a}{r^2}$$

From these shifts the stresses can easily be found and the solution analysed after the method of our Art. 1497: see Boussinesq's pp. 81–92 and 291–5.

Case (d). Take $X = Y = Z = 0$, $A = B = 0$, and C a function for which $\nabla^2\nabla^2 C = 0$, then we have as a solution, if

$$\phi = \frac{\lambda + \mu}{\lambda + 2\mu} C :$$

$$u = -\frac{d^2\phi}{dx\,dz}, \quad v = -\frac{d^2\phi}{dy\,dz}, \quad w = -\frac{d^2\phi}{dz^2} + \frac{\lambda + 2\mu}{\lambda + \mu} \nabla^2\phi, \qquad \Bigg\} \dots\text{(lvi)} \textit{ bis.}$$

where $\qquad \theta = \frac{\mu}{\lambda + \mu} \frac{d}{dz}(\nabla^2\phi)$, and $\nabla^2\nabla^2\phi = 0$

Further: $\qquad \widehat{zx} = \mu \frac{d}{dx} \left(\frac{\lambda + 2\mu}{\lambda + \mu} \nabla^2\phi - 2\frac{d^2\phi}{dz^2} \right),$

$$\widehat{yz} = \mu \frac{d}{dy} \left(\frac{\lambda + 2\mu}{\lambda + \mu} \nabla^2\phi - 2\frac{d^2\phi}{dz^2} \right),$$

$$\widehat{zz} = \mu \frac{d}{dz} \left(\frac{3\lambda + 4\mu}{\lambda + \mu} \nabla^2\phi - 2\frac{d^2\phi}{dz^2} \right).$$

Since ϕ satisfies an equation of the *fourth* order Boussinesq suggests that it might be possible to find a value of ϕ, for which $\widehat{zx} = \widehat{yz} = 0$, or

$$\frac{\lambda + 2\mu}{\lambda + \mu}\, \nabla^2\phi = 2\,\frac{d^2\phi}{dz^2}\,, \text{ when } z = \pm\, a,$$

and $\widehat{zz}/\mu = $ a given function of x and y when $z = \pm\, a$. Thus we should solve the problem of an infinite plate of any thickness subjected to any given system of purely normal loading on its faces. This problem had been solved by Lamé and Clapeyron by aid of quadruple integrals, but their solution does not really exhibit any laws of the phenomena shewn by such a plate (pp. 278–281) : see our Arts. 1020*–21*.

[1520.] We now pass to the last section of the text of Boussinesq's *Treatise*. This is entitled : *Sur les perturbations locales dans la théorie de l'élasticité, et sur la possibilité, pour le géomètre, de remplacer des forces données, s'exerçant sur une petite partie d'un solide, par d'autres forces statiquement équivalentes, appliquées à la même région très petite en tous sens.* It occupies pp. 296–318 and deals with the important principle of the elastic equivalence of statically equipollent systems of loading at small distances from the loaded element of surface. We have frequently had occasion to refer to this principle, remarking how it is practically assumed in all the usual solutions for torsion, flexure and even extension, and appealing to Saint-Venant's experimental arguments in favour of it : see our Arts. 8, 9, 21, etc.

The principle to be demonstrated is stated by Boussinesq in the following words (p. 298) :

Des forces extérieures, qui se font équilibre sur un solide élastique et dont les points d'application se trouvent tous à l'intérieur d'une sphère donnée, ne produisent pas de déformations sensibles à des distances de cette sphère qui sont d'une certaine grandeur par rapport à son rayon.

There are two classes of external forces to be considered, namely body- and surface-forces.

[1521.] *Body-Forces. Case* (i). Let there be two parallel and opposite forces $Z_1 d\varpi$ and $-Z_1 d\varpi$, and let c be the distance between their points of application, supposed small. Let the first be supposed to act at the origin, and let the polar coordinates r, a determine the position of any point with regard to it ; but let the second act at the point 0, 0, c on the axis of z, and let r', a' be the coordinates of a point with regard to 0, 0, c^1. Then for points not in the immediate neighbourhood of the origin we have, if n be any integer:

$$\frac{1}{r'^n} = \frac{1}{r^n}\left(1 + \frac{nc}{r}\cos a\right), \quad \cos a' = \cos a - \frac{c}{r}\sin^2 a, \quad \sin a' = \sin a + \frac{c}{2r}\sin 2a.$$

Hence by aid of (lvi) we easily find, U and w being the radial and axial shifts:

$$\left.\begin{aligned}
\theta &= -\frac{Z_1 cd\varpi}{4\pi\,(\lambda+2\mu)}\,\frac{1-3\cos^2 a}{r^3}, \quad U = \frac{Z_1 cd\varpi}{16\pi\mu(1-\eta)}\,\frac{1-3\cos^2 a}{r^2}\sin a, \\
w &= \frac{Z_1 cd\varpi}{16\pi\mu\,(1-\eta)}\,\frac{1-3\cos^2 a - 2\,(1-2\eta)}{r^2}\cos a
\end{aligned}\right\}\ \text{...(lvii)}.$$

The stresses are easily obtained from these values of the shifts and obviously vary inversely as r^3, or: *the stresses decrease inversely as the cube of the distance from the centre of the region of perturbation:* compare our Art. 1487, (c),

Case (ii). Let there be two parallel forces $Z_1 d\varpi$ and $-Z_1 d\varpi$ acting at the points 0, 0, 0 and c, 0, 0 or a couple of moment $C = cZ_1 d\varpi$. We have then to consider the influence of a couple of small arm applied to an infinitely great elastic solid. Let β be the angle U makes with the plane of the couple, and let V be the shift tending to increase β and perpendicular to both U and w; then we find from (lvi) by a method similar to that of *Case* (i),

$$\left.\begin{aligned}
U &= \frac{C}{16\pi\mu\,(1-\eta)}\,\frac{\cos a\cos\beta}{r^2}(1-3\sin^2 a), \quad V = -\frac{C}{16\pi\mu\,(1-\eta)}\,\frac{\cos a\sin\beta}{r^2}, \\
w &= -\frac{C}{16\pi\mu\,(1-\eta)}\,\frac{\sin a\cos\beta}{r^2}(3-4\eta+3\cos^2 a), \\
\theta &= \frac{C}{8\pi\,(\lambda+2\mu)}\,\frac{3\sin 2a\cos\beta}{r^3}
\end{aligned}\right\}$$

$$\text{............(lvii) }bis.$$

We see that the stresses produced by such a couple again decrease inversely as the cube of the distance from the sphere of perturbation (pp. 303–4).

The results of these two cases compared with those of our Art. 1441 shew us that the influence of such body-forces in an infinite elastic medium does not produce stresses which decrease with the distance anything like so rapidly as in the case of bodies having one or two dimensions small and subjected to surface loading with a zero statical resultant. To such bodies Boussinesq now turns.

---

[1] The two forces are clearly *pushing* and not pulling with the sign we have chosen for Z_1.

[1522.] *Surface-Forces.* Boussinesq points out that in the case of surface-forces we may expect a solution involving exponentials with negative indices and refers to the problem discussed in our Art. 1507 as suggesting this. The earliest solution for a system of forces in equilibrium on the edge of a plate is due to Thomson and Tait[1], and somewhat later a more complete solution has been given by Maurice Lévy[2]: see our Arts. 397 and 1441. Boussinesq discusses the work of these authors on pp. 306–18, and we will indicate the general lines of his investigation here.

Consider a plane plate whose faces are given by $z = 0$ and $z = a$, and let it be bounded laterally by any cylinder whose generators are parallel to the axis of z. We shall suppose the radius of curvature of this cylinder at any generator to be very large as compared with the thickness of the plate a, so that the tangent plane to the cylinder at any generator may be taken to coincide with the boundary of the plate for a distance considerably greater than a. This tangent plane will be taken as the plane of yz, the generator of the cylinder being taken as axis of z, and y being the tangent to the contour of the lower face of the plate. To the faces of the plate we shall suppose no load applied, or

$$(\widehat{zz}, \ \widehat{zx}, \ \widehat{yz}) = 0, \text{ for } z = 0 \text{ and } z = a.$$

On the lateral boundary of the plate, we have the stresses $\widehat{xx}$, $\widehat{xy}$ and $\widehat{zx}$, which it is proposed to analyse, subject to the condition that their *mean* values from $z = 0$ to $z = a$ shall be zero, or that the surface load has a zero statical resultant.

A solution of the body-shift equations suitable to this case is given by

$$u = -\frac{d\psi}{dy}, \quad v = \frac{d\psi}{dx}, \quad w = 0,$$

and therefore $\theta = 0$, where ψ is a function satisfying the equation

$$\nabla^2 \psi = 0.$$

Suppose we take with Lévy

$$\psi = \Sigma \phi_n (x, y) \cos \frac{n\pi z}{a},$$

where n is an integer; then the conditions at the faces of the plate

---

[1] *Treatise on Natural Philosophy.* §§ 724–9. Oxford, 1867.
[2] *Journal de mathématiques.* T. III., pp. 219–306. Paris, 1877.

are clearly satisfied, and further the mean values of the stresses over the lateral boundary will also be zero. We must have

$$\frac{d^2\phi_n}{dx^2} + \frac{d^2\phi_n}{dy^2} = \frac{n^2\pi^2}{a^2}\,\phi_n \quad\ldots\ldots\ldots\ldots\text{(lviii)}$$

as an equation to determine ϕ_n.

For the stresses over $x = 0$, we have:

$$\left.\begin{array}{l}\widehat{xx} = -2\mu\Sigma\left(\dfrac{d^2\phi_n}{dx\,dy}\right)_{x=0}\cos\dfrac{n\pi z}{a}, \\[12pt] \widehat{xy} = \mu\Sigma\left(\dfrac{d^2\phi_n}{dx^2} - \dfrac{d^2\phi_n}{dy^2}\right)_{x=0}\cos\dfrac{n\pi z}{a}, \\[12pt] \widehat{zx} = \mu\Sigma\left(\dfrac{n\pi}{a}\dfrac{d\phi_n}{dy}\right)_{x=0}\sin\dfrac{n\pi z}{a}\end{array}\right\}\ldots\ldots\ldots\text{(lix)}.$$

Now these equations will enable us to give $\widehat{xy}$ any value we please along a generator from $z = 0$ to $z = a$, that is to say they allow us to select at our will the shearing stresses on the edge of the plate which produce a torsional couple M round the normal to the plate. They further allow of this couple or system of shearing forces varying from generator to generator, since ϕ_n is also an undetermined function of y.

[1523.] If we take ϕ_n independent of y, we have $\widehat{xx}$ and $\widehat{zx}$ both zero and we have Thomson and Tait's solution for a distribution of shearing stress along the edge of a plate parallel to the contour of the face and the *same* along each generator.

In this case (lviii) gives us

$$\phi_n = A_n e^{-\frac{n\pi x}{a}},$$

and therefore for the given distribution of shearing stress over $x = 0$:

$$\widehat{xy} = \mu\Sigma A_n\,\frac{n^2\pi^2}{a^2}\cos\frac{n\pi z}{a}.$$

If $\widehat{xy} = \chi(z)$ be the given distribution, we find at once by Fourier's series:

$$A_n\,\frac{n^2\pi^2}{\mu^2} = \frac{2}{a}\int_0^a \chi(z)\cos\frac{n\pi z}{a}\,dz = k_n,\text{ say.}$$

The only finite stresses will then be:

$$\widehat{xy} = \mu\Sigma k_n e^{-\frac{n\pi x}{a}}\cos\frac{n\pi z}{a},$$

$$\widehat{yz} = \mu\Sigma k_n e^{-\frac{n\pi x}{a}}\sin\frac{n\pi z}{a}.$$

Thomson and Tait have shewn (*Treatise on Natural Philosophy*, § 729) that for $x = 2a$, or at a distance equal to twice the thickness

from the edge of the plate, the values of these stresses are only about
·002 of their values at the edge, or we see that the local perturbation
has small influence at slight distances from the edge, supposing the
distribution of shearing stress to be the same along every generator of
the bounding cylinder. See our Arts. 1440-1.

[1524.] Returning to Lévy's more general solution, we notice as
before indicated that :

$$\int_0^a \widehat{xx}\, dz = 0, \qquad \int_0^a \widehat{xy}\, dz = 0.$$

Further we have for the *moment of flexure M'* round the tangent to
the contour of the face $z = 0$, and for the *moment of torsion M* about the
normal to that contour :

$$\left.\begin{aligned}
M' &= \int_0^a \widehat{xx}\, z\, dz = 2\mu\Sigma \left(\frac{d^2\phi_n}{dx\,dy}\right)_{x=0} \frac{a^2}{n^2\pi^2}(1 - \cos n\pi), \\
M &= \int_0^a \widehat{xy}\, z\, dz = -\mu\Sigma \left(\frac{d^2\phi_n}{dx^2} - \frac{d^2\phi_n}{dy^2}\right)_{x=0} \frac{a^2}{n^2\pi^2}(1 - \cos n\pi)
\end{aligned}\right\} \dots \text{(lx)}.$$

The total shearing action F on the edge parallel to a generator
per unit length of rim is given by

$$F = \int_0^a \widehat{zx}\, dz = \mu\Sigma \left(\frac{d\phi_n}{dy}\right)_{x=0} (1 - \cos n\pi).$$

But by (lviii) we may write

$$M = -\mu\Sigma \left(\phi_n - \frac{2a^2}{n^2\pi^2}\frac{d^2\phi_n}{dy^2}\right)_{x=0} (1 - \cos n\pi).$$

Hence, if s be an element of the contour of the edge of the plate,
we have, since $ds = dy$:

$$\frac{dM}{ds} = -F + 2\mu\Sigma \left(\frac{a^2}{n^2\pi^2}\frac{d^3\phi_n}{dy^3}\right)_{x=0} (1 - \cos n\pi)\dots\dots\dots\text{(lxi)}.$$

Now in this second approximation ϕ_n will depend upon y, but *if the
variation of the edge stresses with y be slow*, it is clear that although
we do not as in Thomson and Tait's first approximation take $d\phi_n/dy$
zero, still $\dfrac{d^3\phi_n}{dy^3}$ will be small as compared with $\dfrac{d\phi_n}{dy}$, and $\dfrac{d\phi_n}{dy}$ as compared
with ϕ_n. Hence we see from (lx) that M' the moment of flexure
is negligible as compared with M the moment of torsion, and from
(lxi) that very approximately :

$$F = -\frac{dM}{ds} \dots\dots\dots\dots\dots\dots\dots\dots\dots \text{(lxii)}.$$

Since in this case the magnitude of the shifts and stresses in the material of the plate will decrease as we pass from the edge at least as rapidly as in the first approximation (Art. 1523), we conclude with Boussinesq that: *If the edge of a thin plate be subjected to shearing forces F perpendicular to the faces, and to torsional couples M round normals to the edge, the relation* (lxii) *holding between them, then these actions neutralise each other at a small distance from the contour. In other words torsional couples M and shearing forces dM/ds perpendicular to the faces produce the same effects at a very small distance from the edge of the plate* (p. 313).

This is Thomson and Tait's reconciliation of the Kirchhoff and Poisson boundary-conditions for thin plates. It was first given by them in 1867 and independently by Boussinesq in 1871: see our Arts. 488*, 394, 1438 and 1440. The above investigation shews very clearly the nature of the local action at the edge of the plate and measures the area over which that action is sensibly spread.

Boussinesq concludes his discussion by remarking that it does not seem probable that the local perturbations which present themselves in other cases, in which the principle of the elastic equivalence of statically equipollent loads is applied, will allow of being investigated with the same ease as in this particular case of the boundary conditions at the edge of a thin plate (pp. 317–8).

[1525.] The remainder of Boussinesq's volume is occupied with *Notes complémentaires*, several of which are concerned with results of great value for the theory of elasticity. We will briefly refer to those of importance for the history of our subject.

[1526.] *Note* I. (pp. 318–56) deals with a potential of four variables, or what Boussinesq terms a *spherical potential*. It contains some interesting results for the theory of potentials, but its only value for elasticity is the integration of the equations for the vibration of an isotropic elastic medium (pp. 351–6). The solution takes the form previously given by Stokes: see our Arts. 1268–75. The substance of this *Note* appeared in the *Comptes rendus*, T. xciv. pp. 1465–8 and 1648–50; T. xcv. pp. 479–82. Paris, 1882.

[1527.] *Note* II. (pp. 357–664) deals with a new method of integrating an important class of partial differential equations, and with applications of the method to elastic and other problems. Portions of this

Note were published in the *Comptes rendus*, T. xcIV. pp. 33–6, 71–4, 127–30, 514–7, 1044–7, 1505–8; T. xcv. pp. 123–5; T. xcvII. pp. 154–7, pp. 843–4, 897–900, 1131–2. Paris, 1882 and 1883.

[1528.] § I. (pp. 357–403, 652–5) is occupied with a method of integrating the differential equation

$$A \frac{d^n\phi}{dt^n} + \left(\frac{d^2}{dx^2} + \frac{d^2}{dy^2} + \text{to } p \text{ terms} \right)^n \phi = 0$$

by means of definite integrals of arbitrary functions. The equation is obviously an extremely general one and the solution admits of being modified so as to suit various types of "initial conditions." The results can be applied to a great variety of physical problems, of which for our present purposes it suffices to note the transverse vibrations of bars and plates. Space does not admit of our reproducing in general outline Boussinesq's suggestive analysis and conclusions, but some of his results will be indicated in our discussion of his application of them to the special elastic problems with which we are more closely concerned.

§ II. (pp. 404–34) applies the method to the theory of heat and to the friction of fluids; § III. (pp. 435–577, 655–64) deals with elastic problems; while § IV. (pp. 578–651) discusses applications of the solutions obtained to the theory of liquid waves. It is § III., therefore, with which we shall be occupied in the following articles.

[1529.] The first problem dealt with by Boussinesq is that of a uniform rod or thin prism, the central line of which (coinciding with the axis of *x*) is infinitely extended from the origin in the positive direction. Any forces are supposed to act on the extremity *x* = 0, provided they cause only transverse vibrations in a principal plane of inertia of the prism. Initially the rod is supposed at rest throughout its entire length.

The equations for the motion of such a rod are given with a slightly different notation in our Arts. 343–5, and are the following:
Equation for transverse shift w:

$$\frac{d^4w}{dx^4} + \frac{d^2w}{d(at)^2} = 0,$$

where $a^2 = E\kappa^2/\rho$ in the notation of our work.

Further, $w = 0$ for $t = -\infty$, and $w = 0$ for $x = \infty$ always. The conditions at $x = 0$ may be of the following types:

(a) *Geometrical constraint varying with the time,* i.e.

$$w = F(at), \quad dw/dx = F_1(at), \text{ for } x = 0.$$

(b) *Total shear and the flexural couple varying with the time*, i.e.

$$d^3w/dx^3 = K\,F\,(at), \quad d^2w/dx^2 = -\,K_1F_1\,(at)\ \text{for}\ x = 0,$$

where $\quad\quad\quad K = 1/(E\omega)$ and $K_1 = 1/(E\omega\kappa^2).$

Boussinesq takes instead of these forms the more general ones

$$d^3w/dx^3 = K\{F\,(at) - w\}, \quad d^2w/dx^2 = K_1\{dw/dx - F_1\,(at)\},$$

which he considers might be realised when the definite movements $F\,(at)$ and $F_1\,(at)$ are communicated to the end of the bar by means of springs (*par l'intermédiaire d'un ressort et d'un encastrement élastique*, p. 437).

(c) *Infinitely long bar carrying a load M at its centre*, $x = 0$.

$$dw/dx = 0, \quad \text{and}\ \ \tfrac{1}{2}M\frac{d^2w}{dt^2} = -\,E\omega\kappa^2\frac{d^3w}{dx^3} + \tfrac{1}{2}F\,(at), \quad \text{for}\ \ x = 0,$$

where $F\,(at)$ is the force exerted at time t on M (pp. 481–491). On pp. 438–9 Boussinesq demonstrates the uniqueness of the solution for cases (a) and (b).

[1530.] The solution of the above equations is obtained in the following manner (pp. 360–8, etc.):

Consider the quantity

$$\phi = \int_0^\infty f\left(\frac{a^2}{2}\right) \psi\left(\frac{s^2}{2a^2}\right) da \quad\quad\quad\text{.....................(i):}$$

we have

$$\frac{d\phi}{ds} = \int_0^\infty f\left(\frac{a^2}{2}\right) \psi'\left(\frac{s^2}{2a^2}\right)\frac{sda}{a^2}$$

$$= \int_0^\infty f\left(\frac{s^2}{2a'^2}\right) \psi'\left(\frac{a'^2}{2}\right) da' \quad\quad\text{...................(ii),}$$

if $a' = s/a.$

Similarly

$$\frac{d^2\phi}{ds^2} = \int_0^\infty f'\left(\frac{a''^2}{2}\right) \psi'\left(\frac{s^2}{2a''^2}\right) da'' \quad\text{.....\ ....(iii),}$$

where $a'' = s/a'.$

We may evidently drop the dashes in a' and a'' in (ii) and (iii), and the law of the successive differentials is then obvious.

The above investigation depends for its exactness on the limits being no functions of s, otherwise we should have to introduce special terms depending on the differentiation of the limits. We can get over this difficulty, however, by taking the limits $1/\epsilon$ and ϵs for a instead of ∞ and 0, where ϵ is a vanishingly small quantity. The limits for a' will then be ϵs and $1/\epsilon$, for a'', $1/\epsilon$ and ϵs and so on. Differentiating with regard to s the special term introduced by the limit differentiation will be

$$\left[-\,\epsilon f\left(\frac{a^2}{2}\right) \psi\left(\frac{s^2}{2a^2}\right)\right]_{a=0} \quad \text{or}\ -\frac{1}{s}\left[af\left(\frac{a^2}{2}\right) \psi\left(\frac{s^2}{2a^2}\right)\right]_{a=0}$$

for the first differentiation. Hence we must have

$$af\left(\frac{a^2}{2}\right)\psi\left(\frac{s^2}{2a^2}\right)=0,\text{ for }a=0.$$

Similarly, for the second differentiation :

$$af\left(\frac{s^2}{2a^2}\right)\psi'\left(\frac{a^2}{2}\right)=0,\text{ for }a=0,\quad\Bigg\}\quad\dots\dots\dots\dots(iv),$$

and for the third,

$$af'\left(\frac{a^2}{2}\right)\psi'\left(\frac{s^2}{2a^2}\right)=0,\text{ for }a=0$$

and so on, the law followed by these products being clear. From the above results we can easily deduce a solution of the differential equation :

$$\frac{d^4w}{dx^4}+\frac{d^2w}{d(at)^2}=0.$$

Assume $$w=\int_0^\infty f\left(at\mp\frac{a^2}{2}\right)\psi\left(\frac{x^2}{2a^2}\right)da,$$

then as in (iii): $$\frac{d^4w}{dx^4}=\int_0^\infty f''\left(at\mp\frac{a^2}{2}\right)\psi''\left(\frac{x^2}{2a^2}\right)da,$$

while $$\frac{d^2w}{d(at)^2}=\int_0^\infty f''\left(at\mp\frac{a^2}{2}\right)\psi\left(\frac{x^2}{2a^2}\right)da.$$

Hence, if we take $\psi\left(\frac{x^2}{2a^2}\right)=\cos\frac{x^2}{2a^2}$ or $\sin\frac{x^2}{2a^2}$, we have a solution of the equation.

Noting the interchangeable nature of $\frac{a^2}{2}$ and $\frac{x^2}{2a^2}$ we see that

$$w=\int_0^\infty f\left(at\mp\frac{x^2}{2a^2}\right)\psi\left(\frac{a^2}{2}\right)da$$

is also a solution.

Thus finally we have

$$w=\int_0^\infty f\left(at\mp\frac{a^2}{2}\right)\left(\cos\frac{x^2}{2a^2}\text{ or }\sin\frac{x^2}{2a^2}\right)da\dots\dots\dots(v),$$

or, $$=\int_0^\infty f\left(at\mp\frac{x^2}{2a^2}\right)\left(\cos\frac{a^2}{2}\text{ or }\sin\frac{a^2}{2}\right)da\dots\dots\dots(vi)$$

(p. 439).

Boussinesq points out that for the special case of a rod infinite in one direction only, we must take the upper sign, and have $f(-\infty)=0$, in order to satisfy the conditions that $w=0$ for $t=-\infty$ with any value of x, and for $x=\infty$ with any value of t (pp. 440–1). Further (iv) will hold, since ψ is a sine or cosine, if $f(at)$ and its differentials be supposed finite for all values of t.

[1531.] We can now easily satisfy the special terminal conditions of our Art. 1529. We may write w in the form:

$$w = \int_0^\infty \left[f_1\left(at - \frac{a^2}{2}\right)\cos\frac{x^2}{2a^2} + f_2\left(at - \frac{a^2}{2}\right)\sin\frac{x^2}{2a^2} \right.$$
$$\left. + f_3\left(at - \frac{x^2}{2a^2}\right)\cos\frac{a^2}{2} + f_4\left(at - \frac{x^2}{2a^2}\right)\sin\frac{a^2}{2} \right] da \ldots\ldots(\text{vii}),$$

where we must remember that:

$$\int_0^\infty \cos\frac{a^2}{2} \, da = \int_0^\infty \sin\frac{a^2}{2} \, da = \tfrac{1}{2}\sqrt{\pi} \ldots\ldots\ldots(\text{viii}).$$

Case (*a*). When $w = F(at)$ and $dw/dx = F_1(at)$ for $x = 0$, we take only f_2 and f_4, or put:

$$w = \frac{2}{\sqrt{\pi}} \int_0^\infty \left[F\left(at - \frac{x^2}{2a^2}\right)\sin\frac{a^2}{2} + F_1\left(at - \frac{a^2}{2}\right)\sin\frac{x^2}{2a^2} \right] da \ldots.(\text{ix}).$$

Case (*b*). When $d^3w/dx^3 = KF(at)$, $d^2w/dx^2 = -K_1F_1(at)$ for $x = 0$, we take only f_1 and f_3 or put:

$$w = \frac{2}{\sqrt{\pi}} \int_0^\infty \left[KF^{-1}\left(at - \frac{a^2}{2}\right)\cos\frac{x^2}{2a^2} + K_1F_1^{-1}\left(at - \frac{x^2}{2a^2}\right)\cos\frac{a^2}{2} \right] da \ldots (\text{x})$$

(p. 444).

Here, $\dfrac{d}{d\,(at)}\,F^{-1}(at) = F(at)$, and $\dfrac{d}{d\,(at)}\,F_1^{-1}(at) = F_1(at)$.

Hence $F^{-1}(at) = \displaystyle\int_{-\infty}^{at} F(at)\,dt = \frac{1}{K}\int_{-\infty}^{at} \frac{d^3w}{dx^3}\,dt = $ the total shearing impulsive force applied to the end $x = 0$ of the bar up to time t.

Similarly $F_1^{-1}(at) = $ total flexural impulsive couple applied up to the time t. (p. 447.)

Case (*c*): see our Art. 1539.

[1532.] Two additional cases (*d*) and (*e*) are considered by Boussinesq on pp. 445—6. They are the following:

Case (*d*). When $x = 0$, let $w = F(at)$ and $d^2w/dx^2 = K_1F_1(at)$, then:

$$w = \frac{1}{\sqrt{\pi}} \int_0^\infty \left[F\left(at - \frac{x^2}{2a^2}\right)\left(\cos\frac{a^2}{2} + \sin\frac{a^2}{2}\right) \right.$$
$$\left. + K_1F_1^{-1}\left(at - \frac{x^2}{2a^2}\right)\left(\cos\frac{a^2}{2} - \sin\frac{a^2}{2}\right) \right] da \ldots\ldots(\text{xi}).$$

Case (*e*).   When $x = 0$, let $d^3w/dx^3 = KF'(at)$ and $dw/dx = F_1(at)$, then :

$$w = \frac{1}{\sqrt{\pi}} \int_0^\infty \left[KF^{-1}\left(at - \frac{a^2}{2}\right)\left(\cos\frac{x^2}{2a^2} + \sin\frac{x^2}{2a^2}\right) \right.$$
$$\left. - F_1\left(at - \frac{a^2}{2}\right)\left(\cos\frac{x^2}{2a^2} - \sin\frac{x^2}{2a^2}\right) \right] da \ldots \text{(xii)}.$$

[1533.]   Boussinesq now deals with special subcases of these results (pp. 448–9).

Subcase (*f*).   Suppose the bar to be continuous in both directions but all shifts symmetrical with regard to $x = 0$, then we must have $dw/dx = 0$ at $x = 0$ for all values of t.

Take $at_1 = at - \dfrac{x^2}{2a^2}$ or $at - \dfrac{a^2}{2}$ as the case may be. We have from (ix), if $w = F(at)$ for $x = 0$:

$$w = \frac{x}{\sqrt{2\pi a}} \int_{-\infty}^t \frac{F(at_1)}{(t - t_1)^{\frac{3}{2}}} \sin\frac{x^2}{4a(t - t_1)} dt_1 \ldots \ldots \ldots \text{(xiii)}.$$

From (xii), if $d^3w/dx^3 = KF'(at)$ for $x = 0$:

$$w = \sqrt{\frac{a}{2\pi}} \int_{-\infty}^t \left[\cos\frac{x^2}{4a(t-t_1)} + \sin\frac{x^2}{4a(t-t_1)} \right] \frac{KF^{-1}(at_1)\, dt_1}{(t-t_1)^{\frac{1}{2}}} \ldots \text{(xiv)}.$$

Similarly Boussinesq treats (*Subcase* (*g*)) the problem when the flexural couple vanishes at $x = 0$, i.e. $d^2w/dx^2 = 0$, while either w or d^3w/dx^3 for $x = 0$ are arbitrary functions of the time ; and (*Subcase* (*h*)) when the end $x = 0$ is pivoted, i.e. $w = 0$ while either dw/dx or d^2w/dx^2 for $x = 0$ are arbitrary functions of the time. The reader will find it easy to write down the integral solutions in these cases as we have done for *Subcase* (*f*).

Subcase (*i*).   The particular problem of a bar infinitely long in one direction to which during a very short interval ($t = -\tau$ to $t = \tau$) a definite inclination χ to its unstrained central line is given at a pivoted terminal, is discussed at considerable length on pp. 449–56. Boussinesq finds the following solution :

$$w = 2\tau\chi \sqrt{\frac{2a}{\pi t}} \left(\frac{4at^2}{\tau x^2} \sin\frac{\tau x^2}{4at^2}\right) \sin\frac{x^2}{4at} \ldots \ldots \ldots \text{(xv)}.$$

If t/τ be very large as compared with $\dfrac{x^2}{4at}$, i.e. if a considerable interval of time has elapsed since the inclination was given and if the points considered be not at an immensely great distance from $x = 0$:

$$w = 2\tau\chi \sqrt{\frac{2a}{\pi t}} \sin\frac{x^2}{4at}.$$

The solution obtained in this as in other cases of the transverse vibrations of a rod differs very considerably from the usual type of wave-motion :

la barre élastique ne transmet le mouvement transversal qu'en le disséminant et le rendant insensible, contrairement à ce qui arrive pour le mouvement longitudinal, régi, comme on sait, par l'équation de d'Alembert (ou des cordes vibrantes), laquelle exprime une transmission intégrale, sans altération, c'est-à-dire sans condensation ni dispersion (p. 456).

[1534.] Boussinesq next deals (pp. 456–63) with the case in which the initial shifts and speeds are given at each point of an infinitely long bar. He finds a solution corresponding in form to those obtained by Fourier and Poisson for similar cases (see our Arts. 207–11 and 425), namely :

$$w = \frac{1}{\sqrt{2\pi}} \int_{-\infty}^{\infty} [F(x + 2a\sqrt{at})(\sin a^2 + \cos a^2)$$
$$+ F_1(x + 2a\sqrt{at})(\sin a^2 - \cos a^2)]\, da,$$

where $w = F(x)$, $dw/dt = aF_1''(x)$ when $t = 0$. Fourier in his *Théorie analytique de la chaleur*, § 411–12, Boussinesq states, had obtained this result for the transverse vibrations of a bar for the case of $F_1(x) = 0$; but I think Fourier had really obtained a more general solution : see our Arts. 207–11 and 1462.

For the special case of $F_1(x) = 0$, and $F(x) = 0$ except for a small length dx_1 of the bar about x_1 we easily find by changing the variable of integration to x_1 and writing $F(x_1)\, dx_1 = dq$:

$$w = \frac{dq}{2\sqrt{2\pi at}} \left(\sin \frac{(x_1 - x)^2}{4at} + \cos \frac{(x_1 - x)^2}{4at} \right),$$

which gives the displacement at time t due to a small displacement at x_1.

[1535.] The exact limits within which solutions of the above type are legitimate are discussed by Boussinesq at some length, not only for the case of the rod, but for the infinitely extended elastic plate. In the latter case the discussion occupies pp. 464–80. The evaluation of the integrals involved is treated by a somewhat complex method. An error on p. 465 in the determination of the quantity S is corrected in the memoir referred to in our Art. 1462 : see p. 643 of the memoir. The most important results of Boussinesq's present discussion can, however, be deduced from the conclusion of that article.

Let us consider the case where a definite movement is given at the origin to the infinite plate, everything being symmetrical round the origin. Further, let the initial velocities be zero, or f_1 of Art. 1462 be zero. We easily find that when a definite shift $w = f$ is given at time

$t = 0$ to a small area σ at the origin, then the shift w at distance r from the origin at time t :

$$= \frac{f\sigma}{4\pi bt} \sin \frac{r^2}{4bt}.$$

Hence transferring the epoch to t_1, writing $\psi(t_1)\, dt_1 = f\sigma/(8b)$ and taking the effects of all shifts from $t_1 = -\infty$ to t, we have:

$$w = \frac{2}{\pi} \int_{-\infty}^{t} \frac{\psi(t_1)}{t - t_1} \sin \frac{r^2}{4b(t - t_1)} \, dt_1.$$

Now change the variable from t_1 to ξ where

$$\xi = r^2 / \{2b(t - t_1)\},$$

then we find (p. 470) :

$$w = \frac{2}{\pi} \int_0^{\infty} \psi\left(t - \frac{r^2}{2b\xi}\right) \sin \frac{\xi}{2} \frac{d\xi}{\xi} \quad\ldots\ldots\ldots\ldots(\text{xvi}).$$

This may be shewn directly to satisfy the shift-equation for the transverse vibrations of a plate, i.e. :

$$d^2 w/dt^2 + b^2 \nabla^2 \nabla^2 w = 0 \quad\ldots\ldots\ldots\ldots\ldots\ldots(\text{xvii}),$$

where in this case

$$\nabla^2 = \frac{d^2}{dr^2} + \frac{1}{r}\frac{d}{dr}.$$

See pp. 472–5.

When $r = 0$, we have from (xvi) $w = \psi(t)$. We have thus found a solution giving an arbitrary displacement at the origin at each instant of time. We see further that if we take ψ so that $\psi(-\infty) = 0$, we have $w = 0$ for $t = -\infty$ whatever be r, and also for $r = \infty$ whatever be t (p. 470). It remains to shew that $dw/dr = 0$ for $r = 0$, in order that there may not be an abrupt change of curvature at the origin. This is investigated by Boussinesq on pp. 471–2. The result is not directly obvious on differentiation of w, because the subject of integration becomes infinite at one of the limits.

[1536.] The equation (xvi) also solves the case of given normal impulses applied to the plate (thickness 2ϵ and density ρ) at the origin of coordinates.

This problem requires a solution of the equation (xvii) subject to the conditions $w = 0$ for $t = -\infty$ and for $r = \infty$, $dw/dr = 0$ for $r = 0$ (all these we have stated are satisfied by (xvi)), and further the total shear[1] round a circumference of radius r, or $2\epsilon\rho b^2 \times 2\pi r \dfrac{d\nabla^2 w}{dr}$, must be a given function $F(t)$ of the time for $r = 0$.

By differentiating (xvi) and rearranging we find :

$$2\pi r \frac{d\nabla^2 w}{dr} = \frac{4}{b}\left[2\psi'(t) + \int_0^{\infty} \left\{ \psi'\left(t - \frac{r^2}{2b\xi}\right) - \psi'(t) \right\} \sin \frac{\xi}{2} \, d\xi \right] \ldots(\text{xviii}).$$

---

[1] This follows easily from the value of R', given in our Art. 393, remembering that $b^2 = H\epsilon^2/(3\rho)$.

This gives for $r = 0$, $F(t) = 16\epsilon\rho b\psi'(t)$, so that $\psi(t)$ is fully determined and the problem accordingly solved.

Since dw/dt for $r = 0$ is equal to $\psi'(t)$, we find at once

$$(dw/dt)_{r=0} = \frac{1}{16\epsilon\rho b} F(t),$$

or the speed of the disturbed centre is always proportional to the disturbing force. It follows that the shift of this centre is at each instant proportional to the total impulse up to that instant. On pp. 477–80 Boussinesq draws a number of interesting conclusions with regard to the equation (xvi).

[1537.] The next section (pp. 480–505) of Boussinesq's *Treatise* is of special interest. It is entitled: *Problème de la résistance dynamique des barres et des plaques, notamment de leur résistance au choc, traité par les mêmes procédés : extension d'une loi de Young au cas du choc transversal.*

We shall deal briefly with several cases discussed by Boussinesq.

Case (i). Consider a bar of infinite length in one direction the general expression for the shift of which, when subjected to any kind of action at the end $x = 0$, is given by equation (vii) of our Art. 1531.

We easily deduce the following system of differentials at the origin, *remembering results* (i) *to* (iii) *of our Art. 1530*:

$$w_0 = \tfrac{1}{2}\sqrt{\pi}\left\{f_3(at) + f_4(at)\right\} + \int_0^\infty f_1(at - \tfrac{1}{2}a^2)\,da,$$

$$(dw/dt)_0 = \frac{a}{2}\sqrt{\pi}\left\{f_3'(at) + f_4'(at)\right\} + a\int_0^\infty f_1'(at - \tfrac{1}{2}a^2)\,da,$$

$$(dw/dx)_0 = -\tfrac{1}{2}\sqrt{\pi}\left\{f_1(at) - f_2(at)\right\} - \int_0^\infty f_3'(at - \tfrac{1}{2}a^2)\,da,$$

$$(d^2w/dx^2)_0 = \tfrac{1}{2}\sqrt{\pi}\left\{f_3'(at) - f_4'(at)\right\} - \int_0^\infty f_2'(at - \tfrac{1}{2}a^2)\,da,$$

$$(d^3w/dx^3)_0 = \tfrac{1}{2}\sqrt{\pi}\left\{f_1'(at) + f_2'(at)\right\} + \int_0^\infty f_4''(at - \tfrac{1}{2}a^2)\,da.$$

Now if $(dw/dt)_0 = 0$ for all values of t, we must have

$$f_1 = f_2, \quad \text{and} \quad f_3' = 0,$$

whence we find at once

$$(dw/dt)_0 = -a\,(d^2w/dx^2)_0.$$

Similarly we deduce, if $(d^3w/dx^3)_0 = 0$ for all values of t:

$$(dw/dt)_0 = a\,(d^2w/dx^2)_0.$$

Now $(dw/dt)_0$ is the velocity V taken by the bar at the origin, and if h be the distance of the 'extreme fibre' from the neutral axis, and s the corresponding stretch, we have $s = \pm h\,(d^2w/dx^2)_0$ at the origin, or remembering the value of a:

$$V = \Omega \times s \times \frac{\kappa}{h} \quad\dots\dots\dots\dots\dots\dots\text{(xix)},$$

where Ω is the velocity of longitudinal waves of sound ($= \sqrt{E/\rho}$).

In the case of a circular section $\kappa/h = \tfrac{1}{2}$, of a rectangular section $\kappa/h = 1/\sqrt{3}$, etc.

At the instant of a blow,—for example, a blow at the centre of a rod infinitely long in both directions (i.e. when $(dw/dx)_0 = 0$)—, V will be the velocity of the impinging body, hence if s be the maximum safe-stretch of the material all velocities greater than that given by (xix) will damage the material locally.

It is not however necessary to consider the bar infinitely long; the above results will still hold in the first instant of an impact and before there is time for reflection of the disturbance from fixed or supported ends. We have appealed to this result in our Art. 371. It is an extension of the corresponding result obtained by Young for longitudinal impact (i.e. $V = \Omega \times s$)[1]: see our Art. 1068 (Boussinesq: pp. 480–6, 498–9).

[1538.] *Case* (ii). We can deduce a somewhat similar result for the case of a plate from the result (xvi) of our Art. 1535.

We easily find:

$$\frac{d^2w}{dr^2} - \frac{1}{r}\frac{dw}{dr} = -\frac{4}{\pi b} \int_0^\infty \psi'\left(t - \frac{r^2}{2b\xi}\right) d\left(\frac{\sin\frac{\xi}{2}}{\xi}\right)$$

$$= \frac{2}{\pi b}\,\psi'(t), \quad \text{when } r = 0.$$

But $(dw/dt)_0 = \psi'(t)$, thus

$$\frac{\pi b}{2}\left(\frac{d^2w}{dr^2} - \frac{1}{r}\frac{dw}{dr}\right)_{r=0} = \left(\frac{dw}{dt}\right)_0.$$

Hence, if 2ϵ be the thickness of the plate, s_1 and s_2 the stretches corresponding to the two principal curvatures $\dfrac{d^2w}{dr^2}$ and $\dfrac{1}{r}\dfrac{dw}{dr}$ at the origin, and V the velocity of impact:

$$V = \pm\,\frac{\pi b}{2\epsilon}\,(s_1 - s_2).$$

But $$b^2 = \frac{4\mu\,(\lambda + \mu)}{3\rho\,(\lambda + 2\mu)}\,\epsilon^2 :$$

see our Arts. 385 and 323.

---

[1] Since κ/h is always less than unity, we see that the velocity of the impact which will suffice to damage a bar locally is always less in the case of transverse than in the case of longitudinal impact (Boussinesq: pp. 501–2).

Now $\dfrac{4\mu\,(\lambda+\mu)}{\rho\,(\lambda+2\mu)}$ is the square of the velocity Ω_1 with which 'spreads' are propagated through the plane of the plate (see our Art. 595* and equations (iii) of Art. 389). Hence we have finally :

$$V = \frac{\pi}{2\sqrt{3}} \times \Omega_1 \times (s_1 \sim s_2) \dots\dots\dots\dots\dots(\mathrm{xx}).$$

Unfortunately this does not tell us like (xix) the maximum normal velocity of impact. We see however that any velocity equal to the product of the velocity of spread-propagation into the maximum safe-stretch into $\dfrac{\pi}{2\sqrt{3}}\,(= \cdot9069)$ will on the greatest strain theory damage the plate[1].

[1539.] *Case* (iii). Boussinesq now (pp. 490–6) returns to the problem of an infinite rod to which a mass M is attached at some point of its length. If the mass be subjected to the force $F\,(at)$ we must by (c) of our Art. 1529 satisfy for $x = 0$ the conditions :

$$\tfrac{1}{2}M\,\frac{d^2w}{dt^2} = -E\omega\kappa^2\left(\frac{d^3w}{dx^3}\right)_0 + \tfrac{1}{2}F\,(at),$$

and $\qquad\qquad\qquad\qquad dw/dx = 0.$

[1] Boussinesq's conclusions as to the limit to the velocity with which a body of any mass, however small, can impinge upon a bar or plate without damaging its elasticity seem to me of special physical importance. They indicate how light bodies moving at great speeds may be used to destroy, cut or shape harder and more massive bodies. Thus they are full of suggestion for the science of gunnery and the mechanical arts. One of the most interesting mechanical processes illustrated by Boussinesq's theoretical results is that of the sand-blast. In this case the velocity of the blast ranges from 100 to 2000 feet per second, the blast of air or steam carrying with it 'sand', which term may be used to denote small grains or particles of which quartz sand is a type, but which may include globules of cast-iron or even fine shot. Corundum can be cut by the less hard quartz sand, and quartz rock by fine lead shot, while the hardest steel can be cut by a stream of quartz sand. Sand-blast machines are in use for cutting, perforating, obscuring, or engraving glass, for sharpening files, for cleaning iron and steel castings, for cutting letters, etc., in marble and stone, and so forth. A further example of the same principle is probably to be found in the experiments referred to in Art. 836 (h), in which steel and quartz were cut by soft-iron, and in the copper wheel of 3″ diameter which may be seen cutting glass at the Crystal Palace.

Some idea of the necessary velocity of the sand-blast may be approximately obtained from equation (xx). Assuming uni-constant isotropy of the plate, we have $\Omega_1 = \dfrac{4}{\sqrt{15}}\,\Omega$, and hence $V = \cdot9366\,\Omega \times (s_1 - s_2)$, nearly. For the case of steel taking round numbers, $\Omega = 17{,}000$ feet per second, and $s_1 = \cdot04$ as a maximum for untempered steel (Art. 1134). Thus we see that a blast of 640 feet per second would certainly suffice to cut the steel. For tempered and annealed steel s_1 reduces to $\cdot004$ (Art. 1134) and hence a blast of 64 feet per second would suffice. That something *considerably less* than this might suffice would appear to be indicated by the 34 feet per second of Davier and Colladon's experiments : see our Art. 836 (h). The velocities we have calculated, however, approach nearer to those used in the sand-blast machine. The whole subject is deserving of careful experimental investigation.

The latter condition by Case (i) of Art. 1537 is satisfied by $f_1 = f_2$ and $f_3 = 0$. The former condition will be satisfied by taking:

$$f_4(at) = -\nu f_1(at),$$

and[1] $f_1''(at) - \frac{2}{\nu^2} f_1'(at) = -\frac{2}{Ma^2\nu\sqrt{\pi}}\, F(at) = -\frac{1}{a^2\nu^2\sqrt{\pi}}\frac{F(at)}{\omega\rho},$

where $\nu = \frac{1}{2}$ the ratio of the mass of the central load to the mass of unit length of the rod. Writing $\zeta = at$ and $1/\beta = \dfrac{1}{a^2\nu^2\omega\rho\sqrt{\pi}}$, we have to solve the differential equation:

$$f_1''(\zeta) - \frac{2}{\nu^2} f_1'(\zeta) = -\frac{1}{\beta} F(\zeta).$$

Remembering that $F(-\infty) = 0$, we find as the solution of this

$$f_1(\zeta) = Ce^{2\zeta/\nu^2} + \frac{\nu^2}{2\beta}\int_{-\infty}^{\zeta} F(\gamma)\{1 - e^{2(\zeta-\gamma)/\nu^2}\}\, d\gamma \,\ldots\ldots(xxi).$$

Boussinesq shews that the term involving the arbitrary constant C disappears from the value of w (p. 492). We thus, so far as the shift is concerned, can put it zero or any finite value we please. Let us take it equal to

$$\frac{\nu^2}{2\beta} \times \int_{-\infty}^{\infty} F(\gamma) e^{-2\gamma/\nu^2}\, d\gamma,$$

then we have:

$$f_1(\zeta) = \frac{\nu^2}{2\beta}\left\{\int_{-\infty}^{\zeta} F(\gamma)\, d\gamma + \int_{\zeta}^{\infty} F(\gamma) e^{2(\zeta-\gamma)/\nu^2}\, d\gamma\right\} \,\ldots(xxii),$$

where the exponential has always a negative index.

Equation (xxii) combined with the value of w from (vii) of Art. 1531, or:

$$w = \int_0^{\infty}\left\{f_1\left(at - \frac{a^2}{2}\right)\left(\cos\frac{x^2}{2a^2} + \sin\frac{x^2}{2a^2}\right)\right.$$
$$\left. -\nu f_1\left(at - \frac{x^2}{2a^2}\right)\sin\frac{a^2}{2}\right\} da\ldots\ldots(xxiii)$$

solves the problem completely.

[1] Instead of the last term on the right of this our second condition Boussinesq has on p. 491 the term:

$$-\frac{2}{a^2\mu^2\sqrt{\pi}} F'(t).$$

His $\mu = $ our ν. Hence his $F'(t)$ ought to be equivalent to our $\dfrac{1}{2}\dfrac{F(at)}{\omega\rho}$. This it in fact is, because he defines $F'(t)$ to be *half* the force applied to the mass M, and takes the mass of unit length of the rod as *unit of mass* (p. 481). It seems clearer to take a perfectly general unit of mass.

Boussinesq deals in detail only with the special case in which the motion of M is due to an impulsive force of magnitude Q acting during the very small period $t = 0$ to $t = \epsilon$. Then we have

$$\int_0^\epsilon F(at)\, dt = Q.$$

Hence by putting $C = 0$ in (xxi) we have:

$$f_1(at) = 0, \text{ if } t < 0,$$

$$= \frac{\nu^2 Q a}{2\beta}\{1 - e^{2at/\nu^2}\}, \text{ if } t > \epsilon.$$

Let w_0 be the shift for $x = 0$, then we have by Case (i) Art. 1537, after a slight transformation, if $\tau = 2at/\nu^2$:

$$w_0 = \frac{\nu^3 Q a}{2\beta}\left\{\int_0^{\sqrt{\tau}}(1 - e^{\tau - a'^2})\, da' - \tfrac{1}{2}\sqrt{\pi}\,(1 - e^\tau)\right\}.$$

Substituting the values of ν and β and writing

$$\chi(\tau) = \frac{2}{\sqrt{\pi}}\int_{\sqrt{\tau}}^\infty e^{\tau - a'^2}\, da',$$

we easily find:

$$w_0 = \frac{\nu Q}{4a\omega\rho}\left\{2\sqrt{\frac{\tau}{\pi}} + \chi(\tau) - \chi(0)\right\} \quad\dots\dots\dots(xxiv).$$

Boussinesq discusses at some length the integral $\chi(\tau)$ which may be written

$$\frac{1}{\sqrt{\pi}}\int_0^\infty \frac{e^{-\beta}\, d\beta}{\sqrt{\tau + \beta}},$$

so that $\chi(\tau)$ is always less than $1/\sqrt{\pi\tau}$ to which value it tends as τ increases indefinitely. Generally (p. 496):

$$\chi(\tau) = e^\tau - \sqrt{\frac{2}{\pi}}\sum_0^\infty\left(\frac{(\sqrt{2\tau})^{2n+1}}{1\,.\,3\,.\,5\dots 2n+1}\right).$$

We can easily find for the shift speed at $x = 0$:

$$(dw/dt)_{x=0} = \frac{Q}{2\nu\omega\rho}\chi(\tau) = \frac{Q}{M}\chi(\tau) \quad\dots\dots\dots\dots(xxv).$$

[1540.] This solution can at once be applied to the case in which a body of mass M impinges on a bar infinitely long in both directions with velocity V, for we have only to take $Q = MV$, and then (xxiv) and (xxv) express the solution. Obviously w_0 increases indefinitely with t, while the speed $(dw/dt)_{x=0}$ diminishes and ultimately vanishes.

En effet la formule (xxv) donnant $(dw/dt)_{x=0} = Q/M$ à l'époque $t = 0$, montre que, pour t infiniment petit une masse M unie à la barre à l'origine $x = 0$ détient presque la totalité de la quantité de mouvement qu'une impulsion brusque y a fait naître, tout comme si cette masse s'était trouvée isolée quand elle a subi l'impulsion ; et il doit être, par suite, à peu près indifférent que le corps heurtant ait reçu sa vitesse initiale V quand il était encore libre ou après s'être joint à la barre. Il n'y a, entre les deux cas, de différence, que dans la manière dont la vitesse V se communique, durant l'instant initial ϵ, au tronçon heurté de la barre, manière plus conforme aux hypothèses ordinaires de la théorie de l'élasticité quand on suppose la masse M déjà en contact avec la barre dès l'instant $t=0$ (p. 497).

[1541.]   *Case* (iv).   Boussinesq on pp. 498–502, deals with the problem of a bar indefinitely long in one direction carrying a weight M at its terminal, and subjected to the longitudinal impact Q during the same interval of time $t = 0$ to $t = \epsilon$. Let $Q = MV$, then Boussinesq finds for the shift $u_{x=0}$ of the terminal and for its speed $(du/dt)_{x=0}$, $\Omega = \sqrt{E/\rho}$ being the velocity of longitudinal vibrations :

$$\left.\begin{aligned} u_{x=0} &= \frac{MV}{\rho\omega\Omega}\left(1 - e^{-\frac{\rho\omega\Omega}{M}t}\right), \\ (du/dt)_{x=0} &= Ve^{-\frac{\rho\omega\Omega}{M}t} \end{aligned}\right\}\dots\dots\dots\dots(\text{xxvi}).$$

Thus the shift $u_{x=0}$ does not tend to increase indefinitely with t but to approach the limit $MV/(\rho\omega\Omega)$.

Since $u_{x=0}$ changes its sign with V but its magnitude remains unchanged, we have only to put two bars, infinitely long in one direction, end to end, each bearing a mass $\frac{1}{2}M$ to obtain the solution for the case in which a mass M attached to the middle of an infinitely long bar receives an impulse in the direction of the bar.

Turning to our Art. 222, putting therein $V_1 = V$, $V_2 = 0$, $k_2 = \Omega$, $M_1 = M$, $M_2 = a_2\rho\omega$, and then making a_2 infinite, we easily find from (2^0) of that article by integrating the stretch and putting $x = 0$:

$$\left.\begin{aligned} u_{x=0} &= \frac{MV}{\rho\omega\Omega}\left(1 - e^{-\frac{\rho\omega\Omega}{M}t}\right), \\ (du/dt)_{x=0} &= Ve^{-\frac{\rho\omega\Omega}{M}t} \end{aligned}\right\}\text{ for } t = 0 \text{ to } \infty .$$

These equations agree entirely with (xxvi) after time $t = \epsilon$, or we see that whether M be attached to the bar initially and receive an impulse MV, or a mass M with momentum MV strike the bar, there will be no difference in the values of $u_{x=0}$ and $(du/dt)_{x=0}$ after time $t = \epsilon$.

[1542.]   *Case* (v).   In an *Addition* (pp. 655–64) Boussinesq works out the extremely interesting and practically valuable case of a bar in the form of an infinitely long truncated right circular cone, subjected at the truncated end (supposed at distance c from the vertex) to the longitudinal impact of a body of mass M moving with velocity V. The investigation of this case had been suggested by Saint-Venant's memoir of 1868 : see our Art. 223.

The legitimacy of the solution seems to me, however, to depend upon the cone being of very small vertical angle, otherwise we have no right to use D'Alembert's elementary theory of rods which supposes the cross-sections to remain plane. This assumption is not, I think, directly stated by Boussinesq, but it ought to be kept in mind.

The equation for the longitudinal vibrations of such a cone on D'Alembert's theory is easily found to be

$$\frac{d^2\{(x+c)\,u\}}{dt^2} = \Omega^2 \frac{d^2\{(x+c)\,u\}}{dx^2},$$

where u is the shift at distance $(x+c)$ from the vertex and $\Omega^2 = E/\rho$. For waves in the direction of x positive we have:

$$(x+c)\,u = f(\Omega t - x),$$

whence we easily find:

$$-\frac{du}{dx} = \frac{1}{\Omega}\frac{du}{dt} + \frac{u}{x+c} \quad\ldots\ldots\ldots\ldots\ldots(\text{xxvii}).$$

Thus initially, when $u = 0$, if s be the stretch, and V the velocity,

$$V = -s \times \Omega,$$

or, if s be the safe elastic stretch (or squeeze), no body can strike the truncated end of the cone with greater velocity than $s \times \Omega$ without damaging it. Young's theorem[1] (see our Art. 1537) is thus extended to such solids of revolution with truncated ends, as may *in the very beginning of the motion* be looked upon as truncated cones.

At the end $x = 0$ of the cone we have the condition

$$M\left(\frac{d^2u}{dt^2}\right)_{x=0} = E\omega\left(\frac{du}{dx}\right)_{x=0} \quad\ldots\ldots\ldots\ldots(\text{xxviii}),$$

which enables us to determine the form of $f(\Omega t)$. In addition we have the conditions that $f(0) = 0$, and $du/dt = V$ when $x = 0$ and $t = 0$. If $\nu = M/(\rho\omega)$ we have from (xxviii) using (xxvii):

$$\nu f''(\Omega t) + f'(\Omega t) + \frac{1}{c}f(\Omega t) = 0,$$

whence we determine:

$$f(\Omega t) = \frac{2\nu c\,V}{k\Omega}\,e^{-\frac{\Omega t}{2\nu}}\left(\sin\frac{k\Omega t}{2\nu}, \text{ or, } \sinh\frac{k\Omega t}{2\nu}\right) \quad\ldots\ldots(\text{xxix}).$$

Here $k^2 = 1 \sim 4\nu/c$ and the natural or the hyperbolic sine is to be used according as $4\nu >$ or $< c$, or according as the impinging mass is greater or less than three-quarters of the mass of the truncated part of the cone.

---

[1] A generalized form of Young's Theorem may be found at once from the result given for the squeeze $(-s)$ of the impelled bar in 2° of our Art. 222, by putting $x = 0$ and $t = 0$. We find

$-s = $ (velocity of impact)/(velocity of sound in impelled bar).

We easily find by aid of (xxviii) and (xxix) :

$$
\left.\begin{array}{l}
u_{x=0} = \dfrac{2\nu V}{k\Omega}\, e^{-\frac{\Omega t}{2\nu}} \left(\sin \dfrac{k\Omega t}{2\nu}\,,\ \text{or},\ \sinh \dfrac{k\Omega t}{2\nu} \right), \\[3mm]
\left(\dfrac{du}{dt}\right)_{x=0} = V e^{-\frac{\Omega t}{2\nu}} \left\{ \begin{array}{l} \cos \dfrac{k\Omega t}{2\nu} - \dfrac{1}{k}\sin \dfrac{k\Omega t}{2\nu}\,, \\[3mm] \text{or},\ \cosh \dfrac{k\Omega t}{2\nu} - \dfrac{1}{k}\sinh \dfrac{k\Omega t}{2\nu} \end{array} \right\}, \\[8mm]
\left(\dfrac{du}{dx}\right)_{x=0} = \dfrac{1}{2}\dfrac{V}{\Omega}\, e^{-\frac{\Omega t}{2\nu}} \left\{ \begin{array}{l} \left(\dfrac{1}{k}-k\right)\sin \dfrac{k\Omega t}{2\nu} - 2\cos \dfrac{k\Omega t}{2\nu}\,, \\[3mm] \text{or},\ \left(k+\dfrac{1}{k}\right)\sinh \dfrac{k\Omega t}{2\nu} - 2\cosh \dfrac{k\Omega t}{2\nu} \end{array} \right\}
\end{array}\right\} \ ..(\text{xxx}).
$$

These equations tell us at once a great deal about the impact. We see that if $k=\tan\gamma$ or $\tanh\gamma$ according as $4\nu >$ or $< c$, then the maximum shift $(u_{x=0})_m$ at the free end is reached when $(du/dt)_{x=0}=0$, or :

$$
t = t_1 = \frac{2\nu}{\Omega}(\gamma \cot\gamma,\ \text{or},\ \gamma \coth\gamma),
$$

and

$$
(u_{x=0})_m = \frac{V}{\Omega}\sqrt{\nu c}\, e^{-(\gamma \cot\gamma,\ \text{or},\ \gamma \coth\gamma)}.
$$

Further we have $(du/dx)_{x=0}=0$, or the action of the mass M on the truncated cone ceases, when $t=t_2=2t_1$. Thus the duration of the blow is equally divided between the periods when the mass is continually increasing the compression of the bar and when it is continually releasing that compression. It is easy to see that the blow ends before the cone returns to its original length by substituting $2t_1$ in the value of $u_{x=0}$. The velocity of rebound of M is given by

$$
- V e^{-2(\gamma \cot\gamma,\ \text{or},\ \gamma \coth\gamma)},
$$

which confirms the result referred to in our Art. 216, namely: that the velocity of rebound depends on the masses and dimensions of the bodies in collision. The termination of the blow when $t=t_2$ is of interest, because in the case of the indefinitely long cylindrical rod, there is no limit to the duration of the blow: see our Art. 1541.

[1543.] Boussinesq next considers what happens after the termination of the blow. Instead of (xxviii) the terminal condition is now $(du/dx)_{x=0}=0$, whence we find $cf'(\Omega t)+f(\Omega t)=0$, or remembering that when $t=t_2$ the two solutions must coincide, we have for $t > t_2$:

$$
f(\Omega t) = f(\Omega t_2)\, e^{-\frac{\Omega(t-t_2)}{c}},
$$

We find for the shift speed $t > t_2$

$$\left(\frac{du}{dt}\right)_{x=0} = -\frac{\Omega}{c^2} f'(\Omega t_2) e^{-\frac{\Omega(t-t_2)}{c}},$$

or, this speed decreases with increase of t, and hence the greatest value is reached for $t = t_2$, or at the end of the blow, thus the impelled and the impinging bodies never come into contact again. The shift at the now free end of the cone decreases gradually and ultimately becomes zero with $t = \infty$ (p. 662).

[1544.] It remains to find the maximum squeeze and the time at which it takes place. Boussinesq easily shows by aid of (xxvii) that the maximum squeeze takes place before the end of the blow and at the impelled end of the cone. In order to obtain the maximum value we have only to differentiate the third equation of (xxx) with regard to t and equate the result to zero. We find

$$\left(\frac{d^2 u}{dx\,dt}\right)_{x=0} = \frac{Ve^{-\frac{\Omega t}{2\nu}}}{4\nu} \left\{ \frac{\sin\left(3\gamma - \frac{k\Omega t}{2\nu}\right)}{\sin\gamma \cos^2\gamma}, \quad \text{or,} \quad \frac{\sinh\left(3\gamma - \frac{k\Omega t}{2\nu}\right)}{\sinh\gamma \cosh^2\gamma} \right\}.$$

Thus the squeeze $-(du/dx)_{x=0}$ will decrease as t increases from 0 to t_2, i.e. from the instant after the impact up to the end of the blow, except in the first case ($\nu > \frac{1}{4}c$) for $3\gamma > \pi$, or $k = \tan\gamma > \sqrt{3}$, or $\nu > c$, or when the mass of the impinging body is greater than three times the mass of the truncated portion of the cone. Should this hold the squeeze becomes a maximum $-s_m$, when $t = t_3$, where

$$t_3 = \frac{2\nu}{\Omega} \frac{3\gamma - \pi}{\tan\gamma} \quad \dots\dots\dots\dots\dots\dots (\text{xxxi}),$$

and by the third result of (xxx)

$$-s_m = \frac{V}{\Omega}\sqrt{\frac{\nu}{c}}\, e^{-(3\gamma-\pi)\cot\gamma} \quad \dots\dots\dots\dots (\text{xxxii}).$$

The exponential will take its minimum value for $\gamma = 1\cdot3027$, about, and it then equals $\cdot8101$ which is slightly less than the maximum value, unity, which it takes for $\gamma = \pi/3$ or $\pi/2$.

Thus except for $\nu > c$, $-s_m$ takes its maximum value, V/Ω, at the instant the blow commences. If $\nu > c$ its maximum value must be found from (xxxii) and then by the preceding remarks does not differ widely from $V/\Omega \times \sqrt{\nu/c}$ (pp. 663–4).

Boussinesq concludes his discussion by remarking that if the thicker end of the cone be cut off at the section $x = l$, and this section be fixed, then we shall have (see our Art. 223) a solution of the form

$$(x + c)\,u = f(\Omega t - x) - f(\Omega t + x - 2l),$$

where the second term on the right is due to the reflected wave; this term will, however, be zero at the impelled terminal until $t = 2l/\Omega$, or we see that the above investigation holds for this new case during the whole of the interval $2l/\Omega$ after the impulse.

[1545.] *Case* (vi). Boussinesq deals on pp. 502–5 with the case of a plate of infinite radius struck normally by a mass M at the origin of coordinates with a velocity V. He replaces this problem by that of a mass M attached to the origin of coordinates and subjected to the normal force $F(t)$. Using the notation of our Art. 1536, we have the expression $2\epsilon\rho b^2 \times 2\pi r \dfrac{d\nabla^2 w}{dr}$ for the total shear round a cylinder of radius r about the origin, and therefore for $r = 0$:

$$M \left(\frac{d^2 w}{dt^2}\right)_0 + 2\epsilon\rho b^2 \left(2\pi r \frac{d\nabla^2 w}{dr}\right)_{r=0} = F(t),$$

or, by the results of our Art. 1536 :

$$\nu\psi''(t) + 8b\psi'(t) = \frac{F(t)}{m},$$

m being the mass of the plate per unit of area and $\nu = M/m$.

Solving this equation in the same manner as that for $f(\zeta)$ in our Art. 1539, we have :

$$\psi(t) = \frac{1}{8mb} \int_{-\infty}^{t} F(\zeta)\left(1 - e^{\frac{8b(\zeta - t)}{\nu}}\right) d\zeta.$$

If we consider the special case of the blow produced by the mass M moving with velocity V we have :

$$\int_{-\infty}^{t} F(t)\, dt = 0 \text{ for } t < 0, \text{ and } \int_{-\infty}^{t} F(t)\, dt = MV$$

for t slightly greater than zero. Hence

$$\psi(t) = 0 \ (t < 0), \text{ and } \psi(t) = \frac{\nu V}{8b}\left(1 - e^{-\frac{8bt}{\nu}}\right) (t > 0).$$

Whence we easily find from (xvi) for $t > 0$:

$$w_{r=0} = \frac{\nu V}{8b}\left(1 - e^{-\frac{8bt}{\nu}}\right), \quad (dw/dt)_{r=0} = Ve^{-\frac{8bt}{\nu}}$$

Thus we see that in this case the shift tends to the finite limit $\nu V/(8b)$, and the plate acts in this manner quite differently from the bar of our Arts. 1539–40.

[1546.] The next section (pp. 505–46) of Boussinesq's work is entitled : *Comment il faut modifier ces lois du choc, dans le cas de barres dont la longueur est finie.*

It opens with some remarks on impact generally, noticing that the results obtained in the previous articles hold for finite bodies only, if the velocity of impact be above a certain magnitude and thus damage be done to the body at the instant of the blow. For velocities less than this limiting velocity no damage need be done to the body unless the ratio of the mass of the impinging to that of the impelled body exceeds a certain value, and for such velocities the maximum strain will not be reached at the instant of the impact.

Turning to Saint-Venant's results for the transverse impact of rods given in the table in our Art. 371, (iv), Boussinesq remarks that they may be thrown into the form $s_0 = \dfrac{h}{\kappa} \dfrac{V}{\Omega} \sqrt{\beta \dfrac{Q}{P}}$, where β is a factor depending on the ratio Q to P and the notation is that of our Art. 371. Boussinesq compares this with his condition for damage due to immediate impact, i.e. $s_0 = \dfrac{h}{\kappa} \dfrac{V}{\Omega}$, and notices that when $Q/P =$ or $> 1/\beta$, this latter condition replaces Saint-Venant's. He remarks (p. 508) that β seems to be *roughly* 3 : see our Art. 371, (iii), where $s_0 = h/\rho$. Hence Boussinesq's condition would come into play when $Q/P =$ or $> \frac{1}{3}$. In Art. 371 (p. 254) I have suggested that the critical value of Q/P lies between $\frac{2}{5}$ and $\frac{1}{3}$.

[1547.] The remaining portion of this section deals with the longitudinal impact of bars. Two cases are considered : when the impelled bar has the non-impelled end (i) fixed, (ii) free. The latter case corresponds to that discussed by Saint-Venant in 1868 : see our Art. 221; the former case presents the analytical solution which Saint-Venant and Flamant discussed graphically in their memoir of 1883 : see our Art. 401 *et seq.*

If the impulse occur at the end taken for the origin of x, then, l being the length of the bar, we must have for the first case the shift $u = 0$, when $x = l$, and for the second case $du/dx = 0$ when $x = l$. If[1] $a^2 = E/\rho$, the solution must therefore be of the form

$$u = f(at - x) \mp f(at + x - 2l) \quad \dots\dots\dots\dots\dots\dots(i),$$

the upper sign referring to the first case.

The condition at the impelled end, or for $x = 0$, is

$$v \frac{d^2u}{dt^2} = \frac{F(t)}{m} + \frac{a^2}{l} \frac{du}{dx} \quad \dots\dots\dots\dots\dots\dots(ii),$$

[1] a is here used for the Ω of our Arts. 1541-6, so that the results may at once be compared with those of our Arts. 401-7.

where $\nu = Q/P$ the ratio of the weights of the impinging mass and the bar, $m = P/g$ and $F(t)$ is the force on Q at time t and vanishes for $t < 0$.

Substituting (i) in (ii) putting $x = 0$, integrating and writing $\int_{-\infty}^{t} F(t)\,dt = F^{-1}(t)$ we find, if $at = \zeta$:

$$f'(\zeta) + \frac{1}{\nu l} f(\zeta) = \frac{F^{-1}\left(\frac{\zeta}{a}\right)}{\nu a m} \pm f'(\zeta - 2l) \mp \frac{1}{\nu l} f(\zeta - 2l),$$

or,

$$f(\zeta) = e^{-\frac{\zeta}{\nu l}} \int_{0}^{\zeta} e^{\frac{\zeta}{\nu l}} \left\{ \frac{1}{\nu a m} F^{-1}\left(\frac{\zeta}{a}\right) \pm f'(\zeta - 2l) \mp \frac{1}{\nu l} f(\zeta - 2l) \right\} d\zeta \ldots \text{(iii)}.$$

This holds for $\zeta > 0$. But f and f' vanish for negative arguments. Hence (iii) enables us to write down first the value of $f(\zeta)$ for $\zeta = 0$ to $\zeta = 2l$, and then, from this value of $f(\zeta)$ substituted on the right under the integral, to write down the value of $f(\zeta)$ from $\zeta = 2l$ to $4l$ and so on. Thus $f(\zeta)$ is entirely determined in finite terms. Hence by (i) the problem is analytically solved. The solution involves a novel and valuable method capable of application to a number of problems in impact.

[1548.] For the case of an impact by the mass $M (= Q/g)$ with velocity V we have $F^{-1}(t) = MV$. Hence we find:

$$f(\zeta) = \nu l \frac{V}{a}\left(1 - e^{-\frac{\zeta}{\nu}}\right) \pm e^{-\frac{\zeta}{l}} \int_{0}^{\zeta} \left\{ f'(\zeta - 2l) - \frac{1}{\nu l} f(\zeta - 2l) \right\} e^{\frac{\zeta}{\nu l}} d\zeta \ldots \text{(iv)}.$$

which again completely determines $f(\zeta)$.

Properly the time from $t = 0$ to $t = \tau$, the small interval during which the blow is given, or from $\zeta = 0$ to $\zeta = a\tau$, or ϵ', ought to be excluded from the value of $f'(\zeta)$, for we cannot differentiate $f(\zeta)$ *at* the origin (since $f(\zeta) = 0$ abruptly, for $\zeta < 0$) but only slightly to the positive side of it, i.e. when ζ has any vanishingly small positive value. In fact it will be found that $f'(\zeta)$ increases by jumps (cf. our Diagram IV. p. 278) whenever ζ increases by $2l$ (pp. 515–6).

Boussinesq gives (pp. 513–15) the general solution.

$$\left[f(\zeta)\right]_{\zeta=0}^{\zeta=2l} = \nu l \frac{V}{a}\left(1 - e^{-\frac{\zeta}{\nu l}}\right), \qquad \left[f'(\zeta)\right]_{\zeta=\epsilon'}^{\zeta=2l} = \frac{V}{a} e^{-\frac{\zeta}{\nu l}},$$

$$\left[f(\zeta)\right]_{\zeta=2l}^{\zeta=4l} = \left[f(\zeta)\right]_{\zeta=0}^{\zeta=2l} \pm \nu l \frac{V}{a}\left\{-1 + e^{-\frac{\zeta-2l}{\nu l}}\left(1 + 2\frac{\zeta-2l}{\nu l}\right)\right\},$$

$$\left[f'(\zeta)\right]_{\zeta=2l+\epsilon'}^{\zeta=4l} = \left[f'(\zeta)\right]_{\zeta=\epsilon'}^{\zeta-2l} \pm \frac{V}{a} e^{-\frac{\zeta-2l}{\nu l}}\left(1 - 2\frac{\zeta-2l}{\nu l}\right),$$

$$\left[f(\zeta)\right]_{\zeta=4l}^{\zeta=6l} = \left[f(\zeta)\right]_{\zeta=2l}^{\zeta=4l} + \nu l \frac{V}{a}\left\{1 - e^{-\frac{\zeta-4l}{\nu l}}\left(1 + 2\frac{(\zeta-4l)^2}{\nu^2 l^2}\right)\right\},$$

$$\left[f'(\zeta)\right]_{\zeta=4l+\epsilon'}^{\zeta=6l} = \left[f'(\zeta)\right]_{\zeta=2l+\epsilon'}^{\zeta=4l} + \frac{V}{a}e^{-\frac{\zeta-4l}{\nu l}}\left(1-4\frac{\zeta-4l}{\nu l}+2\frac{(\zeta-4l)^2}{\nu^2 l^2}\right),$$

$$\left[f(\zeta)\right]_{\zeta=6l}^{\zeta=8l} = \left[f(\zeta)\right]_{\zeta=4l}^{\zeta=6l}$$

$$\pm\nu l\frac{V}{a}\left[-1+e^{-\frac{\zeta-6l}{\nu l}}\left(1+2\frac{\zeta-6l}{\nu l}-2\frac{(\zeta-6l)^2}{\nu^2 l^2}+\tfrac{4}{3}\frac{(\zeta-6l)^3}{\nu^3 l^3}\right)\right],$$

$$\left[f'(\zeta)\right]_{\zeta=6l+\epsilon'}^{\zeta=8l} = \left[f'(\zeta)\right]_{\zeta=4l+\epsilon'}^{\zeta=6l}$$

$$\pm\frac{V}{a}e^{-\frac{\zeta-6l}{\nu l}}\left(1-6\frac{\zeta-6l}{\nu l}+6\frac{(\zeta-6l)^2}{\nu^2 l^2}-\tfrac{4}{3}\frac{(\zeta-6l)^3}{\nu^3 l^3}\right)\dots\dots(v).$$

Boussinesq does not calculate these functions to larger values of the variable ζ. The above results generally suffice to determine the maximum strain and the end of the impact. The end of the impact will be reached for the least value of t for which $du/dx=0$ for $x=0$, or by (i) for the least value of ζ for which $f'(\zeta) = \mp f'(\zeta-2l)$.

[1549.] Pp. 517–22 are occupied with the second case viz. that in which the non-impelled end of the bar is free, or we must take the above equations (v) with their *lower* signs. Boussinesq's results are in agreement with Saint-Venant's (see our Art. 221) but his method is easier and his conclusions somewhat more complete.

We will briefly resume the results given by Boussinesq.

(a) *End of the Impact.* This is reached for $t=\dfrac{2l}{a}+\dfrac{\epsilon''}{a}$ where ϵ'' is a very small quantity, or immediately after the wave of impact has travelled to the free end of the bar and back again. After this time the bar and the mass M separate further and further, or the impact is definitely concluded. The velocity of the impelled end of the bar is at this instant $Ve^{-2/\nu}$ It then increases rapidly to $2V$, after which it returns to $Ve^{-2/\nu}$ with every change of time $2l/a$. On the other hand the mass M continues to move with the less of these velocities, i.e. $Ve^{-2/\nu}$ (pp. 518–9).

(b) *Kinetic Energy.* The velocity of the centroid of the bar after the impact is over $= V\nu(1-e^{-2/\nu})$, and therefore the kinetic energy K_1 of translation of the bar $=\tfrac{1}{2}MV^2\nu(1-e^{-2/\nu})^2$. Remembering that the energy of the mass M after the impact is over $=\tfrac{1}{2}MV^2e^{-4/\nu}$, we easily find for K_2 the kinetic energy of vibrations in the bar due to the impact[1]:

$$K_2 = \tfrac{1}{2} M V^2 \left(1 - e^{-4/\nu}\right) \left(1 - \nu \tanh \frac{1}{\nu}\right).$$

Hence we see that if the mass of the impinging body be very great as compared with the mass of the bar (ν very great), the energy lost in vibrations is very small, while if the bar have a large mass as compared with that of the impelling body, almost all the energy is absorbed in vibrations (p. 520).

Obviously, $K_2/K_1 = \dfrac{1}{\nu} \coth \dfrac{1}{\nu} - 1.$

To obtain the case of a rod impelled against a rigid wall, we have only to make $\nu = \infty$ and impress equal velocities, $-V$, on both impinging body and bar after the impact is entirely over (p. 521). We see at once that the bar rebounds with the velocity of impact, and without vibratory energy: see our Art. 205.

(c) *Maximum Strain.* The greatest squeeze is equal to V/a and occurs at points distant not more than $\tfrac{1}{2}\epsilon'$ from the free end at time not greater than $3\epsilon'/a$, i.e. close to the free end immediately after the beginning of the impact. This maximum squeeze is the same as that given by Young's Theorem (see our Art. 1542 *ftn.*) at the instant the impact begins. The maximum stretch equals $\dfrac{V}{a} (1 - \tfrac{1}{2}e^{-2/\nu})$ and occurs close to the impelled end immediately after the end of the blow ($t = 2l/a$). In most cases it will be expedient to take this last strain as that of safe loading, stretch being more important in respect of safety than squeeze.

[1550.] On pp. 522–534 we have the first case treated, the non-impelled end being now fixed, or the upper sign in (v) being taken. The solution in this case has been discussed at considerable length in our Arts. 401–7, and we refer the reader to these articles. We note one or two additional points occurring on pp. 535–46.

(a) On pp. 535–46 Boussinesq shews that to a second approximation we may neglect the inertia of the bar concentrating *one third* its mass at the impelled end[2]. The shift at this end will then be given by:

$$u_0 = \frac{lM}{m} \frac{V}{a} \sqrt{\frac{m}{M + \tfrac{1}{3}m}} \sin \left(\frac{at}{l \sqrt{M/m + 1/3}}\right).$$

[1] This of course neglects any loss of energy due to thermal action, etc.

[2] The use of this mass-coefficient of resilience (see our Vol. I., Appendix Note E (b) and Vol. II., Arts. 367–71, 1450 *et seq.*) is attributed to Saint-Venant; it is, however, as we have pointed out due to Homersham Cox and Hodgkinson.

This expression will give the shift u_0 with a considerable degree of accuracy even without M/m being large, but the assumption does not lead to an accurate expression for the maximum squeeze: see our Art. 406, (2) and footnote. This squeeze is investigated by Boussinesq in an approximate manner on pp. 542–4, and he finds that it is expressed, for M/m large, by $\dfrac{V}{a}\left(\sqrt{\dfrac{M}{m}}+1\right)$: see our Arts. 406, (2) (a) and 407 (3).

(b) A somewhat more elaborate series of values for the maximum squeeze than those of our Art. 406, (2) (a)—(c), are given by Boussinesq on p. 545. For practical purposes, however, those of our Art. 406 would be sufficiently accurate.

(c) In a footnote, pp. 541–3, Boussinesq deals with the interesting case of the mass-coefficient of resilience (see Vol. I., p. 894, (b)) for a thin circular plate of radius a, either built-in at its edge or simply supported. Let us apply the formula of our Art. 368 to this case, first calculating the value of

$$\gamma = \int f^2\,\frac{dP}{P} \quad\text{.............................(i)},$$

where P is the weight of the plate and f the ratio of the statical deflection at the element dP to that at the centre, where the impact of the weight Q is supposed to take place. For the case of isotropy we find from the value of w in (xi) of our Art. 330 that[1]:

$$f = 1 - \frac{r^2}{a^2} + \beta\,\frac{r^2}{a^2}\log\frac{r^2}{a^2},$$

β having the value unity for a built-in edge and $(3\lambda+2\mu)/(7\lambda+6\mu)$ for a simply supported edge. Whence by (i): $\gamma = \tfrac{1}{3}(1 - \tfrac{5}{6}\beta + \tfrac{2}{9}\beta^2)$. Let us put $Q' = Q + \gamma P$, then it only remains to find the $\sqrt{f_s/g}$ of our Art. 368. This is given by putting $r_0 = 0$, and $\gamma^2 = 0$ in (i) and (ii) of our Art. 334. We find after some reductions

$$\sqrt{f_s/g} = \frac{1}{2}\,\frac{a^2}{2\epsilon\Omega_1}\,\sqrt{\frac{3}{\beta}\frac{Q}{P}},$$

where $\Omega_1^2 = H/\rho$. Hence we have

$$w_0 = \frac{1}{2}\,\frac{V}{\Omega_1}\,\frac{a^2}{2\epsilon}\,\sqrt{\frac{3}{\beta}\frac{Q^2}{PQ'}}\,\sin\left(t\,\frac{4\Omega_1\epsilon}{a^2}\,\sqrt{\frac{\beta}{3}\frac{P}{Q'}}\right).$$

This gives us very accurately the depression at the centre of the plate, due to the blow of a body of weight Q at its centre.

---

[1] For both cases since the plate is thin we take the γ^2 of Art. 330 zero and put $z=0$; then for both, $dw/dr=0$ for $r=0$, involves $B=0$, while $w=0$ for $r=a$, determines the central deflection C. When the edge is simply supported $1/\rho=0$, but when the edge is built in $1/\rho$ is determined easily from $dw/dr=0$ for $r=a$.

[1551.] Section 23 *bis* of Boussinesq's work occupying pp. 546–77 is entitled: *Sur les deux problèmes d'un choc par compression faisant fléchir la barre heurtée, supposée très légère, et du mouvement rapide d'une charge roulante le long d'une telle barre horizontale, appuyée à ses deux bouts.* This section really deals with two interesting problems much simplified, however, by neglecting the vibrations of the elastic bodies considered. We shall deal with these in the following two articles.

[1552.] In our Art. 407 (2) we have referred to the possibility of a bar buckling under longitudinal impulse, and have given a not very satisfactory condition against buckling suggested by Saint-Venant and Flamant in their memoir. It is this point which Boussinesq discusses at considerable length on pp. 546–60, on the supposition, however, that *the weight of the bar is negligible as compared with that of the impinging mass.*

Let l be the unstrained length of the bar, l' its strained length; let f be the central deflection on buckling, c the chord, F the longitudinal compressive force, $E\omega\kappa^2$ the flexural rigidity, and $m^2 = F/(E\omega\kappa^2)$. Then if the origin be taken at the centre of the chord and y be the deflection at distance x; we easily find

$$l' - c = \int_0^{c/2} \left(\frac{dy}{dx}\right)^2 dx,$$

$$= m^2 f^2 \int_0^{c/2} \sin^2 mx\, dx,$$

since $y = f \cos mx$, giving $mc = \pi$ to a first approximation, is all that is necessary in order to obtain the value of $l' - c$ to a *second* approximation. Integrating out we have

$$l' - c = \tfrac{1}{4}\pi m f^2 \dots\dots\dots\dots\dots\dots\dots\dots\dots(i).$$

Referring to our Art. 110* for the value[1] of l', and retaining only the first two terms of the bracket we find:

$$l' = \frac{\pi}{m} + \frac{\pi}{16}\, m f^2 \dots\dots\dots\dots\dots\dots\dots\dots(ii).$$

From (ii) we see that l' must be $> \pi/m$, and therefore $l > \pi/m$ or $F > \pi^2 E\omega\kappa^2/l^2$ for there to be any buckling.

Finally, since the squeeze $(l - l')/l$ is due to F is:

$$l' = l \left(1 - \frac{F}{E\omega}\right) \dots\dots\dots\dots\dots\dots (iii).$$

---

[1] l' is the a of that Article, F the P, $E\omega\kappa^2$ the K, and we must put the m of that Article equal to unity.

In the terms in mf^2 of (i) and (ii) we may obviously replace m by its value as given by a first approximation, or by π/l. Let δ be the total shift $l-c$ of the impelled end, then we easily find from (i) and (iii) that:

$$f/l = \frac{2}{\pi}\sqrt{\frac{\delta}{l} - \frac{F}{E\omega}} \quad\dots\dots\dots\dots (iv),$$

a result also holding when $f = 0$.

If $-s_m$ be the maximum squeeze we have:

$$-s_m = \frac{F}{E\omega} \pm h\left(\frac{d^2y}{dx^2}\right)_{x=0},$$

when h is the distance from the central axis of the 'extreme fibre', or we find by (iv):

$$-s_m = \frac{F}{E\omega} \pm 2\pi\frac{h}{l}\sqrt{\frac{\delta}{l} - \frac{F}{E\omega}}.$$

Before flexure the radical on the right will be zero. After flexure we have $F/E\omega = \pi^2\kappa^2/l^2$ nearly. Hence

$$-s_m = \frac{\pi^2\kappa^2}{l^2} \pm 2\pi\frac{h}{l}\sqrt{\frac{\delta}{l} - \left(\frac{\pi\kappa}{l}\right)^2} \quad\dots\dots\dots\dots(v).$$

We are now in a position to measure the action of the impinging body Q. We have very approximately:

$$\frac{Q}{g}\frac{d^2\delta}{dt^2} = -F = \left\{ \begin{array}{l} -\dfrac{E\omega\delta}{l}, \quad \text{for } \delta < l\dfrac{\pi^2\kappa^2}{l^2}, \\[2mm] -E\omega\dfrac{\pi^2\kappa^2}{l^2}, \text{ for } \delta > l\dfrac{\pi^2\kappa^2}{l^2} \end{array} \right\}.$$

Hence the motion of Q will be pendulous until the bar buckles, but after buckling there will be a simple retardation of Q till the initial velocity be destroyed, provided this destruction of the velocity takes place before the deflection ceases to be very small as compared with the length of the bar.

The maximum deflection and strain occur by (iv) and (v) when δ is a maximum, or when the energy $\dfrac{QV^2}{2g}$ has been absorbed by the bar, i.e. when:

$$\frac{QV^2}{2g} = \int_0^\delta F d\delta$$

$$= \frac{E\omega\delta^2}{2l}, \text{ if } F \text{ remain} < F_1. \text{ i.e. } E\omega\frac{\pi^2\kappa^2}{l^2},$$

$$= F_1\left(\delta - \frac{lF_1}{2E\omega}\right). \text{ if } F \text{ exceed } F_1.$$

Putting $E/\rho = a^2$, we find for the maximum terminal shift δ_m :

$$\delta_m = l \frac{V}{a} \sqrt{\frac{Q}{P}}, \text{ if } \frac{V}{a} \sqrt{\frac{Q}{P}} < \frac{\pi^2 \kappa^2}{l^2},$$

$$= \frac{l}{2} \left[\frac{\pi^2 \kappa^2}{l^2} + \frac{V^2}{a^2} \frac{Q}{P} \frac{l^2}{\pi^2 \kappa^2} \right], \text{ if } \frac{V}{a} \sqrt{\frac{Q}{P}} > \frac{\pi^2 \kappa^2}{l^2}.$$

Substituting these values of δ in (iv) and (v), we find :

$$f_m = 0, \quad -s_m = \frac{V}{a} \sqrt{\frac{Q}{P}}, \text{ if } \frac{V}{a} \sqrt{\frac{Q}{P}} < \frac{\pi^2 \kappa^2}{l^2},$$

$$\left. \begin{array}{l} f_m = \dfrac{l\sqrt{2}}{\pi} \sqrt{\dfrac{V^2}{a^2} \dfrac{Q}{P} \dfrac{l^2}{\pi^2 \kappa^2} - \dfrac{\pi^2 \kappa^2}{l^2}}, \\[3mm] -s_m = \dfrac{\pi^2 \kappa^2}{l^2} + \dfrac{\pi h \sqrt{2}}{l} \sqrt{\dfrac{V^2}{a^2} \dfrac{Q}{P} \dfrac{l^2}{\pi^2 \kappa^2} - \dfrac{\pi^2 \kappa^2}{l^2}} \end{array} \right\}, \text{ if } \frac{V}{a} \sqrt{\frac{Q}{P}} > \frac{\pi^2 \kappa^2}{l^2}.$$

The condition $\left(\dfrac{V}{a} \sqrt{\dfrac{Q}{P}} < \dfrac{\pi^2 \kappa^2}{l^2} \right)$ for the non-buckling of the bar agrees with that of our Art. 407, (2), if it be remembered that Boussinesq supposes P very small as compared with Q.

[1553.] The second problem dealt with by Boussinesq is *Willis' Problem* of the rolling load: see our Art. 1419 *. The equation at the bottom of our p. 764 (Vol. I.) may be written

$$\frac{y}{S} = \left(1 - \frac{x'^2}{a^2} \right)^2 \left(1 - \frac{V^2}{g} \frac{d^2 y}{dx'^2} \right),$$

the origin being at the centre and not at the end of the bar, i.e. writing $x' + a$ for x. Boussinesq gives this equation on p. 562 and occupies pp. 562–77 with the discussion of a solution of it. It does not seem to me that Boussinesq's value for the deflection is in a simpler form, or one more capable of readily giving numerical results, than the solutions obtained and discussed by Sir G. G. Stokes in §§ 3–10 of his memoir of 1849 : see our Art. 1279 *.

Boussinesq finds (p. 569) if y be the deflection at distance x' from the centre :

$$\frac{1}{2\beta} \frac{y}{S} = T + \sum_{m=1}^{m=\infty} \left(\frac{1 \cdot 2}{9 + k^2} \cdot \frac{3 \cdot 4}{25 + k^2} \cdots \cdots \frac{2m(2m+1)}{(2m+1)^2 + k^2} \left(1 - \frac{x'^2}{a^2} \right)^{m+1} \right),$$

where for x positive, $T = 0$, and for negative x :

$$T = \pi \frac{1 + k^2}{k} \frac{(\sin, \text{ or, } \sinh) \left(\dfrac{k}{2} \log \dfrac{a + x'}{a - x'} \right)}{(\cosh, \text{ or, } \cos) \dfrac{k \pi}{2}} \sqrt{1 - \frac{x'^2}{a^2}},$$

and the upper sign, with the *sin* and *cosh* in T, are to be taken if $\beta > \frac{1}{4}$, and the lower sign, with the *sinh* and *cos* in T if $B < \frac{1}{4}$. The load is supposed to start from the end $x = -a$. Further, as in our Art. 1419*, S is the statical deflection due to the rolling load concentrated at the middle of the bar, $2a$ is the length of the bar and $\beta = ga^2/(4V^2S)$, V being the velocity of the travelling load. $\pm k^2$ denotes the difference $4\beta - 1$, k being always taken positive.

Boussinesq draws from his form of the solution conclusions similar to those of Sir G. G. Stokes summarised in our Art. 1282*, but I do not think he adds any novel results.

[1554.]   Before leaving this section we must refer to the following important practical problem dealt with by Boussinesq in a footnote on pp. 552–5 : see also our Art. 1556.   Consider a thin cylindrical belt or ring of radius R, thickness τ and breadth b, and suppose it subjected to a uniform pressure p on its outer surface.   What is the least pressure which can cause it to collapse or lose its circular form ?

Let the belt be supposed to have collapsed or bent, so that $r = R(1 + e)$ is the new radius-vector, e being a function of the radial angle θ, then as in our Art. 585 the bending moment at the point defined by θ is given by :

$$M = \frac{E\omega\kappa^2}{R}\left(e + \frac{d^2e}{d\theta^2}\right) = \frac{Eb\tau^3}{12R}\left(e + \frac{d^2e}{d\theta^2}\right).$$

Suppose AC an axis of symmetry of the strained central line and consider the portion AP of the ring.   Take moments about the point P.

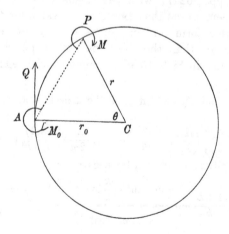

The total thrust Q at A will equal $r_0 p b$, the moment about P of the pressures on the portion of the ring cut off by the chord AP will obviously $= pb\,AP \times \frac{1}{2}AP$.   Hence if M_0 be the moment at A we have:

$$M = M_0 - \tfrac{1}{2}pb\,AP^2 + Q\,(r_0 - r\cos\theta)$$
$$= M_0 - pbR^2\,(e - e_0),$$

Thus:
$$\frac{d^2e}{d\theta^2} + e\left(1 + \frac{12pR^3}{E\tau^3}\right) = \text{a constant.}$$

If R be taken as the compressed radius immediately before collapse, it is easy to see that this constant must be zero, for the mean value of e will be zero.   Hence we find, since at A $de/d\theta = 0$:

$$e = e_0 \cos\left(\theta\sqrt{1 + \frac{12pR^3}{E\tau^3}}\right).$$

But $e = e_0$ when θ increases by 2π, hence if i be an integer we must have

$$\sqrt{1 + \frac{12pR^3}{E\tau^3}} = i.$$

The least collapsing value of p will arise when $i = 2$, thus for collapse we must have

$$p = \text{or} > \tfrac{1}{4}E\,\frac{\tau^3}{R^3}.$$

[1555.]   If the ring become a curved plate, we ought at least, I think, to replace E by the plate modulus $H\,(= \tfrac{16}{15}E$ for uniconstant isotropy).   We should then have for the collapsing pressure on a flue *with unsupported ends*:

$$p = \text{or} > \tfrac{1}{4}H\,\frac{\tau^3}{R^3}.$$

It is noteworthy that this is exactly the form of the old Prussian Government formula for the strength of flues, the origin of which formula is unknown : see our Art. 986.   On the other hand Fairbairn's experiments with flues having cast-iron ends maintained at a *fixed distance* seem to shew that p varies inversely as the length of the flue : see our Art. 984.   When the ends are not thus fixed, the pressure does not vary so exactly as the inverse of the length : see our Art. 982.

[1556.]   The contents of this footnote in the *Treatise* had been previously published by Boussinesq in a memoir entitled : *Résistance d'un anneau à la flexion, quand sa surface extérieure supporte une pression normale, constante par unité de longueur de sa fibre moyenne.   Comptes rendus*, T. xcvii., pp. 843–4.   Paris, 1883.

The problem appears to have been previously discussed by M. Lévy in a paper entitled : *Sur un nouveau cas intégrable du problème de l'élastique et l'une de ses applications (Comptes rendus, T. xcvii., pp. 694–7), which* dealt with the above case and gave somewhat complex results in terms of elliptic integrals. Lévy found: that if

$$p < \frac{9}{4}\frac{E\omega\kappa^2}{bR^3}, \text{ i.e. } < \frac{3}{16}\frac{E\tau^3}{R^3},$$

the belt would not change its circular form or be liable to buckle. But he did not shew that if p be greater than this, the belt would buckle.

Boussinesq in the memoir of which we have given the title above, proved as in our previous article that we must have

$$p < 3\frac{E\omega\kappa^2}{bR^3} \text{ or } < \frac{1}{4}\frac{E\tau^3}{R^3},$$

if the belt be not to lose its circular form.

Lévy replied to Boussinesq in a somewhat inconsequential note to be found in the same volume of *Comptes rendus*, pp. 979–80, remarking in particular that his own result was deduced from a solution for finite changes of shape and that he had previously noticed Boussinesq's conclusion. Boussinesq terminated the discussion on pp. 1131–2 of the same volume by the remark that the ring must pass through an infinitely small change of shape before it can take a finite one.

The problem really involves the same paradoxes (and the same solutions of them) which occur in the case of the buckling of struts or the collapse of flues.

According to Lévy both Boussinesq and he had been anticipated by Résal. I may notice that they had also been anticipated by Bresse, who shewed in 1859 that a flue (? ring) of *any* slight ellipticity would not collapse under external pressure unless $p > \frac{1}{4}E\tau^3/R^3$: see our Art. 537, (d).

[1557.] The last portion of Boussinesq's treatise which deals with the theory of elasticity is *Note* III. (pp. 665–98). It is entitled: *Extension, aux solides hétérotropes les plus simples, c'est-à-dire aux solides isotropes déformés, des lois d'équilibre et des lois les plus importantes de mouvement démontrées dans cette étude pour les solides isotropes.*

The type of aeolotropy for which Boussinesq generalises his results is given by stress-strain relations of the following kind :

$$\left.\begin{aligned}
\widehat{xx} &= as_x + f's_y + e's_z, & \widehat{yz} &= d\sigma_{yz}, \\
\widehat{yy} &= f's_x + bs_y + d's_z, & \widehat{zx} &= e\sigma_{zx}, \\
\widehat{zz} &= e's_x + d's_y + cs_z, & \widehat{xy} &= f\sigma_{xy},
\end{aligned}\right\} \dots\dots\dots \text{(i).}$$

subject to the conditions :

$$d/d' = e/e' = f/f',$$

and $\quad 2d + d' = \sqrt{bc}, \quad 2e + e' = \sqrt{ca}, \quad 2f + f' = \sqrt{ab}$

The latter are the well-known relations of the ellipsoidal kind, true probably for all amorphic bodies. The former are the dangerously near approach to the rari-constant conditions, which in our Art. 140 we have described as but a doubtful sop to the multi-constant Cerberus. The substance indeed of the portion of Saint-Venant's memoir of 1863 considered in that article forms the basis of Boussinesq's note.

[1558.] Adopting the above conditions, with due reservation however, we may throw the stress-strain relations into the following form by taking five new constants a, β, γ, λ, μ such that :

$$d/d' = e/e' = f/f' = \mu/\lambda, \quad ad = \beta e = \gamma f = \mu a \beta \gamma, \quad a/a^2 = b/\beta^2 = c/\gamma^2 = \lambda + 2\mu,$$

we have

$$\begin{aligned}
\widehat{xx} &= a\lambda\chi + 2\mu a^2 s_x, & \widehat{yz} &= \mu\beta\gamma\sigma_{yz}, \\
\widehat{yy} &= \beta\lambda\chi + 2\mu\beta^2 s_y, & \widehat{zx} &= \mu\gamma a\sigma_{zx}, \\
\widehat{zz} &= \gamma\lambda\chi + 2\mu\gamma^2 s_z, & \widehat{xy} &= \mu a\beta\sigma_{xy}
\end{aligned} \right\} \quad \ldots\ldots\ldots\ldots \text{ ... (ii),}$$

where $\chi = a s_x + \beta s_y + \gamma s_z$.

These results become those for bi-constant isotropy, if we take $a = \beta = \gamma = 1$.

Now take new variables such that :

$$\left. \begin{aligned}
x' &= x/\sqrt{a}, & y' &= y/\sqrt{\beta}, & z' &= z/\sqrt{\gamma}, \\
u' &= u\sqrt{a}, & v' &= v\sqrt{\beta}, & w' &= w\sqrt{\gamma}
\end{aligned} \right\} \quad \ldots\ldots\ldots\ldots \text{ (iii),}$$

where u, v, w are the shifts.

Then

$$\chi = \frac{du'}{dx'} + \frac{dv'}{dy'} + \frac{dw'}{dz'} = \theta',$$

and the stress-strain relations become of the form :

$$\widehat{xx}/a = \widehat{x'x'} = \lambda\theta' + 2\mu s_{x'}, \quad \widehat{yz}/\sqrt{\beta\gamma} = \widehat{y'z'} = \mu\sigma_{y'z'}.$$

Thus the body-stress equations of equilibrium will be of the type

$$\frac{d\widehat{x'x'}}{dx'} + \frac{d\widehat{x'y'}}{dy'} + \frac{d\widehat{x'z'}}{dz'} + \rho X' = 0,$$

where $X' = X/\sqrt{a}$, etc.

Thus any solution for an isotropic solid bounded by the surface $f(x', y', z') = 0$ becomes one for the aeolotropic solid bounded by $f(x/\sqrt{a}, y/\sqrt{\beta}, z/\sqrt{\gamma}) = 0$, provided the body-forces applied to the latter be $X'\sqrt{a}$, $Y'\sqrt{\beta}$, $Z'\sqrt{\gamma}$, and the shifts u, v, w at x, y, z be taken as $u'/\sqrt{a}, v'/\sqrt{\beta}, w'/\sqrt{\gamma}$. To obtain the stresses internally or at the surface at x, y, z we must take the system $a\widehat{x'x'}$, $\beta\widehat{y'y'}$, $\gamma\widehat{z'z'}$, $\sqrt{\beta\gamma}\widehat{y'z'}$, $\sqrt{\gamma a}\widehat{z'x'}$, $\sqrt{a\beta}\widehat{x'y'}$ (pp. 665–71).

In a long footnote pp. 665–8, Boussinesq refers to his memoir of 1868 (see our Art. 1467) and deduces the stress-strain relations (ii) above *de novo*.

[1559.] The greater portion of Boussinesq's *Note* III. (pp. 672–98) deals with the vibratory motion which may be set up by a region of disturbance in an aeolotropic medium of the elastic nature defined by (ii). The equations of motion, if there be no body-forces, will be of the type

$$\rho \frac{d^2 u}{dt^2} = a(\lambda + \mu)\frac{d\chi}{dx} + \mu a\left(a\frac{d^2 u}{dx^2} + \beta\frac{d^2 u}{dy^2} + \gamma\frac{d^2 u}{dz^2}\right),$$

when as before $\qquad \chi = a\dfrac{du}{dx} + \beta\dfrac{dv}{dy} + \gamma\dfrac{dw}{dz}.$

Taking as before the new variables given by (iii) the type becomes:

$$\frac{\rho}{a}\frac{d^2 u'}{dt^2} = (\lambda + \mu)\frac{d\theta'}{dx'} + \mu\left(\frac{d^2 u'}{dx'^2} + \frac{d^2 u'}{dy'^2} + \frac{d^2 u'}{dz'^2}\right) \ldots\ldots\ldots\ldots (\text{iv}).$$

Thus these equations do not reduce, like those for equilibrium, to the equations for an isotropic solid. They reduce to the system of equations which have been considered by Sarrau and Boussinesq to hold for the ether, the elastic constants being supposed the same for all directions, but the density of the ether being taken to have the values ρ/a, ρ/β, ρ/γ for vibrations in the directions of x, y and z respectively : see our Art. 1476.

L'intégration de ces équations de mouvement ne se réduit donc pas à une simple application des potentiels sphériques considérés dans le mémoire des p. 319 à 356 ci-dessus (see our Art. 1526). Peut-être deviendrait-elle effectuable, avec moins de complication qu'elle ne l'a été jusqu'ici par la méthode de Blanchet ou par celle de Cauchy basée sur le calcul des résidus (see our Arts. 1166*–1178*), si l'on pouvait généraliser d'une manière convenable la notion de ces potentiels sphériques.

Heureusement, les seuls résultats concrets qu'on ait pu déduire des intégrations difficiles effectuées par Blanchet et par Cauchy, résultats relatifs à la propagation du mouvement autour d'une région d'ébranlement infiniment petite prise pour origine des coordonnées x, y, z, peuvent, à peu près tous, se démontrer directement (pp. 672–3).

[1560.] Boussinesq then discusses at length the solution of the equations of type (iv) above. He first supposes $a = \beta = \gamma = 1$, which leads to the well-known solution. Then he takes on the basis of this solution a second approximation, supposing a, β, γ to differ slightly from each other and from unity. The results, which he obtains, are principally of importance for the elastic theory of light, and may be looked upon as a partial development of those of Sir G. G. Stokes to an elastic medium of the aeolotropic character assumed. Boussinesq applies them on pp. 694–7 to the problem of the lateral limitation of disturbances in the form of light or sound 'rays.'

[1561.] Boussinesq concludes his *Treatise* (pp. 705–12) with a reproduction of the memoir on earthwork to which we have

referred in our Art. 1624. The work as a whole is a remarkable
contribution to our subject, suggesting a wide range of new
analytical methods and a variety of directions for valuable experimental investigations in elasticity. It forms one of the most
important contributions to our subject, published since the *Annotated Clebsch:* see our Art. 298.

SECTION IV.

Memoirs on Plasticity and Pulverulence.

[1562.] *Lois géométriques de la distribution des pressions,
dans un solide homogène et ductile soumis à des déformations
planes. Comptes rendus,* T. LXXIV., pp. 242–6. Paris, 1872.
This is an investigation by aid of what Boussinesq terms
cylindres isostatiques, or what in English are generally spoken
of as *conjugate functions,* of the uniplanar equations of plasticity.
Cylindres isostatiques are to orthogonal curvilinear coordinates
in two dimensions what Lamé's *surfaces isostatiques* are to
those in three dimensions. In a footnote, Boussinesq referring
to Lamé uses the words cited in our Art. 1152.* Boussinesq's
isostatic cylinders in plasticity are, however, a case for which
Lamé's theorem holds.

[1563.] If xy be the plane of symmetry, we shall have to consider
only the stresses $\widehat{xx}$, $\widehat{yy}$, $\widehat{xy}$, $\widehat{zz}$ and these will be functions solely of x and
y. If a be the angle the normal n to an elementary plane makes with
the positive direction of x, we have for the traction and shear across
this plane :

$$\widehat{nn} = \tfrac{1}{2}(\widehat{xx}+\widehat{yy}) + \tfrac{1}{2}R\sin(2a-\psi), \quad \widehat{nt} = \tfrac{1}{2}R\cos(2a-\psi), \Bigg\}\ldots(i),$$
$$\text{where} \quad \cos\psi = \frac{2\widehat{xy}}{R}, \quad \sin\psi = \frac{\widehat{yy}-\widehat{xx}}{R}, \quad R = +\sqrt{(\widehat{yy}-\widehat{xx})^2 + 4\widehat{xy}^2}$$

see our Arts. 248 and 465, (b).

The principal tractions will therefore be determined by the angles a_1
and a_2, where $2a_1-\psi=90°$ and $2a_2-\psi=270°$. Now construct the
cylinders of which the normals are inclined at each point at angles a_1, a_2
respectively to the axis of x and let them be represented by the families
$\rho_1=f_1(x,y)$, $\rho_2=f_2(x,y)$. These are Boussinesq's *isostatic* cylinders.

These surfaces give at each point by their normals the directions of the principal tractions T_1 and T_2. In this case we shall have $R = T_1 - T_2$, and according to Tresca and Saint-Venant (see our Arts. 248 and 259):

$$T_1 - T_2 = 2K \dotfill \text{(ii)}.$$

Boussinesq takes the curvilinear rectangle bounded by two pairs of adjacent curves of the above two families. If dn_1 and dn_2 be the normal distances between the members of the pairs at a given point, we have:

$$dn = d\rho/h, \text{ where } h = \sqrt{\left(\frac{d\rho}{dx}\right)^2 + \left(\frac{d\rho}{dy}\right)^2},$$

for either normal distance.

Considering the equilibrium of the curvilinear rectangle by resolving the principal tractions which act across its faces along the normals dn_1 and dn_2, Boussinesq easily finds

$$\frac{dT_1}{d\rho_1} = (T_1 - T_2) \frac{d(\log h_2)}{d\rho_1}, \quad \frac{dT_2}{d\rho_2} = (T_2 - T_1) \frac{d(\log h_1)}{d\rho_2} \dotfill \text{(iii)}.$$

From (ii) and (iii) we have:

$$T_1 = K \left(2 \log \frac{h_2}{\chi_2(\rho_2)} + 1 \right), \; T_2 = -K \left(2 \log \frac{h_1}{\chi_1(\rho_1)} + 1 \right), \Bigg\} \dots \text{(iv)},$$
$$h_1 h_2 = \chi_1(\rho_1) \chi_2(\rho_2)$$

where χ_1 and χ_2 are arbitrary functions of ρ_1 and ρ_2 respectively.

Now replace ρ_1 and ρ_2 by two new parameters, ρ_1' and ρ_2', determined by $d\rho_1 = \chi_1(\rho_1) d\rho_1'$ and $d\rho_2 = \chi_2(\rho_2) d\rho_2'$, then we see that

$$T_1 = K(1 - \log h_1'^2), \quad T_1 - T_2 = 2K_1, \Bigg\} \dotfill \text{(v)}.$$
$$h_1' h_2' = 1$$

Here ρ_1' and ρ_2' may be treated as the curvilinear coordinates and the dashes may be dropped. We see that equations (v) suffice to determine any three of the quantities T_1, T_2, h_1 and h_2 when the fourth is known.

The third equation of (v) gives us $dn_1 \times dn_2 = d\rho_1 \times d\rho_2$, or if we draw the two families of curves for equal variations $d\rho_1$ and $d\rho_2$ of the parameters, these curves will divide up the plane of xy into curvilinear rectangles of equal area.

[1564.] Boussinesq shews (p. 245) how to construct graphically these families of curves, if one of the quantities T_1, T_2, h_1 and h_2 is given at every point of the whole length of an isostatic line. The construction resembles that adopted by Maxwell in dealing with lines of flow, etc.: see our Art. 1556* and also the *Treatise on Electricity and Magnetism*, Vol. I., Chapter XII.

Since the systems ρ_1 and ρ_2 are orthogonal, Boussinesq easily deduces that both ρ_1 and ρ_2 must satisfy the differential equation :

$$\left\{\left(\frac{d\rho}{dx}\right)^2 - \left(\frac{d\rho}{dy}\right)^2\right\}\left(\frac{d^2\rho}{dx^2} - \frac{d^2\rho}{dy^2}\right) + 4\,\frac{d\rho}{dx}\frac{d\rho}{dy}\frac{d^2\rho}{dxdy} = 0 \;\ldots\ldots\text{(vi)}.$$

Solutions of (vi) give the only possible isostatic cylinders for the deformation of a plastic solid.

[1565.] *Sur l'intégration de l'équation aux dérivées partielles des cylindres isostatiques produits dans un solide homogène et ductile. Comptes rendus*, T. LXXIV., pp. 318–21. Paris, 1872.

In this memoir Boussinesq solves in a rather complicated form equation (vi) of the previous article. Writing that equation in the well-known symbols adopted by treatises on differential equations, we have :

$$(p^2 - q^2)(r - t) + 4pqs = 0.$$

Further taking $h = +\sqrt{p^2 + q^2}$, $p^2 - q^2 = h^2 \cos 2a$, $2pq = h^2 \sin 2a$, Boussinesq finds :

$$x = \frac{d\rho}{dp}, \quad y = \frac{d\rho}{dq},$$

where
$$\rho = \frac{h}{\pi}\int_0^\infty dm \int_{-\infty}^\infty \left\{\frac{\sin(a\sqrt{1+m^2})}{\sqrt{1+m^2}}F_1(\xi)\right.$$

$$+ \cos(a\sqrt{1+m^2})\,e^{-\xi}\left[\int_0^\xi e^{h'}F(h')\,dh'\right]\right\}\cos m\,(h' - \xi)\,d\xi,$$

where $F(h')$ and $F_1(h')$ are the functions of $h' \equiv \log h$ to which x and y reduce when $a = 0$.

The solution appears far too complicated to be of much practical value.

[1566.] *Équation aux dérivées partielles des vitesses, dans un solide homogène et ductile déformé parallèlement à un plan. Comptes rendus*, T. LXXIV., pp. 450–3. Paris, 1872.

Before discussing this memoir we may in the first place refer to a footnote on p. 452, somewhat generalising equation (iii) of our Art. 1563. If T_1 be given as a function of T_2 by some relation other than the plastic relation, $T_1 - T_2 = 2K$, of that article, these equations are still integrable. Boussinesq suggests for example the condition for the limiting equilibrium of loose earth, or $T_1/T_2 = (1 - \sin\phi)/(1 + \sin\phi)$, ϕ being the angle of friction. In this case the third equation of (v) in our Art. 1563 becomes

$$h_1{}^{1-\sin\phi} \times h_2{}^{1+\sin\phi} = 1.$$

[1567.] Boussinesq proposes in this memoir to investigate the uniplanar motion of a plastic mass, and he obtains the components of the velocity perpendicular to the isostatic surfaces from the following considerations:

(a) That the cylinder on the curvilinear rectangle (see our Art. 1563) will have the same volume at times t and $t + dt$.

(b) That the matter situated on one side of an element of an isostatic surface has no slide relative to the matter situated on the other, because the stress is entirely normal.

If U_1 and U_2 be the velocities in the direction of the normals dn_1, dn_2 (see our Art. 1563), Boussinesq finds:

$$U_2 = h_2 \frac{d\psi}{d\rho_2}, \quad U_1 = -h_1 \frac{d\psi}{d\rho_1},$$

where

$$h_1{}^2 \frac{d^2\psi}{d\rho_1{}^2} = h_2{}^2 \frac{d^2\psi}{d\rho_2{}^2}$$ .... (i),

and $h_1 \times h_2 = 1$ as before.

The last equation may be written

$$\frac{d^2\psi}{d\rho_2{}^2} = h_1{}^4 \frac{d^2\psi}{d\rho_1{}^2}$$ .... (ii),

but its integration is rendered extremely difficult by the presence of h_1, a function of both ρ_1 and ρ_2 (p. 453).

[1568.] *Sur une manière simple de déterminer expérimentalement la résistance au glissement maximum dans un solide ductile, homogène et isotrope. Comptes rendus*, T. LXXV., pp. 254–7. Paris, 1872.

Boussinesq draws attention to the formula deduced by Saint-Venant from Tresca's researches in plasticity, namely

$$K = \tfrac{1}{2} (T_1 - T_3),$$

where $T_1 - T_3$ is the greatest difference between the three principal tractions (T_1, T_2, T_3), and K is taken by Saint-Venant to be a constant: see our Arts. 248 and 259. Boussinesq raises the question whether K is an *absolute* constant. He considers that it is not sensibly variable with the small relative velocities of the parts of a plastic mass, and that it cannot really vary with a uniform normal pressure round any element of volume, because such a pressure not sensibly increasing the density would not render the molecular equilibrium more stable. It is only possible he thinks for the variation of K to depend on the manner in which T_2 is comprised between the other two principal tractions or upon the ratio

$$(T_1 - T_2)/(T_2 - T_3).$$

He therefore proposes to take K, or

$$\tfrac{1}{2}(T_1 - T_3) = \text{some function of } \frac{T_1 - T_2}{T_2 - T_3}$$

$$= \chi\left(\frac{T_1 - T_2}{T_2 - T_3}\right).$$

Thus it is possible for K to vary from one case of plastic motion to a second. This variation of K is more fully discussed in the memoir of 1876 : see our Arts. 1586 and 1594.

[1569.] Boussinesq further suggests pure traction experiments as the best means to obtain K, supposing it to be constant. In this case he considers that $T_2 = T_3 = 0$, and therefore that $K (= \tfrac{1}{2}T_1)$ can be found at once.

If K be variable he suggests that its variation could be determined in the following manner. Let a rectangular bar, subjected to a longitudinal traction S, be placed between two parallel polished plates covered with oil or grease, and subjected to a uniform pressure P applied through these plates. Let the stress S requisite to produce plasticity be noted for each value of P. Then we shall have

$$T_2 = 0, \quad T_3 = -P, \quad T_1 = S,$$

and hence :

$$K = \tfrac{1}{2}(S + P) \text{ and } \frac{T_1 - T_2}{T_2 - T_3} = \frac{S}{P}.$$

It would then be easy to determine, whether K is an absolute constant, and, if not, how it varies with the ratio S/P.

It is difficult to see, however, considering the phenomenon of *local stricture* which occurs with ductile metals (see our Vol. I., p. 891) how the proposed arrangement could practically produce the required system of stress, even if the oil or grease really prevented the friction of the plates having any sensible influence.

[1570.] *Intégration de l'équation aux dérivées partielles des cylindres isostatiques qui se produisent à l'intérieur d'un massif ébouleux soumis à de fortes pressions. Comptes rendus,* T. LXXVII., pp. 667–71. Paris, 1873.

Taking the equations (iii) of our Art. 1563 and replacing (ii) of the same article by:

$$\frac{T_1 - T_2}{T_1 + T_2} = -\sin\phi,$$

where ϕ is the angle of friction, Boussinesq obtains the solution of these equations for a mass of loose earth. He finds:

$$T_1 - T_2 = h_2^{-\frac{2\sin\phi}{1-\sin\phi}}, \quad h_1^{1-\sin\phi} \times h_2^{1+\sin\phi} = 1.$$

See our Art. 1566.

As in our Art. 1564, Boussinesq then proceeds to determine the equation which must be satisfied by ρ_1. If $k = \sqrt{\dfrac{1-\sin\phi}{1+\sin\phi}}$, he finds with the notation of our Art. 1565 that:

$$(p^2 - k^2 q^2)\, r + 2\,(1 + k^2)\, pqs + (q^2 - k^2 p^2)\, t = 0.$$

This equation he solves in the same manner as the differential equation of that article. See pp. 669–70 of his memoir.

[1571.] *Essai théorique sur l'équilibre d'élasticité des massifs pulvérulents comparé à celui de massifs solides et sur la poussée des terres sans cohésion. Mémoires couronnés et mémoires des savants étrangers publiés par l'académie......de Belgique*, Tome XL., pp. 1–180. Bruxelles, 1876. See also *Comptes rendus*, Tome LXXVII., pp. 1521–5. Paris, 1873.

Previous memoirs dealing with loose earth had been more especially devoted to the limit of its equilibrium, without reference to its *elasticity*, and from this standpoint they do not properly fall within the limits of our subject. We have, however, referred by title to one or two such memoirs in the course of this history. Thus the researches of Rankine are cited in our Arts. 453 and 465 (*a*), those of Holtzmann in Art. 582 (*b*) and those of Lévy, Saint-Venant and Boussinesq in Art. 242. Rankine's researches were afterwards thrown into a geometrical form by Flamant in the *Annales des ponts et chaussées*, 2ᵉ *Semestre*, pp. 242–68. Paris, 1872. (An interesting elementary discussion of the stability of earth will be found on pp. 111–40 of Flamant's *Stabilité des constructions: Résistance des matériaux*. Paris, 1886.) Boussinesq in his Introduction after referring (pp. 3–4) to the previous history of this branch of the subject continues:

Mais il y a un autre genre d'équilibre également important à considérer: c'est celui que présente une masse sablonneuse en repos, soutenue par un mur assez ferme pour n'éprouver aucun ébranlement. Dans cet état, le frottement mutuel des couches est généralement moindre que dans le précédent, tout comme, à l'intérieur d'un solide en équilibre d'élasticité, les tensions restent partout inférieures à celles qui altéreraient d'une manière permanente la structure du corps : les particules sont donc moins retenues par leurs actions mutuelles que dans le cas où le mur de soutènement les fuirait en cédant sous leur pression, et elles exercent sur ce dernier une poussée supérieure à celle qu'indiquent les formules de Rankine. C'est surtout ce genre d'équilibre que je me propose d'étudier ici : je l'appelle *équilibre d'élasticité*, car je considère les pressions qui s'y trouvent effectivement exercées comme dépendant des petites déformations qu'éprouverait la masse, supposée d'abord homogène et sans poids, si elle devenait ensuite pesante comme elle l'est en effet (p. 5).

More attention must therefore be devoted to Boussinesq's memoirs on pulverulence than to earlier memoirs, because (i) they deal with the *elastic* equilibrium of a pulverulent mass, and (ii) they appear to contain the most complete scientific theory yet given of the stability of such a mass.

[1572.] Boussinesq bases his theory of pulverulence on the hypothesis that masses of pulverulent material stand midway between solid and fluid bodies, and act like fluids when not subjected to pressure, but when subjected to pressure gain an elasticity of form as well as of bulk, and act like solids. Boussinesq considers the slide-modulus to be proportional to the mean pressure. Supposing the dilatation θ to be zero or negligible as compared with the individual stretches s_x, s_y, s_z, we have for the mean pressure of an isotropic medium $\frac{1}{3}(\widehat{xx} + \widehat{yy} + \widehat{zz})$ or $-p$, say. Further the slide-modulus μ is to be proportional to p, or mp say ; hence he finds as stress types :

$$\widehat{xx} = -p\,(1 - 2ms_x), \quad \widehat{yz} = pm\,\sigma_{yz}\ldots\ldots\ldots\ldots(\text{i}).$$

[1573.] On p. 7 of his *Introduction* Boussinesq discusses certain difficulties which arise in the boundary conditions of a pulverulent mass. At a free boundary clearly the pressure is to be zero. At a perfectly rough fixed boundary, as in the case of certain sustaining walls the shifts ought to be zero, or the couch of material along the wall remain unmoved. This fixity of the particles along the wall is, however, generally incompatible with their fixity in the same positions when in the 'primitive' or 'natural' state, i.e. when no body-force such as weight is supposed to act on the mass, but this is the state from which the shifts u, v, w are supposed to be measured.

The remainder of the Introduction is a *résumé* of the conclusions reached in the memoir.

[1574.] § I. of the memoir is entitled : *Formules des pressions principales exercées à l'intérieur des milieux élastiques, solides, fluides ou pulvérulents, dont la constitution est la même en tout sens.* It occupies pp. 11–22. The first two or three pages recite some well-known kinematic properties of strain, concluding with the consideration of the three principal tractions and three principal stretches. If s_1, s_2, s_3 be the latter, T_1, T_2, T_3 the former quantities, Boussinesq puts $-p$ or $\frac{1}{3}(T_1 + T_2 + T_3)$ and

$$\tfrac{1}{2}(T_3 - T_2), \quad \tfrac{1}{2}(T_1 - T_3), \quad \tfrac{1}{2}(T_2 - T_1)$$

equal to functions of s_1, s_2, s_3, which he says, if s_1, s_2, s_3 are sufficiently small, can be expanded by Maclaurin's theorem in rapidly converging series of integral positive powers of s_1, s_2, s_3 (p. 14). This is the same sort of assumption as we have had to criticise in the investigations referred to in our Arts. 928 * and 299.

Accepting it with this qualifying remark, it is easy to follow the considerations of symmetry by which Boussinesq deduces that :

$$-p = A + B\theta + C\theta^2 + D\left\{(s_2 - s_3)^2 + (s_3 - s_1)^2 + (s_1 - s_2)^2\right\} \ldots\ldots\text{(ii)},$$

where A, B, C and D are constants (p. 15).

Write the right-hand side of (ii) as K and its value when the constants A, B, C, D are replaced by dashed letters A', B', C', D' as K'; then Boussinesq shews that the following are the forms of the principal traction differences :

$$\left.\begin{aligned}
\tfrac{1}{2}(T_2 - T_3) &= \{K' + (B'' + C''\theta)\,s_1\}\,(s_2 - s_3),\\
\tfrac{1}{2}(T_3 - T_1) &= \{K' + (B'' + C''\theta)\,s_2\}\,(s_3 - s_1),\\
\tfrac{1}{2}(T_1 - T_2) &= \{K' + (B'' + C''\theta)\,s_3\}\,(s_1 - s_2)
\end{aligned}\right\} \ldots\ldots\ldots\text{(iii)}.$$

These formulae can be applied to all isotropic bodies, and Boussinesq proceeds to apply them to various types (pp. 17–22).

(a) *Elastic solids.* In this case the strains represented by $s_2 - s_3$, $s_3 - s_1$, $s_1 - s_2$ become sensible only when the stress-differences $T_2 - T_3$, $T_3 - T_1$, $T_1 - T_2$ are themselves sensible. Hence for small strains, we have

$$-p = A + B\theta,$$

$$\frac{\tfrac{1}{2}(T_2 - T_3)}{s_2 - s_3} = \frac{\tfrac{1}{2}(T_3 - T_1)}{s_3 - s_1} = \frac{\tfrac{1}{2}(T_1 - T_2)}{s_1 - s_2} = A' + B'\theta \ \ldots\ldots\text{(iv)}.$$

Hence,
$$T_1 \equiv \tfrac{1}{3}(T_1 + T_2 + T_3) + \tfrac{1}{3}(T_1 - T_2) - \tfrac{1}{3}(T_2 - T_1)$$
$$= A + B\theta + \tfrac{2}{3}A'\,(2s_1 - s_2 - s_3),$$
$$= A + (B - \tfrac{2}{3}A')\,\theta + 2A's_1,$$

i.e. is of the form :
$$T_1 = A + \lambda\theta + 2\mu S_1 \dots\dots\dots\dots\dots\dots (v),$$

or the usual expression for a principal traction of an elastic solid, if A be the initial stress, which is supposed to be uniform in all directions : see our Art. 616*.

　　(b) *Fluid Bodies.* Here we have finite strain-differences $s_2 - s_3$, $s_3 - s_1$, $s_1 - s_2$ for vanishingly small stress-differences $T_2 - T_3$, $T_3 - T_1$, $T_1 - T_2$. If these finite strains are produced whatever be the value of p, then we have a fluid and
$$T_1 = T_2 = T_3 = -p = A + B\theta \quad\dots\dots\dots\dots (vi).$$

　　(c) *Pulverulent Bodies.* In such bodies finite strains $s_2 - s_3$, $s_3 - s_1$, $s_1 - s_2$ are produced by vanishingly small stress-differences $T_2 - T_3, T_3 - T_1$, $T_1 - T_2$, only when p is vanishingly small. Hence terms of the type $K' + (B'' + C''\theta)\,s_1$ must be divisible by p, or $= mp$, say. If we neglect terms of the third order in the strain, this can only be realised if $K' = mK$, $B'' = 0$ and $C''\theta$ be of the second order in the strain. Further, since p is to vanish with the strain, we must have $A = 0$. We easily deduce :
$$\frac{\tfrac{1}{2}(T_2 - T_3)}{s_2 - s_3} = \frac{\tfrac{1}{2}(T_3 - T_1)}{s_3 - s_1} = \frac{\tfrac{1}{2}(T_1 - T_2)}{s_1 - s_2} = mp,$$

whence
$$T_1 = -p\,(1 - 2ms_1), \quad T_2 = -p\,(1 - 2ms_2), \quad T_3 = -p\,(1 - 2ms_3),\dots(vii),$$
where　　$-p = B\theta + C\theta^2 + D\,\{(s_2 - s_3)^2 + (s_3 - s_1)^2 + (s_1 - s_2)^2\}.$

　　If we neglect the squares and higher powers of the strains m may be considered a constant ; further, since when p is vanishingly small, θ is known from physical considerations to be vanishingly small even for finite strain-differences $s_2 - s_3$, $s_3 - s_1$, $s_1 - s_2$, it follows that θ must at least be of the order of their squares. Hence we may neglect $C\theta^2$ and put
$$-p = B\theta + D\,\{(s_2 - s_3)^2 + (s_3 - s_1)^2 + (s_1 - s_2)^2\}\dots\dots\dots(viii).$$

Boussinesq for pulverulent bodies neglects the squares of the strains and accordingly puts $\theta = 0$, while $-p\,(= B\theta)$ is supposed to be finite. He thus obtains equations (vii) in conjunction with $\theta = 0$ as the fundamental equations for such media.

　　[1575.]　§ II. of the memoir (pp. 23–7) is entitled : *Expressions générales des forces élastiques, à l'intérieur des corps d'élasticité constante, solides ou pulvérulents.* This opens with a discussion of the transformation of strain from any system of axes to the axes of principal stretch. Boussinesq gives results of the types :

$$s_x = a_1^2 s_1 + b_1^2 s_2 + c_1^2 s_3, \quad \sigma_{yz} = 2\,(a_2 a_3 s_1 + b_2 b_3 s_2 + c_2 c_3 s_3),$$

where a_1, b_1, c_1 are the direction-cosines which the new axis of x makes with the axes of principal stretch, a_2, b_2, c_2 those of y and a_3, b_3, c_3 those of z. These results are identical with those of Maxwell : see our Art. 1539*.

Boussinesq next gives corresponding formulae for the resolution of stress, these are of the type :

$$\widehat{xx} = a_1^2 T_1 + b_1^2 T_2 + c_1^2 T_3, \quad \widehat{yz} = a_2 a_3 T_1 + b_2 b_3 T_2 + c_2 c_3 T_3$$

See our Art. 133.

From (vii) by aid of these results he easily deduces (i), or :

$$\widehat{xx} = -p\,(1 - 2ms_x), \quad \widehat{yz} = pm\sigma_{yz}\ldots\ldots\ldots\ldots(ix),$$

with the condition $\qquad s_x + s_y + s_z = 0,$

as the general type of stresses in pulverulent bodies (p. 27).

[1576.] § III. of the memoir (pp. 28–37) is entitled: *Équations différentielles de l'équilibre d'élasticité des massifs pulvérulents.* Boussinesq here limits the scope of his investigation :

Je m'occuperai principalement, dans la suite de cette étude, de l'équilibre de massifs pesants, tels qu'un monceau de sable, formés de très-petits grains solides juxtaposés sans cohésion, mais se comprimant mutuellement (p. 28).

He supposes the atmosphere to penetrate into the conglomeration of such grains and, pressing round each grain individually, to have no influence on their mutual action, *i.e.* to contribute nothing to the value of p in equations (ix) above. He terms the natural state (*état naturel*) of the mass that in which it is free from its own weight and the pressure p is zero at each point. He takes x, y, z as the coordinates of a particle in this state, and u, v, w for the components of its shift, when the mass is supposed to become heavy, and accordingly to take up a new position of equilibrium. Boussinesq further considers the limit of elasticity of the pulverulent mass not to be passed, and confines his discussion to a uniplanar distribution of strain.

Suppose the plane of xy to be this plane, and the axis of z to be horizontal. Then, if gravity make an angle a with the axis of y, the body stress equations reduce to

$$\frac{d\widehat{xx}}{dx} + \frac{d\widehat{xy}}{dy} + \rho g \sin a = 0,$$
$$\frac{d\widehat{xy}}{dx} + \frac{d\widehat{yy}}{dy} + \rho g \cos a = 0 \left.\right\} \quad \dots\dots\dots\dots\dots(x),$$

where none of the stresses are functions of z.

If we wish to investigate the stresses only we must have a third relation between $\widehat{xx}$, $\widehat{yy}$ and $\widehat{xy}$. This is easily found from (ix) by aid of the identity

$$\frac{d^2\sigma_{xy}}{dxdy} = \frac{d^2 s_y}{dx^2} + \frac{d^2 s_x}{dy^2}$$

(see our Art. 1420), to be

$$2\frac{d^2}{dxdy}\left(\frac{\widehat{xy}}{p}\right) = \left(\frac{d^2}{dx^2} - \frac{d^2}{dy^2}\right)\left(\frac{\widehat{yy} - \widehat{xx}}{2p}\right), \left.\right\} \quad \dots\dots\dots(xi).$$

where
$$p = -\tfrac{1}{2}\left(\widehat{xx} + \widehat{yy}\right)$$

Equations (x) and (xi) are the uniplanar stress-equations for a pulverulent body.

[1577.] In a footnote, p. 31, Boussinesq gives the interesting equivalent to (xi) in the case of uniplanar strain in an elastic solid. It is, if η be the stretch-squeeze ratio :

$$2\frac{d^2\widehat{xy}}{dxdy} = \frac{d^2\widehat{yy}}{dx^2} + \frac{d^2\widehat{xx}}{dy^2} - \eta\left(\frac{d^2}{dx^2} + \frac{d^2}{dy^2}\right)(\widehat{xx} + \widehat{yy})\dots\dots\dots(xii).$$

Solutions of (x) and (xii) for the case of a heavy elastic solid are given by :

$$\widehat{xx} = C + \rho g\left(\frac{d^2\phi}{dy^2} - x\sin a\right), \quad \widehat{yy} = C + \rho g\left(\frac{d^2\phi}{dx^2} - y\cos a\right),$$

$$\widehat{xy} = -\rho g\frac{d^2\phi}{dxdy},$$

where ϕ is any function of x, y which satisfies the equation :

$$\left(\frac{d^2}{dx^2} + \frac{d^2}{dy^2}\right)^2 \phi = 0,$$

and C is an arbitrary constant : see our Art. 1583.

[1578.] Pp. 32–4 of this section of the memoir are occupied with certain supplementary formulae. Thus if $\widehat{nn}$ be the traction and $\widehat{nt}$ the shear across a plane the normal to which makes an angle β with the axis of x, Boussinesq shews that for pulverulent bodies :

$$\widehat{nn} = -p - R\cos 2(\beta - \beta_0), \left.\right\}$$
$$\widehat{nt} = R\sin 2(\beta - \beta_0) \left.\right\} \dots \dots\dots\dots\dots(xiii),$$

where R (taken positive) and β_0 are defined by :

$$R\sin 2\beta_0 = -\widehat{xy}, \quad R\cos 2\beta_0 = \tfrac{1}{2}\left(\widehat{yy} - \widehat{xx}\right).$$

The formulae (xiii) are easy corollaries from those of Rankine : see our Arts. 465 (*b*), and 1563.

If T_1, T_2, T_3 be the principal tractions :

$$T_1 = -p + R, \quad T_2 = -p - R, \quad T_3 = -p \dots\dots\dots\dots(\text{xiv}),$$

where the algebraically least traction, T_2, coincides with the direction which makes an angle β_0 with the axis of x.

If x' be the direction of lines parallel to the plane of xy which make initially an angle β with the axis of x, and y' that of lines perpendicular to x' we have :

$$\left.\begin{aligned} s_{x'} = -s_{y'} = -\frac{R}{2mp}\cos 2\,(\beta - \beta_0), \\ \sigma_{x'y'} = \frac{R}{mp}\sin 2\,(\beta - \beta_0) \end{aligned}\right\} \dots\dots\dots\dots(\text{xv}),$$

where, $R \sin 2\beta_0 = -mp\sigma_{xy}, \quad R \cos 2\beta_0 = mp\,(s_y - s_x).$

The principal stretches are

$$s_1 = \frac{R}{2mp}, \quad s_2 = 0, \quad s_3 = -\frac{R}{2mp}\dots\dots\dots\dots\dots(\text{xvi}).$$

[1579.] Pp. 34–37 deal with the conditions at the boundaries of a pulverulent mass.

At a free surface we must have the stress across the surface zero, or in the case of uniplanar strain :

$$\widehat{xx}\cos\gamma + \widehat{xy}\sin\gamma = 0, \quad \widehat{xy}\cos\gamma + \widehat{yy}\sin\gamma = 0 \dots\dots(\text{xvii}),$$

where γ is the angle the normal to the free surface at any point makes with the axis of x.

For a rigid boundary as a sustaining wall, Boussinesq considers the extreme cases of perfect roughness or perfect smoothness.

For perfect roughness :

$$u = 0, \quad v = 0 \dots\dots\dots\dots\dots\dots\dots(\text{xviii}).$$

For perfect smoothness :

$$u \cos\gamma + v \sin\gamma = 0,$$

and $\widehat{n} = R \sin 2\,(\gamma - \beta_0) = 0,$

γ being the angle the normal to the rigid boundary makes with the axis of x.

Since R will not as a general rule be zero, we may write these conditions :

$$u \cos\gamma + v \sin\gamma = 0, \quad \sin 2\,(\gamma - \beta_0) = 0 \dots\dots\dots\dots(\text{xix}).$$

Boussinesq remarks that the conditions for perfect roughness, or

$$u = 0, \quad v = 0,$$

suppose that the particles of the mass which in the 'natural state'
(i.e. weightless state : see our Art. 1573) were in contact with the rigid
boundary remain so after the mass assumes the strained condition. This
cannot in practice be the real state of the case. Sustaining walls, he
remarks, undoubtedly do not allow of the motion of the particles in
contact with them, but these particles will often be in other positions
than those of the 'natural state'. u and v at such walls may be
theoretically considered as given, but they are in practice unknown
functions of the coordinates of position.

Avec les données dont dispose l'ingénieur, l'équilibre qui se produit dans un
massif, au moment même où on le forme en déchargeant successivement de la
terre sur le sol ou contre un mur de soutènement, ne paraît donc pas susceptible
d'une détermination précise, et il doit être fort complexe ou affecté d'un grand
nombre d'anomalies locales. Mais ce qu'il importe de connaître, c'est le mode
d'équilibre *définitif* qui subsistera, lorsque les petits ébranlements que tout
massif éprouve presque à chaque instant auront fait disparaître les irrégularités
et amené un tassement complet, ou groupé tous les grains sablonneux de la
manière en quelque sorte la moins forcée. Un tel mode d'équilibre, par le fait
même qu'il s'établit de préférence à tout autre, doit être, *de tous les modes
compatibles avec les circonstances, celui qui assure le mieux la stabilité intérieure
du massif en l'écartant le moins possible de l'état naturel* (pp. 36-7).

[1580.] § IV. pp. 37–45 of the memoir solves the equations of our
Art. 1576 for the case of an infinite mass of pulverulent matter bounded
by a plane sloping at an angle ω to the horizon. Boussinesq takes as
plane of xy any vertical plane perpendicular to the bounding plane, and

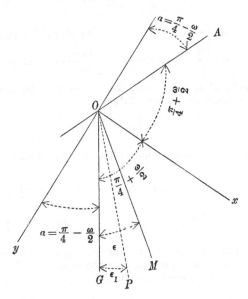

as axis of x the line bisecting the angle between the talus OA and the vertical OG. The magnitudes of the angles will then be those indicated in the accompanying figure. The values of the shifts and stresses in the plane of the figure can now only be functions of the primitive distance l from the line OA. We have

$$l = x \cos a + y \sin a,$$

and equations (x) become:

$$\frac{d}{dl}\{\widehat{xx} \cos a + (\widehat{xy} + g\rho l) \sin a\} = 0,$$

$$\frac{d}{dl}\{(\widehat{xy} + g\rho l) \cos a + \widehat{yy} \sin a\} = 0 \quad\dots\dots\dots\dots(\text{xx.})$$

These equations, having regard to (xvii) which must be satisfied along the bounding surface or for $l = 0$, lead to:

$$\widehat{xx} \cos a + (\widehat{xy} + g\rho l) \sin a = 0, \quad (\widehat{xy} + g\rho l) \cos a + \widehat{yy} \sin a = 0,$$

or, remembering

$$\omega = \frac{\pi}{2} - 2a \text{ and } p = -\tfrac{1}{2}(\widehat{xx} + \widehat{yy}),$$

after some reductions to:

$$\tfrac{1}{2}(\widehat{xx} - \widehat{yy}) - p \sin \omega = 0, \quad \widehat{xy} + \rho g l - p \cos \omega = 0 \dots\dots.(\text{xxi}).$$

Hence by (ix) of Art. 1575, we find:

$$s_x = -s_y = \frac{\sin \omega}{2m}.$$

Thus, if ϕ and ψ be arbitrary functions of y and x respectively:

$$u = \frac{\sin \omega}{2m}[x + \phi(y)], \quad v = \frac{\sin \omega}{2m}[-y + \psi(x)].$$

The properties of σ_{xy}, however, which can only be a function of l, and therefore must satisfy the relation

$$\frac{d\sigma_{xy}}{dx}\frac{1}{\cos a} = \frac{d\sigma_{xy}}{dy}\frac{1}{\sin a},$$

lead to the easy determination of the forms of ϕ and ψ. Boussinesq finds:

$$u = \frac{\sin \omega}{2m}[\ x + cy^2 \sin a + (c' + c'')y + c_1'],$$

$$v = \frac{\sin \omega}{2m}[-y + cx^2 \cos a + (c' - c'')x + c_1''] \left.\right\}\ \dots\dots(\text{xxii}),$$

where c, c', c'', c_1', c_1'' are five arbitrary constants. Obviously the terms in c_1', c_1'' answer to a displacement of the mass as a whole and those in c'' to a rotation of the mass as a whole.

For the strains and stresses we deduce the values :

$$s_x = -s_y = \frac{\sin \omega}{2m}, \qquad \sigma_{xy} = \frac{\sin \omega}{m}(c' + cl) \dots\dots\dots(\text{xxiii}),$$

$$\widehat{xx} = -p(1 - \sin \omega), \quad \widehat{yy} = -p(1 + \sin \omega), \quad \widehat{xy} = p(c' + cl)\sin \omega, \;\Bigg\}$$

$$p = \frac{\rho g l}{\cos \omega - (c' + cl)\sin \omega}$$

$$\dots\dots\dots(\text{xxiv}).$$

If the motion of the mass as a whole be disregarded, Boussinesq (pp. 41–42) shews that any system of initially parallel straight lines in the plane xy becomes a system of concentric and similar conics with their axes parallel to those of x and y. This conic system reduces to a system of circles when the straight lines are parallel to the bounding plane, and to a system of straight lines when $c = 0$.

[1581.] Boussinesq now proceeds to find the stress across any plane from equations (xiii) of Art. 1578. Let the plane pass through the axis of z and make an angle ϵ_1 with the vertical (see figure in Art. 1580), then its trace on the plane of xy being OP, we easily find that β of equations (xiii) is given by

$$\beta = \frac{3\pi}{4} + \frac{\omega}{2} - \epsilon_1.$$

Whence using (xxiv) we have :

$$\tan 2\beta_0 = c' + cl.$$

Take an auxiliary angle ϵ given by $2\beta_0 = \omega - 2\epsilon$, or

$$c' + cl = \tan(\omega - 2\epsilon) \dots\dots\dots\dots\dots\dots(\text{xxv}).$$

We then easily deduce :

$$p = \frac{\rho g l \cos(\omega - 2\epsilon)}{\cos 2(\omega - \epsilon)}, \;\Bigg\}$$

$$\widehat{nn} = -\frac{\rho g l}{\cos 2(\omega - \epsilon)}\{\cos(\omega - 2\epsilon) + \sin \omega \sin 2(\epsilon_1 - \epsilon)\}, \;\Bigg\}\dots(\text{xxvi}).$$

$$\widehat{nt} = \frac{\rho g l}{\cos 2(\omega - \epsilon)}\sin \omega \cos 2(\epsilon_1 - \epsilon) \;\Bigg\}$$

Further, since

$$\widehat{nn} = -p(1 - 2ms_n) \text{ and } \widehat{nt} = pm\sigma_{nt},$$

we deduce :

$$s_n = -\frac{\sin \omega \sin 2(\epsilon_1 - \epsilon)}{2m\cos(\omega - 2\epsilon)}, \;\Bigg\}$$

$$\sigma_{nt} = \frac{\sin \omega \cos 2(\epsilon_1 - \epsilon)}{m\cos(\omega - 2\epsilon)} \;\Bigg\}\dots\dots\dots\dots(\text{xxvii}).$$

Equations (xxv–xxvii) contain the fundamental results of Boussinesq's theory for the elastic equilibrium of pulverulent masses. We see at once that when $\epsilon_1 = \epsilon$, or $= \epsilon - \dfrac{\pi}{2}$, the value of $s_n = 0$; that is to say there is no stretch (or squeeze) in directions making an angle ϵ with the vertical or horizontal, or in directions given by the solution of (xxv) for ϵ. The principal axes of stretch must bisect the angles between these directions of no stretch, and the magnitudes of the principal stretches are given by:

$$s = \pm \frac{\sin \omega}{2m \cos (\omega - 2\epsilon)} \qquad\qquad (\text{xxviii}).$$

[1582.] Boussinesq remarks that the practically useful cases are those in which the constant c of (xxv) is zero, or ϵ is a constant angle, and sums up for such cases as follows:

If two systems of equidistant parallel straight lines be drawn in the unstrained mass, the one inclined at an angle ϵ to the vertical, and the other at an angle ϵ to the horizontal,—thus dividing the transverse section of the mass into a system of equal squares,—then this double system remains after strain a double system of parallel straight lines, the squares being converted into rhombuses having the same sides as the squares, but adjacent sides rotated relative to each other through the small angle $\sin \omega / \{m \cos (\omega - 2\epsilon)\}$. The whole strain therefore reduces to a slide upon each other of parallel slices of the mass inclined at an angle ϵ to the vertical (p. 45).

[1583.] In a footnote on p. 45 Boussinesq deals with the strain in an infinite heavy elastic mass bounded by a plane inclined at an angle ω to the horizon.

Taking: $\quad \widehat{xx} = -p\,(1 - \sin \omega), \quad \widehat{yy} = -p\,(1 + \sin \omega),$

$\qquad\qquad\qquad \widehat{xy} = p \cos \omega - \rho g l \qquad \left.\right\} \quad \dots\dots (\text{xxix}),$

we see from our (xii) that p must be of the form:

$$p = f_1 + f_2\, l \qquad\qquad\qquad (\text{xxx}).$$

I find from $\qquad \widehat{xx} = \lambda\theta + 2\mu s_x, \quad \widehat{yy} = \lambda\theta + 2\mu s_y,$

that the shifts are given by[1]:

$$u = f_0 + \tfrac{1}{2} \sec a \left(f_1 l + \frac{f_2\, l^2}{2} \right) \left(\frac{\sin \omega}{\mu} - \frac{1}{\lambda + \mu} \right),$$

$$v = f_0' - \tfrac{1}{2} \operatorname{cosec} a \left(f_1 l + \frac{f_2\, l^2}{2} \right) \left(\frac{\sin \omega}{\mu} + \frac{1}{\lambda + \mu} \right) \qquad \left.\right\} \quad \dots\dots (\text{xxxi}).$$

---

[1] This is the case of an infinite solid with a free plane surface. Boussinesq's results as expressed in (xxix) do not seem general enough to enable us to deal with a *rigid* plane boundary as well, which we can do in the case of results (xxiv) for a pulverulent mass.

Whence from $\widehat{xy} = \mu\sigma_{xy}$, we have, if $\nu = (\lambda + 2\mu)/(\lambda + \mu)$:

$$\cos\omega - \frac{\rho g l}{p} = \frac{\cos^2\omega - \nu}{\cos\omega}.$$

Now Boussinesq equates the expression on the left-hand side of this result to $\tan(\omega - 2\epsilon)\sin\omega$, and speaks of this auxiliary angle ϵ becoming constant for great values of l. It seems to me that it must always be constant, and that we must have :

$$p = \frac{\rho g l \cos\omega}{\nu} \quad\dots\dots\dots\dots\dots\dots\text{(xxxii)},$$

or, $f_1 = 0$, and $f_2 = (\rho g \cos\omega)/\nu$.

The maximum stretch and squeeze are then given by :

$$\left.\begin{array}{c} s = -\dfrac{p}{2\mu}\{\nu - 1 \mp \sqrt{(\nu - 1)^2 + \nu^2\tan^2\omega}\}, \\[2mm] \text{the angles } \gamma_1 \text{ and } \gamma_2, \text{ they make with the axis of } x \\ \text{being determined as roots of :} \\[1mm] \tan 2\gamma = (\nu - 1)\cot\omega + \nu\tan\omega \end{array}\right\} \dots\dots\text{(xxxiii),}$$

where the squeeze corresponds to the value of $\gamma < \pi/2$.

These results, a slight extension of Boussinesq's, seem to me of possible application to geological problems. For example, supposing rupture to take place perpendicular to the directions of greatest stretch, we find, that a massive slope of rock at an angle of 45°, would under its own weight rupture in planes making an angle of about 80° to 81° with the downward direction of the slope. The planes of rupture thus fall *below* the internal planes perpendicular to the surface. This numerical result supposes uniconstant isotropy to hold for the material of the rock.

That the strains and stresses become infinite with l is only to be expected from the nature of the theoretical problem, which suppose a heavy mass, infinite in size.

[1584.] § v. of the memoir (pp. 46–53) deals with the modifications necessary in the results of the previous section when the pulverulent mass is bounded by a sloping wall. Let this wall slope at an angle i to the vertical (see fig. below), then the angle χ it makes with the axis of x is given by $\chi = \dfrac{\pi}{4} + \dfrac{\omega}{2} - i$. Boussinesq considers two cases, namely, when the wall is either (i) perfectly rough or (ii) perfectly smooth.

Case (i). *Wall perfectly rough.* In this case we take (see our Art. 1579) u and v as given by equations (xxii) equal to zero, when

$$x = r\cos\chi, \quad y = r\sin\chi$$

for all values of r. We easily find that :

$$c = c_1' = c_1'' = 0,$$
$$c'' = -\operatorname{cosec} 2\chi = -\sec(\omega - 2i),$$
$$c' = -\cot 2\chi \quad = \quad \tan(\omega - 2i)$$
$$\left.\right\}\dots\dots\dots\text{(xxxiv)}.$$

Since $c = 0$, we have from (xxv) :

$$\epsilon = i,$$

or, the parameter ϵ is now constant and the line OM of the figure in our Art. 1580 coincides with the direction of the supporting wall. Thus by Art. 1582 the strain of the pulverulent mass consists of a slide of magnitude

$$\sin \omega / \{m \cos(\omega - 2i)\}$$

parallel to the supporting wall.

Suppose a plane in the pulverulent mass perpendicular to that of xy to make an angle ϕ with the fixed wall in the unstrained, and the angle $\phi - \delta\phi$ in the strained condition. Then we easily find $\delta\phi / \sin^2 \phi = \sigma$, the slide parallel to the fixed wall, or

$$\delta\phi = \frac{\sin \omega \sin^2 \phi}{m \cos(\omega - 2i)} \dots\dots\dots\dots\text{(xxxv)}.$$

Boussinesq takes two special cases. Namely, when $\delta\phi = \zeta$, the change in angle of the talus itself, and when $\delta\phi = \zeta'$, the change in angle of a plane making an angle of $45°$ with the wall in the unstrained condition. In the former case $\phi = \pi/2 + \omega - i$, and in the later $\phi = \pi/4$. Hence :

$$\zeta = \frac{\sin \omega \cos^2(\omega - i)}{m \cos(\omega - 2i)}, \qquad \zeta' = \frac{\sin \omega}{2m \cos(\omega - 2i)},$$
$$\text{and} \qquad \zeta - \zeta' = \frac{\sin \omega \cos 2(\omega - i)}{2m \cos(\omega - 2i)}$$
$$\left.\right\}\dots\text{(xxxvi)}.$$

Case (ii). *Wall perfectly smooth.* In this case the $\widehat{m}$ of equation (xxvi) is zero for $\epsilon_1 = i$; this leads to $2(i - \epsilon) = \pi/2$, or from (xxv) :

$$c' + cl = -\cot(\omega - 2i).$$

We have further to make the shift perpendicular to the wall vanish, or

$$v \cos \chi - u \sin \chi = 0,$$

for all values of r when

$$x = r \cos \chi, \quad y = r \sin \chi.$$

This leads by (xxii) to $c = 0$, and, remembering that at the wall

$$\chi + i = \pi/4 + \omega 2, \text{ also to } c'' = 0.$$

Further we find $c_1' / \cos \chi = c_1 \quad \sin \chi = c_0$, say. Boussinesq takes $c_0 = 0$, or the axes of x and y through the strained position of the element at the origin O. Thus we have finally :

$$c = c'' = c_1' = c_1'' = 0, \quad c' = -\cot(\omega - 2i),$$
and
$$\epsilon = i - \pi/4 \qquad \Big\} \dots \dots \text{(xxxvii)}.$$

On se fait une idée nette du tassement qui se produit dans le cas actuel d'un mur poli, ayant OM' pour face postérieure, en concevant, au lieu de ce mur

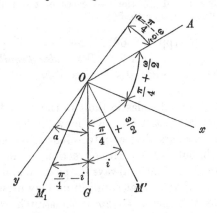

poli, un mur rugueux OM_1, incliné sur celui-ci de 45°, ou faisant avec Oy l'angle $yOM_1 = a + i - \dfrac{\pi}{4}$, et en considérant le tassement, parallèle à OM_1, qui se produirait alors. Ce tassement, à une distance D de OM_1, sera égal à $D \sin \omega \Big/ \Big\{ m \cos \Big[\omega - 2 \Big(i - \dfrac{\pi}{4} \Big) \Big] \Big\} = -D \sin \omega / \{ m \sin (\omega - 2i) \}$[1]. Pour amener le massif AOM' à son état définitif, il suffira de concevoir ensuite qu'il tourne en bloc autour de l'origine O, dans le sens de Oy vers Ox, de la petite quantité $\zeta' = \sin \omega \Big/ \Big\{ 2m \cos \Big[\omega - 2 \Big(i - \dfrac{\pi}{4} \Big) \Big] \Big\} = -\sin \omega / \{ 2m \sin (\omega - 2i) \}$, en vue d'annuler la rotation égale et contraire éprouvée dans ce tassement fictif, d'après (xxxvi), par la ligne matérielle primitivement couchée contre le mur réel OM' et qui ne reçoit effectivement aucune rotation autour de O (pp. 50-1).

This follows since the term in c'' in the first solution corresponds as we have noted in Art. 1580 to a rotation of the mass as a whole of magnitude $\dfrac{\sin \omega}{2m} c''$ [: see equation (xxii)], or by (xxxiv) to

$$\sin \omega \Big/ \Big\{ 2m \cos \Big[\omega - 2 \Big(i - \dfrac{\pi}{4} \Big) \Big] \Big\},$$

i.e. to the value of ζ' in (xxxvi).

[1] The angle GOM_1 is to be reckoned *negative* when substituted for the i of case (i), for OM_1 now falls on the opposite side of OG to OM'.

The change in inclination of the talus to the smooth wall will be given by the third formula of (xxxvi), or by :

$$\zeta - \zeta' = \frac{\sin \omega \cos 2 \left(\omega - \left(i - \frac{\pi}{4} \right) \right)}{2m \cos \left(\omega - 2 \left(i - \frac{\pi}{4} \right) \right)} = \frac{\sin \omega \sin 2 (\omega - i)}{2m \sin (\omega - 2i)} \dots \text{(xxxviii)}.$$

The solutions found for the two cases of the rough and smooth supporting walls are unique (p. 50).

[1585.] § VI. (pp. 53–68) is an interesting, if somewhat hypothetical one. It is entitled : *Des modes d'équilibre qui cessent d'être possibles, par suite des limites d'élasticité de la matière pulvérulente.*

Boussinesq terms the elastic limit in the case of a pulverulent solid the *état ébouleux*, which we may render as the *state of collapse*. He considers that a pulverulent mass may withstand a positive stretch, but not a positive traction. Thus from equation (vii), it follows that p must always be positive and the maximum stretch $s < 1/2m$ in order that equilibrium may exist. But it does not result from this that $s = 1/2m$ is the stretch which marks the point of collapse. A less stretch than this, to be determined by experiment, may be sufficient. Boussinesq accordingly takes for the conditions of non-collapse or stability

$$p > 0 \text{ and } s < \frac{\sin \phi}{2m} \dots\dots\dots\dots\dots \text{(xxxix)},$$

where ϕ is an angle between 0 and 90° (for its physical meaning : see Art. 1587) and s is the greatest positive stretch.

[1586.] We may, according to Boussinesq, look at the condition of stability of any isotropic elastic solid in the following manner (pp. 57–9). Suppose s_1 and s_2 the principal stretches in the case of *uniplanar* strain, then for the limit of elastic stability we must have s_1 some function of s_2 (including a constant as such a function), or we may write, he holds :

$$s_1 - s_2 = f (s_1 + s_2).$$

Now if the corresponding principal tractions be T_1, T_2, we have $T_1 - T_2 = 2\mu (s_1 - s_2)$, $T_1 + T_2 = 2 (\lambda + \mu) (s_1 + s_2)$, and therefore :

$$T_1 - T_2 = 2\mu f \left(\frac{T_1 + T_2}{2 (\lambda + \mu)} \right) \dots\dots\dots\dots\dots \text{(xl)}.$$

Now for the special case of a plastic solid the dilatation-modulus $\frac{1}{3}(3\lambda + 2\mu)$, or—since resistance to change of bulk in such a solid is great as compared to resistance to change of shape—λ must be very great as compared with μ. Hence if $T_1 + T_2$ be not very great as compared with $T_1 - T_2$, we have, *expanding by Maclaurin's theorem and retaining only the lowest terms*:

$$T_1 - T_2 = 2K - K_1 (T_1 + T_2) \quad \dots\dots\dots\dots\dots (\text{xli})$$

where K and K_1 are independent of T_1, T_2.

If we retain only the first term on the right, we have Saint-Venant's fundamental hypothesis for the state of plasticity: see our Arts. 247 and 260. If we retain only the second term on the right, Boussinesq's second condition of (xxxix) that a pulverulent body shall not reach the point of collapse, will be found to coincide with it. For, since

$$T_1 = - p (1 - 2ms_1) \text{ and } p = - \tfrac{1}{2} (T_1 + T_2)$$

we have $\quad (T_1 - T_2)/(T_1 + T_2) = - 2ms_1$, or for stability :

$$- \frac{T_1 - T_2}{T_1 + T_2} < \sin \phi \quad \dots\dots\dots\dots (\text{xlii}).$$

This agrees with (xli), if we take $K = 0$ and $K_1 = \sin \phi$.

Thus $K = 0$ corresponds to pulverulence and $K_1 = 0$ to plasticity.

Here, as in Arts. 1568 and 1594, the discussion seems to trench on ground which much needs accurate physical investigation. Boussinesq cites no experimental evidence, and the appeal to Maclaurin's Theorem is far from convincing. Boussinesq's treatment of the elastic, plastic and pulverulent limits may be suggestive, but it is certainly not final.

[1587.] We may look at condition (xlii) from another stand-point. From (xiii) we easily find the angle the stress across any plane makes with the normal n to that plane. Let this angle be χ. Then it will be found that χ is a maximum when :

$$\left. \begin{aligned} \cos 2 (\beta - \beta_0) &= - R/p = \sin \phi, \\ \text{and that in this case,} \qquad & \\ \sin^2 \chi &= \sin^2 \phi \end{aligned} \right\} \dots\dots\dots\dots (\text{xliii}).$$

Premising that ϕ is termed the angle of *internal friction*, we may interpret this result as follows :

The inclination of the stress across any plane to the produced normal to that plane ought for equilibrium to be possible to be less than, or at most equal to, the internal angle of friction (p. 56).

Equation (xliii) also gives us:

$$\beta - \beta_0 = \pm \left(\frac{\pi}{4} + \frac{\chi}{2} \right),$$

or, we conclude that: The planes, across which the stress makes the greatest possible angle with the production of the normal, make with the plane submitted to the maximum pressure an angle equal to $\frac{1}{4}\pi + \frac{1}{2}\chi$.

This result is due to Rankine, who deduced it, however, from a discussion not involving the same principles.

[1588.]   Using (xxviii) and the second of (xxxix) we easily find that:

$$\cos^2(\omega - 2\epsilon) > \sin^2 \omega / \sin^2 \phi \dots\dots\dots\dots\dots (xliv).$$

Hence ω the slope of the talus must be less than, or at most equal to, the angle of friction ϕ.

Boussinesq gives some details from a memoir of Saint-Guilhem (*Annales des ponts et chaussées*, T. xv. pp. 319-50. Paris, 1858) as to the angle of friction. It varies from 24° or 26° for small shot or mustard seed to about 55° for very dense earth. The natural talus, for example that observed at the foot of steep rocks, etc., is about 31° for fine and dry sand, from 32° to 33° for marl, limestone and earth recently thrown from the wheelbarrow, about 37° for chalky earth, about 38° for moist quartz sand and about 45° for moist gypseous sand.

[1589.]   The relation (xliv) is easily shewn to involve the condition that $\cos^2 \omega > \sin^2 \omega \tan^2(\omega - 2\epsilon)$, or, looking at (xxiv) and (xxv), the condition that $p > 0$. Thus the first condition of (xxxix) is satisfied if the second be.

Suppose $\cos \tau = \sin \omega / \sin \phi$, where τ is an angle between 0 and $\pi/2$, then we easily find from (xliv) that

$$\epsilon > \tfrac{1}{2}(\omega - \tau) \text{ and } < \tfrac{1}{2}(\omega + \tau).$$

In the case of a rough revetment wall its inclination i to the vertical is equal to ϵ: see our Art. 1584. The limits within which i, or the direction of the face of the revetment wall must lie thus form what Boussinesq terms a Maltese cross, and for all values outside the arms of this cross slipping takes place for the given slope of the talus[1].

---

[1] Boussinesq in a footnote, p. 63, returns to the solution for the elastic solid, which we have dealt with in Art. 1583. He appears to think that the principal stretches can both be negative, but I have been unable to find any special application of the solution with possible boundary conditions in which this would occur. He states that it would be necessary for the plane boundary of such a solid to have a less slope than that given by $\sin \omega = \dfrac{\mu}{\lambda + \mu}$, or $= \frac{1}{2}$ for a uniconstant elastic solid, i.e. $\omega < 30°$. I do not think this is true for an infinite elastic solid with a free plane surface.

We easily find from (xliv) that the two limiting values of the slope of the talus for a given value i of the inclination of the revetment wall are the roots of:

$$\tan \omega = \frac{\cos 2i}{\pm \operatorname{cosec} \phi - \sin 2i} \quad \text{......(xlv)}.$$

Boussinesq remarks that in actual practice this only gives the slope of the tangent plane to the talus at the revetment wall, the talus itself generally not being plane and having at a distance from the wall any declivity not greater than the angle of friction ϕ. Thus we have the cases:

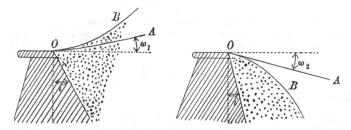

in the first of which ω_1 corresponds to the upper, and in the second ω_2 to the lower sign in (xlv).

The remainder of this section of the memoir is devoted to numerical details of the relations between ω and i for the cases of perfectly rough and perfectly smooth revetment walls.

[1590.] § VII. (pp. 68–81) is entitled: *Calcul de la pression exercée sur tout élément de surface normal au plan des déformations, et de la poussée totale que supporte un mur plan de soutènement.*

In this section Boussinesq first discusses at some length (pp. 68–74) the values of the parameter ϵ (which characterises the mode of equilibrium: see our Art. 1581 and *fig.* Art. 1580) for which the stresses $-\widehat{nn}$ and $\widehat{nt}$ across the plane given by ϵ_1 (see equation (xxvi) of Art. 1581) take their maximum and minimum values. Let ϕ_1 be the angle the resultant stress makes with the normal to the plane given by ϵ_1 then:

The extreme values of the two components $-\widehat{nn}$, $\widehat{nt}$ of the stress,—or those which correspond to a mass with its talus at an angle ω to the horizon on the point of collapse—are given by the equations:

$$\left.\begin{aligned}
&\sin(\omega + 2\psi) = \sin \omega / \sin \phi, \\
&\tan(\phi_1 + \epsilon_1 + \psi) = \tan(\epsilon_1 + \psi) \Big/ \tan^2\left(\frac{\pi}{4} - \frac{\phi}{2}\right), \\
&\widehat{nt} = \frac{\sin \phi \cos \psi \sin 2(\epsilon_1 + \psi)}{2 \cos^2\left(\frac{\pi}{4} - \frac{\phi}{2}\right) \cos(\omega + \psi)}\, g\rho l, \quad -\widehat{nn} = \frac{\widehat{nt}}{\tan \phi_1}
\end{aligned}\right\} \text{......(xlvi)}.$$

The auxiliary angle ψ, which is to be calculated from the first of these equations, ought to be so chosen that the absolute value of $\omega + 2\psi$ is less than $\pi/2$, if we require the limit when $-\widehat{nn}$ takes its least value; and it is to be taken between $\pi/2$ and π if we require the limit when $-\widehat{nn}$ takes its greatest value. Since the inclination ϕ_1 of the resultant stress to the produced normal to the elementary plane must lie between $-\pi/2$ and $\pi/2$, its value is completely determined by the second of the equations (p. 74).

Boussinesq points out that the limiting equilibrium when $-\widehat{nn}$ is a minimum is that which has been studied by Maurice Lévy in his memoir of 1867–9 : see our Art. 242, while Rankine in 1856 (see our Art. 453) had considered both cases of limiting equilibrium for $\epsilon_1 = 0$.

[1591.] For the special case of a rough supporting wall $\epsilon = i$, or the inclination of the wall (see our Art. 1584), whence from (xxvi) we have at the wall where $\epsilon_1 = i$:

$$-\widehat{nn} = \frac{\rho g l}{\cos 2\,(\omega - i)} \cos\,(\omega - 2i), \quad \widehat{nt} = \frac{\rho g l}{\cos 2\,(\omega - i)} \sin \omega\, ;$$

whence, if R be the resultant stress :

$$\left.\begin{aligned} \tan \phi_1 &= \sin \omega/\cos\,(\omega - 2i), \\ R &= \frac{\rho g l \sin \omega}{\cos 2\,(\omega - i)\sin \phi_1} \end{aligned}\right\} \dots\dots\dots\dots(\text{xlvii}).$$

We see at once that the stress across each element of the wall is the same in direction. Taking a strip of the wall of unit horizontal length and depth *d measured along its sloping face* from the talus, we easily find for the resultant action P :

$$P \equiv \int_0^L R\,dL = \frac{\rho g L^2 \sin \omega \cos\,(\omega - i)}{2 \cos 2\,(\omega - i)\sin \phi_1} \dots\dots\dots\dots(\text{xlviii}),$$

and this acts at the hydrostatic centre of pressure of the strip, i.e. at a distance $\frac{2}{3}L$ from the top of the wall.

If $i = 0$, or the sustaining wall be vertical :

$$\tan \phi_1 = \tan \omega, \quad P = \frac{\rho g L^2}{2} \frac{\cos \omega}{\cos 2\omega} \dots\dots\dots\dots(\text{xlix}).$$

In this last case we have, from (xliv), $\tan \omega = \sin \phi$, whence we see that the angle of friction between the wall and the pulverulent mass (which ought always to be greater than ϕ_1) must have its tangent greater than the sine of the internal angle of friction of the mass (pp. 80–1).

Boussinesq also deals with the corresponding problems for the *smooth* wall; these having less practical importance, I do not reproduce.

[1592.] § VIII. (pp. 81–95) is entitled : *Résolution des problèmes d'équilibre les plus importants dans les applications, au moyen d'une condition de stabilité qui tient lieu des relations spéciales aux parois.*

This chapter is occupied with a discussion of rules for practically finding the resultant action between a sustaining wall and a pulverulent mass which will: (i) just support the pulverulent mass, or (ii) which will give the greatest possible stability to the mass.

Boussinesq first remarks that having regard to the manner in which earthwork is generally put together, it is impossible to consider that u and v, the shifts from the "natural state," are zero at the rough wall: see our Art. 1578. He considers that the pulverulent mass is in reality put in with many finite slippings and shakings, so that it finally takes up a form of equilibrium totally different from that obtained by using the conditions $u = v = 0$, the wall.

Ce mode doit être le plus favorable possible à la stabilité intérieure du massif, c'est-à-dire, celui pour lequel la dilatation maxima s_1 acquiert aux divers points ses plus petites valeurs compatibles avec le degré de résistance que le mur peut opposer : car le mode d'équilibre ainsi défini, s'il n'était pas déjà complètement réalisé un instant après que l'on a déposé les dernières couches de terre ou de sable, ne tarderait pas à s'établir par l'effet des petits ébranlements, dus à mille causes diverses, que le massif éprouve presque à tout instant, et qui permettent aux grains sablonneux de se grouper de la manière en quelque sorte la moins forcée (p. 82).

Excluding the material in the immediate neighbourhood of the wall where the wall's influence produces perhaps local disturbances, we see that all the modes of equilibrium realisable are given, subject to the limitation of the value of ϵ in Art. 1589, by our equations (xxv)–(xxviii). The last equation, however, shews us that the principal stretch s_1 will be a minimum, ω being a constant, when $\epsilon = \frac{1}{2}\omega$, and this corresponds to the most stable mode of equilibrium, or that nearest to the "natural state" for which $s_1 = 0$. This mode of equilibrium will therefore be set up if the wall be able to carry the corresponding resultant action.

In the case of most stable equilibrium the resultant action, taking place at the hydrostatic centre of pressure, is given by:

$$P = \frac{\rho g L^2}{2} \frac{\tan \omega \cos (\omega - i) \cos (2i - \omega)}{\sin \phi_1} \quad\dots\dots\dots\dots(1),$$

where ϕ_1, the angle P makes with the normal to the wall, is determined by:

$$\tan\left(\phi_1 + i + \frac{\pi}{4} - \frac{\omega}{2}\right) = \frac{\tan\left(i + \frac{\pi}{4} - \frac{\omega}{2}\right)}{\tan^2\left(\frac{\pi}{4} - \frac{\omega}{2}\right)} \dots\dots\dots\dots\text{(li)}.$$

For $i = 0$, $P = \dfrac{\rho g L^2}{2} \cos\omega$, and $\phi_1 = \omega$.................(lii).

If the sustaining wall will support this thrust then the pulverulent mass has the maximum of stability possible.

Boussinesq next turns to cases in which the wall will not support this thrust, and discusses the moment of the forces tending to capsize it, and the relative degree of stability corresponding to these cases (pp. 86–95). He shews that, when the sustaining wall is vertical or $i = 0$, then the pulverulent mass will still be in equilibrium, if the wall will just withstand a thrust given by:

$$P_1 = \frac{\rho g L^2}{2} \frac{\cos\omega \sin 2\psi}{\sin 2(\omega + \psi)},$$

where ψ is the least root of

$$\sin(\omega + 2\psi) = \sin\omega/\sin\phi$$

$$\left.\right\}\dots\dots\dots\dots\text{(liii)}.$$

The values of P and P_1, as given by (lii) and (liii) respectively, differ very considerably. Thus for $\omega = 20°$, $\phi = 45°$, we have $P_1 : P :: \cdot1935 : \cdot9397$. There is thus a wide range of values from the just stable stage to that of maximum stability (pp. 91–2).

[1592 *bis.*] The reader must carefully bear in mind the different character of the solutions obtained in our Art. 1590 and again in our Art. 1592. The solutions of Art. 1590 give the resultant action of the pulverulent mass on the sustaining wall, *provided we make the physically incorrect assumption that* $u = v = 0$, *at the wall.* The solutions of Art. 1592 do not involve this assumption (which in the case of a rough wall leads to $\epsilon = i$: see our Art. 1584, Case (i)) but, *disregarding the condition of affairs in the immediate neighbourhood of the wall* (i.e. giving u, v definite, but undetermined values there), give the resultant actions on the wall for those values of ϵ which correspond to (i) the maximum stability in the pulverulent mass, and (ii) the least resultant action on the wall consistent with equilibrium at all.

[1593.] § IX. (pp. 95–133) is entitled: *Sur l'équilibre-limite en général. Étude particulière de l'état ébouleux qui se produit dans un massif pulvérulent, au moment où un mur de soutènement commence à se renverser.* This section deals with the stability of a pulverulent mass on the point of collapse as a limiting case of a pulverulent mass in motion.

In obtaining the equations for the motion of a pulverulent mass Boussinesq supposes :

(i) that the mass remains homogeneous and isotropic,

(ii) that the strains are produced so slowly that the inertia of the elements of the mass is negligible [1],

(iii) that the stresses in consequence do not sensibly differ from the maximum elastic stresses,

(iv) that the elastic stretch in any direction is proportional to the set stretch which occurs in the same direction in a small interval of time *dt*.

These assumptions lead Boussinesq to equations equivalent to (x) of our Art. 250, which hold for either a plastic or a pulverulent mass. The equation (ix) of Art. 250 resulting from (iii) of Art. 247 will not, however, be the additional relation peculiar to the case of a pulverulent mass. Indeed Boussinesq criticises the form of that relation as stated by Saint-Venant and Lévy even for a plastic mass.

If s_1, s_2, s_3 be the principal elastic stretches in descending order of magnitude, Boussinesq, generalising the results of Art. 1586, considers that $s_1 - s_3$ takes at the elastic limit a definite value for each value of the dilation $(s_1 + s_2 + s_3)$, and for each value of $(s_1 - s_2)/(s_2 - s_3)$. Thus he considers the elastic limit given by a relation of the form :

$$s_1 - s_3 = f\left(s_1 + s_2 + s_3, \frac{s_1 - s_2}{s_2 - s_3}\right) \quad \ldots\ldots\ldots\ldots\text{(liv)},$$

or, in the case of both plastic and pulverulent masses, for which $s_1 + s_2 + s_3 = 0$, by

$$s_1 - s_3 = f\left(\frac{s_1 - s_2}{s_2 - s_3}\right) \quad \ldots\ldots\ldots\ldots\ldots\ldots\text{(lv)},$$

---

[1] The equations for the small elastic vibrations of a pulverulent mass are not linear and cannot be even approximately satisfied by sine and cosine terms involving the time. This is the analytical equivalent of the effectiveness of sawdust or sand in checking sound vibrations.

or,
$$T_1 - T_3 = 2\mu f\left(\frac{T_1 - T_2}{T_2 - T_3}\right),$$

μ being as usual the slide-modulus.

He remarks that for the cases treated by Saint-Venant:

(a) torsion of right circular cylinder (see our Art. 255), $s_2 = 0$, $s_3 = -s_1$, and therefore $T_1 - T_2 = 2\mu f(1)$,

(b) circular flexure of a prism, $s_2 = s_3$ for the extended fibres, $s_2 = s_1$ for the contracted fibres, and therefore

$$T_1 - T_2 = 2\mu f(\infty), \text{ or } = 2\mu f(0) \text{ respectively.}$$

Hence Saint-Venant's plastic modulus K, which equals μf, will not be the same for all these cases, unless f is an absolute constant (p. 101).

It seems to me that Boussinesq's relation (liv) is really of a very arbitrary character, and that at least some physical evidence in favour of it ought to have been adduced.

[1594.] With regard to the surface-conditions, Boussinesq practically refers to Saint-Venant's treatment (see our Art. 254), and like Saint-Venant he also neglects a possible semi-plastic midzone of material (p. 102): see our Art. 244.

[1595.] On p. 103 Boussinesq makes an important remark with regard to the form of (lv). He notes that for a pulverulent mass, by aid of equation (vii) of our Art. 1574, it may be written in the form:

$$\frac{3}{2}\frac{T_1 - T_2}{T_1 + T_2 + T_3} = mf\left(\frac{T_1 - T_2}{T_2 - T_3}\right).$$

Now this equation, if satisfied for any system T_1, T_2, T_3, will still be satisfied if the principal tractions and therefore the general system of stresses be altered in any definite ratio. Further if the body-forces be negligible as compared with the stresses, equations (x) of our Art. 250 combined with the body-stress equations shew us that the magnitudes of the velocities u, v, w will remain unaltered by this change of stress. Boussinesq applies this result to the case of a reservoir of pulverulent matter with a small hole in its base:

Dans un écoulement de sable par un orifice, la vitesse tend donc vers une limite dès que la hauteur de charge devient un peu grande, et elle se maintient dès lors constante.

Ainsi s'explique l'uniformité d'écoulement qu'obtenaient les anciens avec les sabliers dont ils se servaient pour mesurer le temps (p. 104).

[1596.] In the particular case of uniplanar strain studied by Boussinesq, $\widehat{yz}$, $\widehat{zx}$ and $\widehat{zz}$ are respectively 0, 0, and $-p$, while w, du/dz, dv/dz are all zero. Thus equations (x) of our Art. 250 become:

$$\frac{\widehat{xy}}{u_y + v_x} = \frac{\widehat{yy} + p}{2v_y} = \frac{\widehat{xx} + p}{2u_x} \dots\dots\dots\dots\dots(\text{lvi}).$$

These will be found to be satisfied by (xxii) and (xxiv) of our Art. 1580, remembering that u and v are now *velocities*. Further the solutions of that article satisfy the conditions, which are necessary at the free surface.

But in order that the limit of elasticity may be reached at all points where the mass is commencing to move we must have by (xliii):

$$- R/p = \sin \phi,$$

or, substituting for the values of R and p, we have Rankine's relation:

$$4\widehat{xy}^2 + (\widehat{xx} - \widehat{yy})^2 - (\widehat{xx} + \widehat{yy})^2 \sin^2 \phi = 0 \dots\dots\dots(\text{lvii}).$$

This condition is again satisfied by the solutions for the cases of limiting equilibrium that we have found in Art. 1590.

[1597.] The condition that remains to be satisfied is that at the wall. Its introduction into the modern theory is due to Lévy. Boussinesq writes:

Une dernière relation, spéciale à la face postérieure du mur, ne s'applique qu'autant que les particules contiguës du massif sont sur le point d'y éprouver des glissements finis, circonstance qui semble devoir se produire dès le commencement de renversement du mur, toutes les fois qu'elle ne sera pas en contradiction avec les autres équations du problème. Or sa réalisation exige que l'angle fait en chaque point, avec le prolongement de la normale à la face postérieure du mur, par la poussée qui lui est appliquée, vaille précisément l'angle du frottement maximum du mur et de la matière sablonneuse du massif (p. 107).

The solutions (xlvi) for the limits of stability satisfy as we have seen all but this last condition. In order that they should also satisfy this condition, it would be necessary that the value of ϕ_1, as given for the wall ($\epsilon_1 = i$) by the second equation of (xlvi), should be just equal to the angle of friction between the wall and the pulverulent mass. If it be, then the solutions (xlvi) we have obtained for limiting equilibrium will still continue to give the stresses when the mass begins to collapse owing to the upsetting of the wall (p. 109).

When the angle of friction is greater than the above value of ϕ_1, then a wedge of material adjacent to the wall retains its elastic equilibrium at the instant the wall begins to upset.

[1598.] Now Boussinesq remarks that in practice sustaining walls
are generally sufficiently rough to render a thin stratum of the
pulverulent mass stationary upon them. Hence the angle of friction
between wall and mass really reduces to the angle of friction of the
pulverulent mass upon itself, or to what we have denoted by the
angle ϕ. The second equation of (xlvi) then leads, if $\phi_1 = \phi$, to
$\phi + 2i + 2\psi = \frac{1}{2}\pi$ as the only practical solution, or to:

$$i = \tfrac{1}{4}\pi - \tfrac{1}{2}\phi - \psi \dots\dots\dots\dots\dots\dots\text{(lviii)},$$

for the requisite inclination of the wall.

It will be found in this case that the value of the resultant action
P reduces to (p. 111):

$$P = \tfrac{1}{2}\rho g L^2 \cos (\phi + i) \dots\dots\dots\dots\dots \text{(lix)}.$$

This result is due to Lévy.

Thus we see that the solutions (xlvi) only hold for a special type of
wall.

[1599.] Boussinesq next turns to the case in which the wall
has not the above inclination and proceeds to solve this case by a
method suggested by Saint-Venant (*Comptes rendus*, T. LXX., p.
283: see our Art. 242). It consists in supposing the stresses to
differ by small quantities from the values they have for the case
in which (lviii) is satisfied. The analysis by which a solution is
obtained (pp. 112–124) is too long for reproduction, but we shall
examine it in a later modified form in our Arts. 1613–6. The
solution obtained is exact provided the pulverulent mass instead
of being supposed uniform be considered as slightly heterogeneous,
its angle of internal friction ϕ' being taken slightly greater than
ϕ and varying from point to point of the mass. Boussinesq
shews that the divergence of ϕ' from ϕ necessary to insure the
exactness of the solution only exists in a wedge of the material
bounded by the face of the wall and by a plane through the in-
tersection of the wall and talus and making an angle $\frac{1}{4}\pi - \frac{1}{2}\phi - \psi$
with the vertical (pp. 116 and 122).

Boussinesq further shews (p. 125) that

$$1 < \sin \phi'/\sin \phi < \sec \delta,$$

where δ is $\frac{1}{4}\pi - \frac{1}{2}\phi - \psi - i$, or measures the divergence of true
inclination of wall from that given by (lviii). When the secant of
the angle δ (which is really the angle of the above-mentioned
wedge) is small, then we may regard ϕ' as equal to ϕ and apply
the approximate solution (lix).

For small values of δ Boussinesq finds for the resultant action P on the wall:

$$P = \tfrac{1}{2}\rho g L^2 \tan\left(\frac{\pi}{4} - \frac{\phi}{2}\right) \frac{\cos\psi\,\cos(\phi+\delta)\,\cos(\omega-i)}{\cos(\phi_1-\delta)\,\cos(\omega+\psi)} \quad\ldots\ldots (lx).$$

This will be found to agree for $\phi = \phi_1$, $\delta = 0$ with (lxix).
For a horizontal talus and vertical wall:

$$\omega = 0, \quad \psi = 0, \quad i = 0, \quad \delta = \frac{\pi}{4} - \frac{\phi}{2},$$

and:

$$P = \tfrac{1}{2}\rho g L^2 \frac{\tan^2\left(\dfrac{\pi}{4} - \dfrac{\phi}{2}\right) \cos\left(\dfrac{\pi}{4} - \dfrac{\phi}{2}\right)}{\cos\left\{\phi_1 - \left(\dfrac{\pi}{4} - \dfrac{\phi}{2}\right)\right\}} = \tfrac{1}{2}\rho g L^2 k \sec\phi_1, \text{ say,}\ldots(lxi).$$

These results hold for all cases in which δ being positive, ϕ_1 the external angle of friction nearly satisfies

$$\tan(\phi_1 + i + \psi) = \frac{\tan(i+\psi)}{\tan^2\left(\dfrac{\pi}{4} - \dfrac{\phi}{2}\right)}.$$

See our Arts. 1590 and 1618.

In the remaining pages of this section Boussinesq deals with the cases in which the back of the wall and the talus are no longer plane. The results, owing to their complexity, do not seem likely to lend themselves to practical applications (pp. 127–33).

[1600.] In the present memoir Boussinesq speaks of the results (lx) and (lxi) as *approximations* to the true solutions. In later papers he takes the values of P given in them as providing *a lower or an upper limit* to the value of P according as the heterogeneity of the wedge (Art. 1599) is supposed to be such that the whole mass is either more or less stable than it really is. These limits can be reached by a proper choice of the values of ϕ and ϕ_1 in (lx) or (lxi): see our Arts. 1607 and 1616–8.

[1601.] § x. (pp. 134–56) is entitled: *Étude, en coordonnées polaires, de l'équilibre-limite (par déformations planes) d'une masse plastique ou pulvérulente comprimée. Applications à une masse annulaire, à un massif compris entre deux plans rigides qui se coupent.* This section of the memoir may be looked upon as a supplement to the memoirs dealt with in our Arts. 245–64.

Boussinesq supposes the inertia of the plastic or pulverulent mass as well as the body-forces to be negligible. He further supposes the flow to be uniplanar and a function only of the axial distance r and the longitude ϕ. We have from the equations in the footnote on our p. 79:

$$\left.\begin{array}{l} \dfrac{d\,\widehat{rr}}{dr} + \dfrac{d\,\widehat{r\phi}}{rd\phi} + \dfrac{\widehat{rr} - \widehat{\phi\phi}}{r} = 0, \\[2mm] \dfrac{d\,\widehat{r\phi}}{dr} + \dfrac{d\,\widehat{\phi\phi}}{rd\phi} + \dfrac{2\widehat{r\phi}}{r} = 0 \end{array}\right\} \quad \ldots\ldots\ldots\ldots\ldots \text{(lxii)}.$$

Now if $-p$ be half the sum and R half the difference of the principal tractions, and a the angle r makes with the algebraically least principal traction, then we have:

$$\left.\begin{array}{l} \widehat{rr} = -p - R\cos 2a, \quad \widehat{\phi\phi} = -p + R\cos 2a, \\[2mm] \widehat{r\phi} = -R\sin 2a \end{array}\right\} \quad \ldots\ldots \text{(lxiii)}.$$

Now by equation (xli) for limiting equilibrium

$$R = K + K_1 p \ldots\ldots\ldots\ldots\ldots\ldots\ldots \text{(lxiv)},$$

where for plasticity $K_1 = 0$, and for pulverulence $K = 0$ and $K_1 = \sin\phi$. By means of (lxii–lxiv) Boussinesq finds:

$$\left.\begin{array}{l} \dfrac{1 - (R_p)^2}{R}\dfrac{dp}{dr} = -\dfrac{2}{r}(\cos 2a - R_p)\left(1 + \dfrac{da}{d\theta}\right) - \dfrac{d\cos 2a}{dr}, \\[4mm] \dfrac{1 - (R_p)^2}{R}\dfrac{dp}{d\theta} = -2\sin 2a\left(1 + \dfrac{da}{d\theta}\right) - 2r(\cos 2a + R_p)\dfrac{da}{dr}, \end{array}\right\} \ldots\text{(lxv)}.$$

where $R_p = dR/dp$

Since R_p is a constant, if we differentiate the first equation with regard to θ and the second with regard to r, we shall then, by subtracting one of these equations from the other, obtain a differential equation involving a only. If this be solved the value of p can then easily be found by multiplying the first equation of (lxv) by dr and the second by $d\theta$, adding and integrating after substitution of the value of a.

[1602.] Boussinesq treats in particular the following special cases of *plasticity*:

(*a*) A belt of matter bounded by two coaxial, right-circular cylindrical surfaces subjected to normal pressures and by two smooth rigid planes perpendicular to its axis (pp. 137–9). In this case p and a are independent of θ. For the special case of plastic material Boussinesq's results agree with Saint-Venant's: see our Art. 261.

(*b*) Although the results of (*a*) are rigidly true only for the special conditions stated, Boussinesq considers them as approximately applicable to the case of a like belt placed upon a smooth plane, the interior surface of the belt being subjected to a pressure p_0 tending to extend it and the external surface and the upper face being free (p. 140). In this case we have from (ix) of our Art. 261 :

$$p_0 = 2K \log \frac{r_1}{r_0} \quad \dots \dots \dots \dots \dots \dots \text{(lxvi)},$$

where r_0 is the radius of the inner, r_1, of the outer surface.

(c) A hollow cylinder of which the base and exterior surface are placed in contact with rigid smooth surfaces, but of which the upper face is submitted to a *mean* pressure p_z and the interior surface to a pressure p_0 (pp. 140–2). The treatment is again only approximate. Boussinesq argues :

Alors la matière se dilate à la fois ou se contracte à la fois, et en moyenne presque également, dans deux sens rectangulaires normaux aux rayons r, tandis qu'elle éprouve par suite, suivant les rayons r, une contraction ou une dilatation moyennement doubles. On a donc presque $\widehat{zz} = \widehat{\phi\phi}$, c'est-à-dire que les éléments plans parallèles aux bases de l'anneau supportent des tractions $\widehat{zz}$ (positives ou négatives) assez peu différentes de celles qu'éprouvent, aux mêmes points, les plans méridiens (p. 141).

I see no reason why this should be true. Assuming its truth, however, we have:

$$-\widehat{zz} = p_0 \mp 2K \left(1 + \log \frac{r}{r_0}\right) \quad \dots \dots \dots \dots \text{(lxvii)} :$$

which is equal to the value of $\widehat{\phi\phi}$ given in equation (viii), of our Art. 261, but *not* to that of $\widehat{zz}$.

Hence, from the relation

$$\pi \left(r_1{}^2 - r_0{}^2\right) p_z = \int_{r_0}^{r_1} (-\widehat{zz}) \, 2\pi r dr,$$

we find :
$$p_z = p_0 \mp K \left(1 + \frac{2r_1{}^2}{r_1{}^2 - r_0{}^2} \log \frac{r_1}{r_0}\right) \dots \dots \dots \text{(lxviii)}.$$

If the radius r_0 is decreasing we must take the lower sign (p. 141).

For $p_z = 0$, as in this case r_0 is increasing, we take the upper sign, and find :

$$p_0 = K \left(1 + \frac{2r_1{}^2}{r_1{}^2 - r_0{}^2} \log \frac{r_1}{r_0}\right) \dots \dots \dots \dots \text{(lxix)}.$$

The results (lxvii)–(lxix) were first given by Tresca; his proof, however, is even less satisfactory than the above. See also our Art. 262.

(d) The action of a circular punch (pp. 142–4) Here Boussinesq applies in Tresca's manner, the above doubtful formulae to the two cases where the material to be punched rests on a rigid plane, either (i) with, or (ii) without, a circular orifice of the size of the punch and immediately opposite to its face in the plane.

Tresca's discussion of the cases (c) and (d) will be found on pp. 784–803 of his memoir of 1869: *Sur le poinçonnage des métaux...* published in the *Recueil des savants étrangers*, T. xx. Paris, 1872.

(e) A general solution of equations (lxv), when $da/d\phi = 0$, but not necessarily $dp/d\phi = 0$. No results of practical importance seem to flow from this assumption (pp. 145–6).

[1603.] (f) Case of $da/dr = 0$ in equations (lxv), or the least principal traction making the same angle with a radius-vector at every point of its length (pp. 145–56).

R_p being a constant, we have in this case:

$$\left. \begin{aligned} a + \phi &= C + \int \frac{c - R_p}{c - \cos 2a}\, da, \\ (1 - R_p{}^2) \int \frac{dp}{R} &= C' - (c - Rp) \log \{r^2 (c - \cos 2a)\} \end{aligned} \right\} \quad \ldots\ldots \text{(lxx)},$$

where C, C', and c are arbitrary constants.

Boussinesq applies results (lxx) to the consideration of the special case, where a cylindrical sector of material is placed between two intersecting rigid planes, sufficiently rough to stop all sliding of the particles of the material touching them. The application to the case where the rigid planes are hinged to a common axis and squeeze a wedge of plastic material placed between them might possibly have some practical interest.

[1604.] To the memoir is appended a *Note Complémentaire: Sur la méthode de M. Macquorn-Rankine, pour le calcul des pressions exercées aux divers points d'un massif pesant*......(pp. 157–173). The method dealt with by Boussinesq is that discussed in Rankine's memoir *On the Stability of Loose Earth* (see our Art. 453), but as it does not start from an elastic hypothesis, we have not considered it in our volume. Boussinesq explains and supplements Rankine's investigations, but remarks of the hypothesis by which Rankine solves his fundamental equation:

Peut-être trouvera-t-on un jour quelque ordre de phénomènes auquel l'hypothèse considérée sera plus applicable, et qui réalisera ainsi cette curieuse analogie d'une distribution de pressions avec le mouvement de la chaleur dans une barre (173).

The memoir concludes with two notes, the first of which deduces Hopkins' theorem for the value and direction of the maximum shear (see our Art. 1368*), while the second deduces Saint-Venant's theorem for the direction and magnitude of the maximum slide: see our Arts. 1570* and 4 (δ).

[1605.] Various papers on the stability and thrust of loose earthwork by Boussinesq, briefly resuming, generalising or simplifying the results of the above memoir, will be found in volumes of the *Comptes rendus*.

(*a*) T. LXXVII., pp. 1521–5. Paris, 1873. An extract from the Brussels memoir under the same title as the latter: see our Arts. 1571–99.

(*b*) *Sur les lois de la distribution plane des pressions à l'intérieur des corps isotropes dans l'état d'équilibre limite.* T. LXXVIII., 757–9 and 786–9. Paris, 1874.

In the first part of this paper Boussinesq supposes a conservative system of body-forces applied to a mass under uniplanar strain. The first two body-stress equations then become:

$$\frac{d\widehat{xx}}{dx} + \frac{d\widehat{xy}}{dy} + \frac{d\phi}{dx} = 0, \quad \frac{d\widehat{xy}}{dx} + \frac{d\widehat{yy}}{dy} + \frac{d\phi}{dy} = 0 \dots\dots\dots (i),$$

ϕ being the potential of the body-forces.

Hence we find, as in our Arts. 1576–7, for an elastic body:

$$\frac{d^2\widehat{xx}}{dx^2} + \frac{d^2\widehat{yy}}{dy^2} - 2\frac{d^2\widehat{xy}}{dxdy} - \frac{\lambda}{2\lambda+2\mu}\left(\frac{d^2}{dx^2} + \frac{d^2}{dy^2}\right)(\widehat{xx}+\widehat{yy}) = 0 \dots (ii),$$

and for a pulverulent or plastic mass:

$$\left(\frac{d^2}{dx^2} - \frac{d^2}{dy^2}\right)\left(\frac{\widehat{xx}-\widehat{yy}}{\widehat{xx}+\widehat{yy}}\right) + 4\frac{d^2}{dxdy}\left(\frac{\widehat{xy}}{\widehat{xx}+\widehat{yy}}\right) = 0 \dots\dots\dots (iii).$$

The remainder of the first part of the memoir is devoted to the condition of limiting equilibrium discussed in our Art. 1585.

(*c*) The second part of the memoir is devoted to discussing the integration of equations (i) for the case of limiting equilibrium.

Boussinesq, as in our Art. 1568, takes T_1 and T_2 for the principal tractions, and puts $p = -\frac{1}{2}(T_1 + T_2)$, $q = \frac{1}{2}(T_1 - T_2)$, then for limiting equilibrium q will be a function of p and generally a linear one. We have, if a be the angle the greater traction T_1 makes with the axis of x:

$$\widehat{xx} = -p + q\cos 2a, \quad \widehat{yy} = -p - q\cos 2a, \quad \widehat{xy} = q\sin 2a.$$

Hence from (i), if $P = p - \phi$:

$$\left.\begin{array}{l} \dfrac{d(P-q)}{dx}\cos a + \dfrac{d(P-q)}{dy}\sin a - 2q\left(-\dfrac{da}{dx}\sin a + \dfrac{da}{dy}\cos a\right) = 0, \\[3mm] -\dfrac{d(P+q)}{dx}\sin a + \dfrac{d(P+q)}{dy}\cos a - 2q\left(\dfrac{da}{dx}\cos a + \dfrac{da}{dy}\sin a\right) = 0 \end{array}\right\}\dots(iv).$$

Boussinesq remarks that (iv) can be solved when 1° q is a constant, i.e. $2K$: see our Art. 1863, and 2° when $\phi = 0$, or the weight of the material is negligible as compared with the pressures to which it is subjected. In both these cases q is a given function (very approximately linear) of the sole variable $p - \phi$ or P, and (iv) contains therefore only the variables P and a.

Boussinesq now proceeds to take P and a as the independent variables, or solves the equations (iv) for x and y in terms of P and a. In other words he finds the point at which there is a given condition of stress.

The solution is an interesting piece of analytical investigation, but the results seem too complicated to be of very great practical service. They are used however in the investigations of the following papers (d).

(d) *Sur les modes d'équilibre limite les plus simples que peut présenter un massif sans cohésion fortement comprimé.* T. LXXX., pp. 546-9 and pp. 623-7. Paris, 1875. These papers discuss at considerable length of analysis the cases dealt with in Arts. 52-3 of the Brussels memoir. The processes adopted in the later treatment seem very much simpler than those of the *Comptes rendus* investigation.

[1606.] In the *Minutes of Proceedings of the Institution of Civil Engineers*, Vol. LXV., pp. 140-241 (London, 1881), will be found a paper by Sir Benjamin Baker entitled : *The Actual Lateral Pressure of Earth-work*, together with a discussion and correspondence. The paper itself deals with some sixty-five actual examples of retaining walls and of the pressure of earthwork. It points out the defect of Rankine's theory in supplying accurate data for practical construction, but suggests empirical rules rather than an improved theory as the best way out of the difficulty.

In the *Correspondence* will be found (pp. 209-12) an application by Flamant of Boussinesq's theory as published in his *Essai* (see our Art. 1571) to a special case cited by Sir B. Baker. This is followed on pp. 212-23 by some discussion by Boussinesq himself of the case of a level mass of homogeneous material supported by a vertical wall. He follows the lines of his *Essai*, Articles 43-8, see our Arts. 1596-1600, giving as his equation (15) p. 218, equation (lxi) of our Art. 1599 for P. The thickness of the wall is then easily found by the method of Article 41 of the *Essai* (see our Art. 1592). If h be the height, b the minimum breadth, ρ the density of the earth, ρ' that of the wall, we have[1] (p. 219):

$$\frac{h}{b} = \frac{3}{2} \left[\tan \phi_1 + \sqrt{\tan^2 \phi_1 + \frac{4\rho'}{3\rho k}} \right] \dots\dots\dots (i),$$

where k is given by equation (lxi) of our Art. 1599, and ϕ_1 is the angle of friction of the earthwork against the wall: see our Art. 1597.

[1607.] Now in order to make the above solution *exact* Boussinesq supposes that in the neighbourhood of the wall the angle of interior friction ϕ takes slightly higher values than at other parts of the

---

[1] This result follows at once from Art. 1599, if we take moments for all the forces acting on the wall round the axis about which it would rotate if it capsized as a whole.

pulverulent mass, at which it is constant, and he denotes these values by ϕ': see our Art. 1599. It is shewn that the maximum value of ϕ' is given by the formula (p. 221 : see our Art. 1616):

$$\sin^2 \phi' = \sin^2 \phi + (1 - \sin \phi)^2 \tan^2 \phi_1 \dots\dots\dots\dots(ii).$$

Boussinesq further considers that his equations in the actual case slightly exaggerate the influence of the interior friction and so lead to a slightly too small value of b.

To be quite safe and to obtain a limit of b too high, values might be given to ϕ and to ϕ_1, slightly less than the true ones, by calculating these lesser values as if the maximum of the variable angle of interior friction ϕ', supposed by the formulae, were just equal to the true value of the angle of friction of the earthwork in question ; for in that case the latter would be more stable than the theory supposed (p. 221).

After some discussion as to how this may be done, Boussinesq considers that a higher limit will be found for b by taking

$$\sin \phi = \sin \phi_1 = \frac{\sin \phi' + \sqrt{8 + \sin^2 \phi'}}{4} \sin \phi' \dots\dots\dots(iii).$$

Here ϕ' is to be put equal to the known interior angle of friction, and then the values of ϕ and ϕ_1 substituted in the value of k and in (i).

For the particular example, $\rho'/\rho = 1$ and $\phi = 45°$, Boussinesq finds, on his earlier hypothesis :

$$(\phi = 45°, \quad \phi_1 = 45°) \qquad h/b = 6.69 ;$$

on Rankine's hypothesis of a smooth wall :

$$(\phi = 45°, \quad \phi_1 = 0) \qquad h/b = 4.18 ;$$

on his modified hypothesis, using (iii) :

$$(\phi' = 45° \text{ and } \phi = \phi_1 = 39° 49' \text{ by (iii)}) \qquad h/b = 5.79.$$

According to Boussinesq therefore we must have :

$$b \text{ between } \frac{h}{6.69} \text{ and } \frac{h}{5.79},$$

and he says we may take it equal to 1/6. Sir B. Baker's rule as a result of his experience was to take $b = h/3$ (p. 184), or almost *double* the value given by Boussinesq's theory, supposing no factor of safety to be used in the latter. Boussinesq still further modifies this superior limit in a late paper : see our Art. 1618.

It does not seem to me that Boussinesq's modified hypothesis is entirely satisfactory from the theoretical standpoint ; it involves a number of disputable points : see *Proceedings*, pp. 221–3.

[1608.] *Note sur la détermination de l'épaisseur minimum que doit avoir un mur vertical, d'une hauteur et d'une densité données, pour contenir un massif terreux, sans cohésion, dont la surface supérieure est horizontale. Annales des ponts et chaussées,* T. III. pp. 625–43. Paris, 1882. This is perhaps the best and clearest account Boussinesq had yet published of the application of his method of treating earthwork to the special problem of a uniform vertical wall supporting a pulverulent mass with a horizontal surface. It may be looked upon as a slight expansion of the paper contributed to the *Institution of Civil Engineers* (see our Art. 1606), and it practically gives a complete treatment of this case independently of the results reached in the *Essai* (see our Art. 1599). The same objections may of course be raised as to the manner in which ϕ' is determined. We shall pass this memoir by, however, as Boussinesq did not give his final and complete treatment till 1884.

[1609.] *Note on Mr G. H. Darwin's Paper 'On the Horizontal Thrust of a Mass of Sand.' Minutes of Proceedings of the Institution of Civil Engineers,* Vol. LXXII. pp. 262–71. London, 1883. Boussinesq considers that the value, 35°, adopted by Darwin does not represent the angle ϕ of interior friction of sand, but concludes partly from experimental, partly from theoretical considerations, that it should be taken as 40°·5. With this value of ϕ Boussinesq shews that the results of Darwin's Experiments, Series I.—IV., are fairly in accord with the theory developed in our Arts. 1599 and 1607.

Boussinesq next turns (pp. 266–7) to Darwin's Experiments, Series VI., where the talus had a slope equal to 35° or to the angle of friction at the surface. In this case Boussinesq refers to pp. 125–6 of his *Essai* for the expression for the thrust on the wall : see our Art. 1599. The value there given, however, is only an inferior limit. Boussinesq now develops the theory of the *Essai* with a view to the discovery of a superior limit. He considers that this can be obtained by giving ϕ a value derived from

$$\sin \phi / \cos \delta = \sin \phi',$$

ϕ' being the interior angle of friction (= 40°·5 for sand) and δ the angle defined in our Art. 1599. The assumption is defended in the same manner as that for the value of ϕ in the case of a horizontal talus : see our Art. 1607.

It does not seem to me, however, that Boussinesq's discussion of Darwin's experiments in the case of Series VI. can be looked upon as on the whole favourable to Boussinesq's theory, and I can hardly agree with the concluding remarks of the author :

Mr G. H. Darwin's valuable observations appear to confirm as fully as possible the Author's formulas for the thrust of a pulverulent mass in limiting equilibrium. These formulas are due to Rankine's principles, simply developed and completed by the addition of the element of slip of the mass against the wall sustaining it, and constituting in this form the rational and corrected expression of principles due to Coulomb himself. Coulomb's theory, in all cases where it is justifiable to apply his fundamental hypothesis of a plane rupture of the mass, gives identically the same results as Rankine's formulas as has been shewn by M. Maurice Lévy. It will then be found that these instances........are just those in which the author's formulas merge into those of Rankine, in such a way as to represent all that may now be retained of the old theory of Coulomb (p. 270).

The contents of this *Note* by Boussinesq on Darwin's experiments appear also in an article contributed to the *Annales des ponts et chaussées*, T. VI., 2ᵉ *Semestre*, pp. 494–510. Paris, 1883.

[1610.] *Résumé d'articles publiés par la Société des ingénieurs civils de Londres sur la poussée des terres.* Annales des ponts et chaussées, *Mémoires*, T. VI., 2ᵉ *Semestre*, 1883, pp. 477–532. Paris, 1883. This paper is by Flamant. It considers matter to which we have already referred, *i.e.* the memoir of Darwin with the notes of Boussinesq, Gaudard and others.

Boussinesq gives, pp. 510–24, an *Addition relative aux expériences de M. Gobin.* This has reference to a long paper by Gobin on pp. 98–231 of the same volume of the *Annales*. The theoretical basis of Gobin's investigations seems to be very doubtful. His hypotheses are briefly resumed by Boussinesq on p. 511. The experimental part of Gobin's memoir occupies pp. 184–212, and Boussinesq in his paper compares Gobin's results with his own theory :

En résumé, les expériences de M. Gobin s'accordent parfaitement avec celles de M. G. Darwin, pour confirmer la théorie de l'équilibre-limite des terres exposée au § IX. de mon *Essai* publié en 1876 (p. 524).

See our Art. 1609.

[1611.] *Formules simples et très approchées de la poussée des terres, pour les besoins de la pratique.* Comptes rendus, T. XCIX., pp. 1151–3. Paris, 1884. In this paper Flamant, after pointing out that Boussinesq

a établi la parfaite concordance avec les faits d'expériences, constatés surtout en Angleterre par M. Darwin et en France par M. Gobin, de sa théorie de l'équilibre des massifs pulvérulents ou sans cohésion (p. 1151),

remarks that he had formed the idea of preparing tables of the thrust for the cases most commonly occurring in practice, where the mass adheres to the wall (*i.e.* $\phi_1 = \phi$). In the course of his calculations, however, he discovered that for a horizontal talus and vertical wall, the *vertical* component of the thrust, or with the notation of our Art. 1599,

$$k \tan \phi \, \frac{g\rho L^2}{2},$$

is almost a constant for values of ϕ from 20° to 33°, and then

$$= \cdot 16 \, \frac{g\rho L^2}{2},$$

and that for values of ϕ from $33°$ to $45°$, it diminishes so slowly as to be equal to

$$\cdot14\,\frac{g\rho L^2}{2}, \text{ when } \phi = 45°.$$

For the more general case of a wall of height h ($= L\cos i$) with an internal slope of i to the vertical and of a talus sloping at ω to the horizontal, there still exists a direction in which the component of the thrust is sensibly equal to

$$\cdot16\,\frac{g\rho}{2}\left(\frac{h}{\cos i}\right)^2,$$

but this direction varies with ϕ and makes very approximately an angle

$$\chi = \frac{\omega}{2} + \frac{i}{4}\left(\frac{\phi}{10°} - 1\right),$$

with the back of the wall for all values of ϕ (measured in degrees) between $20°$ and $45°$, of i less than $20°$ and of ω less than $\phi - i$. Within these limits this component does not differ from the above constant value by $1/10$ of its value.

For $i > 15°$, Flamant says that a closer approximation to the angle χ will be found from

$$\chi = \frac{\omega}{2} + \frac{i}{4}\left(\frac{\phi}{12°} - 1\right).$$

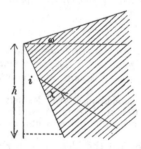

Since the resultant thrust acts at a third of the depth of the wall from its base and makes as a rule the angle of friction $\phi_1 = \phi$, with the normal to the wall it is possible by the simplest graphical construction to obtain from the above known component the resultant thrust.

[1612.]   In the *Comptes rendus*, T. XCVIII. (Paris, 1884), will be found the following memoirs by Boussinesq:

(a) *Sur la poussée d'une masse de sable, à surface supérieure horizontale, contre une paroi verticale ou inclinée,* pp. 667–70.

(b) *Sur la poussée d'une masse de sable, à surface supérieure horizontale, contre une paroi verticale dans le voisinage de laquelle son angle de frottement intérieur est supposé croître légèrement d'après une certaine loi,* pp. 720-3.

(c) *Calcul approché de la poussée et de la surface de rupture, dans un terre-plein horizontal homogène, contenu par un mur vertical,* pp. 790-3.

These papers put in an easy form the approximate integration of the differential equations for a pulverulent mass supported by a vertical wall and having a horizontal talus. They give Boussinesq's theory in its final form. The method of integration had been suggested by Saint-Venant in a report on Lévy's memoir (see our Art. 242) and had been first carried out by Boussinesq for a more general case in a rather complicated manner in his *Essai:* see our Art. 1599. We have already referred to the results of the integration as given in the memoirs discussed in our Arts. 1606-11.

We propose here to consider Boussinesq's method of integration at slightly greater length for this special case.

[1613.] Let the rear-side of the wall make an angle i with the vertical; let the origin O be taken at the intersection of this face with the talus, Oy being the horizontal line perpendicular to the trace of the

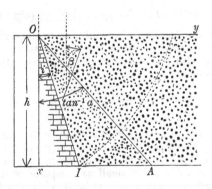

talus on this face, and Ox being vertical. Let ϕ be the interior angle of friction, and let

$$a^2 = (1 - \sin \phi)/(1 + \sin \phi) = \tan^2 (45° - \tfrac{1}{2}\phi).$$

23—2

Now if the wall be about to collapse and the pulverulent mass accordingly in its limiting condition of equilibrium (*l'état ébouleux*) we have:

$$\sin^2 \phi = \frac{(\widehat{xx} - \widehat{yy})^2 + 4\widehat{xy}^2}{(\widehat{xx} + \widehat{yy})^2} \quad \dots\dots\dots\dots \dots\dots\dots\dots(\epsilon),$$

at every point: see our Art. 1596. Hence the equations to be solved become, if ρ be the density of the mass:

$$\left.\begin{array}{c} \dfrac{d\widehat{xx}}{dx} + \dfrac{d\widehat{xy}}{dy} + g\rho = 0, \quad \dfrac{d\widehat{xy}}{dx} + \dfrac{d\widehat{yy}}{dy} = 0, \\[2mm] (\widehat{xx} - a^2\widehat{yy})(\widehat{yy} - a^2\widehat{xx}) = (1 + a^2)^2\,\widehat{xy}^2 \end{array}\right\} \dots\dots\dots\dots(i).$$

Now at the free surface $\widehat{xx} = \widehat{xy} = 0$, and therefore by the third equation of (i): $\widehat{yy} = 0$. At the wall, where we may suppose a thin coating of the pulverulent mass to adhere, we must have the ratio of the tangential component of the reaction to the normal component equal to $\tan \phi$. If no coating of the mass adheres to the wall this ratio must equal $\tan \phi_1$, the angle of friction between the wall and the mass.

Now the following is a solution of (i):

$$\widehat{xx} = -g\rho x, \quad \widehat{xy} = 0, \quad \widehat{yy} = -g\rho a^2 x \dots\dots\dots\dots\dots(ii).$$

This satisfies the surface-condition at the talus, but not that at the wall except for special values of ϕ_1 or i. For example if $i = 0$, we must have $\phi_1 = 0$, or the wall perfectly smooth. Herein lies the inconsistency of Rankine and Lévy's solution for the stability of pulverulent masses: see our Arts. 1596–8.

Now Boussinesq introduces into the values of the stresses as given by (ii), additional small terms with a view of making them exact. Thus let

$$\widehat{xx} = -g\rho(x + t), \quad \widehat{xy} = g\rho s, \quad \widehat{yy} = -g\rho(a^2x + r) \dots\dots\dots(iii),$$

where t, s and r are functions of x and y to be determined. The first two equations of (i) shew us that r, s, t are the three differentials of a single function ϖ with regard to dx^2, $dxdy$, dy^2 respectively. The last equation of (i) leads to

$$r - a^2 t = \frac{(1 + a^2)^2\, s^2}{(1 - a^4)\, x + t - a^2 r} \quad \dots\dots\dots\dots\dots(iv).$$

[1614.] Boussinesq now (p. 669) enters upon an investigation with a view to shewing that the right-hand side of (iv) may be put zero. It seems to me that this follows at once if we neglect the squares of r, s, and t, and consider t and r small as compared with x. The only difficulty which arises occurs when x is itself small, and I do not clearly follow Boussinesq's reasoning as to this point. *Assuming* it to be correct we have:

$$r - a^2 t = 0.$$

Hence

$$r = a^2 (f'' + f_1''), \quad s = -a (f'' - f_1''), \quad t = (f'' + f_1'') \ldots\ldots\ldots(\mathrm{v}),$$

where f and f_1 are arbitrary functions of $y - ax$ and $y + ax$ respectively.

Now $r = s = t = 0$ when $x = 0$ and $y > 0$, hence f'' and $f_1'' = 0$ for all positive values of their variables $y \mp ax$. This includes all possible values of these variables in the pulverulent mass, if $\tan i$ is $> a$. If on the other hand $\tan i$ is $< a$, or $i < 45° - \frac{1}{2}\phi$, there will be a wedge comprised between the rear-side of the wall and the line $y = ax$ where, although f_1'' is still zero, f'' will exist because its variable $y - ax$ is negative.

The more usual case is that in which i is $< 45° - \frac{1}{2}\phi$. In this case r, s and t will be zero all over the plane $y = ax$, and we can take as the solution for all points in the wedge IOA (see figure Art. 1613) between the planes i and $\tan^{-1} a$

$$r = a^2 f'', \quad s = -af'', \quad t = f'' \ldots\ldots\ldots\ldots\ldots(\mathrm{vi}),$$

provided these do not give values of r, s, t comparable with that of x. They hold with decreasing exactness as we pass from the plane $y = ax$ where $r = s = t = 0$ towards the face of the wall (p. 670).

[1615.] Boussinesq remarks (pp. 720-1) that of the two 'surfaces of rupture' through the bottom I of the wall one must lie in the solid angle between the planes i and $\tan^{-1} a$ (for $i < 45° - \frac{1}{2}\phi$); for otherwise, if they both passed out of this angle, they would in the upper portion, where $r = s = t = 0$, become as in Rankine's theory directions parallel to the planes $y = \pm ax$, but these meet without entering the above solid angle wherein I lies. Accordingly Boussinesq considers it natural to suppose that one of the rupture-surfaces takes the line of the wall. The pulverulent mass will then be on the point of slipping at the wall, and if the angle of friction between wall and mass be taken $= \phi_1$, we shall have a condition to determine the arbitrary function $f'' (y - ax)$. All the equations will then be fully satisfied except (iv), which we have supposed to reduce to $r - a^2 t = 0$. In its unreduced form (iv) is not satisfied by a homogeneous mass in the region IOA, (see the figure, Art. 1613). Boussinesq now points out that this is the only equation which involves ϕ, and he suggests that ϕ be given a variable value ϕ' in this region. This value ϕ' will be supposed to differ from ϕ more and more as we pass from the plane OA and approach the wall, being taken to satisfy (iv), or what is the same thing the equation (ϵ) of Art. 1613, which may then be written :

$$(\widehat{xx} + \widehat{yy})^2 \sin^2 \phi' = (\widehat{xx} - \widehat{yy})^2 + 4\widehat{xy}^2 \ldots\ldots\ldots\ldots(\mathrm{vii}).$$

The values of the stresses given by (iii) and (vi) will then be exact for material homogeneous in yOA and heterogeneous in the manner indicated by (vii) in AOI.

Let β be the acute angle, reckoned positive on the side of y positive, which the plane subjected to the least pressure makes with the vertical measured upwards, then by our Art. 1578 :

$$\tan 2\beta = 2\widehat{xy}/(\widehat{yy} - \widehat{xx}).$$

It follows from (vii) by using (vi) and the value of a^2 that :

$$\sin \phi' \cos 2\beta = \sin \phi,$$

ϕ' is thus always greater than ϕ.

Boussinesq then shews (p. 721) that the surface of rupture which starts from the base of the wall will have its several elements inclined to the vertical at an angle a given by

$$a = \frac{\pi}{4} - \frac{\phi'}{2} + \beta.$$

He further demonstrates that the surface of rupture through the bottom of the wall when it ceases at the plane $y = ax$ to be plane (*i.e.* parallel to $y + ax = 0$) becomes concave to the upward vertical through the bottom of the wall.

[1616.]   The remainder of the memoir confines itself to the case of $i = 0$. In this case when $y = 0$, we must have, ϕ_1 being angle of friction between wall and pulverulent mass :

$$\tan \phi_1 = - \widehat{xy} / \widehat{yy},$$

$$= -f'' / \{a (x + f'')\}, \text{ by (iii) and (vi).}$$

Whence we find

$$f'' (-ax) = \frac{-ax}{a + \cot \phi_1},$$

or,

$$f'' (y - ax) = \frac{y - ax}{a + \cot \phi_1}.$$

The normal pressure on the wall is now easily determined to be :

$$g\rho x a^2 \left(\frac{\cot \phi_1}{a + \cot \phi_1} \right) = kg\rho x \dots\dots\dots\dots\dots (\text{viii}),$$

where,

$$k = \frac{a^2}{1 + a \tan \phi_1} \dots\dots\dots\dots\dots\dots (\text{ix}).$$

Supposing $y/x = \tan \theta$, we find by substituting the values of the stresses in (vii) when $\tan \theta < a$:

$$\sin^2 \phi' = \sin^2 \phi + \cos^2 \phi \left(\frac{a - \tan \theta}{\cot \phi_1 + \tan \theta} \right)^2 \dots\dots\dots (\text{x}).$$

Thus ϕ' increases as θ diminishes, or takes its greatest value Φ at the wall, or remembering the value of a :

$$\left. \begin{array}{l} \sin^2 \Phi = \sin^2 \phi + (1 - \sin \phi)^2 \tan^2 \phi_1, \\[2mm] \tan \phi_1 = \dfrac{\sqrt{\sin^2 \Phi - \sin^2 \phi}}{1 - \sin \phi} \end{array} \right\} \dots\dots\dots (\text{xi}).$$

whence,

Boussinesq shews that for a given value of x the least value of the normal pressure in (viii), or of $\dfrac{a^2}{1 + a \tan \phi_1}$, is to be found for a value of $\sin \phi$ lying between $\sin \Phi$ and $\sin^2 \Phi$ (pp. 722–3).

He finishes the second section of the memoir by deducing certain results for the curved part of the surface of rupture on the assumption that it is a circular arc.

[1617.] In the third section (p. 790) Boussinesq points out how the conditions of the equilibrium limit are affected by the presence of a second bounding wall parallel to the axis of z. In this case the function $f_1''(y + ax)$ must be retained, and the results for the special case of two *vertical* and parallel walls are given without analysis.

[1618.] In the fourth section of the memoir, Boussinesq shews how the above formulae,—obtained for the case of a mass for which the interior angle of friction ϕ' increases slightly towards the wall from the value ϕ to Φ,—may be applied in practice to obtain to a sufficient degree of approximation the thrust of a homogeneous mass with a constant angle of friction.

In the first place Boussinesq points out that the normal component of the thrust on the wall, upon which the overturning effect really depends, will be decreased if the angle of internal friction is increased and *vice versa*.

Il suffit donc d'imaginer deux massifs hétérogènes constitués conformément à la formule (x), dans l'un desquels ϕ' et ϕ_1 soient égaux ou un peu supérieurs à l'angle de frottement du massif homogène donné, tandis qu'ils lui seront un peu inférieurs dans l'autre, pour que le moment de la poussée soit moindre, dans le premier, et plus grand, dans le second, qu'il n'est dans le massif proposé (p. 791).

We thus obtain inferior and superior limits k_1 and k_2 of k in (ix), and the mean of them if the limits be taken sufficiently near will constitute a close practical solution of the problem.

If we want as high an inferior limit k_1 of k as possible, then, since k increases when ϕ or ϕ_1 decreases, the first of the heterogeneous masses should be obtained by making ϕ and ϕ_1 as small as possible, consistent with their being equal to or greater than the real value of the angle of internal friction of the real mass, or by making them both equal to ϕ. We then find (see equation (lxi) of our Art. 1599):

$$k_1 = \frac{\tan(45° - \tfrac{1}{2}\phi)\sin(45° - \tfrac{1}{2}\phi)\cos\phi}{\cos(\tfrac{3}{2}\phi - 45°)} \quad\dots\dots\dots\dots(\text{xii}).$$

On the other hand for the superior limit k_2, we must make the interior angle of friction ϕ' of the second heterogeneous mass less than the true ϕ, and this will be done by putting the maximum value of ϕ', i.e. Φ, equal to the true ϕ, and therefore ϕ' will be less than this real value ϕ at other points. Accordingly we make the value of k, as given by (ix) with $\tan \phi_1$ substituted from (xi), the least possible for variations in ϕ. Afterwards the maximum Φ must be replaced by the angle of interior friction ϕ. Thus to obtain k_2 we seek the minimum value of k as given by :

$$\frac{1}{k} = \frac{1 + \sin \phi}{1 - \sin \phi} + \sqrt{\frac{1 + \sin \phi}{1 - \sin \phi}} \, \frac{\sqrt{\sin^2 \Phi - \sin^2 \phi}}{1 - \sin \phi} \quad \ldots\ldots(\text{xiii}).$$

Boussinesq (p. 792) finds that $\frac{1}{k}$ in (xiii) is a maximum for

$$\tan \phi = \frac{2 \sqrt{2} \tan \Phi}{\sqrt{9 + \tan^2 (45° - \tfrac{1}{2}\phi)}} \quad \ldots\ldots\ldots\ldots(\text{xiv}).$$

The solution of (xiv) for ϕ, Boussinesq says, is very nearly

$$\tan (\phi + 2') = \frac{2 \sqrt{2} \tan \Phi}{\sqrt{9 + \tan^2 (45° - \tfrac{1}{2}\Phi)}} \quad \ldots\ldots\ldots(\text{xv}).$$

After some reductions we have from (xiii) and (xiv) :

$$k_2 = \tfrac{1}{4} \tan^2 (45° - \tfrac{1}{2}\phi) \{3 + \tan^2 (45° - \tfrac{1}{2}\phi)\} \quad \ldots\ldots(\text{xvi}),$$

where ϕ is to be given its value as obtained from (xiv), or approximately from (xv), Φ being put equal to the angle of friction of the homogeneous mass. The true normal thrust will then be

$$\tfrac{1}{2}(k_1 + k_2) g\rho x \ldots\ldots\ldots\ldots\ldots\ldots(\text{xvii}).$$

The calculation of the superior limit of k is thus clearly a complicated matter, and the real accuracy of Boussinesq's theory will depend upon the closeness to each other of k_1 and k_2.

Boussinesq concludes by stating that the numbers obtained from these results are in agreement with the experiments of Darwin and Gobin : see our Arts. 1609–10.

[1619.] On pp. 850–2 of the *Comptes rendus*, T. XCVIII. (Paris, 1884), will be found a notice by Saint-Venant of Boussinesq's theory of the thrust of loose earth. After shewing how Boussinesq's investigation improves on those of Rankine and Lévy, and pointing out how much closer it agrees with experiment, Saint-Venant concludes:

Les nouvelles recherches de M. Boussinesq rendent à l'art des constructions, où les économies possibles et sans danger ont tant d'importance, un service réel, et on peut les regarder comme fournissant aux ingénieurs des moyens de calcul qui répondront, pour bien longtemps, à ce qui était désiré dans la question énoncée (p. 852).

[1620.] *Sur le principe du prisme de plus grande poussée, pose par Coulomb dans la théorie de l'équilibre-limite des terres. Comp-*

tes rendus, T. XCVIII., pp. 901–4. Paris, 1884. In this memoir
Boussinesq deduces from his theory of the limiting equilibrium of
pulverulent matter—

...la propriété suivante de maximum, qui est comme l'expression
développée du principe du prisme de plus grande poussée, émis et si
ingénieusement utilisé par Coulomb en 1773 (*Savants étrangers*, T. VII.,
Essai sur une Application des règles de Maximis et Minimis *à quelques
Problèmes de statique*, pp. 343–82 : see p. 359) : la poussée exercée effectivement sur la paroi mobile continuerait à s'y exercer si le massif
pulvérulent se terminait à la surface de rupture la plus éloignée de la
paroi, toute la masse sous-jacente devenant solide, et elle est la plus
forte des poussées qui ont lieu, à l'état d'équilibre-limite, quand le
massif se trouve limité ainsi par une surface rugueuse quelconque allant
du bas de la paroi mobile à la surface libre (p. 901).

Unfortunately we do not know the surface of rupture, and to
try all possible surfaces would be a more complicated process than
integrating approximately the differential equations for the equilibrium limit as Boussinesq has done. Coulomb assumed that the
surface was a plane, which is *à priori* arbitrary as well as inexact.
The thrust determined by this assumption can therefore only be
considered roughly approximate in default of any better method.

[1621.] *Complément à de précédentes Notes sur la poussée des terres.*
Annales des ponts et chaussées. Mémoires, T. VII., 1ᵉʳ *Semestre*, pp. 443–81.
Paris, 1884.

Boussinesq begins by citing the results of his *Essai* (see our Art.
1599) and then applies them to the case of a horizontal talus.

Let k be the coefficient which occurs in the expression for the
normal pressure on the wall (see equation (lxi) of our Art. 1599), then
Boussinesq points out that the method of the *Essai* and of the memoir
of 1881 (see our Art. 1607) does not give a sufficiently close superior
limit to the value of k.

[1622.] Let the superior limits to k as given by the values of ϕ
from (iii) of our Art. 1607 substituted in (xii) of our Art. 1618, and
from (xiv) of our Art. 1618 substituted in (xvi) of the same article be
k_2' and k_2 respectively, and let the inferior limit as given by (xii) of our
Art. 1618 with ϕ its true value be k_1.

Boussinesq states (p. 451) that he has found by taking a sufficient
number of values of Φ from $20°$ to $50°$ that very approximately :

$$k_2 - k_1 = \tfrac{9}{11}(k_2' - k_1)\dots\dots\dots\dots\dots\dots\dots\text{(xviii)}.$$

---

[1] There appears to be an almost verbal reprint of this memoir on pp. 975–78 of
the same volume of the *Comptes rendus*.

Hence the mean, $\frac{1}{2}(k_1 + k_2)$, adopted by Boussinesq as the value of k (see our Art. 1618):

$$= k_1 + \tfrac{9}{22}(k_2' - k_1) \dots\dots\dots\dots\dots\text{(xix)}.$$

Or, to obtain the value of the coefficient k, we need not for practical purposes calculate the difficult k_2, but can deduce the approximate solution from the easily found k_1 and k_2'. Obviously this new approximation consists in subtracting $\frac{1}{11}(k_2' - k_1)$ from the old approximation $\frac{1}{2}(k_1 + k_2')$ obtained by the method of Art. 1607.

[1623.]   On pp. 453–5 of the memoir, Boussinesq applies (xix) to M. Gobin's experiments (see our Art. 1610), and on pp. 455–7 to G. H. Darwin's experiments (see our Art. 1609).   In both cases theory now approaches closer to experiment, but it may be questioned whether the agreement is still a sufficient one.

Boussinesq next turns to some experiments of Audé recorded in the *Mémorial du génie*, No. 15, p. 269, 1848, and also cited in Note v of a memoir by M. Saint-Guilhem, *Annales des ponts et chaussées. Mémoires*, T. xv. pp. 340–5.   Paris, 1858.   The discussion occupies pp. 457–69, but the results are not in my opinion more decisive with regard to Boussinesq's theory than those of Darwin and Gobin.   On pp. 469–73 we have a discussion of the angle given by theory for the inclination of the surface of rupture to the vertical and comparison with the experimental value determined in certain cases by Gobin (see our Art. 1615).

The remainder of the memoir (pp. 473–81) is occupied with a discussion of Coulomb's prism of greatest thrust, after the manner of the *Comptes rendus* article : see our Art. 1620.

[1624.]   *Sur l'intégration, par approximations successives d'une équation aux dérivées partielles du second ordre, dont dépendent les pressions intérieures d'un massif de sable à l'état ébouleux. Mémoires de la Société des Sciences de Lille*, T. XIII., pp. 705–12.   Lille, 1885. This memoir is also included in the *Application des potentiels... ..*: see our Art. 1561.

We have seen how Boussinesq makes the pressure on a vertical, or moderately inclined wall, of a mass of earth with horizontal talus depend on the solution of the differential equation (see our Art. 1613):

$$r - a^2 t = \frac{(1 + a^2)^2 s^2}{(1 - a^4) x + t - a^2 r} \dots\dots\dots\dots\text{(i)}.$$

We have noted how by neglecting the right-hand side of (i) (if s, r, t are small as compared with x), he obtains as a *first* approximation a solution of the form :

$$\varpi = f(y - ax) + f_1(y + ax).$$

If the forms of f and f_1 are known then by substituting on he right-hand side of (i) we obtain for a *second* approximation the equation

$$r - a^2 t = F(x, y) \quad\ldots\ldots\ldots\ldots\ldots\ldots\ldots\ldots\text{(ii)},$$

where $F(x, y)$ is a known function of x and y.

This equation Boussinesq now proceeds to solve. He writes $a = y + ax$, $\beta = y - ax$, and obtains first

$$\frac{d^2\varpi}{da\,d\beta} = -\frac{1}{4a^2} F\left(\frac{a-\beta}{2a}, \frac{a+\beta}{2}\right),$$

and then

$$\varpi = f(\beta) - \frac{1}{4a^2} \int_0^a da \int_0^\beta F\left(\frac{a-\beta}{2a}, \frac{a+\beta}{2}\right) d\beta \ldots\ldots\ldots\text{(iii)},$$

where f must be determined by the condition for the slipping of the mass against the wall.

Boussinesq then proceeds to calculate F on the basis of his first approximation in the case of a *vertical* supporting wall: see our Art. 1616. He obtains a somewhat complicated value of ϖ to a second approximation for this case on p. 709. From this he determines r, s, t and the stresses at the wall. He finds for the value of the constant k (see our Art. 1616) determining the thrust on the wall:

$$k = \frac{a^2}{1 + a\tan\phi_1} \left[1 + (-1 + 2\log 2)\frac{1 + a^2}{1 - a^2}\left(\frac{a}{a + \cot\phi_1}\right)^2\right].$$

In the case of ordinary sand resting against a rough wall, when $\phi = \phi_1 = 34°$, we find for k by the first approximation:

$$k = \cdot 2081,$$

while by the second approximation $k = \cdot 2181$.

Like the first approximation the second may be looked upon as an *exact* answer to a problem in which the angle of internal friction varies from its value ϕ at $y = ax$ and increases up to ϕ_1 at the wall: see our Art. 1615. The law of variation would, however, be difficult to express. Boussinesq contents himself with a tentative investigation, from which he concludes that a very close value of k is $\cdot 2308$ for the above special case. His practical method of the arithmetical mean between a too great and a too small value (see our Art. 1618) had given $\cdot 2309$. Hence he considers that his practical method of the mean gives close results. It will be observed that it gives closer results than the second approximation, the rate of approximation being apparently not sufficiently rapid.

[1625.] *Tables numériques pour le calcul de la poussée des terres. Annales des ponts et chaussées. Mémoires*, T. IX., 1er *Semestre*, pp 515–40. Paris, 1885. This is a statement by Flamant

of the final form of Boussinesq's theory with numerical tables for
the value of the thrust calculated for various angles of the talus,
of internal friction and of the rear-side of the supporting wall.
From these tables Flamant deduces the results we have referred
to in our Art. 1611. The tables ought to be useful as giving
with sufficient accuracy all that is obtainable from Boussinesq's
theory.

[1626.] *Summary.* Among the many pupils of Saint-Venant
few have dealt with such a wide range of elastic problems as Boussinesq, or contributed more useful work to elastic theory. Belonging in a more marked manner than Saint-Venant, Phillips or
Maurice Lévy to the mathematical as distinguished from the
technical group of elasticians, he has illustrated our subject by
ingenious analysis rather than by special solutions of mechanical
and physical problems. If in several cases his researches have been
anticipated by those of other investigators, he has yet managed to
throw new light on old problems, and where he has not succeeded
in giving final solutions, he has greatly added to our appreciation
of yet unsurmounted difficulties. Thus his investigations on plates
and rods were preceded by those of Kirchhoff and Clebsch, yet
Boussinesq's methods, if scarcely final, are at least clearer and
more concise than Kirchhoff's. If Thomson and Tait reconciled
Poisson's and Kirchhoff's contour conditions some years before
Boussinesq, yet the latter, especially in his work on potentials, has
thrown the whole matter into a more concise and simple form.
If Cerruti in the application of potentials in some respects
anticipated Boussinesq's results, yet the latter's great treatise will
remain for many years to come a classic of our subject, every page
almost of which is fascinating to the mathematical physicist.
Much the same may be said of Boussinesq's contributions to
the theory of impact; he followed Stokes, Saint-Venant and
Hertz, but he was able to follow them without loss of individuality
in either method or results. Putting aside his contributions to the
theory of potentials, perhaps the most original part of his work
lies on the border-land of elasticity proper,—namely, in his contributions to the elastic theory of light and to the theory of
pulverulence. These theories must of course be judged by the
physicist and the engineer; yet if they be not final they are still

the best which have hitherto been propounded from the elastic standpoint; indeed, they are perhaps the limit to what elastic theory can provide in these directions. Like the majority of the leading French mathematicians, Boussinesq is as a rule lucid in his analysis, if occasionally wanting in physical touch.

CHAPTER XIV.

SIR WILLIAM THOMSON (LORD KELVIN[1]), THOMSON AND TAIT.

[1627.] ONE of the many fields in which Sir William Thomson has laboured with much profit to physical science is that of elasticity, and the concluding chapter of this volume may not unfitly be devoted to a *résumé* of his contributions to our subject.

His first papers on elasticity will be found in the *Cambridge and Dublin Mathematical Journal*, Vol. II., pp. 61–4, and Vol. III., pp. 87–9. Cambridge, 1847 and 1848. They are entitled:

(*a*) *On a Mechanical Representation of Electric, Magnetic and Galvanic Forces* (*M. P.*, Vol. I., pp. 76–80)[2].

(*b*) *Note on the Integration of the Equations of Equilibrium of an Elastic Solid* (*M. P.*, Vol. I., pp. 97–99).

[1628.] Consider the body-shift-equations of an isotropic elastic medium subjected to no body force. These are of the type[3]:

$$(\lambda + \mu)\frac{d\theta}{dx} + \mu \nabla^2 u = 0 \dots\dots\dots\dots\dots\dots\dots \text{(i)}.$$

Then if C, l, m, n be arbitrary constants, the following are solutions:

$$\left.\begin{aligned}
u &= C \left\{ \frac{\lambda + \mu}{2(\lambda + 2\mu)}\frac{d}{dx}\left(\frac{lx + my + nz}{r}\right) - \frac{l}{r}\right\}, \\
v &= C \left\{ \frac{\lambda + \mu}{2(\lambda + 2\mu)}\frac{d}{dy}\left(\frac{lx + my + nz}{r}\right) - \frac{m}{r}\right\}, \\
w &= C \left\{ \frac{\lambda + \mu}{2(\lambda + 2\mu)}\frac{d}{dz}\left(\frac{lx + my + nz}{r}\right) - \frac{n}{r}\right\}
\end{aligned}\right\} \dots\dots\dots\text{(ii)},$$

where
$$r = (x^2 + y^2 + z^2)^{\frac{1}{2}}.$$

---

[1] The manuscript of this chapter was completed before Sir William Thomson became Lord Kelvin.

[2] The letters *M. P.* stand throughout this chapter for the *Mathematical and Physical Papers* by Sir William Thomson, Vols. I.–III. Cambridge, 1882–1890.

[3] Sir William Thomson speaks of Sir G. G. Stokes' memoir of 1845 as being the only work in which the "true formulae" (*i.e.* bi-constant formulae for isotropy), had in 1847 been given. This is hardly exact: see our Arts. 614* and 1267*.

It will be found that:

$$\theta = C \frac{\mu}{\lambda + 2\mu} \frac{lx + my + nz}{r^3} \quad\quad\quad\text{......................(iii)}.$$

This is clearly a special case of results, which Boussinesq much later dealt with under the title of 'potential solutions'. Sir William Thomson remarks that the most general solution can be expressed in terms of particular solutions of the type (ii). He adds (b), p. 89, (M. P. Vol. I., p. 99) that u, v, w can be easily shewn to be the shifts produced at the point x, y, z of an infinite isotropic elastic medium due to a force applied at the origin of coordinates in the direction l, m, n: see our Art. 1519 (b).

[1629.] In the preceding article we have given the results of the second paper, those of the first hold only for an *incompressible* isotropic solid. Putting $\lambda/\mu = \infty$ in (iii) we find $\theta = 0$ and results (ii) become of the type:

$$u = C \left\{ \frac{1}{2} \frac{d}{dx} \frac{lx + my + nz}{r} - \frac{l}{r} \right\}.$$

In this case the twists are of the form

$$\tau_{yz} = \frac{C}{2} \left\{ \frac{ny - mz}{r^3} \right\}, \quad \tau_{zx} = \frac{C}{2} \left\{ \frac{lz - nx}{r^3} \right\}, \quad \tau_{xy} = \frac{C}{2} \left\{ \frac{mx - ly}{r^3} \right\}.$$

These Sir William compares with the expressions for the components of the force which an infinitely small element of a galvanic current at the origin, in the direction of l, m, n, produces on a unit magnetic pole at the point x, y, z. ((a) p. 64, M. P., Vol. I., p. 80.)

[1630.] In the first paper the author further shews that the following systems of shifts are also solutions in the case of an incompressible solid:

(I). $u = C \dfrac{x}{r^3}, \quad v = C \dfrac{y}{r^3}, \quad z = C \dfrac{z}{r^3};$

and   (II). $u = C \dfrac{mz - ny}{r^3}, \quad v = C \dfrac{nx - lz}{r^3}, \quad w = C \dfrac{ly - mx}{r^3}.$

The first set of shifts are compared with the components of the force exerted at x, y, z by a charge of electricity at the origin; and the twists corresponding to the second set of shifts are compared to the components of the magnetic force of a small magnet placed at the origin with its axis in the direction l, m, n upon an ideal magnet pole at x, y, z: see our Art. 1813.

[1631.] *On the Thermo-elastic, Thermo-magnetic and Pyro electric Properties of Matter.* This paper appeared under the title: *On the Thermo-elastic and Thermo-magnetic Properties of Matter, Quarterly Journal of Mathematics,* Vol. I. pp. 57–77, Cambridge,

1857. It was reprinted with additions in the *Philosophical Magazine*, Vol. 5, pp. 4–27, London, 1878, and appears as Part VII. of the article: *On the Dynamical Theory of Heat, M. P.*, Vol. I., pp. 291–316. We are solely concerned with pp. 291–313 of this latter paper, and our reference will be first to the pages of the *Philosophical Magazine* and secondly to those of Vol. I. of the *Papers*.

This is one of the most important of Sir William Thomson's contributions to our subject. Its object, so far as we are concerned, is to obtain from the *Second Law of Thermo-dynamics* the most general possible theory of elasticity for unmagnetised and unelectrified bodies.

The author defines the " intrinsic energy of a body in a given state" to be, "the mechanical value of the whole agency" that would be required to bring the body from a standard state to the given state. This agency may be spent in overcoming the resistances of the body or in exciting thermal motions. The intrinsic energy, e, can depend only on the standard and given states, if we are to deny the possibility of perpetual motion.

[1632.] Sir William Thomson now assumes that six independent variables can fully express " the mechanical condition of a homogeneous solid mass, homogeneously strained in any way"[1]. The words *homogeneous strain* are not in this paper further defined: see our Art. 1672. Let these six variables be denoted by the letters $\alpha, \alpha', \alpha'', \sigma, \sigma', \sigma''$, and let t be the temperature in the given state, the standard state being denoted by $\alpha_0, \alpha_0', \alpha_0'', \sigma_0, \sigma_0', \sigma_0''$ and t_0. Then we have :

$$e = \phi \,(\alpha,\, \alpha',\, \alpha'',\, \sigma,\, \sigma',\, \sigma'',\, t)\dots\dots\dots\dots\dots\dots(\text{i}),$$
$$\text{and } 0 = \phi \,(\alpha_0,\, \alpha_0',\, \alpha_0'',\, \sigma_0,\, \sigma_0',\, \sigma_0'',\, t_0),$$

where ϕ denotes a certain function depending on the nature of the substance.

Now suppose the body strained so as to pass from the state $\alpha_0, \alpha_0', \alpha_0'', \sigma_0, \sigma_0', \sigma_0''$ to the state $\alpha, \alpha', \alpha'', \sigma, \sigma', \sigma''$ without change of the temperature t, and let H be the quantity of heat that must be supplied to it during this process to prevent its temperature being

---

[1] Here there is a footnote referring to Rankine's introduction of the word *strain*, and calling the corresponding forces *straining tensions or pressures*. This is of historical interest as shewing that the word *stress* had not yet (1855) come into general use (pp. 6, 293).

lowered. Now let the body be brought back to its first mechanical condition through the same or any other of the infinitely numerous possible sequences of states, the temperature being still always kept at t, and let H' be the heat supplied. Then, by the Second Law of Thermodynamics:

$$\frac{H}{t} + \frac{H'}{t} = 0, \text{ or } H = -H'.$$

It follows therefore that the amount of heat absorbed by the body in passing from one state to the other must be independent of the sequence of states through which the body passes, and depend only on the initial and final states, provided the temperature remain constant throughout. Accordingly we have :

$$H = \psi\,(\alpha, \alpha', \alpha'', \sigma, \sigma', \sigma'', t) - \psi\,(\alpha_0, \alpha_0', \alpha_0'', \sigma_0, \sigma_0', \sigma_0'', t)...\text{(ii)},$$

where ψ denotes a function of the variables.

Hence, if ϵ denote the whole augmentation of mechanical energy which the body experiences, *i.e.* the change in its intrinsic energy from one state to the second, or

$$\epsilon = \phi\,(\alpha, \alpha', \alpha'', \sigma, \sigma', \sigma'', t) - \phi\,(\alpha_0, \alpha_0', \alpha_0'', \sigma_0, \sigma_0', \sigma_0'', t)...\text{(iii)},$$

we have : $$\epsilon = w + JH ....\text{(iv)},$$

where w denotes the work done by applied forces in compelling the body to pass from one condition to the other, and J is "Joule's equivalent".

It is clear from this that the work, w, required to strain the body from one to another of two given states, keeping it always at the same temperature, is independent of the particular succession of mechanical states through which the body passes; it depends only on the initial and final conditions. This theorem Sir William Thomson attributes to Green: see our Art. 918*.

He adds :

It is now demonstrated as a particular consequence of the Second General Thermodynamic Law. It might at first sight be regarded as simply a consequence of the general principle of mechanical effect; but this would be a mistake, fallen into from forgetting that heat is in general evolved or absorbed when a solid is strained in any way; and the only absurdity to which a denial of the proposition could lead would be the possibility of a self-acting machine going on continually drawing heat from a body surrounded by others at a higher temperature, without the assistance of any at a lower temperature, and performing an equivalent of mechanical work (p. 7, *M. P.*, Vol. I., p. 295).

[1633.] To obtain the most complete results available from the Second Law of Thermodynamics, which is expressible in the form[1]: $\Sigma\left(H_t/t\right) = 0$, Sir William Thomson now supposes the body to pass through the following reversible cycle from and back to its primitive condition as to strain and temperature:

(i) Without altering the strain (a_0, a_0', a_0'', σ_0, σ_0', σ_0'') raise the temperature from t to t'. If $t' - t$ be small, this requires the quantity of heat represented by

$$\frac{1}{J}\frac{de_0}{dt}(t' - t),$$

where e_0 denotes the value of e (see Equation (i)) for a_0, a_0', a_0'', σ_0, σ_0', σ_0'', t.

(ii) Keeping the temperature at t', pass from the strain a_0, a_0', a_0'', σ_0, σ_0', σ_0'' to a, a', a'', σ, σ', σ''. This requires the heat (see Equation (ii)):

$$H + \frac{dH}{dt}(t' - t).$$

(iii) Without altering the strain, lower the temperature to t. This requires the heat

$$-\frac{1}{J}\frac{de}{dt}(t' - t).$$

(iv) Return to the primitive strain without altering the temperature. This involves the heat

$$-H.$$

Hence we find

$$\Sigma\left(\frac{H_t}{t}\right) = (t' - t)\left\{\frac{d}{dt}\left(\frac{H}{t}\right) - \frac{1}{Jt}\frac{d\epsilon}{dt}\right\},$$

where $\epsilon = e - e_0$: see Equation (iii). Hence it follows that

$$\frac{d}{dt}\left(\frac{H}{t}\right) - \frac{1}{Jt}\frac{d\epsilon}{dt} = 0 \quad\dots\dots\dots\dots\dots\dots\dots (v).$$

From Equations (iv) and (v) we easily find:

$$H = -\frac{t}{J}\frac{dw}{dt} \quad\dots\dots\dots\dots\dots\dots\dots (vi),$$

$$e = e_0 + w - t\frac{dw}{dt} \quad\dots\dots\dots\dots\dots\dots\dots (vii).$$

These are the fundamental thermo-elastic equations: see p. 9 (*M. P.*, Vol. I., p. 297).

<hr>

[1] See J. H. Parker: *Elementary Thermodynamics*, p. 139, Cambridge, 1891, or Sir William Thomson, *Math. Papers*, Vol. I. p. 236.

[1634.] Let N be the specific heat at constant strain for any temperature t, and let K be the specific heat for any temperature t when the body is allowed or compelled to alter its strain with the temperature in any fixed manner. Then we find

$$JN = \frac{de}{dt} \quad\text{.........................} \quad\text{(viii)},$$

$$JK = \frac{de}{dt} + \Sigma \frac{d(JH)}{d\zeta} \frac{d\zeta}{dt} \quad\text{.........................} \text{(ix)},$$

where ζ is to be taken successively equal to each component of strain.

The last equation by aid of (iv) may be written :

$$JK = \frac{De}{dt} - \Sigma \frac{dw}{d\zeta} \frac{d\zeta}{dt} \quad\text{.........................} \text{(x)},$$

De/dt denoting the total differential of e.

[1635.] On pp. 12–15 ($M. P.$, Vol. I., pp. 300–304) Sir William Thomson supposes the strain to be small and he practically takes as his six strain components the three stretches and slides from the standard state, $i.e.$ we may put :

$$a - a_0 = s_x, \qquad a' - a_0' = s_y, \qquad a'' - a_0'' = s_z,$$

$$\sigma - \sigma_0 = \sigma_{yz}, \qquad \sigma' - \sigma_0' = \sigma_{zx}, \qquad \sigma'' - \sigma_0'' = \sigma_{xy}.$$

He then *assumes* that w may be expanded by Maclaurin's Theorem, whence retaining only the expressions up to the squares and products of the strains, he easily finds for the stresses expressions of the types:

$$
\begin{aligned}
\widehat{xx} &= \left(\frac{dw}{da}\right)_0 + \left(\frac{d^2w}{da^2}\right)_0 s_x + \left(\frac{d^2w}{dada'}\right)_0 s_y + \left(\frac{d^2w}{dada''}\right)_0 s_z \\
&\quad + \left(\frac{d^2w}{dad\sigma}\right)_0 \sigma_{yz} + \left(\frac{d^2w}{dad\sigma'}\right)_0 \sigma_{zx} + \left(\frac{d^2w}{dad\sigma''}\right)_0 \sigma_{xy}, \\
\widehat{yz} &= \left(\frac{dw}{d\sigma}\right)_0 + \left(\frac{d^2w}{d\sigma da}\right)_0 s_x + \left(\frac{d^2w}{d\sigma da'}\right)_0 s_y + \left(\frac{d^2w}{d\sigma da''}\right)_0 s_z \\
&\quad + \left(\frac{d^2w}{d\sigma^2}\right)_0 \sigma_{yz} + \left(\frac{d^2w}{d\sigma d\sigma'}\right)_0 \sigma_{zx} + \left(\frac{d^2w}{d\sigma d\sigma''}\right)_0 \sigma_{xy}
\end{aligned}
\left.\vphantom{\begin{aligned}x\\x\\x\\x\end{aligned}}\right\} \dots \text{(xi)}.
$$

These are clearly the usual expressions for the stresses in a multi-constant solid when there are initial stresses in the standard state. If these initial stresses be zero, we have the usual stress-strain relations with twenty-one constants.

We have already pointed out that such important physical conclusions as those which flow from the linearity of the stress-strain relations seem to demand a basis in physical experiment rather than in a mathematical theorem: see our Arts. 928* and 299.

[1636.] Sir William Thomson then turns (pp. 16–18; *M. P.*, Vol. I., pp. 304–7) to the problem of rari- and multi-constancy and the inter-constant relations. He remarks:

Whether or not it may be true that such relations do hold for natural crystals, it is quite certain that an arrangement of actual pieces of matter may be made, constituting a homogeneous whole when considered on a large scale (being, in fact, as homogeneous as writers adopting the atomic theory in any form consider a natural crystal to be), which shall have an arbitrarily prescribed value for each one of these twenty-one coefficients. No one can legitimately deny for all natural crystals, known and un-known, any property of elasticity, or any other mechanical or physical property, which a solid composed of natural bodies artificially put to-gether may have in reality. To do so is to assume that the infinitely inconceivable structure of the particles of a crystal is essentially restricted by arbitrary conditions imposed by mathematicians for the sake of shortening the equations by which their properties are expressed (p. 16).

It is, perhaps, somewhat hard to accuse the rari-constant elasti-cians of being actuated by a desire to shorten mathematical equa-tions, when certainly one of their objects was to get over the difficulty of a purely mathematical deduction of the generalised Hooke's Law by appealing to a general physical principle of intermolecular action : see our Arts. 192 and 300–6. At the same time the appeal to the existence of a 21-constant model is a valid argument *pro tanto*. The exact nature of this mechanical model was not de-scribed for many years (see our Art. 1771), and I cannot say that when described it carries conviction to my mind.

The further arguments cited against rari-constancy are the stock examples of cork, jelly, india-rubber (see our Arts. 924*, 930*, 1322*, 192 (*b*) and 610) and the values of the stretch-squeeze ratio as determined by Wertheim, Everett, Clerk-Maxwell and Sir William Thomson himself. The materials above cited may be fairly excluded from the list of elastic bodies to which the rari-constant theory applies; while the group of experiments referred to were in several cases made on bodies the isotropy of which was more than doubtful.

In none of these cases were any investigations made as to whether *two* constants would really suffice to describe the elastic properties of the material, or whether the actual elastic system was not represented better by some suitable distribution of elastic homogeneity than by bi-constant isotropy : see our Arts. 925*, 932*, 192, 1201 and 1272.

[1637.] On p. 18 (*M. P.*, Vol. I., p. 307) Sir William Thomson notes that some of the rari-constant relations lead to three principal axes of elasticity. Many natural crystals certainly have complete symmetry of form with regard to three rectangular axes and "therefore probably possess all their physical properties symmetrically with reference to these axes". But it is further noted that many natural crystals do not exhibit this symmetry of form in reference to rectangular axes, and the instance of Iceland spar is cited with three cleavage planes inclined at equal angles to one another and to the 'optic axis' of the material. Then Sir William Thomson adds:

If, as probably must be the case, the elastic properties within the limits of elasticity have correspondence with the mechanical properties on which the brittleness in different directions depends, the last-mentioned class of crystals cannot have three principal axes of elasticity at right angles to one another (p. 18 ; *M. P.*, Vol. I., p. 307).

Now exception must, I think, be taken to both the principles enunciated above. It does not appear that all the physical properties of crystals with three rectangular axes of symmetry of form are symmetrically arranged about these axes : see our Arts. 683–7, 1218–20. Further, if the distribution of hardness has relation to a system of rectangular axes differing from those of form, it does not seem *a priori* certain that we should expect distributions of elasticity and brittleness to be symmetrical about the same system of rectangular axes. In fact without experimental investigation it does not seem legitimate to assert that the *shape* of the crystal, as determined by its planes of cleavage, defines in any way the nature of its elastic distribution. In particular it may be observed that the elastic constants of a material are frequently insensibly altered by large sets ; it is very probable, however, that such sets may materially influence the cohesive powers of the material. It does not appear, therefore, improbable that distributions of cohesion and brittleness may follow different laws or systems of axes from the distribution of elasticity : see our Arts. 683–7 and 1218–9.

[1638.] From Equation (vi) of our Art. 1633, by supposing w expanded by Maclaurin's theorem and the first terms only retained, we have :

$$H = -\frac{t}{J}\frac{d}{dt}\left\{\left(\frac{dw}{da}\right)_0 s_x + \left(\frac{dw}{da'}\right)_0 s_y + \left(\frac{dw}{da''}\right)_0 s_z\right.$$

$$\left. + \left(\frac{dw}{d\sigma}\right)_0 \sigma_{yz} + \left(\frac{dw}{d\sigma'}\right)_0 \sigma_{zx} + \left(\frac{dw}{d\sigma''}\right)_0 \sigma_{xy}\right\}$$

$$= -\frac{t}{J}\left\{\frac{\widehat{dxx}}{dt} s_x + \frac{\widehat{dyy}}{dt} s_y + \frac{\widehat{dzz}}{dt} s_z + \frac{\widehat{dyz}}{dt} \sigma_{yz} + \frac{\widehat{dzx}}{dt} \sigma_{zx} + \frac{\widehat{dxy}}{dt} \sigma_{xy}\right\} \ \ldots\ldots (\text{xii}),$$

where the usual relations between the stresses and the strain-energy have been assumed.

From this result Sir William Thomson draws an important series of physical conclusions which are embraced in the following sentences :

We conclude that cold is produced whenever a solid is strained by opposing, and heat when it is strained by yielding to, any elastic force of its own, the strength of which would diminish if the temperature were raised —but that, on the contrary, heat is produced when a solid is strained against, and cold when it is strained by yielding to, any elastic force of its own, the strength of which would increase if the temperature were raised (p. 19 ; *M. P.*, p. 308).

This may be expressed otherwise thus : If the strain remaining constant, an increase of temperature is marked by increase of the stress required to maintain the strain, then the body will give off heat when the strain is produced.

The following are given as examples of these statements :

(i) The cubical compression of any elastic fluid or solid in any ordinary condition would cause an evolution of heat. This follows at once from the fact that most elastic bodies require increased pressure to maintain their volume constant when the temperature is raised[1].

(ii) A twisted wire, if further twisted within its elastic limits, will produce cold, and if it be allowed to suddenly untwist will evolve heat. This follows from the fact that $d\mu/dt$ is negative: see our Art. 754.

(iii) Spiral springs, as we have seen in Arts. 1382*–3* and 1284 (c), act principally by torsion ; hence when suddenly drawn out they will cool and when suddenly released they will rise in temperature. This result was confirmed experimentally by Joule: see our Arts. 689–690.

(iv) A bar, rod or wire if suddenly stretched by terminal traction is cooled, and warmed when the traction is suddenly removed.

Sir William Thomson's next case is that of india-rubber, which in the early version of the paper he supposed would be cooled, if suddenly

---

[1] The dilatation-modulus for an isotropic material is given by $G=\frac{1}{3}(3\lambda+2\mu)$: see our Vol. I., p. 885. This may be put in the form

$$G=\frac{1}{3}\frac{\mu E}{3\mu - E}.$$

Hence

$$\frac{1}{3}\frac{1}{G^2}\frac{dG}{dt} = \frac{3}{E^2}\frac{dE}{dt} - \frac{1}{\mu^2}\frac{d\mu}{dt}.$$

Now dE/dt and $d\mu/dt$ are generally negative. Thus in the notation of our Arts. 752—4, we have :

$$\frac{1}{3}\frac{1}{G^2}\frac{dG}{dt} = \frac{\beta_\tau}{\mu} - \frac{3\beta_f}{E}.$$

We must therefore have $\beta_\tau/\beta_f > 3\mu/E$, if compression is to be accompanied by the evolution of heat. For the case of uni-constant isotropy this reduces to $\beta_\tau/\beta_f > 6/5$, and appears to be satisfied as far as Kupffer's numbers allow of any real comparison.

drawn out[1]. The cooling effect was only found for low temperatures, but at a higher temperature, 15° C., a pull was shown by Joule to produce a heating effect: see our Art. 689. This led Sir William Thomson to predict that a vulcanised india-rubber band with a weight attached at one end would shorten on being heated. The phenomenon may be termed the " Gough-effect ".

This is an experiment which anyone can make with the greatest ease by hanging a few pounds weight on a common india-rubber band, and taking a red-hot coal, in a pair of tongs, or a red-hot poker, and moving it up and down close to the band (p. 20 ; *M. P.*, Vol. I., p. 309).

[1639.]   The remainder of the memoir deals with certain properties of elastic crystals.

For a regular crystal Sir William Thomson obtains stress-strain relations agreeing with those of Neumann given in our Art. 1203 (*d*), except that he writes them in forms of the type:

$$\widehat{xx} = \lambda\theta + 2\mu s_x, \quad \widehat{yz} = (\mu + \kappa)\, \sigma_{yz}.$$

Here, if $\kappa = 0$, the regular crystal becomes an isotropic solid. Hence κ expresses the "crystalline quality" in the elasticity of a crystal of the cubic class (p. 22 ; *M. P.*, Vol. I., p. 311).

[1640.]   We have next and lastly a suggestion that the state of strain of an elastic body, which can be expressed by any six independent variables which describe the changes of shape, should be indicated by the "six edges of a tetrahedron enclosing always the same part of the solid ". In the case of a regular crystal Sir William Thomson takes this tetrahedron with its edges parallel to the diagonals of the faces of the cube.   He obtains an expression for the strain-energy in terms of three crystalline constants and the stretches of the six edges of this tetrahedron.   He further obtains expressions for quantities corresponding to the stresses, which are in this case the tractions normal to the faces of the dodecahedron with unit facial area, obtained by drawing planes perpendicular to the edges of the tetrahedron.   The investigation is ingenious, but it has not I believe been made the basis of any further investigations.

[1641.]   Sir William Thomson's memoir placed the principles of thermo-elasticity on a firm foundation, and advanced that branch of our subject much beyond the theories of Duhamel and Neumann: see our Arts. 868*–896*, 1196–7 and 1200.

It opened up the path of accurate investigation into the difficult borderland of thermo-dynamics and elasticity, wherein

---

[1] The correct statement of the thermo-elastic properties of india-rubber had been given by Gough in 1805 ; but his paper had been forgotten : see our Vol. I., p. 386, footnote.

more than one distinguished physicist had gone astray: see our Arts. 716, 717, 725 and 745. The memoir may fairly be said to give the first really legitimate proof of the existence of a strain-energy function depending only on the strain from a standard state and not on the manner in which the strain is reached.

[1642.] *On Thermo-electricity in Crystalline Metals, and in Metals in a state of Mechanical Strain.* This forms § III. of a memoir entitled: *Experimental Researches in Thermo-electricity. Proceedings of the Royal Society*, Vol. VII., pp. 56–58. London, 1856. (*M. P.*, Vol. I., pp. 467–8.) Sir William Thomson had been led to believe, by the analogy of strain as influencing the optical properties of transparent bodies, that the application of stress to a mass of metal would give it the thermo-electric properties of a crystal. The present paper announces the results of experiments on copper and iron wires. Let a portion of a circuit of copper wire be stretched within the elastic limits and let an extremity of this portion be heated, then a current sets from the stretched to the unstretched part through the hot junction. If the wire be alternately stretched and unstretched on the two sides of the heated portion, the current is reversed at each change. In the case of iron wire the current flows from the unstretched to the stretched portion through the hot junction, *i.e.* the reverse of the case for copper wire.

[1643.] *On the Effects of Mechanical Strain on the Thermo-electric Qualities of Metals. British Association Report*, Glasgow Meeting, 1855, *Transactions*, pp. 17–18. London, 1856. (*M. P.*, Vol. II., pp. 173–4.) This paper announces further results similar to those stated in the previous article. The experiments were extended to other metals than copper and iron and to set as well as elastic strain. Fuller details are given in a later memoir: see our Art. 1645.

[1644.] *On the Electro-dynamic Qualities of Metals.* (*a*) *Philosophical Transactions*, Vol. 146, pp. 649–751. London, 1856. (*b*) *Proceedings of the Royal Society*, Vol. VIII., pp. 546–50. London, 1857. (*c*) *Philosophical Transactions*, Vol. 166, pp. 693–713. London, 1876. (*d*) *Proceedings of the Royal Society*, Vol. XXIII., pp. 473–6. London, 1875. (*e*) *Philosophical Transactions*, Vol. 170, pp. 55–85. London, 1879. Abstracts of these memoirs will further be found in the *Proceedings of the Royal Society;*

(*f*) Vol. VIII., pp. 50–5, 1856; (*g*) Vol. XXIII., pp. 445–6, 1875; (*h*) Vol. XXVII., pp. 439–43, 1878. The whole series forms an *Article* under the above title divided into seven parts and an Appendix in the *Mathematical and Physical Papers*, Vol. II., pp. 189–407.

The parts of this *Article* which directly concern us are Parts III., IV., VI., VII. and portions of the *Appendix*.

[1645.]   Part III. is entitled: *Effects of Mechanical Strain and of Magnetization on the Thermo-electric Qualities of Metals.* (*a*) pp. 709–36; (*f*) pp. 52–4 (*M. P.,* Vol. II., pp. 267–97), and it is the first portion concerned with our present subject. It gives fuller details of the experiments referred to in our Arts. 1642 and 1643. Pp. 709–27 (*M. P.,* Vol. II., pp. 267–86) deal with the action of elastic strain and set in the production of thermo-electric effects.

It is well known that if a circuit be formed of two *different* metals, one junction being maintained at a higher temperature than the second, then a current will flow in the circuit. Let it be from metal *A* to metal *B* through the hot junction. The metal *B* is then said to be higher in the thermo-electric scale than the metal *A*. At the bottom of such a scale stands bismuth, near the top iron and above iron antimony. No thermo-electric effect has been found in an unequally heated circuit of the *same* metal, if that metal be all in the same condition as to strain[1]. The object of the present memoir is to ascertain what thermo-electric effects elastic strain and set have on portions of the same metal forming a circuit. The effects of a uniform dilatation and compression are not ascertained, but Sir William discusses in a series of ingenious experiments the effects of longitudinal traction and lateral contraction in the case of both elastic and set strains in differentiating a metal into classes (*i.e.* the strained and unstrained) which do not coincide in the thermo-electric scale. Thus Sir William found:

A.   *For elastic strain.*

(i)   That a longitudinal traction caused a deviation in copper wire from its position in the unstrained state towards bismuth, but in iron wire towards antimony. (Strain was found also to shift the position in the thermo-electric scale of platinum wire.)

---

[1] The section of the conductor must not change suddenly: see Maxwell, *Electricity and Magnetism.* 3rd Ed. Vol. I., p. 371, *ftn.*

(ii)   That a lateral contraction caused a deviation in iron wire
towards antimony.   Hence Sir William argues that a lateral *traction*
would cause a deviation towards bismuth, or that it would have an
effect the reverse of that produced by a longitudinal traction[1].   The
crystalline characteristic is therefore established for the thermo-electric
effect of mechanical stress applied to iron, if it be true that traction produces the reverse temporary effect to that of pressure in the same direction ((*a*) p. 715 ; *M. P.*, Vol. II., p. 275).

Sir William cites an ingenious experiment to shew that iron under
a simple longitudinal stress has "different thermo-electric qualities in
different directions" ((*a*) pp. 715–7 ; *M. P.*, Vol. II., pp. 275–8.)

B.   *For set.*

(iii)   That set produced by a longitudinal traction in both copper
and iron wire causes a deviation from the thermo-electric position in the
reverse direction to that caused by an elastic strain of the same kind ((*a*)
pp. 712–3 ; *M. P.*, Vol. II., pp. 270–2).

(iv)   That set produced by a lateral contraction in iron wire causes
a deviation in the reverse direction of the elastic strain of the same kind
((*a*) pp. 717–18 ; *M. P.*, Vol. II., pp. 278–9).

The combination of these results (iii) and (iv) leads to Magnus' conclusion that drawn wire, *i.e.* wire subjected to longitudinal stretch set
and lateral squeeze set, differs in position from the unstrained wire in
the thermo-electric scale.   Magnus stated his results for iron in the
words "the current is from hard to soft though hot".   This Sir William
Thomson shews is not an exact description of all thermo-electric currents
produced by set.   He constructed a conductor of 24 little iron cylinders
set end to end, alternate cylinders having been compressed to set.   By
an ingenious system alternate junctions were heated and cooled.   A
current was then found to pass from unstrained to strained through hot,
i.e. from "soft to hard through hot".   Thus it appears that it is not the
hardening of the iron, but the *direction of the strain* which is the determining element.   Copper and tin wires were found, like iron, to give
the same thermo-electric effects in the cases of set due to longitudinal
traction and to lateral contraction.   The whole series of phenomena point
to strain producing a crystalline character in the metal so far as its
thermo-electric action is concerned.

[1646.]   Further experiments were made on the thermo-electric effect
in the cases of coils, parts of which were hammered and parts not, of coils
parts of which were annealed and parts unannealed ; and of coils parts
of which had torsional set and parts not.   In the first case the current
for iron was from hammered to unhammered through hot, but for steel,
copper, tin, brass, lead, cadmium, platinum and zinc this direction was
reversed.   In the second case for iron and steel the current was from
unannealed to annealed through hot ; this direction was reversed for

---

[1] The terms 'longitudinal' and 'lateral' are here applied to directions along
and perpendicular to the current.

copper and brass. In the third case the current was from torted to
untorted (brittle to soft) in iron, and the reverse for copper. In these
experiments the wire was first uniformly torted to set and then the
set in parts of it removed by annealing ((*a*) pp. 720-2 ; *M. P.*, Vol. II.,
pp. 283-6).

[1647.] The next part of the memoir which is of interest for our
present purposes is entitled : *Methods for comparing and determining
Galvanic Resistances, illustrated by Preliminary Experiments on the
Effects of Tension...on the Electric Conductivity of Metals.* ((*a*) pp. 730-6;
M. P., Vol. II., pp. 298-306). Pp. 733-4 (*M. P.*, Vol. II., pp. 301-6) are
all that concern us. Here a single experiment is given to shew that
equal longitudinal stretches, whether elastic or set, in iron and copper
wires alter their relative electric conductivities. The resistance of
the iron had increased relatively to that of the copper; the author had
not then determined the absolute effect on the conductivities of the
strain, but had been led by a partial investigation to believe that it
diminished in both metals.

The remainder of the series of memoirs cited in our Art. 1644
will be found dealt with in our Arts. 1727-1736.

[1648.] *Elements of a Mathematical Theory of Elasticity.
Philosophical Transactions*, Vol. 146, pp. 481-98. London, 1856.
This memoir is incorporated in the *Encylopaedia* Article on
Elasticity: see our Art. 1741.

[1649.] *On the Stratification of Vesicular Ice by Pressure.
Royal Society, Proceedings*, Vol IX., pp. 209-13. London, 1859.
*Note on Professor Faraday's Recent Experiments on 'Regelation.'
Royal Society, Proceedings*, Vol. XI., pp. 198-204. London, 1862.

These papers deal with the melting of ice under pressure, and
on the nature of the motion of a plastic solid like ice under stress.
The discussion is general and unaccompanied by mathematical
analysis, but to enter into it would lead us too far beyond our
present limits.

[1650.] *Note on Gravity and Cohesion. Proceedings of the
Royal Society of Edinburgh*, Vol. IV., pp. 604-6. Edinburgh, 1862.
(*Popular Lectures and Addresses*, Vol. I., pp. 59-63. London,
1889).

This is an attempt to shew that gravitation will suffice to
explain cohesive force, provided only that the ratio of the space
occupied by matter to the space unoccupied by matter in any
finite body is sufficiently great. Sir William Thomson refers to

woven and fibrous structures as exemplifying this position and adds:

> ...it is clear that the same result would be produced by any sufficiently intense heterogeneousness of structure whatever, provided only some appreciable proportion of the whole mass is so condensed in a continuous space in the interior that it is possible, from any point of this space as centre, to describe a spherical surface which shall contain a very much greater amount of matter than the proportion of the whole matter of the body which would correspond to its volume (p. 606).

I do not feel convinced by the arguments used, especially if matter be not treated as continuous as it is in the case of fibrous or woven structure. The hypothesis of *un tessuto fibroso o reticolare* has been dealt with by Belli (see our Art. 756*.) Sir William Thomson does not seem to have been acquainted with Belli's memoir, nor does he, I think, meet such arguments as those of Belli.

[1651.] *Dynamical Problems regarding Elastic Spheroidal Shells and Spheroids of Incompressible Liquid. Philosophical Transactions*, Vol. 153, pp. 583–616. London, 1864 (*M. P.*, Vol. III., pp. 351–94). This paper was read November 27, 1862. It contains a solution of *Lamé's Problem* by means of *solid spherical harmonics*. The introduction of these harmonics seems to be due independently to Sir William Thomson and Clebsch: see our Art. 1397. In a note added to the memoir in December, 1863, Sir William Thomson refers to Lamé's memoir of 1854 (see our Art. 1111*), which he had only discovered after the communication of his own paper to the Royal Society.

The form in which the analysis has been applied in the present paper is very different from that chosen by Lamé (who uses throughout polar coordinates); but the principles are essentially the same, being merely those of spherical harmonic analysis, applied to problems presenting peculiar and novel difficulties (p. 616 ; *M. P.*, Vol. III., p. 394).

Whether it is easier to deal with the strain of elastic spherical bodies by means of polar or cartesian coordinates will, perhaps, be always a matter of opinion, and depends very much on the method in which the student has first approached the problem. At the same time the solutions of a considerable number of interesting problems concerning the physics of the earth depend, assuming perfect elasticity, only on harmonics of the second order, and the discussion can in these cases be carried out in an especially easy and elementary manner by aid of polar-coordinates,—which, indeed, give the results in the form most convenient for geometrical interpretation.

As we have already dealt at length with *Lamé's Problem* in our first volume (see Arts. 1112*–1148*) and there put on record the general forms required for special investigations we shall content ourselves here by referring to the principal results of Sir William Thomson's treatment.

[1652.]   Taking the body-shift-equations of the type:

$$\mu \nabla^2 u + (\lambda + \mu) \frac{d\theta}{dx} + \rho X = 0 \dots\dots\dots\dots \text{(i)},$$

if we write, $p = -(\lambda + \frac{2}{3}\mu)\,\theta$, we change the type to:

$$\mu \nabla^2 u - \frac{\lambda + \mu}{\lambda + \frac{2}{3}\mu} \frac{dp}{dx} + \rho X = 0 \dots\dots\dots\dots\text{(ii)}.$$

Here p is the mean normal pressure per unit of surface of a small portion of the solid.   Put $\lambda = \infty$, and $\theta = 0$, and we have for an incompressible solid three equations of the type:

$$\mu \nabla^2 u - \frac{dp}{dx} + \rho X = 0,$$

and

$$\frac{du}{dx} + \frac{dv}{dy} + \frac{dw}{dz} = 0$$

to find the four unknowns u, v, w and p (§§ 4–5).   See our Arts. 1215 and 1217.

[1653.]   To remove the body-forces assume

$$
\begin{aligned}
u &= u' + u_0 = u' + \frac{1}{\mu}\left(U - \frac{\lambda + \mu}{\lambda + 2\mu} \frac{d\chi}{dx}\right), \\
v &= v' + v_0 = v' + \frac{1}{\mu}\left(V - \frac{\lambda + \mu}{\lambda + 2\mu} \frac{d\chi}{dy}\right), \\
w &= w' + w_0 = w' + \frac{1}{\mu}\left(W - \frac{\lambda + \mu}{\lambda + 2\mu} \frac{d\chi}{dz}\right),
\end{aligned}
\quad \Bigg\} \dots\dots \text{(iii)}.
$$

where $U = -\nabla^{-2}(\rho X)$, $V = -\nabla^{-2}(\rho Y)$, $W = -\nabla^{-2}(\rho Z)$,

$$\chi = \nabla^{-2}\left(\frac{dU}{dx} + \frac{dV}{dy} + \frac{dW}{dz}\right)$$

and therefore U, V, W, χ can theoretically be found[1].

[1] Thomson and Tait's *Natural Philosophy*, Part II., Arts. 730–1: see our Art. 1715.

On substitution the body-shift-equations reduce to the type :

$$\mu \nabla^2 u' + (\lambda + \mu)\, d\theta'/dx = 0 \dotfill \text{(iv)},$$

which is the form from which Sir William starts his investigation (§ 2 and §§ 38–44).

When a force-function exists, (iii) can be much simplified : see § 42 and our Arts. 1658 and 1716.

When the conditions of the problem are that the surface-shifts are given for the spherical shell, then the above values of u_0, v_0, w_0 must be subtracted from these given surface-shifts before the problem is stated in its reduced form.

When the conditions of the problem are that the surface-stresses are given, then the stresses which result from these shifts at the surface must be deducted from the given surface-stresses before the problem is solved from (iv). Special cases of this are dealt with in §§ 42–3.

[1654.] §§ 7–13 give the general solutions of the equations of the type

$$\mu \nabla^2 u + (\lambda + \mu)\, d\theta/dx = 0 \dotfill \text{(v)}.$$

These are :

$$u = \sum_{i=0}^{i=\infty} \left\{ u_{i+1} + u'_{i-1} r^{-2i+1} - \frac{(\lambda+\mu)\, r^2}{2} \frac{d}{dx} \left[\frac{\psi_i}{(\lambda+3\mu)\, i + \mu} \right. \right.$$
$$\left. \left. - \frac{\psi'_i r^{-2i-1}}{(\lambda+3\mu)\, i + (\lambda+2\mu)} \right] \right\},$$

$$v = \sum_{i=0}^{i=\infty} \left\{ v_{i+1} + v'_{i-1} r^{-2i+1} - \frac{(\lambda+\mu)\, r^2}{2} \frac{d}{dy} \left[\frac{\psi_i}{(\lambda+3\mu)\, i + \mu} \right. \right.$$
$$\left. \left. - \frac{\psi'_i r^{-2i-1}}{(\lambda+3\mu)\, i + (\lambda+2\mu)} \right] \right\},$$

$$w = \sum_{i=0}^{i=\infty} \left\{ w_{i+1} + w'_{i-1} r^{-2i+1} - \frac{(\lambda+\mu)\, r^2}{2} \frac{d}{dz} \left[\frac{\psi_i}{(\lambda+3\mu)\, i + \mu} \right. \right.$$
$$\left. \left. - \frac{\psi'_i r^{-2i-1}}{(\lambda+3\mu)\, i + (\lambda+2\mu)} \right] \right\},$$
$$\dotfill \text{(vi)},$$

where
$$\psi_i = \frac{du_{i+1}}{dx} + \frac{dv_{i+1}}{dy} + \frac{dw_{i+1}}{dz},$$

and
$$\psi'_i \, r^{-2i-1} = \frac{d\,(u'_{i-1} r^{-2i+1})}{dx} + \frac{d\,(v'_{i-1} r^{-2i+1})}{dy} + \frac{d\,(w'_{i-1} r^{-2i+1})}{dz},$$

u_i, v_i, w_i, u'_i, v'_i, w'_i, denoting six solid harmonics of degree i.

I am inclined to think the separation of the solution into two elements one of which depends on the twist terms pure and simple—after the manner of Clebsch : (see our Arts. 1394–5)—would have given to (vi) a more concise form.

[1655.] §§ 14–18 determine the values of the six typical solid harmonics u_i, v_i, w_i, u'_i, v'_i, w'_i in terms of the spherical surface harmonics which determine the values of the shifts at the inner and outer surfaces of the shell for the particular problem of given surface-shifts.   Thus for $r = a$, and $r = a'$, we have

$$(u)_{r=a} = \Sigma A_i, \qquad (u)_{r=a'} = \Sigma A'_i,$$
$$(v)_{r=a} = \Sigma B_i, \qquad (v)_{r=a'} = \Sigma B'_i,$$
$$(w)_{r=a} = \Sigma C_i, \qquad (w)_{r=a'} = \Sigma C'_i,$$

and the problem is to find the six solid harmonics u_i, v_i, w_i, u'_i, v'_i, w'_i in terms of A_i, B_i, C_i, A'_i, B'_i, C'_i. The problem presents little difficulty beyond rather cumbersome algebraical expressions, the length of which prevents their being reproduced here.   For the case of the first one or two harmonics, which are really those of chief practical and physical interest, the reader will find it easy to reproduce a simple form of the investigation for himself.

[1656.]   §§ 21–30 deal with the case when the two surfaces of the shell are subjected to given surface stresses.   Here the components P, Q, R of load parallel to the axes of x, y, z on an element of the surface of the shell are shewn to be given by the type :

$$Pr = \lambda\theta x + \mu\left\{\left(r\frac{d}{dr} - 1\right)u + \frac{d\zeta}{dx}\right\},$$

where

$$r\frac{d}{dr} = x\frac{d}{dx} + y\frac{d}{dy} + z\frac{d}{dz},$$

and

$$\zeta = ux + vy + wz$$

$$\qquad\qquad\qquad\qquad\qquad\text{......... (vii).}$$

The values of u, v, w and θ as given by (vi) have then to be substituted in (vii), and P, Q, R reduced to proper solid harmonic form.

Sir William Thomson shews that the surface stresses are given by the type (Equation (43), § 28):[1]

$$Pr = \mu \sum_{i=-\infty}^{i=\infty}\left\{(i-1)u_i - 2(i-2)M_i r^2 \frac{d\psi_{i-1}}{dx}\right.$$
$$\left. - N_i r^{2i+1}\frac{d(\psi_{i-1}\,r^{-2i+1})}{dx} - \frac{1}{2i+1}\frac{d\phi_{i+1}}{dx}\right\},$$

where

$$\psi_{i-1} = \frac{du_i}{dx} + \frac{dv_i}{dy} + \frac{dw_i}{dz},$$

$$\phi_{i+1} = r^{2i+3}\left\{\frac{d(u_i\,r^{-2i-1})}{dx} + \frac{d(v_i\,r^{-2i-1})}{dy} + \frac{d(w_i\,r^{-2i-1})}{dz}\right\},$$

and

$$M_i = \tfrac{1}{2}\frac{\lambda+\mu}{(\lambda+3\mu)i - (\lambda+2\mu)}, \quad N_i = \frac{(\lambda-\mu)i + (2\lambda+3\mu)}{(2i+1)\{(\lambda+3\mu)i - (\lambda+2\mu)\}}$$

$$\qquad\qquad\qquad\qquad\qquad\qquad\text{(viii).}$$

---

[1] Sir William here replaces the double series of terms in (vi) by a single series.   The terms in $u'_{i-1}\,r^{-2i-1}$ and $\psi'_i\,r^{-2i-1}$ are clearly harmonics of degree $-(i+1)$.   Thus we may drop them in (vi), if we note that the summation is to be from $i = \infty$ to $-\infty$.   This I think should, perhaps, have been been more clearly explained in §§ 28—9, where the range of the summations is not directly stated.

As in the case of given shifts, the surface stresses will give us six surface-harmonics of each degree, *e.g.*

$$(P)_{r=a} = \Sigma A_i, \qquad (P)_{r=a'} = \Sigma A'_i,$$

$$(Q)_{r=a} = \Sigma B_i, \qquad (Q)_{r=a'} = \Sigma B'_i,$$

$$(R)_{r=a} = \Sigma C_i, \qquad (R)_{r=a'} = \Sigma C'_i.$$

These six individual surface-harmonics must then be equated to the terms in the values of P, Q, R in (viii) which lead to surface-harmonics of the ith degree for the two values respectively of $r = a$ and $r = a'$. These surface-harmonics will arise partly from positive and partly from negative values of $.i$ in the expressions for the stresses. The method by which this may be accomplished is indicated rather than carried out in §§ 29–30, and for the general case would require the addition of a large amount of algebraical work which is only suggested. Even Lamé, who carried the solution further than Sir William Thomson, still leaves it in the form of linear equations for the undetermined constants : see our Arts. 1133* and 1141*.

[1657.] The method in which the terms of (vi) in u'_i and ψ'_{i-1} are dropped in § 27 and reintroduced with a different notation in § 29 is not a little likely to puzzle the reader. Here as elsewhere in the discussion of this problem, the method of the general solution does not seem the readiest to reach the simpler cases, which are after all those most frequently occurring in physical applications.

[1658.] The interesting general case, when the force-function is a harmonic, W_{i+1}, of the $(i+1)$th degree is worked out by Sir William Thomson in §§ 44–7. He takes $\rho X = -dW_{i+1}/dx$, $\rho Y = -dW_{i+1}/dy$, $\rho Z = -dW_{i+1}/dz$, and he indicates, without fully determining all the constants, the solution for the case of a spherical shell subjected to no surface-loading (§§ 45–6).

For the particular case of a solid sphere with no surface-forces, he does fully determine all the constants. The shifts are then given by (§ 47) :

$$(u, v, w) = G_{i+1}\left(\frac{d}{dx}, \frac{d}{dy}, \frac{d}{dz}\right) W_{i+1} + G'_{i+1}\left(\frac{d}{dx}, \frac{d}{dy}, \frac{d}{dz}\right)(W_{i+1} r^{-2i-3}),$$

where

$$G_{i+1} = -\frac{(i+1)\left[(\lambda + \mu)(i+3) - \mu\right]a^2}{2\mu\{(\lambda + \mu)\left[2(i+2)^2 + 1\right] - \mu(2i+3)\}}$$

$$+ \frac{\left[(i+2)(2i+5)(\lambda + \mu) - (2i+3)\mu\right]r^2}{2\mu(2i+3)\{(\lambda + \mu)\left[2(i+2)^2 + 1\right] - \mu(2i+3)\}},$$

$$G'_{i+1} = \frac{(i+1)(\lambda + \mu)r^{2i+5}}{\mu(2i+3)\{(\lambda + \mu)\left[2(i+2)^2 + 1\right] - \mu(2i+3)\}}.$$

As a corollary we may note the case of chief physical interest for which $i = 1$, we then have :

$$(u, v, w) = G_2 \left(\frac{d}{dx}, \frac{d}{dy}, \frac{d}{dz} \right) W_2 + G_2' \left(\frac{d}{dx}, \frac{d}{dy}, \frac{d}{dz} \right) (W_2 r^{-5}),$$

where
$$G_2 = \frac{-10 (4\lambda + 3\mu) a^2 + (21\lambda + 16\mu) r^2}{10\mu (19\lambda + 14\mu)},$$

$$G_2' = \frac{4 (\lambda + \mu) r^7}{10\mu (19\lambda + 14\mu)}.$$

Sir William Thomson calculates W_2 for the case of the disturbing force due to the tides raised in the solid earth by a distant body. If m be the mass of the tide raising body, c its distance and ρ the density of the earth, he finds (§§ 49–51) :

$$W_2 = -\frac{m\rho}{2c^3} (2x^2 - y^2 - z^2).$$

The application of this has been discussed in the other works by our author dealt with in our Arts. 1663–4, and 1720–6.

[1659.] § 54 gives the value of the shifts of a *solid* sphere for given surface displacements, and indicates the like results for a spherical cavity in an infinite elastic solid.

§§ 55–8 deal with the oscillations of shape in a gravitating liquid sphere. A simple harmonic normal displacement of the ith order has for period

$$2\pi \sqrt{\frac{a}{g} \frac{2i+1}{2i(i-1)}},$$

where a is the radius of the sphere and g gravity at its surface. For the case of $i = 2$, or an ellipsoidal deformation, the length of the isochronous pendulum at the sphere's surface is $\frac{5}{4}a$. If the liquid globe were homogeneous and $5\frac{1}{2}$ times the density of water, and of the size of the earth, the period would be 1 hr. 34 m. 24 s. We may compare this with the result for a homogeneous elastic sphere given an ellipsoidal deformation of the type $u = A Y_2 \cos kt$. Lamb finds for a globe of the size of the earth and of the density and rigidity of steel a period of 1 hr. 18 m. A difference of less than 2 minutes is made in the result whether we suppose steel incompressible or of uniconstant isotropy. Thus the earth if it were as rigid as steel would oscillate more rapidly than if it were made of a liquid $5\frac{1}{2}$ times as dense as water : see *Proceedings, London Mathematical Society*, Vol. XIII., pp. 211–2. London, 1882.

Sir William Thomson in his paper on the rigidity of the earth (*M. P.* Vol. III.. p. 313) says (§ 3):

A steel globe of the same dimensions [as the earth], without mutual gravitation of its parts, could scarcely oscillate so rapidly [as 1 hr. 34 m. 24 s.], since the velocity of plane waves of distortion in steel is only about 10,140 feet per second, at which rate a space equal to the earth's diameter would not be travelled in less than 1 hr. 8 m. 40 s.

As a matter of fact Lamb finds, if τ be the time a wave of distortion would take to traverse the earth's diameter, and P the period of oscillation: $P = \tau / \cdot 848$ if the material be incompressible and $= \tau / \cdot 840$ if it possess uniconstancy. Thus Sir William's minimum estimate based on the liquid sphere is about 16 *per cent.* in excess.

[1660.] The memoir besides dealing with spherical shells points out that the problem of an infinite plane plate of homogeneous isotropic material, with given shifts or stresses at its plane faces might be treated as a limiting case of the spherical shell (§§ 19–20 and §§ 31–4). To work out the plate, however, as a limiting case of a spherical shell would involve, for the general case some rather formidable analytical difficulties.

Sir William Thomson in §§ 32–4 briefly sketches a different method of solution.

The following system of shifts will be found to satisfy the body-shift-equations of elasticity:

$$u = U - \frac{(\lambda + \mu)\,x}{\lambda + 3\mu}\,\psi,$$

$$v = V - \frac{(\lambda + \mu)\,x}{\lambda + 3\mu}\,\frac{d}{dy}\left(\int \psi dx\right),$$

$$w = W - \frac{(\lambda + \mu)\,x}{\lambda + 3\mu}\,\frac{d}{dz}\left(\int \psi dx\right),$$

where U, V, W satisfy $\nabla^2 \phi = 0$, while:

$$\psi = \frac{dU}{dx} + \frac{dV}{dy} + \frac{dW}{dz},$$

and $\int \psi dx$ is to be so taken that it also satisfies $\nabla^2 \phi = 0$.

Sir William Thomson now remarks that if we take

$$U = (fe^{-px} + f'e^{px}) \sin(sy) \sin(tz),$$

$$V = (ge^{-px} + g'e^{px}) \cos(sy) \sin(tz),$$

$$W = (he^{-px} + h'e^{px}) \sin(sy) \cos(tz),$$

subject to the condition $p^2 = s^2 + t^2$, we have a solution capable of giving over the faces of a plate (taken as $x = 0$ and $x = a$)

$$\widehat{xx_0} = A \sin (sy) \sin (tz),$$
$$\widehat{xy_0} = B \cos (sy) \sin (tz),$$
$$\widehat{xz_0} = C \sin (sy) \cos (tz),$$

and three like expressions for $\widehat{xx_a}$, $\widehat{xy_a}$, $\widehat{xz_a}$ with A', B', C' for A, B, C.

Hence by a series of such terms we have the most general solution according to Fourier's principles.

As a matter of fact, if the solution were completed, we should merely reach a somewhat extended form of Lamé and Clapeyron's rather unwieldy results in quadruple integrals, of which since their statement in 1828 no practical use has, so far as I am aware, ever been made: see our Arts. 1020*–1*.

[1661.]   §§ 59–71 are occupied by an Appendix entitled: *General Theory of the Equilibrium of an Elastic Solid*. This appendix was reprinted in the *Treatise on Natural Philosophy*: see Part II., pp. 461–8.

§§ 59 and 60 point out that the quantities ϵ_x, ϵ_y, ϵ_z, η_{yz}, η_{zx}, η_{xy} of our Art. 1619* (in Sir William Thomson's notation $\frac{1}{2}(A-1)$ $\frac{1}{2}(B-1)$, $\frac{1}{2}(C-1)$, a, b and c) suffice to determine the most general form of strain which can be given to a body in the neighbourhood of a point. The temperature being kept constant the strain-energy w is a function of ϵ_x, ϵ_y, ϵ_z, η_{yz}, η_{zx}, η_{xy}. Hence by the method of variation Sir William Thomson deduces equations which are identical with those of C. Neumann, or with those which flow from Kirchhoff's memoir of 1852: see our Arts. 670–1 and 1250.

In § 63 the possibility of a solution of these generalised equations of elasticity for any type of elastic body subjected to a given system of surface-shifts is indicated, and it is shewn that under certain conditions there can be only one solution of the elastic equations for this case. § 64 is a brief reference to similar results for the case of surface-stress.

§§ 65–6, 69–71 contains a short theory of elasticity for *small* strains, giving the usual results of Green's investigation of the strain-energy.

[1662.]   § 67 proves in a manner differing slightly from that of Neumann, Clebsch and Kirchhoff the uniqueness of the solution in the case of small strains, when the surface-shifts are given: see our Arts. 1198, 1255, 1278 and 1331.

§ 68 turns to the like problem when the surface data are those of load not shift, or when a force acts on the interior of the material. In this case the solution is not in general unique—configurations of unstable equilibrium occurring even with infinitely small shifts.

For instance, let part of the body be composed of a steel bar magnet; and let a magnet be held outside in the same line, and with a pole of the same name in its end nearest to one end of the inner magnet. The equilibrium will be unstable, and there will be positions of stable equilibrium with the inner bar slightly inclined to the line of the outer bar, unless the rigidity of the rest of the body exceed a certain limit.

This conclusion as to the want of uniqueness in the solution appears to be deduced from physical considerations and not from the analysis of the problem. It depends on the system of applied force itself changing its characteristics owing to the shifts of a portion of the body, e.g. from a simple pressure in an unstable position to, perhaps, a force and a couple in the stable positions of equilibrium. Such a dependence of the system of applied force on the shifts is supposed not to exist in a proof like that by which Clebsch demonstrates the uniqueness of the solution of the elastic equations : see our Art. 1331.

[1663.] *On the Rigidity of the Earth.* Royal Society Proceedings, Vol. XII., pp. 103–4. London, 1863. *Philosophical Transactions*, Vol. 153, pp. 573–82. London, 1864. *Glasgow Philosophical Society Proceedings*, Vol. V. pp. 169–70. Glasgow, 1864. *British Association Report* (Glasgow Meeting, 1876), *Transactions* pp. 1–12. London, 1877. §§ 21–32 of the *Phil. Trans.* memoir were withdrawn by the author and in the reprint of the memoir in the *Mathematical and Physical Papers*, Vol. III., pp. 312–36, these sections are replaced by the opening address to Section A in the British Association *Report* referred to above. Thus we may look upon the final form of this memoir as Art. XCV. of the collected *Papers: On the Rigidity of the Earth; Shiftings of the Earth's Instantaneous Axis of Rotation; and Irregularities of the Earth as a Time-keeper.*

Most of the important results of the memoir are embodied in the *Treatise on Natural Philosophy* and will be found pretty fully discussed in our Arts. 1719–26.

[1664.]   After some remarks on Hopkins's view that the earth
cannot be a liquid mass enclosed in a thin shell of solidified matter—a
view with which Sir William Thomson agrees—the memoir passes at
once to the consideration of "the relative values of gravitation and
elasticity in giving rigidity to the earth's figure." A formula is now
cited which may be obtained from that of our Art. 1724 (c) in the
following manner.   Put $\lambda = \infty$ in the value of ϵ', or make the elastic
mass incompressible, then we have by Art. 1724, (b) :

$$\epsilon' = \frac{\epsilon_g}{1 + \frac{19}{2}\frac{\mu}{g\rho a}} = \frac{\epsilon_g}{1 + 9 \cdot 5 \frac{\mu}{g\rho a}} .$$

If Sir William had taken uni-constant isotropy the result would
have been very nearly

$$\epsilon' = \frac{\epsilon_g}{1 \cdot 03 + 9 \cdot 17 \frac{\mu}{g\rho a}} .$$

Then follow investigations corresponding to those of our Art. 1725.
See the memoir §§ 4–7.

§ 5–15 of the memoir cover in a less concise and lucid manner the
results of the *Natural Philosophy* epitomised in our Art. 1724–5.   It
will be noticed that in that article we neglect the self-attraction of
the superficial coating of water.   This neglect is defended in § 12 of
the memoir, which thus refers to the result for *e* in our Art. 1725 :

It may be regarded as a better expression of the true tidal tendency on the
actual ocean, than the slightly different result calculated with allowance for
the effect of the attraction of the altered watery figure constituting the
equilibrium spheroid, and its influence on the figure of the elastic solid ; since
the impediments of land and the influence of the sea bottom render the actual
ocean surface altogether different from that of the equilibrium spheroid.

I do not quite follow the argument here.   The neglect of the self-
attraction of the ocean may be justifiable considering the hypothetical
and rough character of the approximation, but I do not clearly follow
why it should necessarily give a *better* result than the treatment which
includes the self-attraction.

§§ 16–20 contain suggestions for determining the amount of rigidity
of the solid earth by means of the fortnightly tide.   But to enter
into the details here would carry us beyond our limits.

[1665.]   §§ 21–33—forming part of the *British Association* address,—
deal with the *Effects of Elastic Yielding on Precession and Nutation.*
Arguments are here cited against "the geological hypothesis of a thin
rigid shell full of liquid", and the theory of a mainly solid mass, containing small hollows or vesicles filled with liquid, is supported.   A
number of results are cited with regard to the effect of interior liquidity

on the tides and on nutation (§§ 24–6) the mathematical analysis of
which has not yet been published. The general conclusions are thus
resumed in § 28 :

The state of the case is shortly this :—The hypothesis of a perfectly rigid
crust containing liquid, violates physics by assuming preternaturally rigid
matter, and violates dynamical astronomy in the solar semi-annual and lunar
fortnightly nutations ; but tidal theory has nothing to say against it. On the
other hand, the tides decide against any crust flexible enough to perform the
nutations correctly with a liquid interior, or as flexible as the crust must be
unless of preternaturally rigid matter.

§§ 34–8 deal with the irregularity of the earth as a time-keeper, and
although of much interest, do not touch on the topics of our *History*.
§§ 39–40 are appendices, the latter bearing upon the formula cited in
§ 4 : see our Art. 1664. Further Appendices deal with the *Tidal
Retardation* and the *Thermodynamic Acceleration* of the Earth's rotation. These are taken respectively from the *Philosophical Magazine*,
Vol. xxxi., pp. 533–7 (London, 1866), and the *Proceedings of the Royal
Society* (Edinburgh), Vol. xi., pp. 396–405. Edinburgh, 1882.

[1666.]   *On the Elasticity and Viscosity of Metals. Proceedings
of the Royal Society*, Vol. xiv., pp. 289–97. London, 1865. This
memoir is incorporated in the *Encylopaedia* article on Elasticity :
see our Art. 1741.

[1667.]   *On the Fracture of Brittle and Viscous Solids by
Shearing. Proceedings of the Royal Society*, Vol. xvii., pp. 312–13.
London, 1869. *Philosophical Magazine*, Vol. xxxviii., pp. 71–3.
London, 1869. The author noted on a visit to Kirkaldy's testing
works in Southwark that the rupture of bars of circular cross-
section by torsion took place in two different manners. The
rupture surface of bars of hardened steel

shewed complicated surfaces of fracture, which were such as to
demonstrate, as part of the whole effect in each case, a spiral fissure
round the circumference of the cylinder at an angle of about 45° to the
length.

On the other hand in softer or more viscous solids there was
a tendency to break right across perpendicular to the axis of the
bar.

These experiments of Kirkaldy's were confirmed by the rupture
surfaces of sealing-wax and hard steel bars, which gave spiral
fractures, while those of steel tempered to various degrees of soft-

ness, brass, copper and lead were planes perpendicular to the axis
of torsion (Compare our Art. 810). It was thus demonstrated:

that continued "shearing" parallel to one set of planes of a viscous solid,
developes in it a tendency to break more easily parallel to these planes
than in other directions, or that a viscous solid, at first isotropic, acquires
"cleavage planes" parallel to the planes of shearing (*Proc. R. S.*, p. 313).

Clearly in a hard elastic solid with small strain the direction
of greatest stretch would be an angle of 45° to the axis of the bar,
and hence the spiral fissure tends so far to confirm the maximum
stretch theory of rupture. On the other hand in the case of a
material which passes through the plastic stage before rupture, we
know that it will begin to flow when the maximum shear reaches
a certain value (see our Arts. 236, 247 and 1586), and this flow
may lead as Sir William Thomson suggests to the formation of
planes of cleavage.

The paper concludes by noticing Forbes' and Hopkins's views as
to the manner of rupture in the case of glaciers, and their reconciliation by means of the above distinction between two kinds of
rupture.

[1668.] *Treatise on Natural Philosophy* by Sir William
Thomson and Peter Guthrie Tait. Vol. I., Oxford, 1867 (pp. xxiii.
+ 727).

A new edition of this first volume, Part I. (pp. xvii. + 508),
1879, and Part II. (pp. xxv. + 527), 1883, has been issued by the
Cambridge University Press. Our references will be to the pages
of this edition[1]. A smaller work, *Elements of Natural Philosophy*,
by the same authors appeared at Oxford, 1873, and at Cambridge,
1879, in a new edition. It will not be necessary to refer, however,
to this popular *résumé* of the more important treatise. Although
only the first volume of the *Natural Philosophy* has been published
and the authors announce in the preface to *Part* II. of the second
edition that the completion of the work is definitely abandoned,
still the theory of elasticity and many of its applications naturally
fall into this first volume, and the non-appearance of the later
volumes, regretable as it is, does not inflict such a severe loss on

---

[1] A German edition of the work with a preface by von Helmholtz appeared in
Braunschweig, 1871—4, entitled: *Handbuch der theoretischen Physik übersetzt von
Helmholtz und Wertheim.*

the elastician as it does on students of other branches of mathematical physics.

The following are the portions of the *Treatise* dealing with our subject: *Part* I., §§ 119–190, 300–6, and *Part* II., §§ 573–741, 829, 832–48 and *Appendix* C. The paragraph numbers are the same in both editions, but the second edition has been largely modified and extended.

[1669.] Part I. discusses our subject from the standpoint of strain only. In §§ 119–27 we have a discussion of the curvature and tortuosity of flat bars or rods. The following definitions are of interest:

A bent or straight rod of circular or any other form of section being given, a line through the centres, or any other chosen points of its sections, may be called its *axis*. Mark a line on its side all along its length, such that it shall be a straight line parallel to the axis when the rod is unbent and untwisted. A line drawn from any point of the axis perpendicular to this side line of reference is called the *transverse* of the rod at this point.

* * * * * * * * *

The twist (t) of a curved, plane or tortuous, rod at any point is the rate of component *rotation of its transverse round its tangent line, per unit of length along it* (§ 120).

By the tangent line in the last definition is meant the tangent to the *axis* at the given point. *Integral twist* over any length s of the axis $= \int t ds$.

The following proposition is then shewn to hold for the twist in any part of a bar:

Let a point move uniformly along the axis of the bar: and parallel to the tangent at every instant, draw a radius of a sphere cutting the spherical surface in a curve, the hodograph of the moving point. From points of this hodograph draw parallels to the transverses of the corresponding points of the bar. The excess of the change of direction from any point to another of the hodograph, above the increase of its inclination to the transverse, is equal to the twist in the corresponding part of the bar (§ 123).

If the hodograph be a closed curve and the sphere be of unit radius the change in direction of the hodograph is simply the area enclosed by it.

[1670.] Some instructive examples of the 'Dynamics of twist in kinks' are given in § 123, rather by way of suggestion than proof at this stage. Thus a piece of steel pianoforte-wire being free from stress when straight is given any degree of twist and then bent into a circle, its ends being securely joined. This circle can then be twisted into a figure of 8, the two parts being tied together at the crossing.

The circular form, which is always a figure of free equilibrium, may be stable or unstable, according as the ratio of torsional to flexural rigidity is more or less than a certain value depending on the actual degree of twist. The tortuous 8 form is not (except in the case of whole twist $=2\pi$, when it becomes the plane elastic lemniscate of Fig. 4, § 610 [see our Art. 1694]), a continuous figure of free equilibrium, but involves a positive pressure of the two crossing parts on one another when the twist $>2\pi$, and a negative pressure (or a pull on the tie) between them when twist $<2\pi$: and with this force it is a figure of stable equilibrium (§ 123, p. 98).

[1671.] After some examples of tortuosity and twist of a geometrical character, the authors pass to the curvature of surfaces, define *anticlastic* and *synclastic* (or 'saddle-back' and 'dome') curvature (§ 128) and have some remarks of special interest for our subject on flexible and inextensible surfaces and the conditions for their development into plane surfaces. Cases of inextensibility in two directions only (those of the warp and woof) are pointed out as existing in woven materials. In this case theoretically a stretch from 0 up to $\sqrt{2}-1$ can be given in a diagonal accompanied by a squeeze from 0 to -1 in the perpendicular diagonal. It is pointed out how the grace of drapery largely depends on this power of extensibility in certain directions (§§ 142–3).

[1672.] § 154–90 deal at considerable length with the geometry of strain and form a novel and lucid discussion of a somewhat trite topic. The authors commence with a definition of *strain* and then pass to homogeneous strain, which they define as follows :

If when the matter occupying any space is strained in any way, all pairs of points of its substance which are initially at equal distances from one another in parallel lines remain equidistant, it may be at an altered distance; and in parallel lines, altered, it may be, from their initial direction; the strain is said to be homogeneous (§ 155).

The magnitude of the strain is thus not in any way limited. The analytical expressions for the coordinates x_1, y_1, z_1 of the point x, y, z after such a strain are :

$$\left.\begin{aligned}
x_1 &= [xx]\,x + [xy]\,y + [xz]\,z, \\
y_1 &= [yx]\,x + [yy]\,y + [yz]\,z, \\
z_1 &= [zx]\,x + [zy]\,y + [zz]\,z
\end{aligned}\right\} \quad \dots\dots\dots\dots\dots(i),$$

where $[xx]$, $[xy]$, etc., are nine arbitrary constants.

[1673.] Clearly any plane remains after strain a plane, any line a line, and any ellipsoid an ellipsoid. As a special case of the last result a sphere will become an ellipsoid after strain. This is Cauchy's ellipsoid : see our Art. 617*. It is termed the *strain-ellipsoid* (§ 160). Its axes are the *principal axes of the strain*.

Let the lengths of the semi-axes of this ellipsoid be a, β, γ, the radius of the unstrained sphere being unity. Then $a-1$, $\beta-1$, $\gamma-1$

are in our terminology the principal stretches s_1, s_2, s_3; the authors term them the *principal elongations*. They demonstrate the following propositions:

(*a*)   The stretch s_r of the body in the direction l, m, n is given by

$$s_r = (a^2 l^2 + \beta^2 m^2 + \gamma^2 n^2)^{\frac{1}{2}} - 1. \qquad (\S\ 164).$$

(*b*)   The angle ϕ after strain between two directions with initial direction-cosines l, m, n and l', m', n' is given by

$$\cos\phi = \frac{a^2 l l' + \beta^2 m m' + \gamma^2 n n'}{\{a^2 l^2 + \beta^2 m^2 + \gamma^2 n^2\}^{\frac{1}{2}} \{a^2 l'^2 + \beta^2 m'^2 + \gamma^2 n'^2\}^{\frac{1}{2}}}. \qquad (\S\ 164).$$

(*c*)   The angle χ after strain between two planes, the equations of which are $lx + my + nz = 0$ and $l'x + m'y + n'z = 0$ before strain, is given by:

$$\cos\chi = \frac{l l'/a^2 + m m'/\beta^2 + n n'/\gamma^2}{\{l^2/a^2 + m^2/\beta^2 + n^2/\gamma^2\}^{\frac{1}{2}} \{l'^2/a^2 + m'^2/\beta^2 + n'^2/\gamma^2\}^{\frac{1}{2}}}. \qquad (\S\ 165).$$

(*d*)   There are two systems of parallel planes in which there is no distortion or the strain is a uniform spread (see our Art. 595* and Vol. I., p. 882). These are parallel to the circular sections of the strain-ellipsoid (§ 167).

[1674.]   The authors now (§ 169–76) deal by an elegant geometrical analysis with the special case of the strain specified by $a - 1$, 0, and $1/a - 1$ as principal stretches. This strain corresponds to the distortion of a lozenge into an equal lozenge by squeezing its greater axis till it is of length equal to the initially less axis and stretching the less till it is of length equal to the initially greater axis. It is shewn that this strain corresponds to the sliding of one plane in the material parallel to a second, or to what we term in this *History* a *slide*. The authors term it a *simple shear*. This is unfortunate, for that word was introduced by George Stephenson to denote the transverse *stress* in rivets, and has been consistently used in this sense of stress by Rankine and the majority of engineers since. Its present confused use partly for stress and partly for strain has been avoided in our own work by the introduction of the term *slide* for shearing strain.

The principal axes of a slide are defined (§ 173) to be the axes of maximum stretch and maximum squeeze. a is the *ratio of the slide*, and the amount of relative motion per unit distance between the planes of no distortion is *the amount of the slide*. It is shewn to equal $a - 1/a$, or the excess of the maximum stretch over the maximum squeeze (§§ 174–5).

[1675.]   An interesting problem appears, I think, for the first time in the history of our subject in § 177.   It is shewn that a pure stretch, a simple slide and a dilatation combine to form the most general homogeneous strain.   Thus if that strain be denoted by a, β, γ, it may be considered as compounded of: (i) a uniform dilatation denoted by a stretch $\sqrt{a\gamma}$ in *all* directions, superimposed on (ii) a pure stretch $\beta/\sqrt{a\gamma}$ in the direction of the principal axis β, superimposed on a simple slide of amount $\sqrt{a/\gamma} - \sqrt{\gamma/a}$ in the plane of the other two principal axes.

[1676.]   In § 181 the authors carry out an analytical investigation of formulae (i) of our Art. 1672.   They inquire whether there is a line in the body which remains unaltered in direction by strain, or, if values of x, y, z can be found for which $x_1/x = y_1/y = z_1/z = \zeta$, say.   It is easy to see that there results a cubic for ζ, so that one such line always exists. There may, however, be three real solutions, in which case there will be three lines of directional identity, oblique to each other in the most general case.   In the special case, however, when

$$[yz] = [zy], \qquad [zx] = [xz], \qquad [xy] = [yx]\dots\dots\dots\dots(ii),$$

these three lines will be always real and rectangular, coinciding with the principal axes of the strain-ellipsoid.

In the course of the analysis the equation of the *inverse strain-ellipsoid* (or the ellipsoid into which a sphere in the strained condition would change, if the strain were remitted: see Vol. I., p. 882) is given. If $[XX]$, $[YZ]$ etc. represent quantities of the types:

$$[XX] = [xx]^2 + [yx]^2 + [zx]^2, \ [YZ] = [xy][xz] + [yy][yz] + [zy][zz]$$

then the equation is

$$[XX] x^2 + [YY] y^2 + [ZZ] z^2 + 2 ([YZ] yz + [ZX] zx + [XY] xy) = r^2,$$

where r is the radius of the spherical surface (p. 130).

[1677.]   The authors conclude:

that any homogeneous strain whatever applied to a body generally changes a sphere of the body into an ellipsoid, and causes the latter to rotate about a definite axis through a definite angle.   In particular cases the sphere may remain a sphere.   Also there may be no rotation. In the general case, when there is no rotation, there are three directions in the body (the axes of the ellipsoid) which remain fixed; when there *is* rotation, there are generally three such directions but not rectangular. Sometimes, however, there is but one (§ 182).

When the axes of the strain-ellipsoid are the lines which do not change their direction the strain is said to be *pure*, and relations (ii) are the necessary and sufficient conditions for a pure strain (§ 183).

[1678.]   Subject to (ii) of our Art. 1676 the formulae (i) of our Art. 1672 may be written in the form :

$$x_1 = Ax + cy + bz, \quad y_1 = cx + By + az, \quad z_1 = bx + ay + Cz \ldots\ldots(\text{iii}).$$

Let a body thus strained be strained further in the manner

$$x_2 = A_1 x_1 + c_1 y_1 + b_1 z_1, \quad y_2 = c_1 x_1 + B_1 y_1 + a_1 z_1, \quad z_2 = b_1 x_1 + a_1 y_1 + C_1 z \ldots(\text{iv}).$$

Combining (iii) and (iv) we find :

$$\begin{aligned}
x_2 &= (A_1 A + c_1 c + b_1 b)\, x + (A_1 c + c_1 B + b_1 a)\, y + (A_1 b + c_1 a + b_1 C)\, z, \\
y_2 &= (c_1 A + B_1 c + a_1 b)\, x + (c_1 c + B_1 B + a_1 a)\, y + (c_1 b + B_1 a + a_1 C)\, z, \\
z_2 &= (b_1 A + a_1 c + C_1 b)\, x + (b_1 c + a_1 B + C_1 a)\, y + (b_1 b + a_1 a + C_1 C)\, z
\end{aligned} \right\} \ldots(\text{v}).$$

Although (iii) and (iv) express irrotational strains, they give when superimposed a strain (v) which is in general rotational, or two pure strains, if superimposed, may give a pure strain and a rotation.

If the strains be small, we shall have the constants represented by capitals nearly unity, and those represented by small letters small. Hence the squares and products of small quantities being neglected, we have the *pure strain* :

$$\begin{aligned}
x_2 &= A_1 A x + (c + c_1)\, y + (b + b_1)\, z, \\
y_2 &= (c + c_1)\, x + B_1 B y + (a + a_1)\, z, \\
z_2 &= (b + b_1)\, x + (a + a_1)\, y + C_1 C z
\end{aligned} \right\} \ldots\ldots\ldots\ldots\ldots(\text{vi}),$$

arising from the superimposition of the two pure strains (§ 185).

[1679.]   Our authors now turn to discuss what they term the *entire tangential displacement* of a curve taken in a continuous solid or fluid mass. We might speak of it in the terminology of our work as the *integral tangential shift*. Consider any series of physical points forming a curve in the unstrained body. Divide this curve up into small elements, and let the length of each element be multiplied by its shift resolved in the direction of the element. If these products be summed for the curve the sum is the *integral tangential shift for the unstrained curve*. The same reckoning carried out for the strained curve is the *integral tangential shift for the strained curve*. Representing these quantities by I and I' we cite the following propositions :

$$(a) \qquad\qquad I' - I = \tfrac{1}{2}\,(D''^2 - D'^2),$$

where D' and D'' are respectively the shifts at the beginning and end of the curve as determined by the sense in which the arc is measured. Thus it follows that the integral tangential shift for a *closed* curve is the same whether reckoned along the strained or unstrained curve, and that the integral tangential shift is the same reckoned along either of two conterminous arcs.   (§§ 188–9.)

(b) Let τ_{yz}, τ_{zx}, τ_{xy} be the twist-components (see our Vol. I., p. 882) of a homogeneous strain, i.e. $\tau_{yz} = \frac{1}{2}\{[zy] - [yz]\}$, etc. in the notation of formula (i) of our Art. 1672. Then the integral tangential shift round a closed curve is given by

$$2\{\varpi_1\tau_{yz} + \varpi_2\tau_{zx} + \varpi_3\tau_{xy}\},$$

where ϖ_1, ϖ_2, ϖ_3 are the areas of the projections of the closed curve in its initial position on the coordinate planes yz, zx, and xy respectively.

(c) The most general homogeneous strain can be expressed by the shifts [1]:

$$u = \frac{d\psi}{dx} - \tau_{xy}y + \tau_{zx}z,$$

$$v = \frac{d\psi}{dy} - \tau_{yz}z + \tau_{xy}x,$$

$$w = \frac{d\psi}{dz} - \tau_{zx}x + \tau_{yz}y,$$

where

$$\psi = \frac{1}{2}\{(A-1)x^2 + (B-1)y^2 + (C-1)z^2 + 2(ayz + bzx + cxy)\}.$$

Thus for non-rotational homogeneous strain, if the integral tangential shifts be measured from a definite point of the body as origin up to any point x, y, z we have:

$$I = \psi, \quad I' = \psi + \frac{1}{2}\left\{\left(\frac{d\psi}{dx}\right)^2 + \left(\frac{d\psi}{dy}\right)^2 + \left(\frac{d\psi}{dz}\right)^2\right\}.$$

Thus the integral tangential shifts for the strained and unstrained curves depend only on the terminals of the curve (§ 190, (a)).

[1680.] The next stage in our authors' analysis of strain is to consider the strain round any point when a body is submitted to a heterogeneous strain. They shew that " at distances all round any point, so small that the first terms only of the expressions by Taylor's theorem for the differences of displacement are sensible, the strain is sensibly homogeneous (p. 140) ".

In other words if u, v, w be the shifts of x, y, z relative to any axes:

$$x_1' - x' = \frac{du}{dx}x' + \frac{du}{dy}y' + \frac{du}{dz}z',$$

$$y_1' - y' = \frac{dv}{dx}x' + \frac{dv}{dy}y' + \frac{dv}{dz}z',$$

$$z_1' - z' = \frac{dw}{dx}x' + \frac{dw}{dy}y' + \frac{dw}{dz}z',$$

---

[1] The expressions for ϖ, ρ, σ in § 190, (a) have wrong signs.

where x', y', z' are the coordinates relative to the given point and to the selected axial directions of any point in its neighbourhood before, and x_1', y_1', z_1' the coordinates after strain. Clearly we have for the quantities $[xx]$, $[yz]$, etc. of our Art. 1672,

$$[xx] = 1 + \frac{du}{dx}, \quad [yz] = \frac{dv}{dz}, \quad [zy] = \frac{dw}{dy}, \quad \text{etc.}$$

This result obviously assumes that the second shift-fluxions

$$\frac{d^2u}{dx^2}, \quad \frac{d^2u}{dydz}, \quad \text{etc.},$$

can never be infinitely great as compared with the first shift-fluxions

$$\frac{du}{dx}, \quad \frac{du}{dy}, \quad \text{etc.}$$

[1681.] If dS be any element of a surface in the body, l, m, n the direction cosines of its normal, τ_{yz}, τ_{zx}, τ_{xy} the twist-components at the point x, y, z of the surface, we easily find

$$2\iint (l\tau_{yz} + m\tau_{zx} + n\tau_{xy})\, dS = \int (u\,dx + v\,dy + w\,dz)$$
$$= \text{the integral tangential shift round the perimeter of } S.$$

If T be the resultant twist and ϕ the angle its direction makes with the normal to the corresponding element of S, we see that the quantity $\iint T \cos \phi\, dS$ is constant for all surfaces drawn through the same curve.

When the twist vanishes, or the conditions:

$$dv/dz = dw/dy, \qquad dw/dx = du/dz, \qquad du/dy = dv/dx$$

are satisfied, then $u\,dx + v\,dy + w\,dz$ is a perfect differential; or when a strain is irrotational we must have u, v and w of the form:

$$u = \frac{dF}{dx}, \qquad v = \frac{dF}{dy}, \qquad w = \frac{dF}{dz}.$$

In this case $\int (u\,dx + v\,dy + w\,dz)$ may be termed the *shift-function* ("the displacement function") and we see that it represents the entire tangential shift from the fixed point of the body up to the point x, y, z along any curve whatever. (§ 190, (i)–(l).)

[1682.] Some notice must be taken here of §§ 300–6, which deal with the impact of elastic bodies. The authors, objecting strongly to the terminology usually adopted in the discussion of Newton's Law[1] in the text-books, yet appear to give their sanction to the validity of that law in one of the worst forms in which

---

[1] *i.e.* that the velocity of rebound is proportional to the velocity of impact for the same two bodies.

it is often stated. They say that the results of recent experiments[1]
have confirmed Newton's Law, but they do not say that the
results of more recent theory are opposed to it. They speak
of Newton's finding the coefficient of restitution, e, for balls of
compressed wool to be $\frac{5}{9}$, of iron nearly the same and glass $\frac{15}{16}$, but
they fail to point out that e probably depends not only on the
elastic nature of the materials in contact, but also on the masses
of the colliding bodies, their shapes and their dimensions[2]: see
our Arts. 941*, 1183*, 1523*, 209, 213, 217 and 1224.

In § 302 the generalised Hooke's Law (see our Art. 8*) is
cited to demonstrate that Newton's experimental law is consistent
with perfect elasticity, but the argument used is not opposed to
the variation of e with the masses, sizes and shapes of the colliding
bodies.

[1683.] §§ 303–4 deal in a very brief manner with the longitudinal
impact of cylindrical bars. The only case dealt with is that of Case (i)
of our Art. 213, it being noted that e in this case is theoretically the
ratio of the lesser to the greater *mass*. This statement ought to have
saved the writers of elementary text-books, which have been largely
based on the *Natural Philosophy*, from making the erroneous statements
current with regard to the nature of e. Thomson and Tait refer for
further particulars to their discussion of the kinetics of elastic solids.
As that portion of their work has never been written a reference in the
second edition (1879) to Saint-Venant's elaborate memoir of 1867 might
have been helpful to the writers of elementary works.

[1684.] §§ 305–6 refer to the amount of energy lost in vibrations,
and notice that but a small part of the whole kinetic energy can
remain in the form of vibrations after the impact of solid spheres
of glass or ivory. This is the view since taken by Hertz and Boussinesq
of the collision of massive bodies, and although the theory of the
vibrations of solid elastic spheres has not yet been so fully worked
out, that its application to the case of vibrations produced by
impact is possible there is still no doubt that Hertz's theory throws a
large amount of light on this all-important problem: see our Arts. 1515–7.

---

[1] The experiments seem far from conclusive, the influence on e of variation of
mass, size and shape have not yet been investigated with the needful accuracy: see
Encyklopaedie der Naturwissenschaften, Handbuch der Physik, Bd. I., S. 296–301.
[2] Even such a great authority as Dr Routh speaks of e as a constant ratio
depending on the *material* of the balls and does not hint that it may vary with
their mass and size : *Elementary Rigid Dynamics*, 1882, p. 158. In one of the
most recent Cambridge text-books we are told that e depends on "the substances
of which the bodies are made and is independent of the masses of the bodies".
(Loney's *Elementary Dynamics*, p. 203. Cambridge, 1889). All the elementary
books seem to go astray on this point.

It brings out in particular why in the case of a *hollow* sphere much of the kinetic energy of the blow is spent in vibrations, while in the case of the solid sphere this loss is little.

[1685.] We now pass to Part II. of the *Natural Philosophy*, which deals with the dynamical aspect of strain, *i.e.* with stress and the stress-strain relations. The authors pass from the treatment of rigid bodies by the stages: (i) flexible strings, (ii) rods and wires, and (iii) thin plates to the complete elastic equations for any solid body. This arrangement, while certainly carrying the student by a graduated course to the more complex problems of elasticity, fails, I think, to fully emphasize the transcendent difficulties associated with the wire and plate problems, nor does it bring into clear relation the elastic coefficients of wires and plates and those for extended masses of the same material: see our Arts. 383–94, 1236, 1251–67, 1292–1300, 1358–1364 and 1418–40.

[1686.] §§ 573–87 deal with the general theory of catenaries, *i.e.* of flexible and sensibly inextensible cords hanging freely, or constrained to lie on smooth or rough surfaces. There is nothing so closely related to our subject in this discussion that it need detain us here.

[1687.] §§ 588–626 deal with wires and rods and present many points of interest. The authors define a *wire* to be "an elongated body of elastic material...bent or twisted to any degree, subject only to the condition that the radius of curvature and the reciprocal of the twist [see our Art. 1669] are everywhere very great in comparison with the greatest transverse dimension". They suppose that certain constants termed by them "the constants of flexural and torsional rigidity" are *known*. These constants for an isotropic wire are the $E\omega\kappa_1^2$, $E\omega\kappa_2^2$ and $E\omega\chi$ of our Art. 1287. The axial stretch in the wire is neglected throughout the investigation. This is justified in § 592 (see, however, our Arts. 1592*, 1367, 1373 and 1425). I do not think, however, that the "conditional limits", frequently referred to in the discussion as those of § 588, and apparently amounting only to the single one cited above in the definition of *wire* are really sufficient. They do not seem to me to exclude the possibility of set, nor the application of such a system of load that in a small portion of the wire axial

stretch or transverse slide may become of relative importance.
Further it seems practically assumed that the system of load will
solely produce curvature and twist, and that the effects of the
distortion of the cross-sections are *nil* or negligible. This is the
fact, indeed, for the cases dealt with by our authors, but some
word of warning seems very necessary, especially when we
remember that the constant of torsional rigidity can only be
ascertained after the form of the distorted cross-section has been
actually calculated[1].

[1688.] Premising that their wire may be isotropic, crystalline,
fibrous, or laminated in structure, Thomson and Tait state (§ 591) the
following "laws of flexure and torsion".
Suppose the resultant stress of the matter on one side of any cross-
section of the wire on matter on the other side to be reduced to a single
force through any point of the cross-section and a single couple, then:

I.   The twist and curvature of the wire in the neighbourhood of this
section are independent of the force, and depend solely on the couple.

II.   The curvatures and rates of twist, producible by any several couples
separately, constitute, if geometrically compounded, the curvature and rate of
twist which are actually produced by a mutual action equal to the resultant
of those couples.

[1689.] In § 592 the line of centroids of the cross-sections is defined
as the *elastic central line*. This line in our work is spoken of as the
central line, the term *elastic line* being retained especially for its strained
form. The series of points of zero stretch in the plane of the cross-section
form the neutral axis, and the points of section of these neutral axes by
the corresponding osculating planes of the elastic line form the *neutral
line*. Now Thomson and Tait write:

the elastic central line remains sensibly unchanged in length to whatever
stress within our conditional limits [see our Art. 1687] the wire be subjected.
The elongation or contraction produced by the neglected resultant force, if
this is in such a direction as to produce any, will cause the line of *rigorously
no elongation* to deviate only infinitesimally from the elastic central line, in
any part of the wire finitely curved.

This amounts practically to saying that at points of finite curvature
the central and neutral lines deviate only infinitesimally. Such a statement is, however, incorrect. An examination of the figure in our Vol.
I., p. 403, shews that the neutral line may pass at points of finite
curvature to a considerable distance from the central line. *But, as a*

---

[1] That the flexural rigidity theoretically varies with the amount of curvature is
shown in our Arts. 619—20, but this variation is really excluded by our authors'
'conditional limits'.

matter of fact, this deviation does not sensibly affect the flexural effect of the stress-couples, which is the real point upon which our authors' theory depends.

[1690.] A wire "of uniform constitution and figure throughout, and naturally straight" is now taken. Two planes of reference are drawn through its central axis cutting any cross-section at P in the lines PN_1 and PN_2. ν_1 and ν_2 are the component curvatures in two planes perpendicular respectively to PN_1 and PN_2, and τ is the twist at P. The authors then proceed as follows (§ 594):

Considering now the elastic forces called into action, we see that if these constitute a conservative system, the work required to bend and twist any part of the wire from its unstrained to its actual condition, depends solely on its figure in these two conditions. Hence if $w \cdot PP'$ denote the amount of this work, for the infinitely small length PP' of the rod, w must be a function of ν_1, ν_2, τ; and therefore if N_1, N_2, M denote the components of the couple-resultant of all the forces which must act on the section through P' to hold the part PP' in its strained state, it follows...that:

$$N_1 \delta \nu_1 = \delta_{\nu_1} w, \quad N_2 \delta \nu_2 = \delta_{\nu_2} w, \quad M \delta \tau = \delta_\tau w.$$

Law II. of our Art. 1688, or the principle of superimposition, then leads at once to w being a homogeneous quadratic function of ν_1, ν_2, τ, or

$$w = \tfrac{1}{2} \left(A\nu_1^2 + B\nu_2^2 + C\tau^2 + 2a\nu_2\tau + 2b\tau\nu_1 + 2c\nu_1\nu_2 \right) \dots \dots (i),$$

where A, B, C, a, b, c are constants of the wire.

[1691.] Now it seems to me that this investigation is wanting in accuracy in several points. First our authors' definition of twist (see our Art. 1669) when applied to a material rod or curve seems to exclude the possibility of the 'transverse' becoming inclined to the tangent, and being itself distorted by the strain. Once it is recognised that the cross-section of the wire in the cases of both flexure and torsion is distorted, it does not seem to me possible without an investigation such as that of Kirchhoff's (introducing Saint-Venant's results) to assume that the work is a function of ν_1, ν_2, and τ only. To do so appears to be only repeating the old hypothesis of Euler, Bernoulli and Coulomb under a disguised form, *i.e.* the non-distortion of the cross-sections is practically assumed without sufficient discussion under the purely geometrical definition of twist.

It cannot be said that this distortion is negligible, for it plays an important part in the determination of the constants C, a and b, for all but a rod of circular cross-section with elastic isotropy in the plane of the cross-section. The result (i) is, however, deduced

without any limitations of this kind.   That it is practically
correct for a thin wire of any cross-section may, however, be
recognised from the investigations of Kirchhoff, Clebsch and
Boussinesq: see our Arts. 1251–66, 1359–64 and 1418–36.

[1692.]   By the well-known process for reducing a homogeneous
quadratic function, w in (i) may be put into the form:

$$w = \tfrac{1}{2}\left(A_1\theta_1^2 + A_2\theta_2^2 + A_3\theta_3^2\right) \dots\dots\dots\dots\dots\text{(ii)},$$

corresponding to three component couples about three rectangular axes

$$\mathfrak{M}_1 = A_1\theta_1, \quad \mathfrak{M}_2 = A_2\theta_2, \quad \mathfrak{M}_3 = A_3\theta_3,$$

where θ_1, θ_2, θ_3 are linear functions of ν_1, ν_2 and τ.   Hence our authors
conclude:

There are in general three determinate rectangular directions PQ_1, PQ_2,
PQ_3, through any point P of the middle line of a wire, such that if opposite
couples be applied to any two parts of the wire in planes perpendicular to any
one of them, every intermediate part will experience rotation in a plane
parallel to those of the balanced couples.   The moments of the couples
required to produce unit rate of rotation round these three axes are called
the *principal torsion-flexure rigidities* of the wire.   They are the elements
denoted by A_1, A_2, A_3 in the preceding analysis. (§ 596.)

The corresponding rectangular directions are termed the three *principal axes*, and the form taken by the wire when balanced by couples round
any one of the three principal axes is a uniform helix having a line
parallel to the principal axis for axis.   The helices so obtained are the
three *principal helices*[1] (§ 598).

If one of the principal axes coincides with the central line of the
wire[2] then the three principal helices become the axis of the wire corresponding to pure torsion, and two circles corresponding to pure flexure
in either principal plane (§ 599).

[1693.]   Our authors now demonstrate a number of properties of
rods strained into helices, or of helical springs.

(*a*)   Wantzel's theorem of the helical form taken by a straight rod
of which the central line is a principal axis and the flexural rigidities
are equal, when subjected to couples in parallel planes not perpendicular to the central line, is proved in § 601: see our Arts. 1240*
and 1606*

(*b*)   We have already discussed Giulio's memoir of 1841 and Saint-
Venant's of 1844 on helical springs: see our Arts. 1219* and 1608*.

---

[1] Thomson and Tait speak of *helix* in the text and *spiral* in the margin.   The
latter word seems better reserved for a *plane* curve.   A watchspring is the true type
of spiral spring, not the spring of a spring balance, which is a helical spring.
[2] The authors speak of   common metallic wires' being 'sensibly isotropic'—a
somewhat questionable statement: see our Arts. 332*, 831*, 858*, 925* and 1271–3.

In §§ 602 and 605 our authors[1] give Saint-Venant's expressions for the force and couple, *i.e.* the results of § 605 correspond with those of our Art. 1608* and those of § 602 with the same results for the special case when $\beta_0 = \pi/2$. A comparison of Saint-Venant's method with that of Thomson and Tait is of value, as bringing out the terms supposed to be negligible in their investigation, *i.e.* the longitudinal stretch and Saint-Venant's ϵ : see our Arts. 1593*–1608*. Thomson and Tait's results are slightly more general than the corresponding conclusions of Kirchhoff (see our Arts. 1268 and 1283 (c)) which suppose the cross-section of the wire to be circular.

(c) Let l be the length of the helix, x the distance between planes through its two terminals perpendicular to its axis, ϕ the angle between planes through its axis and its two terminals in the strained condition, x_0 and ϕ_0 the corresponding quantities for the unstrained condition. Then with the notation of our Art. 1608*, $x = l \sin \beta$, $\phi = l \cos \beta / r$, and we may write :

$$
\left.
\begin{aligned}
N &= \frac{E\omega\kappa^2}{l^3} \left\{ \sqrt{(l^2 - x^2)}\,\phi - \sqrt{(l^2 - x_0^2)}\,\phi_0 \right\} \sqrt{(l^2 - x^2)} + \frac{2\mu\nu}{l^3}(x\phi - x_0\phi_0)x, \\
P &= \frac{E\omega\kappa^2}{l^3} \left\{ \sqrt{(l^2 - x^2)}\,\phi - \sqrt{(l^2 - x_0^2)}\,\phi_0 \right\} \frac{x\phi}{\sqrt{(l^2 - x^2)}} - \frac{2\mu\nu}{l^3}(x\phi - x_0\phi_0)\,\phi
\end{aligned}
\right\} \text{(i)},
$$

P being the force in the axis tending to *compress* the helix (§ 607). The authors then take $x - x_0$ and $\phi - \phi_0$ small, and deduce various results bearing on the practical use of helical springs. For example :

$$
P = \frac{1}{l^3}\left(E\omega\kappa^2 \frac{x_0^2}{l^2 - x_0^2} + 2\mu\nu \right)\phi_0^2 (x_0 - x) - \frac{1}{l^3}(E\omega\kappa^2 - 2\mu\nu)\,x_0\phi_0(\phi_0 - \phi).
$$

Hence if the spiral be of very small inclination to the axis, or x_0/l be small we have approximately :

$$
P = \frac{2\mu\nu\phi_0^2}{l^3}(x_0 - x).
$$

Thus (i) the load is proportional to the compression in the axis, a property first determined by Hooke at a much earlier period from experiment : see our Arts. 7* and 250*; (ii) helical springs act chiefly by torsion, a property first stated by Binet and after him by Giulio, J. Thomson and Kirchhoff : see our Arts. 175*, 1382* and 1283 (c); (iii) if the number of coils be n, $l = 2\pi r \times n$ nearly, if r be the radius of the helix, and $\phi_0 = l/r$, whence :

$$
P = \frac{1}{\pi}\frac{\mu\nu}{nr^3}(x_0 - x),
$$

or, the total compression, $(x_0 - x)$, for a given load varies directly as the number of coils and as the cube of the radius of the helical spring.

[1] Two misprints of the first edition α', r' for α, r and L for G in § 605 remain in the second edition.

This result agrees after proper changes of notation with that of Giulio
cited as (vi) in our Art. 1220*.

[1694.] Our authors next refer to Kirchhoff's elastico-kinetic
analogy (see our Arts. 1267, 1283 and 1364) and cite as a special case
of it the *Elastic Curve* of James Bernoulli (see our Arts. 18*–25*).
A straight wire having one set of principal axes of its cross-sections
coplanar is bent in this plane by the action of two equal and opposite
forces, F, $-F$, acting in any line in the plane taken as the axis of x,
and connected with the wire, if needful, by rigid bars. The corresponding elastico-kinetic analogy is that of a rigid body swinging on an axis
under the action of gravity (§ 613). If $1/\rho$ be the curvature we easily
find $E\omega\kappa^2/\rho = Fy$, whence the equation to the curve is $\rho y = a^2$, a being a
constant. Thomson and Tait suggest that the elastic line for this case
might be found by drawing successive arcs of circles whose radii vary
inversely as the ordinate y (§ 611). They discuss somewhat briefly
the types of solution of the differential equation $\rho y = a^2$, and depict some
of the forms[1] (traced experimentally from a flat steel spring) which the
solution may take (§ 611). These forms are of very great physical
interest and their comparison with various cases of pendulum motion is
instructive.

A conclusion worthy of remark is, that the rectification of the elastic curve
is the same analytical problem as finding the time occupied by a pendulum in
describing any given angle (§ 613).

[1695.] § 614 gives general equations for the equilibrium of a bent
and twisted rod, in some respects slightly more comprehensive than those
of Kirchhoff and in other respects slightly less luminous as to form
than those of Clebsch. The points to be considered are of a very difficult
and delicate kind, and the difficulty and delicacy are both increased by
the manner in which Thomson and Tait obtain their expression for the
strain-energy of a bent rod: see our Arts. 1687–91. According to
both Kirchhoff and Clebsch there is a certain principle involved in
the discussion of the equilibrium of elastic bodies with one or two
dimensions indefinitely small: see *Kirchhoff's Principle* referred to in
our Art. 1253. Kirchhoff on the ground of this principle neglects
the body-forces on the rod, he further supposes surface-load to be
applied only at the terminal cross-sections: see our Art. 1259. Clebsch
also supposes no surface-load except at terminal cross-sections, but he
introduces body-forces: see our Art. 1363. The absence of surface-load
seems essential to the treatment of both Clebsch and Kirchhoff, for the

---

[1] Forms drawn *directly from the equation* for a spring loaded in a great variety
of modes will be found in L. Saalschütz: *Der belastete Stab unter Einwirkung einer
seitlichen Kraft*, Leipzig, 1880, a most interesting work. Saalschütz's curves agree
closely with Thomson and Tait's experimental forms, only he gives as a rule a
smaller piece of the curve placed in a somewhat different situation. Thus compare
his Fig. 10 with their Fig. 1, his Fig. 21 with their Fig. 3, his Fig. 11 (12) with
their Fig. 5, his Fig. 30 with their Fig. 2, etc., etc.

former bases his expressions for the shifts (Art. 1360) and the latter
his expression for the strain-energy (Art. 1287) on the assumption
of Saint-Venant that $\widehat{yy} = \widehat{zz} = \widehat{yz} = 0$ (Arts. 1262 and 1286), and accordingly that the stress at the surface of the rod vanishes. Now Thomson
and Tait after taking a, β, γ as "the components of the mutual force,
and ξ, η, ζ as those of the mutual couple acting between the matter on
the two sides of the normal section through x, y, z"—x, y, z being axes
fixed in space,—proceed to take $X\delta s$, $Y\delta s$, $Z\delta s$ and $L\delta s$, $M\delta s$, $N\delta s$ as
"the components of the applied force, and applied couple, on the
portion δs of the wire" between the normal sections at x, y, z and
$x + \delta x$, $y + \delta y$, $z + \delta z$. There seems to me great difficulty about this.
Do X, Y, Z, L, M, N refer to surface load or to body-force, or to both[1]?
If they refer to surface-load, we cannot fall back on Kirchhoff and
Clebsch's discussion for the exactness of the expression of our Art. 1690
adopted for the strain-energy. If the above quantities, however,
represent merely body-forces, the generality is not greater than that of
Clebsch's investigation and the treatment seems in many respects less
luminous. Thomson and Tait's Equation (i) becomes Kirchhoff's (xxiii)
in our Art. 1265, if X, Y, Z be put zero; their Equations (i) and
(ii), supposing X, Y, Z, L, M, N to refer to body-forces, ought to be
contained in Clebsch's (viii) in our Art. 1363, but the analysis necessary
to prove the identity by transformation would be complicated.

    Thomson and Tait do not refer, like Clebsch, their force- and couple-
components (a, β, γ, ξ, η, ζ) of the total stress on a cross-section to axes
fixed in the element of the rod, but to axes fixed in space. Thus since
the rod in the general case has finite shifts their method of treating the
question does not directly bring the bending moments, shears etc., into
the equations of equilibrium. Further in the application of many
systems of applied force, the system would be known relative to axes
fixed in the element, and therefore X, Y, Z, L, M, N would be not
given directly, but only in terms of the unknown strained form of the
rod ; the Equations (4) and (5) of Thomson and Tait would thus be
still more complicated in application than they at first sight appear.
Comparing the two methods with the corresponding equations for
the elastico-kinetic analogy, we may say that Thomson and Tait's
method leads to differential equations corresponding to the motion of a
rigid body referred to axes fixed in space, while Kirchhoff and Clebsch's
method leads to differential equations more nearly corresponding with
Euler's equations for the motion of a body referred to its principal axes.

[1696.] In the following sections, §§ 615–9, the results are
applied to special cases of naturally straight wires, but the authors
pass by a somewhat abrupt transition to " to a uniform bar, beam or

---

[1] The conditions under which a surface-load may be practically replaced by a
body-force are of great importance. One investigation for a special case of continuous loading has been given by the Editor : *Quarterly Journal of Mathematics*, Vol.
xxiv., pp. 87 and 106. Cambridge, 1889.

plank" and even to continuous beams. The need here of a preliminary investigation as to the form of the distorted cross-section, and
as to the real limits within which the strain-energy of this distortion
may be neglected, becomes very manifest. In § 619 it is shewn that
the problem of a continuous beam is determinate, but the method
sketched for its general solution becomes in most practical cases
far too laborious to be workable, and a reference might have been
expected to Clapeyron's Theorem: see our Arts. 603, 607 and 893.

[1697.] We now pass to, perhaps, the most interesting part of
our authors' treatment of wires, namely, problems relating to wires
of equal or unequal flexibility rotated round their central line.
This occupies §§ 621-6. We note the following points:

(a) A wire of equal flexibility, straight when unstrained, offers
when bent and twisted in any manner no resistance to being turned
round its central line. This is the principle of the *equable elastic rotating
joint*, which admits of the rotation about any axis of one body being
transferred equably to a second body rotating about any other axis.
The wire which acts as the joint must have the tangents at the
terminals of its central line exactly in the axes of rotation of the two
bodies. If the wire be not accurately of equal flexibility there will be a
periodic inequality in the rotations of the two bodies having for period
half a turn of either; if it be not absolutely straight an inequality of
period equal to a whole turn of either (§§ 621-2).

(b) Consider a piece of wire or ribbon which in the unstrained
state has its central line a circular arc of radius a; the plane of greatest
flexural rigidity at each point being inclined at an angle a to the plane
of the central line. Let its central line be strained into a complete
circle of radius r, and let a couple $L\delta s$ applied to each element δs of
the wire in the normal plane of the central line be required to hold the
wire, so that its planes of greatest flexural rigidity make an angle ϕ
with the plane of the central line. Then the expression (i) of our Art.
1690 for the strain-energy per unit length of central line, since there is
no twist, reduces to the form:
$$w = \tfrac{1}{2}\left(Av_1{}^2 + Bv_2{}^2 + 2cv_1v_2\right),$$
or, transforming this expression by reference to the planes of principal
flexural rigidities (A_1 and A_2, $A_1 > A_2$), to:
$$w = \tfrac{1}{2}\left\{A_1\left(\frac{\cos\phi}{r} - \frac{\cos a}{a}\right)^2 + A_2\left(\frac{\sin\phi}{r} - \frac{\sin a}{a}\right)^2\right\} \quad\ldots\ldots\ldots\ldots(i).$$
Clearly:
$$L = \frac{dw}{d\phi} = -\frac{A_1\sin\phi}{r}\left(\frac{\cos\phi}{r} - \frac{\cos a}{a}\right) + \frac{A_2\cos\phi}{r}\left(\frac{\sin\phi}{r} - \frac{\sin a}{a}\right) \quad\ldots\text{(ii)},$$
$$\text{and}\quad \frac{d^2w}{d\phi^2} = -A_1\left(\frac{\cos 2\phi}{r^2} - \frac{\cos a \cos\phi}{ra}\right) + A_2\left(\frac{\cos 2\phi}{r^2} + \frac{\sin\phi\sin a}{ra}\right)\ldots\text{(iii)}.$$

Further let C be the couple in the plane of the central line, or plane of bending, which acts between the matter on either side of a cross-section then:

$$C \cos \phi = A_1 \left(\frac{\cos \phi}{r} - \frac{\cos \alpha}{a} \right),$$
$$C \sin \phi = A_2 \left(\frac{\sin \phi}{r} - \frac{\sin \alpha}{a} \right) \left.\right\} \quad\cdots\cdots\cdots\cdots\cdots \text{ (iv)}.$$

This follows from our Art. 1692, for clearly

$$C \cos \phi = N_1 \text{ and } C \sin \phi = N_2.$$

Thomson and Tait now consider most suggestive special cases of these results.

Case (i). *Rotation of a straight wire bent into the form of a hoop round its central line.*

Here $a = \infty$, therefore:

$$L = -\frac{A_1 - A_2}{2r^2} \sin 2\phi, \text{ and } \frac{d^2w}{d\phi^2} = -\frac{A_1 - A_2}{r^2} \cos 2\phi.$$

Hence when $\phi = 0$, or when planes of maximum flexural rigidity coincide with the plane of the central line, w is a maximum, and the equilibrium is unstable. When $\phi = \pi/2$, we have again equilibrium, but, as w is a minimum, it is stable (§ 623).

Case (ii). *A wire equally flexible in all directions is strained from a circular arc of radius a to a ring of radius r and then turned round its central line.*

Here $A_1 = A_2$, $a = 0$, therefore:

$$L = \frac{A_1 \sin \phi}{ar}, \quad \frac{d^2w}{d\phi^2} = \frac{A_1 \cos \phi}{ar}.$$

Hence $\phi = 0$ and $\phi = \pi$ are positions of equilibrium, the former being stable and the latter unstable (§ 624).

Case (iii). Suppose $A_2 = \infty$, which corresponds closely to the case of a flat band or metal ribbon—for example "a common hoop of thin sheet-iron fitted upon a conical vat, or on either end of a barrel of ordinary shape".

Here if the strain-energy is not to be infinite we must have $r^{-1} \sin \phi = a^{-1} \sin \alpha$, or the plane of inflexibility must make an angle $\sin^{-1}(ra^{-1} \sin \alpha)$ with the plane of the central line, when the band is bent to a radius r. We have from (iv):

$$C = \frac{A_1}{\cos \phi} \left(\frac{\cos \phi}{r} - \frac{\cos \alpha}{a} \right).$$

Hence if ϕ approaches near to $\pi/2$, or if the plane of inflexibility approaches the plane of the central line C gets extremely large and the band must snap across (§ 626).

The many suggestive points with regard to problems of stability which arise from this interesting discussion may justify the space here devoted to its reproduction.

[1698.] The next portion of Thomson and Tait's *Treatise* deals with the bending of plates (§§ 627–57). Of this §§ 627–49 give a general theory of such plates, containing much that is of a most instructive character. At the same time certain assumptions are made which it is needful for us to notice here, and which will, perhaps, be brought out best by the independent discussion of a special case.

If the plane of xy be taken as the tangent plane to a surface at the point $x = y = z = 0$, then in the neighbourhood of the origin the form of the surface is given by:

$$z = \tfrac{1}{2}\left(rx^2 + 2sxy + ty^2\right) \quad\dots\dots\dots\dots\dots (i),$$

where $r = d^2z/dx^2$, $s = d^2z/dxdy$, and $t = d^2z/dy^2$. As in the case of the rod, Thomson and Tait take the strain-energy per unit area of the plate's mid-plane to be a homogeneous quadratic function of the r, s, and t of the bent mid-plane at any point. The constants of this function are not expressed at this stage in terms of the thickness of the plate and the usual elastic coefficients (see, however, our Art. 1713), and it does not appear from the discussion how far the general equations of elasticity are satisfied by the assumptions made. I take it that, $z = 0$ being the unstrained mid-plane of the plate, §§ 634–5 really amount to the Saint-Venant-Boussinesq hypothesis that:

$$\widehat{zx} = \widehat{yz} = \widehat{zz} = 0\dots\dots\dots\dots\dots\dots(ii),$$

throughout the material of the plate.

Assuming this to be so, I propose to find an expression for the strain-energy at a given point of a plane plate with three rectangular planes of elastic symmetry, one being the mid-plane, when (ii) holds and the mid-plane at the point in question is *slightly* bent to the form (i).

[1699.] The stress-strain relations are of the form (see our Art. 117):

$$\begin{aligned}
\widehat{xx} &= as_x + f's_y + e's_z, & \widehat{yz} &= d\sigma_{yz}, \\
\widehat{yy} &= f's_x + bs_y + d's_z, & \widehat{zx} &= e\sigma_{zx}, \\
\widehat{zz} &= e's_x + d's_y + cs_z, & \widehat{xy} &= f\sigma_{xy},
\end{aligned} \Bigg\} \quad\dots\dots\dots (iii).$$

Assume the following values for the shifts:

$$\begin{aligned}
w &= \tfrac{1}{2}\left(rx^2 + 2sxy + ty^2\right) + \tfrac{1}{2}Cz^2, \\
u &= -rxz - syz + f_1(x, y), \\
v &= -sxz - tyz + f_2(x, y),
\end{aligned} \Bigg\} \quad\dots\dots\dots\dots(iv),$$

where r, s, t and C are arbitrary constants, and f_1, f_2 arbitrary functions. It will be found that u and v in (iv) have the most general values consistent with the value chosen for w and with $\widehat{zx} = \widehat{yz} = 0$. To satisfy $\widehat{zz} = 0$, we must further have $C = \dfrac{e'r + d't}{c}$ and $e'\dfrac{df_1}{dx} + d'\dfrac{df_2}{dy} = 0$.

The value chosen for w gives:

$$w_0 = \tfrac{1}{2}\left(rx^2 + 2sxy + ty^2\right)$$

for the form of the distorted mid-plane, $i.e.$ the mid-plane of the plate is bent at the origin to the form suggested by Thomson and Tait. The term $\tfrac{1}{2}Cz^2$ enables us to satisfy the relation $\widehat{zz} = 0$ at all points of the plate. Further f_1 and f_2 clearly refer only to strains in the plane of the plate uniform throughout the thickness, and these terms are therefore not due to bending at all. We may therefore neglect them from the outset[1]. Thus we conclude:

$$\left. \begin{aligned} u &= -rxz - syz, \\ v &= -sxz - tyz, \end{aligned} \qquad w = \tfrac{1}{2}\left(rx^2 + 2sxy + ty^2\right) + \tfrac{1}{2}\frac{e'r + d't}{c}z^2 \right\} \dots \text{(v)}.$$

Further:

$$\left. \begin{aligned} s_x &= -rz, \quad s_y = -tz, \quad s_z = \frac{e'r + d't}{c}z, \\ \sigma_{yz} &= 0, \quad \sigma_{zx} = 0, \quad \sigma_{xy} = -2sz \end{aligned} \right\} \dots\dots\dots\dots \text{(vi)},$$

whence:

$$\left. \begin{aligned} \widehat{xx} &= -z\left\{\left(a - \frac{e'^2}{c}\right)r + \left(f' - \frac{d'e'}{c}\right)t\right\}, \\ \widehat{yy} &= -z\left\{\left(f' - \frac{d'e'}{c}\right)r + \left(b - \frac{d'^2}{c}\right)t\right\}, \\ \widehat{xy} &= -2zfs; \\ \widehat{zz} &= \widehat{yz} = \widehat{zx} = 0 \end{aligned} \right\} \dots\dots\dots \text{(vii)}$$

These stresses will be found to satisfy the body stress-equations. Forming the expression for W the strain-energy per unit area of the mid-plane of the plate at the origin we have:

$$W = \tfrac{1}{2}\int_{-\epsilon}^{+\epsilon}\left(\widehat{xx}\,s_x + \widehat{yy}\,s_y + \widehat{xy}\,\sigma_{xy}\right)dz,$$

2ϵ being the thickness of the plate. Thus

$$W = \frac{\epsilon^3}{3}\left\{\left(a - \frac{e'^2}{c}\right)r^2 + 2\left(2f + f' - \frac{d'e'}{c}\right)rt + \left(b - \frac{d'^2}{c}\right)t^2\right.$$
$$\left. + 4f(s^2 - rt)\right\}\dots\dots\text{(viii)},$$

---

[1] Their most general forms as determined from the body stress-equations and $\widehat{zz} = 0$, are $f_1 = Gd'x + D_1 y$ and $f_2 = -Ge'y + D_2 x$, where G, D_1 and D_2 are arbitrary constants.

or with abbreviated expressions for the constants:

$$W = \frac{\epsilon^3}{3}\{H_1 r^2 + 2Grt + H_2 t^2 + 4f(s^2 - rt)\} \quad \text{...(viii bis)}.$$

Now let us find the couples acting on matter *inside* an element round lines in the mid-plane of the plate parallel respectively to the axes of x and y on strips of unit breadth and height 2ϵ perpendicular to these axes:

$$\left. \begin{aligned}
L &= -\int_{-\epsilon}^{+\epsilon} \widehat{xx}\,z\,dz = \frac{2\epsilon^3}{3}\{H_1 r + (G-2f)t\} = \frac{dW}{dr}, \\
M &= -\int_{-\epsilon}^{+\epsilon} \widehat{yy}\,z\,dz = \frac{2\epsilon^3}{3}\{(G-2f)r + H_2 t\} = \frac{dW}{dt}, \\
N &= -\int_{-\epsilon}^{+\epsilon} \widehat{xy}\,z\,dz = \frac{4\epsilon^3}{3}fs \qquad\qquad = \frac{dW}{ds},
\end{aligned} \right\} \quad \text{...... (ix)}.$$

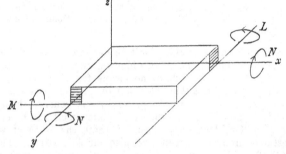

Here the couples L and N acting on an elementary strip parallel to the plane zy, tend to turn x to z and y to z, and the couples M and N acting on an elementary strip parallel to the plane zx tend to turn y to z and x to z respectively. This is indicated in the accompanying figure.

[1700.] Let $x' = x + u$, $y' = y + v$, $z' = z + w$. Then a straight line perpendicular to the mid-plane before strain is given by $x = a$, $y = \beta$. After strain this line becomes by (v):

$$\frac{x'-a}{ra+s\beta} = \frac{y'-\beta}{sa+t\beta} = -\frac{z'-\frac{1}{2}(ra^2 + 2sa\beta + t\beta^2)}{1+\frac{1}{2}Cz}.$$

This is clearly no longer an exact straight line unless $C = 0$, but if C be taken zero then $\widehat{zz}$ will not be zero and *the values of the elastic constants in* (viii) *will be changed.* On the other hand, if the plate be *very* thin, $\frac{1}{2}Cz$ will be negligible in the above result and the transverses will remain very approximately straight lines.

[1701.] We are now in a position to sum up our conclusions for a plate with three planes of elastic symmetry, one coinciding with the mid-plane of the plate:

(i)   The strain-energy per unit area of the mid-plane required to bend without stretching a small portion of the plate to the form (i) is a quadratic function of the curvatures r, s and t.

(ii)   The bending couples are given by the differentials of the strain-energy with regard to r, s and t.

These agree with Thomson and Tait's conclusions for the most general case of aeolotropy in §§ 640–1.   But:

(iii)   A straight line in the material of the plate originally perpendicular to the mid-plane, only remains approximately perpendicular to the mid-plane after bending.   If it be assumed to remain absolutely perpendicular, then the value of the flexural rigidities of the plate will not be given as the proper functions of the elastic constants a, b, c, d', e', f' and f.

This result is stated by Thomson and Tait in (2) of § 633 :

The particles in any straight line perpendicular to the plate when plane, remain in a straight line perpendicular to the curved surfaces into which its sides, and parallel planes of the substance between them, become distorted when it is bent.

Thomson and Tait cite this result as deducible from "the general theory of elastic solids."   The above investigation by the "semi-inverse method" (see our Art. 3) while by no means free from objections, may suffice, perhaps, to suggest that the result in question is an approximation to be justified for very thin plates by the general theory rather than an axiom upon which the theory of plates itself can be based.   It coincides with Kirchhoff's hypothesis of 1850, the truth of which we have questioned in our Art. 1236 (see also our Art. 1412).   The results (2) and (3) of our authors' § 633 may be both true as approximations obtained by the general theory, but they will hardly serve as a basis for an elementary theory of plates, for while (3) would lead us to introduce the term $\frac{1}{2}Cz^2$, (2) would cause us to omit it.   The whole question is, I think, on a par with the Bernoulli-Eulerian theory of rods.   The postulate that the cross-sections of a rod under flexure remain undistorted is not an *a priori* truth.   It received its first justification in Saint-Venant's memoir on flexure (see our Art. 92), which shewed that the cross-sections actually are distorted but that in certain cases this distortion is negligible.   Experiment on a moderate sized iron bar shews clearly the distortion of the cross-sections, and the distortion of the transverses can be exhibited in moderately thick iron or glass plates.

If it be objected that our investigation depends on the assumptions (a) that the sole stress lies in the plane of the plate (*i.e.* Equation (ii)) and (b) that there is only negligible stretching of the mid-plane (see our Art. 1699) we must remark that the same assumptions are made by our authors : see their §§ 634 (1) and 633 (1).   Hence their results may be compared with our investigation based on a more general theory.

The further condition on which Thomson and Tait insist, namely, that :

The deflection is nowhere, within finite distance from the point of reference, more than an infinitely small fraction of the thickness (§ 632, (3)),

does not seem involved in our investigation, and it certainly appears at first sight to exclude from treatment the most useful and ordinary applications of the theory to thin plates[1].

From the above, I think, we may conclude that Thomson and Tait's ultimate conclusions are true, but that their axioms are not absolutely necessary; they would, indeed, if treated as rigidly and not approximately correct, lead to erroneous values for the flexural rigidities. Further their mode of discussion scarcely enables us to clearly realise the nature of the internal stresses in the plate.

[1702.] Some valuable remarks and definitions occupying §§ 637–40 must be referred to here. We have already noticed Rankine's analysis of uniplanar stress (see our Arts. 453 and 465 (b)). Clearly, since L, M, N are only the z-integrals of the products of z and $\widehat{xx}$, $\widehat{yy}$ and $\widehat{xy}$, a precisely similar analysis holds for these couples. Thus let n as in our Art. 465 (b) denote the normal to a plane perpendicular to the mid-plane of the plate, and let this normal make an angle ϕ with the axis of x, and let, as in that article, t be the trace of this plane on the mid-plane; then if $\mathfrak{L}$ be the stress-couple in the plane normal to n, tending to turn n towards z and $\mathfrak{M}$ the stress-couple tending to turn t towards z, we have by integrating the results of that article multiplied by z with regard to z:

$$\mathfrak{L} = L \cos^2 \phi + M \sin^2 \phi + N \sin 2\phi, \atop \mathfrak{M} = \tfrac{1}{2}(L - M) \sin 2\phi - N \cos 2\phi. \Bigg\} \quad \cdots\cdots\cdots(\text{x}).$$

Hence clearly there are two directions determined by:

$$\tan 2\phi = 2N/(L - M)\cdots\cdots\cdots\cdots\cdots(\text{xi}),$$

for which the stress-couple, whose axis is normal to the plane over which we are reckoning the stress, vanishes.

These are termed by Thomson and Tait the *principal axes of bending stress* (§ 637). Let ϕ_0 be a value of ϕ satisfying (xi) then we may write (x):

$$\mathfrak{L} = \tfrac{1}{2}(L + M) + \Omega \cos 2 (\phi - \phi_0), \atop \mathfrak{M} = \Omega \sin 2 (\phi - \phi_0) \Bigg\} \cdots\cdots\cdots(\text{xii}),$$

where $\Omega = \sqrt{N^2 + \tfrac{1}{4}(L - M)^2}$.

---

[1] The pressure of the finger at the centre of the bottom of a round tin canister seems to produce a deflection which is far from being an "infinitely small fraction of the thickness" and which might I think be fairly discussed by the ordinary theory.

Thomson and Tait now term $\frac{1}{2}(L+M)$ a *synclastic* stress. Clearly this term gives the same stress-couple round every line in the plane of the plate. On the other hand they term Ω an *anticlastic* stress; clearly the terms in the stress-couples due to Ω are such that they change sign without alteration of magnitude when ϕ is increased by a right-angle. If the axes of x and y be the principal axes of bending stress then clearly the condition for a pure synclastic stress is that $L = M$, or for a pure anticlastic stress that $L = -M$, *i.e.* the principal stress-couples must be equal with the same or opposite signs respectively. Compare our Arts. 325 and 453.

[1703.] In § 644 our authors deduce from purely statical considerations the general equation connecting the couples L, M, N and the forces applied to any small element of the plate. This amounts in the notation of our Art. 384 to:

$$\frac{d^2L}{dx^2} + 2\,\frac{d^2N}{dx\,dy} + \frac{d^2M}{dy^2} = Z' + \frac{dX''}{dx} + \frac{dY''}{dy}\ldots\ldots\ldots\ldots(\text{xiii}).$$

Thomson and Tait do not distinguish between body-forces and surface-load, nor do they investigate how far the existence of the normal force Z' is consistent or inconsistent with the assumptions: (*a*) that the stress $\widehat{zz}$ is supposed zero (§ 634 (1)), and (*b*) that the whole thickness of the plate remains unchanged (§ 633 (3)). Further they introduce shearing stresses α, β, perpendicular to the mid-plane of the plate, which seem directly excluded by their § 634, (1), from which § 639 and equation (3) of their present investigation indirectly flows.

Assuming W a homogeneous quadratic function of the three curvature-components r, s and t, we can write down at once the equation for the normal deflection of an aeolotropic plate. This is more general than our result (v) of Art. 385, which may be at once obtained from Equations (ix) and (xiii) above. But the authors do not shew how the six flexural rigidities of the aeolotropic plate are to be determined in terms of the 21 elastic constants.

[1704.] In §§ 645—8 we have a discussion of the boundary-conditions for a thin plate. The authors point out the contradiction between Poisson and Kirchhoff, and then proceed to reconcile their conclusions. This is the famous Thomson and Tait "reconciliation" to which we have had repeated occasion to refer when discussing Poisson, Saint-Venant, Kirchhoff and Boussinesq : see our Arts. 488*, 394 and 1239, 1522-4. It has been so fully explained in our Arts. 488* and 394, and the precedence of our authors so fully acknowledged by Boussinesq, that there is no need to discuss the subject further here.

[1705.] §§ 649–653 deal with the case of a finite or infinite plate symmetrically strained round a point. The material of the plate is supposed to be isotropic. The authors only investigate that part of the shift of points on the mid-plane which is perpendicular to the plane. Their results are therefore far less complete than those of our Arts. 328–37. They give the couple L per unit length of a cylindrical surface of radius r, whose generators are perpendicular to the mid-plane, the axis of the couple being a tangent to the trace of the mid-plane on the cylindrical surface: and they further give the value of the total shear S per unit-length of the same surface parallel to the generators. They do not, however, give the radial shift or the stresses in the material of the plate: see our Arts. 329–31.

Let Z' be the force applied normally to the plate per unit area of the mid-plane. Then for an isotropic plate we have by our Art. 385:

$$\frac{1}{r}\frac{d}{dr}\left\{ r\frac{d}{dr}\left[\frac{1}{r}\frac{d}{dr}\left(r\frac{dw}{dr}\right)\right]\right\} = \frac{3}{2H\epsilon^3}Z'\dots\dots\dots(i).$$

Call the right-hand side Z'/A, then the solution is:

$$w = \frac{1}{A}\int\frac{dr}{r}\int rdr\int\frac{dr}{r}\int rZ'dr + \tfrac14 C\left(\log r - 1\right)r^2 + \tfrac14 C'r^2$$
$$+ C''\log r + C''' \dots\dots\dots(ii),$$

where C, C', C'' and C''' are undetermined constants.

Further, L and S are given, if H be the plate-modulus of our Arts. 323 and 385, by:

$$L = \frac{2\epsilon^3}{3}\left(H\frac{d^2w}{dr^2} + (H - 2\mu)\frac{dw}{rdr}\right)$$
$$= A\frac{d^2w}{dr^2} + \mathbf{c}\frac{dw}{rdr}\dots\dots\dots\dots\dots\dots(iv),$$

if $\mathbf{c} = \dfrac{2\epsilon^3}{3}(H - 2\mu)$,—and:

$$S = -A\frac{d}{dr}\left(\nabla^2 w\right)\dots\dots\dots\dots\dots\dots(v).$$

The expression for L follows at once by substituting the isotropic values of the constants in the results (ix) of our Art. 1699 and taking the axis of x to coincide with the radius r. The expression for S has been previously considered in our work (see our Art. 1536). Thomson and Tait do not apparently recognize that with their previous assumptions S ought to be zero. They appeal in fact to the "general theory of elasticity" (see §§ 633–4), but on the basis of that general theory their solution is only accurate provided stresses like $\widehat{zz}$, $\widehat{zx}$, $\widehat{yz}$, or in this case $\widehat{zz}$ and $\widehat{zr}$, are zero, or negligible as compared with the other stresses. The relative order of the stresses and the conditions under which we may neglect certain of them in the case of a thin plate form one of the most delicate investigations in the whole theory of elasticity: see our Arts. 385–8, 1438–40. Yet after an investigation which really depends for its accuracy on the neglect of the shearing stress $\widehat{zr}$ and the normal stress $\widehat{zz}$, we are confronted by the introduction without remark of

$$S = \int_{-\epsilon}^{+\epsilon} \widehat{zr} \, dz,$$

and of Z' equal presumably to $(\widehat{zz})_{+\epsilon} - (\widehat{zz})_{-\epsilon}$. Clearly if $\widehat{zr}$ and $\widehat{zz}$ are not zero the statements § 633 (2) and § 634 (1) are only approximations, and we must shew in calculating the strain-energy W (see our Art. 1699) that the terms due to these stresses are negligible. This in most cases is probably the fact, but the difficulties of the investigation do not seem effectively brought out by our authors' mode of investigation. For the value of $\widehat{zr}$ in a special case : see our Arts. 329–30. Assuming our authors' conclusions to be correct, we have (§ 649) :

$$L = -\frac{A-c}{Ar^2} \int r\,dr \int \frac{dr}{r} \int rZ'dr + \int \frac{dr}{r} \int rZ'dr$$
$$+ \tfrac{1}{2}C \left\{ (A+\mathbf{c}) \log r + \tfrac{1}{2}(A-\mathbf{c}) \right\}$$
$$+ \tfrac{1}{2}C'(A+\mathbf{c}) - C''(A-\mathbf{c})\frac{1}{r^2} \cdots\cdots\cdots\cdots\cdots \text{(vi)},$$

$$S = -\frac{1}{r}\int rZ'dr - C\frac{A}{r}\cdots\cdots\cdots\cdots\cdots\cdots\cdots\cdots\text{(vii)}.$$

[1706.] Our authors give an interesting investigation of the physical meaning of the various terms in the solution expressed by (ii), (vi) and (vii), (§ 651), and then work out the following interesting problems, for the special analysis of which we must refer our readers to the *Treatise* :

(*a*) The symmetrical flexure of a circular annulus acted upon by any given bending couples and shearing forces distributed uniformly round the outer and inner edges (§ 652).

(*b*) The same annulus acted upon in the same manner with the addition of any load symmetrically spread over its area. The solution is indicated for a special case only (§ 653). In § 655 our authors indicate the solution of the circular plate problem for the case of non-symmetrical loading. This problem had been completely solved by Clebsch some years earlier in his *Treatise :* see our Arts. 1380–2.

[1707.] Two further paragraphs (§§ 654 and 656) in the authors' treatment of plates deserve special notice.

Let $Z'(x, y)$ be the load on a plate at the point x, y, then the form of the plate equation for an isotropic material is :

$$\left(\frac{d^2}{dx^2} + \frac{d^2}{dy^2}\right)^2 w = Z'(x, y)/A \quad \text{(see our Art. 399)}.$$

A particular integral, w_0, of this is

$$w_0 = \frac{1}{4\pi^2 A} \iint dx' \, dy' \log R \iint dx'' \, dy'' \, Z' \, (x'', \, y'') \log R',$$

where $R = \sqrt{(x-x')^2 + (y-y')^2}$ and $R' = \sqrt{(x''-x')^2 + (y''-y')^2}$.

The solution is thus thrown back on the complete integral of

$$\left(\frac{d^2}{dx^2} + \frac{d^2}{dy^2} \right)^2 w = 0.$$

[1708.]   Returning to the results (iv) of our Art. 1699, let us suppose the material subjected to a uniform anticlastic curvature (Art. 1702), obtained by putting $L = M = 0$, and therefore $N = \frac{4}{3}\epsilon^3 fs$.

Hence if a rectangle with its sides parallel to the axes of x and y were to have its edges subjected to the uniform couple N (see figure of our Art. 1699), there would be anticlastic curvature and a deflection given ((iv), Art. 1699) by

$$w = \frac{3N}{4f\epsilon^3} xy.$$

Now by Thomson and Tait's 'reconciliation' a couple may be replaced by a distribution of shearing force. N the couple per unit length of the edge may be replaced by shearing forces $\frac{1}{2}P$ and $-\frac{1}{2}P$ at infinitely small distances from each other and such that $P = \frac{1}{2}N$. These will cancel each other except at the corners, where from each edge we shall have a force $\frac{1}{2}P$, or P as a whole. Thus we have the case of a rectangular plate subjected to normal forces P, P at the ends of one diagonal and normal forces $-P$, $-P$ at the ends of the other diagonal. Such a system of force therefore produces uniform anticlastic curvature in the plate and a deflection from the mid-point given by

$$w = \frac{3P}{8f\epsilon^3} xy.$$

In the case of isotropy, $f = \mu$, the slide-modulus.

Clearly the shifts in this case are given by (iv) of our Art. 1699 as :

$$w = \frac{3P}{8f\epsilon^3} xy, \quad u = -\frac{3P}{8f\epsilon^3} yz, \quad v = -\frac{3P}{8f\epsilon^3} xz.$$

Turning to our Art. 29, we see that these results exactly correspond to the torsion of a very thin rectangular prism, *i.e.* if in the results of that article we neglect c/b, write $M = 2bP$, and therefore take $\tau = 3P/(8f\epsilon^3)$, f and ϵ being respectively μ and c. Thus the breadth of our plate being $2b$, we see that uniform anticlastic curvature is produced in a thin plate by torsion. Thomson and Tait return to this identical case of the flexure and torsion problems later (§§ 719–23 : see our Art. 1713) as a means of determining the flexural rigidity of the plate in terms of the slide-modulus,—a determination which is necessary in their mode of

approaching the plate problem : see our Art. 1698. The investigation
is one of very considerable suggestiveness and great physical interest.

In § 657 the authors remark :

Few problems of physical mathematics are more curious than that presented
by the transition from this solution, founded on the supposition that the
greatest deflection is but a small fraction of the thickness of the plate, to the
solution for larger flexures, in which corner portions will bend approximately
as developable surfaces (cylindrical, in fact), and a central quadrilateral part
will remain infinitely nearly plane ; and thence to the extreme case of an infinitely thin perfectly flexible rectangle of inextensible fabric.

Unfortunately, they give no analysis, nor any suggestion for the
mathematical treatment of this case.

[1709.] The next forty-one articles (§§ 658–98) of the *Treatise*
form a luminous discussion of the general equations of elasticity,
which, however, as traversing well-known ground need not detain
us long.

We may draw attention to the following points:

(*a*) If W denote the "whole amount of work done per unit of
volume in any part of the body while the substance in this part
experiences a strain (s_x, s_y, s_z, σ_{yz}, σ_{zx}, σ_{xy}) from some initial state
regarded as a state of no strain", then

$$dW = \widehat{xx}ds_x + \widehat{yy}ds_y + \widehat{zz}ds_z + \widehat{yz}d\sigma_{yz} + \widehat{zx}d\sigma_{zx} + \widehat{xy}d\sigma_{xy}$$

Upon this result our authors make the following weighty statement,
which I believe had not been clearly expressed before, and which has
too often been disregarded since:

This equation, as we shall see later, under Properties of Matter [*alas!*],
expresses the work done in a natural fluid, by distorting stress (or difference
of pressure in different directions) working against its innate viscosity; and
W is then, according to Joule's discovery, the dynamic value of the heat
generated in the process. The equation may also be applied to express the
work done in straining an imperfectly elastic solid, or an elastic solid of
which the temperature varies during the process. In all such applications
the stress will depend partly on the speed of the straining motion, or on the
varying temperature, and not at all, or not solely, on the state of strain
at any moment, and the system will not be dynamically conservative
(§ 671).

An attempt has been made by the Editor of the present volume to
form the generalised equations of elasticity when the speed of the
straining motion is taken into account: see the *Proceedings of the
London Mathematical Society*, Vol. xx., pp. 297–350. London, 1889.

(*b*) There are a number of definitions in these paragraphs which ought
to be regarded. A *perfectly elastic body* is defined as a body which,
"when brought to any one state of strain, requires at all times the

same stress to hold it in this state; however long it be kept strained, or however rapidly its state be altered from any other strain, or from no strain, to the strain in question". Here the effect of variation of temperature is neglected in the theoretical 'perfectly elastic' body, but our authors point out that "by making the changes of strain quickly enough to prevent any sensible equalization of temperature by conduction or radiation," or again "by making them slowly enough to allow the temperature to be maintained sensibly constant", the perfectly elastic body of theory finds close approximations among natural bodies (§ 672.)

The first clear statement of the relation of thermal effect to strain is due to Sir William Thomson: see our Art. 1631.

(c) In § 673 we have the expression of W (see (a) above) as a quadratic function of the strain-components, and the remarks as to Boscovich's theory, which we have criticised in other parts of our *History*: see Arts. 924*, 928*, 276 and 299.

(d) What we have termed the dilatation- and slide- moduli, or the $F \equiv \frac{1}{3}(3\lambda + 2\mu)$, and the μ of an isotropic solid, are defined as the *bulk-modulus* and the *rigidity* in § 680. There seems some objection to the latter word as the term *flexural rigidity* has been widely used in quite a different sense, *i.e.* in the cases of a beam and of a plate, where its value has on the multiconstant theory no direct relation to the slide-modulus μ. The reciprocal of the bulk-modulus is termed the *compressibility*. Thomson and Tait use the letter n for our μ, the letter k for our $F \equiv \frac{1}{3}(3\lambda + 2\mu)$ and the letter m for our $\lambda + \mu$. It follows that our $E = 9nk/(3k + n)$ and our η, the stretch-squeeze ratio, $= \frac{1}{2}(3k - 2n)/(3k + n)$. Moduli expressed in terms of k and n are generally more complex than when expressed in terms of λ and μ, but k has a more direct physical signification than λ. Young's modulus, our stretch-modulus, is identified in § 686 with "what we also sometimes call...*longitudinal rigidity*". This I venture to think completes the confusion which has hitherto been attached to the word rigidity[1].

(e) The criticisms of uniconstancy in §§ 684–5, for the reasons often cited in this *History*, do not seem to me to carry conviction with them: see our Arts. 921*–933*, 192, 196, 1201, 1212 and 1273.

---

[1] In a footnote to this paragraph occurs the slip concerning the stretch-modulus of ice to which I have referred in Art. 372*, footnote. It should be noted that if a perfect fluid *might* be compared with an elastic solid for which the slide-modulus μ is zero, but the dilatation-coefficient λ finite, then the stretch-modulus would also be zero; but if a column of the material were placed in a cylindrical vessel with rigid sides and stretched by a traction T, the corresponding stretch s would be T/λ. In this case, owing to the vanishing of μ, the dilatation-modulus would also be λ. The phenomenon of the *stretching* of such a material is illustrated by the fact that even a column of water will, if air-free, bear a pull of many atmospheres before rupture. If a fluid be compressible in the least degree, then a column of it will resist stretching, if it be given lateral support. The 'water rope' paradox thus finds an elastic analogue.

[1710.]   §§ 699–710 deal with Saint-Venant's *Torsion Problem:* see our Arts. 17–60.   Our authors treat of the application of conjugate functions to the torsion of prisms and indicate their application to the case of cross-sections in the form of annular sectors (§ 710).   The method itself is due to Clebsch (see our Art. 1348 (*e*)), and the suggested application has been later fully worked out by Saint-Venant: see our Art. 285.

In § 705 a hydrokinetic analogue to the torsion problem is given, which differs, however, materially from that of Boussinesq published some years later: see our Art. 1430.

It runs as follows :

Conceive a liquid of density μ completely filling a closed infinitely light prismatic box of the same shape within as the given elastic prism and of length unity, and let a couple be applied to the box in a plane perpendicular to its length.   The *effective* moment of inertia of the liquid will be equal to the correction by which the torsional rigidity of the elastic prism calculated by the false extension of Coulomb's law must be diminished to give the true torsional rigidity.

Further, the actual *shear* [*i.e.* slide] of the solid, in any infinitely thin plate of it between two normal sections, will at each point be, when reckoned as a differential sliding parallel to their planes, equal to and in the same direction as the velocity of the liquid relatively to the containing box (§ 705).

By "*effective* moment of inertia" the authors understand that of a rigid solid fixed within the box, which if the liquid were removed would make the motions of the box the same as when it contained liquid.

The reader will find it of interest to compare Thomson and Tait's analogue with Boussinesq's,—especially in reference to the insight both throw on the position of the fail-point, as in general the point on the contour nearest the axis: see our Arts. 23 and 1430.

[1711.]   Another important matter in our authors' discussion of Saint-Venant's torsion problem is contained in the following words of § 710:

A solid of any elastic substance, isotropic or aeolotropic, bounded by any surfaces presenting projecting edges or angles, or re-entrant angles or edges, however obtuse, cannot experience any finite stress or strain in the neighbourhood of a *projecting* angle (trihedral, polyhedral, or conical); in the neighbourhood of an edge, can only experience simple longitudinal stress parallel to the neighbouring part of the edge; and generally experiences infinite stress and strain in the neighbourhood of a *re-entrant* edge or angle; when influenced by any distribution of force, exclusive of surface-tractions infinitely near the angles or edges in question. An important application of the last part of this statement is the practical rule, well known in mechanics, that every re-entering edge or angle ought to be rounded to prevent risk of rupture, in solid pieces designed to bear stress.

The writers remark that want of space obliges them to leave this statement without formal proof. A certain portion of the proof may be given readily as follows, although the general demonstration in the case of a polyhedral angle might be difficult.

Consider an edge of a solid bounded by two planes meeting in a line taken as axis of z. Further let two parallel planes be taken perpendicular to the edge cutting off a wedge-shaped portion of the edge. If the planes be taken very close together and we deal only with a very small portion of the wedge in the immediate neighbourhood of the edge, the variation of the stresses with z may be neglected as compared with their variations with regard to r and ϕ, polar coordinates in the angular face of the wedge. If $2a$ be the angle of the wedge and ϕ be measured from the angular bisector, the most general expressions for the radial and cross-radial shifts and for the dilatation will be found to be:

$$
\left.
\begin{aligned}
u &= \Sigma \left\{ C_{m'} \cos m'\phi + D_{m'} \sin m'\phi \right\} r^{m'-1} \\
&\quad + \Sigma \left\{ L_{m-2} \cos (m-2)\,\phi + M_{m-2} \sin (m-2)\,\phi \right\} r^{m-1}, \\
v &= \Sigma \left\{ D_{m'} \cos m'\phi - C_{m'} \sin m'\phi \right\} r^{m'-1} \\
&\quad + \Sigma \left\{ v_{m-2} \left(M_{m-2} \cos (m-2)\,\phi - L_{m-2} \sin (m-2)\,\phi \right) \right\} r^{m-1}, \\
\theta &= \Sigma \left\{ (mL_{m-2} - (m-2)\,v_{m-2}L_{m-2}) \cos (m-2)\,\phi \right. \\
&\quad \left. + (mM_{m-2} - (m-2)\,v_{m-2}M_{m-2}) \sin (m-2)\,\phi \right\} r^{m-2}
\end{aligned}
\right\} \dots \text{(i)},
$$

where $v_{m-2} = \{(\lambda + 2\mu)\,m - \mu\,(m-2)\} / \{(\lambda + 2\mu)\,(m-2) - \mu m\},$

and m, m', $C_{m'}$, $D_{m'}$, L_{m-2}, M_{m-2}, are arbitrary constants to be determined by the surface conditions, *i.e.*, the values of the stresses over (i) the surfaces $\phi = a$ and, $\phi = -a$, and over (ii) a cylindrical surface of small radius about the axis of z, giving the internal stresses in the body at small distances from the edge. The latter stresses $\widehat{rr}$, $\widehat{r\phi}$ will be of the form

$$
\left.
\begin{aligned}
\widehat{rr} &= a_0 + \sum_{p=1}^{p=\infty} \left(a_p \cos \frac{p\pi\phi}{a} + b_p \sin \frac{p\pi\phi}{a} \right), \\
\widehat{r\phi} &= a'_0 + \sum_{p=1}^{p=\infty} \left(a'_p \cos \frac{p\pi\phi}{a} + b'_p \sin \frac{p\pi\phi}{a} \right)
\end{aligned}
\right\} \dots\dots\dots \text{(ii)},
$$

where the values of the constants a_p, b_p, a'_p and b'_p are supposed known. The former stresses $\widehat{r\phi}$, $\widehat{\phi\phi}$ are according to Thomson and Tait to be zero in the neighbourhood of the edge, *i.e.* to vanish with r.

By forming from (i) the expressions for $\widehat{rr}$ and $\widehat{r\phi}$ we find from (ii) that m' and $m-2$ will both be of the form $\dfrac{p\pi}{a}$, and that accordingly both $\widehat{rr}$ and $\widehat{r\phi}$ will involve powers of r of the order $\dfrac{p\pi}{a}-2$. Hence in order that the stresses may not become infinite at the angle we must have

$$p\pi > 2a,$$

or, since from (ii) the least value of p will be unity,

$$2a < \pi.$$

Thus the stress at the edge will not be finite if the edge be re-entering. By taking the tangent to any point of a curved edge as axis of z, we may apply the above analysis to any small portion in the immediate neighbourhood of its point of contact. A very similar proof holds in the case of a conical angle.

The condition that $\widehat{\phi\phi}$ and $\widehat{r\phi}$, over $\phi = \pm a$, are to vanish in the neighbourhood of the edge, compels us to give zero values to any constant terms in the expressions for $\widehat{rr}$ and $\widehat{r\phi}$. Hence the strain vanishes in the neighbourhood of the edge, if it be a projecting edge. This is the first part of Thomson and Tait's proposition.

[1712.] §§ 711–718 deal with the problem of flexure. In a marginal note, this treatment is spoken of as "Saint-Venant's solution of flexure problem". But the problem to which reference is made is not that of the great memoir of 1856 (see our Art. 69), but that much simpler case of "circular flexure," or of bending a straight rod into a circular arc by couples, which is dealt with in the memoir on *Torsion* (pp. 299–304) and in the *Leçons de Navier*: see our Arts. 9–13, 170. In this case if the rod be of isotropic material, and if z be the direction of the unstrained axis, xz the plane of bending and $1/\rho$ the curvature after bending:

$$s_x = \eta\frac{x}{\rho}, \quad s_y = \eta\frac{x}{\rho}, \quad s_z = -\frac{x}{\rho},$$
$$\sigma_{yz} = \sigma_{zx} = \sigma_{xy} = 0.$$

Whence we have for the shifts:

$$u = \frac{1}{2\rho}\{z^2 + \eta(x^2 - y^2)\}, \quad v = \frac{1}{\rho}\eta xy, \quad w = -\frac{1}{\rho}xz,$$

and therefore for the stresses:

$$\widehat{zz} = -E\frac{x}{\rho}, \text{ and } \widehat{xx} = \widehat{yy} = \widehat{yz} = \widehat{zx} = \widehat{xy} = 0.$$

The above shifts correspond in the case of a rectangular cross-section to the distorted form depicted in our Art. 1485*, the under-edge of the section corresponding to the outer side of the beam after flexure, and the axis of x being positive when measured towards the upper edge of the section in our figure. The anticlastic nature of the curvature on the faces of the beam perpendicular to the plane of flexure is obvious.

Since the usual mathematical theory of elasticity assumes that measured from the same set of axes, the shifts are small, it is clear that the radius of curvature must be great compared with x and y, *i.e.* with both the depth and breadth of the beam. This is a point to which we have frequently had to refer in cases where the theory of beams has been applied to the case of cylindrical shells: see our Arts. 537 and 1555. Thomson and Tait remark (§ 717) that:

Unhappily mathematicians have not hitherto succeeded in solving, possibly not even tried to solve, the beautiful problem thus presented by the flexure of a broad very thin band (such as a watchspring) into a circle comparable with a third proportional to its thickness and its breadth.

[1713.] An ingenious application of the results in Art. 1708 is made in §§ 719–20 to obtain the flexural rigidities[1] of a plate of isotropic material. Take a square element of the plate of unit side and suppose the thickness of the plate to be 2ϵ. Let pairs of balancing couples N_1 be applied to one pair of opposite sides, and pairs N_2 to the other pair of opposite sides, each tending to produce concavity in the same sense. Then we easily see that $N_1 = \frac{2}{3}E\epsilon^3/\rho$ and $N_2 = \frac{2}{3}E\epsilon^3/\rho'$. Hence by the results of the preceding article, if ν_1 and ν_2 be the total curvatures:

$$\nu_1 = (N_1 - \eta N_2)\Big/\frac{2E\epsilon^3}{3}\,; \qquad \nu_2 = (N_2 - \eta N_1)\Big/\frac{2E\epsilon^3}{3}\,,$$

or
$$N_1 = \frac{2E\epsilon^3}{3(1-\eta^2)}(\nu_1 + \eta\nu_2), \qquad N_2 = \frac{2E\epsilon^3}{3(1-\eta^2)}(\nu_2 + \eta\nu_1).$$

If w be the strain-energy of the plate per unit area of mid-plane, assumed a quadratic function of ν_1, ν_2 (see our Art. 1698) and A' and c' the 'flexural rigidities' we have:

$$w = \tfrac{1}{2}\{A'\,(\nu_1^2 + \nu_2^2) + 2c'\nu_1\nu_2\},$$

whence
$$N_1 = dw/d\nu_1 = A'\nu_1 + c'\nu_2.$$

Thus
$$A' = \frac{2E\epsilon^3}{3(1-\eta^2)} \text{ and } c' = \frac{2\eta E\epsilon^3}{3(1-\eta^2)}.$$

These results agree with those given by Kirchhoff by aid of a very different process: see our Arts. 1237 and 1296.

[1] So termed by our authors, notwithstanding that they have previously defined 'rigidity' with reference to resistance to shearing action: see our Art. 1709, (d).

[1714.] §§ 724–9 deal with what we have termed "the elastic equivalence of statically equipollent systems of load", with special reference, however, to Thomson and Tait's reconciliation of the Kirchhoff and Poisson boundary conditions for a thin plate. It is shewn that systems of forces in equilibrium applied to elementary lengths of the edge of a plate produce only insensible shifts at a distance two or three times the thickness from the edge. These investigations take as their starting-point Saint-Venant's solution for the torsion of a flat prism of rectangular cross-section. They are a necessary part of the Thomson and Tait reconciliation and a valuable contribution to our knowledge of the exact meaning of the above-mentioned principle of elastic equivalence. At the same time we shall not discuss them further here, as the whole matter has been investigated at a later date with rather more complete results by Boussinesq, and these results have been already cited in our Chapter XIII.

[1715.] §§ 730–4 deal with the solution of the general body-shift equations of elasticity with certain special applications. Lamé, as we have pointed out, first introduced the potential solution into the theory of elasticity: see our Arts. 1062* and 1489. But to Thomson and Tait belongs the honour of having indicated its wide applications;—applications, which have been carried out with great ingenuity by Cerruti and Boussinesq: see our Arts. 1486–1524.

Without entering into the elegant analysis by which our authors obtain their solutions we may, in our own notation and terminology, record their results for the important cases with which they deal.

(a) Let a spherical element (radius a) of an infinite homogeneous *isotropic* elastic solid be subjected to the *constant* body-forces ρX, ρY, ρZ, so that the type of body-shift equation is:

$$(\lambda + \mu)\frac{d\theta}{dx} + \mu \nabla^2 u + \rho X = 0.$$

Then Thomson and Tait find shifts of the type:

$$u = \frac{1}{18\mu(\lambda + 2\mu)}\left\{(2\lambda + 5\mu)\,\rho X\,(3a^2 - r^2) \right.$$
$$\left. - \tfrac{2}{5}(\lambda + \mu)\,r^5\frac{d}{dx}\frac{\rho(Xx + Yy + Zz)}{r^3}\right\}\quad (r < a)\ldots\ldots(i),$$

$$u = \frac{a^3}{18\mu\,(\lambda + 2\mu)} \left\{ 2\,(2\lambda + 5\mu)\,\frac{\rho X}{r} \right.$$
$$\left. - (\lambda + \mu)\,(r^2 - \tfrac{3}{5}a^2)\,\frac{d}{dx}\,\frac{\rho\,(Xx + Yy + Zz)}{r^3} \right\} \Bigg\} \quad (r > a)\dots\dots\text{(ii)},$$

where the centre of the spherical element is at the origin and r is its distance from the point at which we are measuring the shift.

(b)   From the second of the above results our authors easily deduce, by making a vanishingly small, expressions for the shifts at x, y, z in an infinite elastic medium subjected to the body-forces X', Y', Z' at x', y', z'. Let $R = \{x - x')^2 + (y - y')^2 + (z - z')^2\}^{\frac{1}{2}}$, and let the body forces be such that $R\sqrt{X'^2 + Y'^2 + Z'^2}$ approaches zero as the point x', y', z' moves to an infinite distance from the origin. Then the type of the shifts is given by:

$$u = \frac{1}{24\pi\mu\,(\lambda + 2\mu)}$$
$$\times \iiint dx'dy'dz \left\{ 2\,(2\lambda + 5\mu)\,\frac{\rho' X'}{R} - (\lambda + \mu)\,R^2\,\frac{d}{dx}\left(\frac{\rho' F'}{R^2}\right) \right\}\dots\dots\text{(iii)},$$

where $\rho' F' = \{\rho' X'(x - x') + \rho' Y'(y - y') + \rho' Z'(z - z')\}/R$, or is the body-force resolved in the line joining x', y', z' to x, y, z. The integration must be extended over the whole portion of the medium to which body-force is applied (§ 731).

[1716.]   (c)   As the authors point out, the above general solutions for an infinite solid enable us to reduce the body-shift equations for a finite solid to the type:

$$(\lambda + \mu)\,\frac{d\theta}{dx} + \mu\nabla^2 u = 0,$$

where the body-forces have been removed by aid of a solution of the above type and by imposing the needful surface-stresses or surface-shifts (§ 732).

(d)   If the body-forces form a conservative system, or

$$\rho\,(Xdx + Ydy + Zdz) = dW,$$

it is pointed out in § 733 that they give rise to a dilatation

$$\theta = - W/(\lambda + 2\mu),$$

and that the shifts which remove the body-forces are then of the type

$$u = \frac{1}{\lambda + 2\mu}\,\frac{d\chi}{dx}$$

where χ is a solution of $\nabla^2\chi = - W$. If $W = \Sigma W_i$, W_i being a solid spherical harmonic of degree i, then $\chi = - \Sigma\,\dfrac{r^2}{2\,(2i + 3)}\,W_i.$ (§ 733).

The remarks in the article cited on conservative systems are of great interest and should be consulted.

[1717.]   In §§ 735–9 two important general cases are solved:

(i)   The strain in a solid sphere or spherical shell subjected either to given surface-stresses or to given surface-shifts: these problems have already been dealt with in our discussion of the memoir of 1862: see our Arts. 1651–5.

(ii)   The general solution for uniplanar strain in terms of polar-coordinates.   The values given for the radial and cross-radial shifts agree with those of our Art. 1711 in other symbols and with other expressions for the four series of constants of the solution.

[1718.]   A few remarks must suffice to indicate the remaining features of this portion of the *Treatise*.

(a)   In § 740 a general proposition of importance is stated.   The authors draw attention to the fact that if two elastic solids of like substance and similar shapes be taken, and by the application of force they be similarly strained, then the stresses across similarly situated elements either of real boundary or of geometrical surface within the substance will be equal.   The total stresses across any similar surfaces are accordingly as the *squares* of the linear dimensions of the two bodies, but any similar body-forces or the mass-accelerations are as the *cubes* of the linear dimensions.   Hence it follows that the greater body will be the more strained.   The strains at similar points will be simply as the linear dimensions, while the shifts at similar points will be as the squares of the linear dimensions.

Analytically we may look at this result in the following manner. Taking the body-stress equations we see, since the body-forces per unit volume are the same, and since the bodies are similar, that the stresses must be as the linear dimensions.   Therefore the strains, since the bodies are of the same elastic substance, are also in the ratio of the linear dimensions, while the shifts are as the squares of those dimensions.

(b)   In § 741 our authors adopt the term *plasticity* for that group of phenomena, wherein bodies change indefinitely and continuously their shape under the action of continued stress.   This is the sense in which the word has been used in our *History*.   They further describe under the term *viscosity of solids* "a distinct *frictional resistance* against every change of shape" which they say has been demonstrated by many experiments ("on metals, glass, porcelain, natural stones, wood, india-rubber, homogeneous jelly, silk fibre, ivory, etc.") and has been "found to depend on the speed with which the change of shape is made." They further state that:

A very remarkable and obvious proof of frictional resistance to change of shape in ordinary solids is afforded by the gradual, more or less rapid, subsidence of vibrations of elastic solids; marvellously rapid in india-rubber, and even in homogeneous jelly; less rapid in glass and metal springs, but still demonstrably much more rapid than can be accounted for by the resistance of the air.

The last statement embodies Kupffer's discovery of 1852 : see our Art. 748.   The reference to *silk* suggests Weber's classical experiments: see our Art. 707*.   Yet in both these cases the reduction of the amplitude of oscillation was attributed to elastic after-strain.   Now it seems to me difficult to identify frictional resistance and elastic after-strain. The "creeping back" to the original shape which goes on, it may be for minutes, hours or even days after the removal of the load (see our Arts. 720*, 817*, 827*, 1224*–6*, and 1431*), can hardly be due to any frictional action[1].   I have previously referred to the danger of masking the real nature of elastic after-strain by the use of the term viscosity : see our Arts. 708* ftn. and 750.   I think it would be better to limit the use of the term viscosity to *after-set*.

[1719.]   The only other portion[2] of Thomson and Tait's great *Treatise* with which we as elasticians are concerned is contained in §§ 832–48, and deals with the earth as a solid elastic body. A great deal of this portion is rewritten with supplementary articles by G. H. Darwin in the second edition, which I follow in this analysis. The problem itself is stated in the following words :

A few years ago [see our Art. 1663], for the first time, the question was raised : Does the earth retain its figure with practically perfect rigidity, or does it yield sensibly to the deforming tendency of the moon's and sun's attractions on its upper strata and interior mass? It must yield to *some* extent, as no substance is infinitely rigid : but whether these solid tides are sufficient to be discoverable by any kind of observation, direct or indirect, has not yet been ascertained [see our Art. 1726].   § 832.

[1720.]   The first point to be dealt with is the limit to the mathematical theory.   This is considered in § 832′, but in a manner with which the Editor of the present volume cannot express himself satisfied.   The following statements should be noted :

(*a*)   Nature, however, does impose a limit on the stresses : if they exceed a limit the elasticity breaks down, and the solid either flows (as in the punching or crushing of metals) or ruptures (as when glass or stone breaks under excessive tension).

[1] A curious example of elastic after-strain, which some of our readers may have remarked, occurs occasionally with razors. A razor which seems by rough or continual usage to have quite lost its sharpness, will frequently, if laid aside for a few weeks, be found quite capable of again performing its functions after this lapse of time.

[2] An Appendix reproduces the Appendix to Sir William Thomson's memoir of 1863: see our Arts. 1661 and 1250.

(*b*) The theory of elastic solids as developed in §§ 658, 663, &c., shews that when a solid is stressed, the state of stress is completely determined when the amount and direction of the three principal stresses are known, or, speaking geometrically, when the shape, size, and orientation of the stress-quadric is given. It is obvious that the tendency of the solid to rupture must be intimately connected with the shape of this quadric.

(*c*) The precise circumstances under which elastic solids break have not hitherto been adequately investigated by experiment. It seems certain that rupture cannot take place without difference of stress in different directions. One essential element therefore is the difference between the greatest and least of the three principal stresses. How much the tendency to break is influenced by the amount of the intermediate principal stress is quite unknown.

Now throughout the investigation the stress-difference is calculated from the *elastic* theory, and therefore the very important assumption appears to be made that the elastic theory holds up to the beginning of plasticity or even to rupture. This is far from being borne out, except for very special materials, by experimental facts : see our Vol. I., pp. 891–3. The stress-quadric as found from "the mathematical theory of elastic solids" can only be used in discussing the limit to perfect elasticity, *i.e.* to a linear stress-strain relation. At the same time we have seen in the course of our work that it is rather a value of stretch than of stress which ought to fix a limit to the application of the mathematical theory. The stretch-quadric may determine the *fail-limit* (see our Arts. 5 (*e*) and 169 (*g*)) but it is very doubtful whether we have any right to associate this fail-limit with the rupture-limit.

In the next place the statements quoted do not seem to me to clearly mark the distinction between materials which flow previously to rupture, and those which do not. Nor further, if the material be one which flows, is it clear that it will in all cases of stress have the power of doing so. It is well known from Tresca's experiments that flow commences in a plastic solid when the maximum shear (or half the difference between the greatest and least of the principal stresses) reaches a certain value : see our Arts. 1368*, 259 and 1586. But the general equations of plasticity (see our Art. 250) are not those of mathematical elasticity, and it is the latter equations which are applied by Thomson and

Tait and Darwin to the present problem. That the equations of
mathematical elasticity hold up to flow is not borne out by the
simple phenomenon of stricture in a bar under longitudinal
traction : see our Vol. I., pp. 889–91. Even if plasticity followed
at once on linear elasticity, it does not seem justifiable to apply
the plastic condition to rupture, which follows, if at all, long after
plasticity has been established and linear elasticity disappeared.
Further, the theory that rupture depends only on the maximum
stress-difference leads us to the conclusion that neither a plastic
nor a brittle material, if subjected to a strain in which the principal
stresses are all three equal will ever give way. It may be inconceivable that any amount of uniform pressure applied to the
surface of a solid sphere of isotropic material would cause it to
rupture, but it is also very difficult to believe that a uniform
tension, if it could be applied to its surface, would not, were it
indefinitely increased, produce rupture. To hold that such a
tension would not produce rupture seems to involve the assertion
that intermolecular force is not only infinitely great at an infinitely
small distance, but also at some finite distance. If this were true,
it would be difficult to grasp how even a shear of a certain
amount could cause the molecules of the material to permanently
separate by sliding over each other ; for such a slide is accompanied
by a finite separation of the molecules in the direction of one of
the principal axes of the slide. To sum up : it seems to me
that we may legitimately find a "fail-limit" by the condition of
maximum stretch, and that when the material is such that it
has a very high-elastic limit (*e.g.* hard steel), we may look upon
the fail-points or fail-surfaces as those at which, in the present
state of our knowledge, rupture will probably take place. I
think, that the maximum stress-difference does not give a limit
which can be safely applied to the mathematical theory of
elasticity, and, if we are to take it as a rupture limit, it ought
only to be applied to plastic materials which are being dealt with
by the general equations of plasticity. This is, I think, the use
which Boussinesq practically makes of it, when applying it to the
problem of loose earth : see our Arts. 1568, 1586 and 1594. The
remainder of § 832 is a *résumé* of G. H. Darwin's memoir: *On the
Stresses caused in the Interior of the Earth by the Weight of
Continents and Mountains. Phil. Trans.*, Vol. 173, Part I., pp. 187–

230. London, 1882. The discussion of the results of this paper
would carry us beyond the scope of the present chapter.

[1721.] Chree in a very valuable memoir entitled: *Some
Applications of Physics and Mathematics to Geology* (*Philosophical
Magazine*, Vol. 32, pp. 233–52 and 342–53. London, 1891) has
dealt with the application of the mathematical theory of elasticity
to the problem of strains in a solid earth. He states some important objections to the application of the theory of an isotropic
elastic solid to physico-geological problems. These must be noted
here, so far as they qualify the problem of the elastic solid tides
of the earth. The following causes have to be considered as
contributing to the deformation of the earth's surface: 1° the
mutual gravitation of its parts, 2° the centrifugal acceleration
produced by the diurnal rotation about its axis, 3° the gravitational
influence of the sun and moon.

We may express the action of the last two causes by the following
force-function:

$$\left. \begin{aligned} F &= \rho \left\{ \tfrac{1}{3}\omega^2 r^2 + \tfrac{1}{2}\omega^2 r^2 \left(\tfrac{1}{3} - \cos^2 \phi \right) - \tfrac{3}{2} \frac{M}{D^3} r^2 \left(\tfrac{1}{3} - \cos^2 \psi \right) \right\}, \\ &= \rho \left\{ \tau_0 r^2 + \tau r^2 \left(\tfrac{1}{3} - \cos^2 \phi \right) - \tau' r^2 \left(\tfrac{1}{3} - \cos^2 \psi \right) \right\} \end{aligned} \right\} \quad \dots (i),$$

where ω = the spin of the earth about its polar axis, ρ = the density of
the earth at distance r from its centre, ϕ = the angle the direction r
makes with the polar axis, ψ = the angle the same direction makes with
the line from the centre of the earth to the tide-raising body, M = the
mass of the latter body, D = its central distance, and τ_0, τ, τ' are written
for $\tfrac{1}{3}\omega^2$, $\tfrac{1}{2}\omega^2$ and $\tfrac{3}{2}M/D^3$ respectively. Clearly if the term $\tfrac{1}{3}\omega^2 r^2$ be put
on one side, as only producing a radial extension, the effects of rotation
and of the tide-producing body can be deduced, the one from the other,
by interchanging τ and $-\tau'$, and the line of centres with the polar axis.
The solution therefore for the case of the body-tides in an isotropic elastic
sphere can be deduced from the results of our Arts. 563 and 568.

The first type of force, that due to mutual gravitation, leads to some
difficulties in the treatment. Suppose the sphere in a state of strain
owing to spin or tide to be converted into the spheroid $r = a (1 + \Sigma Y_i)$,
Y_i being a surface harmonic of degree i. Then the internal stress due
to mutual gravitation will be partly due to "the attraction of the
harmonic inequalities" which produce a potential

$$= 4\pi \rho a^2 \Sigma \left(Y_i \frac{r^i}{a^i} \frac{1}{2i+1} \right),$$

or,

$$3ga\Sigma \left(Y_i \frac{r^i}{a^i} \frac{1}{2i+1} \right),$$

if g be the mean value of gravity over the surface *neglecting the spin*, and partly due to surface-stresses on the spherical surface $r = a$ caused by the action of the harmonic inequalities. In the case of an incompressible viscous fluid these stresses reduce to a surface-traction due to the weight of the harmonic inequalities, *i.e.* to a surface-traction $= -g\rho a \Sigma Y_i$. Darwin has shewn that in the case of an *incompressible* viscous fluid, we may replace this surface-traction together with the potential due to gravitational attraction of the harmonic inequalities by an "effective potential"

$$- 2g\rho a \Sigma \left(Y_i \frac{r^i}{a^i} \frac{i-1}{2i+1} \right) \qquad (\S\ 840').$$

According to Thomson and Tait, Darwin's analysis is "almost *literatim* applicable to the case of an elastic incompressible spheroid." But the hypothesis of incompressibility is scarcely justified in the case of the earth. Further, this result, unlike the above expression (i) for F, necessarily supposes ρ to be a constant.

[1722.] Chree in an important memoir in the *Cambridge Philosophical Transactions* (Vol. xiv., pp. 278—86. Cambridge, 1888) has worked out the shifts produced by the mutual gravitation of a nearly spherical mass, of which the boundary may be represented by $r = a\,(1 + \Sigma Y_i)$. For the purposes of our present discussion it will be sufficient to deal with his results (p. 280) for the case in which $i = 2$, and $Y_2 = \epsilon\,(\frac{1}{3} - \cos^2 \phi)$, *i.e.* the boundary of the gravitating mass is a spheroid of ellipticity ϵ. In our notation (Art. 568) he finds for the shifts, u_1, v_1, w_1 :

$$u_1 = \frac{g\rho a}{10\,(\lambda + 2\mu)} \left\{ \frac{r^3}{a^2} - r\,\frac{5\lambda + 6\mu}{3\lambda + 2\mu} \right\}$$
$$+ \frac{g\rho a Y_2}{5\mu(\lambda + 2\mu)(19\lambda + 14\mu)} \left\{ \frac{r^3}{a^2}(6\lambda^2 - 18\lambda\mu - 12\mu^2) - r(16\lambda^2 + 23\lambda\mu + 6\mu^2) \right\},$$
$$v_1 = \frac{g\rho a\, dY_2/d\phi}{10\mu(\lambda + 2\mu)(19\lambda + 14\mu)} \left\{ \frac{r^3}{a^2}(10\lambda^2 + 3\lambda\mu - 10\mu^2) - r(16\lambda^2 + 23\lambda\mu + 6\mu^2) \right\},$$
$$w_1 = 0$$

$$\dots\dots\dots\dots(ii).$$

The shifts in a perfect sphere due to a force-function of the form :

$$F = \rho\,(\tau_0 r^2 + \tau r^2 Y_2'), \quad \text{if} \quad Y_2' = (\tfrac{1}{3} - \cos^2 \phi),$$

are easily found from our Art. 568, or better still from p. 287 of Chree's *Camb. Phil. Trans.* paper cited above, to be of the form :

$$u_2 = -\frac{\tau_0 \rho a^2}{5\,(\lambda + 2\mu)} \left\{ \frac{r^3}{a^2} - r\,\frac{5\lambda + 6\mu}{3\lambda + 2\mu} \right\}$$
$$- \frac{\tau \rho a^2 Y_2'}{\mu\,(19\lambda + 14\mu)} \left\{ \frac{r^3}{a^2}(3\lambda + 2\mu) - r\,(8\lambda + 6\mu) \right\},$$
$$v_2 = -\frac{\tau \rho a^2\, dY_2'/d\phi}{2\mu\,(19\lambda + 14\mu)} \left\{ \frac{r^3}{a^2}(5\lambda + 4\mu) - r\,(8\lambda + 6\mu) \right\},$$
$$w_2 = 0$$

$$\dots\dots(iii).$$

Here if τ_0 be put zero, $\tau = -\tau'$ and ϕ replaced by ψ, we have the shifts in a perfect sphere due to a tide-raising body.

[1723.]   Now several important conclusions may be drawn from the above results.

(i) Consider only the term in $u_1 + u_2$ tending to produce the same compression of the body along each radius, *i.e.*

$$u' = (g - \tfrac{2}{3}\omega^2 a)\, \frac{\rho a}{10\,(\lambda + 2\mu)} \left\{ \frac{r^3}{a^2} - r\, \frac{5\lambda + 6\mu}{3\lambda + 2\mu} \right\}.$$

Now $g - \tfrac{2}{3}\omega^2 a$ may be practically taken equal to the mean surface value g_0 of gravitational acceleration, and we then have the following results: $s_\phi = u'/r$ is everywhere negative, but $s_r = du'/dr$ will be a positive stretch at the surface, where it equals s_0, say.   Thus we find u_0' being the radial shift at the surface:

(*a*)   Uni-constant isotropy, if E be the stretch-modulus:

$$u_0' = -\frac{g_0 \rho a^2}{10 E}, \qquad s_0 = \frac{g_0 \rho a}{15 E},$$

(*b*)   Incompressible substance, or $\mu/\lambda = 0$:

$$u_0' = -\frac{g_0 \rho a^2}{15 \lambda}, \qquad s_0 = \frac{2 g_0 \rho a}{15 \lambda}.$$

Now the very roughest attempt to turn (*a*) into numbers shews that u_0' and s_0 have quite impossible values, if E be given a value not largely exceeding that of any known mineral.   As Chree (*Philosophical Magazine, loc. cit.* pp. 247–8) has been the first to point out, u_0' becomes a large fraction of the earth's radius and the strain s_0 becomes immense, both suppositions entirely inconsistent with the mathematical theory of perfect elasticity, which supposes the shifts and strains to be both small.   On the other hand for a nearly incompressible substance (for which μ is finite) both the surface shift and strain will be vanishingly small.   It is difficult, however, with our knowledge of the materials which form the terrestrial crust to suppose that at any rate at the crust λ is immensely greater than μ; those materials approximate more closely to iron and stone than to india-rubber in their nature.   It is clear then that the strains produced by gravitation are such that permanent set and probably variations in density would be produced, if the earth were treated simply as an isotropic substance, compressed under the mutual gravitation of its parts.   We are therefore compelled to suppose that mutual gravitation has produced nearly its full effect before we proceed to investigate the effect of a tide-producing body in directly altering the ellipticity of the earth, or in indirectly altering it by altering the form of its mutually attracting parts.   But it will then be at once noticed, that to treat as homogeneous and isotropic the substance of the earth which has consolidated under the enormous stresses

resulting from the mutual gravitation of its parts is by no means a
satisfactory hypothesis. It must only be adopted as a very rough first
approximation, and until our knowledge of the arrangement of density
in bodies consolidating under great stresses has advanced beyond its
present stage. These points do not seem brought out very clearly in
Thomson and Tait's discussion of this matter. Thus in § 834 they
only remark of the term $\rho \tau_0 r^2$, *i.e.* $\frac{1}{3}\rho\omega^2 r^2$, that its effect "is merely a
drawing outwards of the solid from the centre symmetrically all round".
But this term may have very considerable influence on the magnitude
of the stresses. Indeed, the rotational terms as a whole, as Chree has
shewn (*Phil. Mag.* pp. 245–6), lead on Darwin's hypothesis of the
maximum stress-difference to results, under which it is certainly doubtful
whether masses of rock or heterogeneous mineral would remain permanently in equilibrium. It seems, therefore, desirable that the reader
should regard these articles of the *Treatise* on the distortion of the solid
earth as replete with suggestions for future investigation, rather than
as expressing the definite analytical results of an irreproachable physical
investigation.

(ii) So far as the terms measuring the ellipticity produced by
rotation directly and indirectly through the change in the character of
mutual gravitation *i.e.* the terms in τ, are concerned, these do not lead
on the maximum stretch hypothesis to results necessarily incompatible
with the *elastic* straining of an *isotropic* solid. They are, however,
identical in form with those due to tidal action and thus need not detain
us here.

(iii) We now come to the terms due solely to the tidal action[1],
and we note that for $r = a(1 + Y_2)$, the radial shift $u'' - u_1 + u_2$ is then
of the form :

$$u'' = (\beta_0 Y_2 - \beta_1 Y_2 - \beta_2 Y_2') a,$$
$$= (\beta_0 \epsilon - \beta_1 \epsilon - \beta_2) Y_2' a.$$

This gives at once for the ellipticity ϵ :

$$\epsilon = \beta_0 \epsilon - \beta_1 \epsilon - \beta_2,$$

or ϵ is negative, *i.e.* the spheroid prolate, and of ellipticity

$$\epsilon' = \beta_2/(1 - \beta_0 + \beta_1).$$

We easily find on substituting the values of β_0, β_1 and β_2:

$$\epsilon' = \dfrac{\dfrac{\tau'\rho a^2 (5\lambda + 4\mu)}{\mu(19\lambda + 14\mu)}}{1 + \dfrac{g\rho a (30\lambda^3 + 105\lambda^2\mu + 108\lambda\mu^2 + 36\mu^3)}{5\mu(\lambda + 2\mu)(19\lambda + 14\mu)(3\lambda + 2\mu)}} ,$$

$$= \dfrac{\dfrac{\tau'\rho a^2 (5\lambda + 4\mu)}{\mu(19\lambda + 14\mu)}}{1 + \dfrac{3g\rho a (10\lambda^2 + 15\lambda\mu + 6\mu^2)}{5\mu(19\lambda + 4\mu)(3\lambda + 2\mu)}} .$$

[1] They are obtained from (ii) of Art. 1722, and from (iii) of the same article by
putting in the latter $\tau_0 = 0$ and $-\tau'$ for τ.

[1724.] We will consider special cases of this result.

(a) Suppose the indirect gravitational influence to be neglected, then if $\epsilon' = \epsilon_r$,

$$\epsilon_r = \frac{\tau'\rho a^2 (5\lambda + 4\mu)}{\mu (19\lambda + 14\mu)}.$$

If F be the dilatation-modulus $\equiv \frac{1}{3}(3\lambda + 2\mu)$, this may be thrown into the form :

$$\epsilon_r = \frac{5\tau'\rho a^2}{19\mu} \left\{ 1 + \frac{\frac{6}{95}\mu/F}{1 + \frac{4}{57}\mu/F} \right\},$$

which agrees with the second result of (20) in § 834.

(b) Next let us suppose $\lambda = \infty$ and $\mu = 0$, or that we are dealing with a perfect incompressible fluid. In this case if $\epsilon' = \epsilon_g$

$$\epsilon_g = \frac{5\tau'a}{2g}. \qquad \text{(§§ 819 and 839.)}$$

Thus ϵ_r and ϵ_g are respectively the ellipticities due to rigidity without gravitation, and to gravitation without rigidity.

(c) Generally we have :

$$\frac{1}{\epsilon'} = \frac{1}{\epsilon_r} + \frac{1}{\epsilon_g} \cdot \frac{3}{2} \frac{(10\lambda^2 + 15\lambda\mu + 6\mu^2)}{(5\lambda + 4\mu)(3\lambda + 2\mu)}.$$

If we have uni-constant isotropy ($\lambda = \mu$):

$$\frac{1}{\epsilon'} = \frac{1}{\epsilon_r} + \frac{31}{30}\frac{1}{\epsilon_g}.$$

If we have an incompressible substance ($\mu/\lambda = 0$):

$$\frac{1}{\epsilon'} = \frac{1}{\epsilon_r} + \frac{1}{\epsilon_g}.$$

The last relation is stated by Thomson and Tait in § 840 as if it were *universally* true. This is only approximately the fact, as is indicated by the previous case of uniconstant isotropy.

(d) For uniconstant isotropy, $\epsilon_r = \frac{3}{11}\tau\rho a^2/\mu$, and for incompressibility $\epsilon_r = \frac{5}{19}\tau'\rho a^2/\mu$. Since $\frac{3}{11} = \cdot 2727$ and $\frac{5}{19} = \cdot 2632$, we see that the magnitude of the ratio λ/μ does not exercise a very large influence on the value of ϵ_r (§ 837). In § 838 Thomson and Tait give the value of ϵ_r for a steel ball of the size of the earth. They calculate it on the supposition that steel is incompressible and find for its value $77 \times 10^4 \cdot \tau'$. It would, perhaps, be better to suppose the steel mass to possess uniconstant isotropy. In that case a closer value would be $79 \times 10^4 \cdot \tau'$. The value of ϵ_g is found in § 839 to be $162 \times 10^4 \cdot \tau'$. Thus from (c) we see that the tides have somewhat less effect—supposing the earth as rigid as steel—indirectly through the changes they make in gravitational action between the parts of the earth, than directly through the gravitational action of sun or moon. Approximately for steel $\epsilon_g = 2\epsilon_r$, and $\epsilon' = \frac{1}{5}\epsilon_g$. For glass $\epsilon' = \frac{3}{5}\epsilon_g$ about (§ 841).

[1725.] We are now able to estimate the influence of the elastic strain of a solid earth of uniform density on the superficial water-tides.

Disregarding the diurnal rotation the equation to the form of the prolate spheroid that would be assumed by the solid earth is $r = a(1 - \epsilon' Y_2')$. This produces a potential at a point outside itself given by :

$$g\frac{a^2}{r} - \tfrac{3}{5}g\frac{a^4}{r^3}\epsilon' Y_2'.$$

Thus neglecting the self-attraction of the superficial coating of water, it will have for its level surface under tidal attraction :

$$g\frac{a^2}{r} - \tfrac{3}{5}g\frac{a^4}{r^3}\epsilon' Y_2' - \tau' r^2 Y_2' = \text{constant}.$$

This is clearly a prolate spheroid of ellipticity ϵ'' given by

$$\epsilon'' = \tfrac{3}{5}\epsilon' + \frac{\tau' a}{g}.$$

Hence the difference of ellipticity between the solid earth and the superficial fluid is :

$$e = \epsilon'' - \epsilon' = \frac{\tau' a}{g} - \tfrac{2}{5}\epsilon' = \tfrac{2}{5}(\epsilon_g - \epsilon'), \text{ by } (b) \text{ of Art. } 1724.$$

If we write ν for the expression

$$\tfrac{3}{2}(10\lambda^2 + 15\lambda\mu + 6\mu^2)/(5\lambda + 4\mu)(3\lambda + 2\mu),$$

we find by (c) of Art. 1724 :

$$e = \frac{\tau' a}{g}\frac{\epsilon_g + (\nu - 1)\epsilon_r}{\epsilon_g + \nu\epsilon_r}.$$

For the case of an incompressible solid $\nu = 1$, and we have :

$$e = \frac{\tau' a}{g}\frac{\epsilon_g}{\epsilon_g + \epsilon_r},$$

which agrees with Thomson and Tait's result in § 842. In rough numbers (Art. 1724) for steel $\epsilon_g = 2\epsilon_r$, whence

$$e = \tfrac{2}{3}\frac{\tau' a}{g}.$$

Thus, if the earth were as rigid as steel, its elastic yielding would reduce the height of the tide to about $\tfrac{2}{3}$ of its value as calculated from a theory in which the earth is supposed to be absolutely rigid. If the earth had only the rigidity of glass ($\epsilon_g = \tfrac{2}{3}\epsilon_r$ about), then the tide would be decreased by as much as $\tfrac{3}{5}$ of its value on the absolutely rigid theory. Thomson and Tait remark :

Imperfect as the comparison between theory and observation as to the actual height of the tides has been hitherto, it is scarcely possible to believe that the height is in reality only two-fifths of what it would be if, as has been universally assumed in tidal theories, the earth were perfectly rigid. It seems, therefore, nearly certain, with no other evidence than is afforded by the tides, that the tidal effective rigidity of the earth must be greater than that of glass (§ 843).

There is a point here which it is important, however, to bear in mind. The theory really deals with the "equilibrium hypothesis", and on that hypothesis there is an admitted 'lagging' of the tides. It is hardly reasonable to suppose that the water and earth tides will lag at the same rate. There is no reason therefore why the major axes of the two prolate spheroids corresponding respectively to water and earth tides should approximately coincide, unless we are dealing with tides of long period, *i.e.* at least with the fortnightly tides. It is these fortnightly and monthly tides which G. H. Darwin has considered in detail.

[1726.] The remaining sections of the *Treatise*, §§ 844–8, deal with evidence deducible from tidal data in favour of earth tides. The evidence is chiefly due to G. H. Darwin, who does not feel, however, that it justifies any very definite statements. He sums up with the remark :

On the whole we may fairly conclude that, whilst there is some evidence of a tidal yielding of the earth's mass, that yielding is certainly small, and that the effective rigidity is at least as great as that of steel (§ 848, Part II., p. 460 of 2nd Edition).

In a later paper (*Dynamical Theory of the Tides of Long Period : Royal Society's Proceedings.* Vol. 41, pp. 337–42. London, 1886) Darwin raises an objection to Laplace's equilibrium theory, and he concludes from a dynamical theory which neglects friction (p. 342) :

1°   That it is not possible to evaluate the effective rigidity of the earth as attempted in the *Natural Philosophy* from the fortnightly and monthly tides by aid of the equilibrium hypothesis.

2°   That the investigation in that work may however be accepted as confirming Sir William Thomson's view of "the great effective rigidity of the whole earth's mass".

3°   That Laplace's theory would hold for the minute tide of nearly nineteen years' period, but that this tide cannot probably be appreciated.

4°   That "it does not seem likely that it will ever be possible to evaluate the effective rigidity of the earth's mass by means of tidal observations".

With the words cited at the commencement of this article the text of Thomson and Tait's *Treatise* closes. If occasionally the analysis adopted does not seem to the present writer free from difficulties, yet the work as a whole made mathematical elasticity a branch of academic instruction in Great Britain. Few works on elasticity have been published which present so much that is suggestive, and arouse in the reader so great a desire to push further the many inquiries which the authors place before him.

[1727.] [1] *Note on Mr Gore's Paper on Electro-torsion.* *Philosophical Transactions*, Vol. CLXIV., pp. 560-2.   London, 1874. This refers to the twisting observed by Gore, and previously by G. Wiedemann, in an iron wire when magnetised at once longitudinally and circularly.   An explanation of the twisting is derived from the alteration of length in magnetised iron bars observed by Joule (see our Art. 688).   The direction of the resultant magnetisation is inclined to the axis of the wire, and so in accordance with Joule's results (for intensities lower than the critical points found by Shelford Bidwell, *Proceedings of the Royal Society*, Vol. L., pp. 109-133.   London, 1886 and subsequent papers) there is a lengthening of the material in this direction and a contraction in a perpendicular direction.   The two strains are equivalent to a torsional strain round the axis.   This theory had been already given by Maxwell though in a less complete form (*Electricity and Magnetism*, Vol. II., Art. 448.   Oxford, 1873). It does not appear to be accepted by Wiedemann (see *Annalen der Physik*, Bd. 27, S. 381-2), but it fits in well with a number of the facts (see Knott : *Trans. Roy. Soc. Edinburgh*, Vol. XXXVI., p. 507.   Edinburgh, 1892).   At the end of the paper it is inferred from the effects of loading observed by Joule (see our Art. 688, (iv)) that with sufficient longitudinal and torsional stress the direction of the twist would be reversed.   A reversal has in fact been obtained by Shelford Bidwell *in high fields* (*Philosophical Magazine*, Vol. XXII., pp. 251-5.   London, 1886).

[1728.] *Electrodynamic Qualities of Metals* [2] —*Part VI. Effects of Stress on Magnetization.* *Phil. Trans.*, Vol. CLXVI., pp. 693-713. London, 1877 (*M. P.*, Vol. II., pp. 332-53).   An abstract is given in *Proceedings Royal Society*, Vol. XXIII., pp. 445-6.   London, 1875 (*M. P.*, Vol. II., pp. 401-3).   This deals with the influence of longitudinal load on the induced and residual magnetisation of steel and iron wires.   The wire was suspended vertically and magnetised by a current in a surrounding coil, its magnetic changes being observed by the ballistic method.   The earth's vertical magnetic component remained uncompensated during the experiments on residual magnetisation.   The principal results are given

---

[1] I owe the following eleven articles to the kindness of Mr C. Chree whose knowledge of the topics discussed in them is far more extensive than my own.

[2] For the earlier portions of this memoir : see our Arts. 1644-7.

in the abstract in the *Proceedings* and on pp. 712–3 of the *Transactions*. The following are the results given for steel:

(1)   The magnetization is diminished by hanging on weights, and increased by taking the weights off, when the magnetizing current is kept flowing.

(2)   The residual magnetism remaining after the current is stopped is also diminished by hanging on the weights, and increased by taking them off.

(3)   The absolute amount of the difference of magnetization produced by putting on or taking off weights is greater with the mere residual magnetism when the current is stopped, than with the whole magnetism when the magnetizing current is kept flowing.

(4)   The changes of magnetization produced by making the magnetizing current always in one direction and stopping it are greater with the weights on than off.

(5)   After the magnetizing current has been made in either direction and stopped, the effect of making it in the reverse direction is less with the weights on than off.

(6)   The difference announced in (5) is a much greater difference than that in the opposite direction between the effects of stopping the current with weights on and weights off, announced in (4).

(7)   When the current is suddenly reversed, the magnetic effect is less with the weights on than with the weights off.

[1729.]   These results refer apparently only to a single field, 123 C.G.S. units approximately (see p. 696), and to hard steel pianoforte wire under loads from about an eighth to a half of the breaking load. When stating them Sir W. Thomson was not aware of the previous observations of Matteucci (see our Art. 705) and Villari (*Annalen der Physik*, Bd. 126, S. 87–122. Leipzig, 1865). The latter observer had found the induced magnetisation in iron and some specimens of soft steel to be increased or diminished by longitudinal pull according as the field was low or high. Prof. Ewing has found similar phenomena even in hard pianoforte steel wire (*Phil. Trans.*, Vol. CLXXVI., p. 625. London, 1886). Thus (1) is true only in fields above the *Villari critical field* as it is called, and there is a similar limitation with respect to (2). The critical fields or magnetisations, as Ewing has shewn, depend greatly on the nature of the wire, and are lower the larger the load. The phenomena to which conclusions (3)–(7) refer are also largely dependent on the field and the load (see Ewing, *l.c.*, pp. 623–630). In his experiments on iron Sir W. Thomson found what seemed very anomalous results. These find however a satisfactory explanation in the existence of the Villari critical field, which he did not recognise till later.

[1730.] *Effects of Stress on Inductive Magnetism in Soft Iron. Proceedings Royal Society*, Vol. XXIII., pp. 473–6. London, 1875 (*M. P.*, Vol. II., pp. 353–7). This records the rediscovery of the *Villari critical field* for soft iron. Observations were made in a large variety of fields, and the results are shewn in curves whose abscissae represent the fields and ordinates the changes in magnetisation due to load, p. 475 (*M. P.*, Vol. II., p. 356). The exactness of the information as to the fields and the relative magnitudes of the changes in magnetisation in different fields mark a great advance from the somewhat vague data previously existent.

[1731.] *Electrodynamic Qualities of Metals.—Part VII. Effects of Stress on the Magnetization of Iron, Nickel, and Cobalt. Phil. Trans.*, Vol. CLXX., pp. 55–85. London, 1880 (*M. P.*, Vol. II., pp. 358–395). An abstract occurs in *Proceedings Royal Society*, Vol. XXVII., pp. 439–443. London, 1878 (*M. P.*, Vol. II., pp. 403–7). This commences with a reference to Villari's discovery of a critical field. It then describes, pp. 56–63 (*M. P.*, Vol. II., pp. 359–69), experiments determining how the effect of tension on a soft iron wire depends on the temperature. The wire, ·75 mm. in diameter, received a small permanent stretch under a load of 18 lbs. and was then subjected to cycles of load on and off with a load of 14 lbs. Experiments were made in a series of fields up to about 40 C.G.S. units. In each field loading and unloading were repeated until the changes of magnetisation became cyclic, and it is this cyclic change that is dealt with. Observations were taken at the ordinary temperature and at 100°C. The results are shewn in plate 3 and on p. 61 (*M. P.*, plates II. and III.). The position of the Villari point was practically the same at both temperatures, but the magnitude of the cyclic change of magnetisation in fields both above and below the Villari point was greater at the lower temperature. P. 62 and plate 4 (*M. P.*, pp. 367–8, plates IV. and V.) describe similar results when the load was 7 lbs. or 21 lbs. The statement on p. 62 that the Villari field was much greater for 7 lbs. than for 14 lbs. is in accordance with the general conclusion of Ewing (*Phil. Trans.*, Vol. CLXXVI., pp. 621–3. London, 1886). The result however that the Villari field was higher for 21 lbs. than for 14 lbs. seems anomalous, unless perhaps the elastic qualities of the wire were altered by the greater weight.

432   SIR WILLIAM THOMSON.   [1732—1734

[1732.]   Pp. 62–3 and plate 5 (*M. P.*, Vol. II., pp. 368–9, and plates
VI. and VII.) describe some experiments of the following character.
A load of 14 lbs. or 21 lbs. was applied and removed 10 times, and
with it off the magnetising current was made and the throw t_1 of the
ballistic galvanometer observed.   Then while the current continued to
flow 14 lbs. or 21 lbs. was applied and removed 10 times, and with it
off the current was broken and the galvanometer throw t_2 observed.
Both t_1 and t_2 were considerably greater at an ordinary temperature
than at 100° C. for all the fields tried.

[1733.]   Pp. 64–7 (*M. P.*, Vol. II., pp. 370–4) treat of the effects of
"transverse stress" on the longitudinal magnetisation of iron.   The
inner surface of a gun barrel of "tolerably soft iron" was subjected to
applications and removals of a hydrostatic pressure 1000 lbs. per sq.
inch, and the (cyclic?) changes of magnetisation were observed by the
magnetometric method.   The effect was found to be the exact opposite
of that of longitudinal pull, *i.e.* pressure diminished or increased the
magnetisation according as the field was below or above a critical field.
The data on p. 65, and in curves (2), plate 7 (*M. P.*, Vol. II., p. 371 and
plate X.), seem to prove that this Villari field was much lower for the
material near the middle of the barrel than for that at the ends.   This
may be accounted for in part by the probable hypothesis that the
intensity of magnetisation was greatest near the middle.   It would be
desirable, however, to know the distribution of strain in the barrel, as
the validity of interpretations of the phenomena may depend largely on
this.   Pp. 65–7 and plates 8 and 9 (*M. P.*, Vol. II., pp. 371–4, plates XI.
and XII.) deal with the changes in the induced and residual magnetisations of the gun barrel produced in each case by 10 pressure cycles.
These changes measure what may be called the *non-cyclic* effects of
pressure.   In weak fields the pressure cycles caused a marked increase
in induced magnetisation; and the general effect on the residual
magnetisation was a marked decrease (see Wiedemann *Lehre von der
Elektricität*, Bd. III., S. 666–7).   The phenomena were however complicated
by the uncompensated action of the earth's vertical magnetic component.

[1734.]   P. 67 (*M. P.*, Vol. II., pp. 373–4) propounds the theory of
the development of an "aeolotropic property of different magnetic inductive susceptibility in different directions" by all systems of stress
other than uniform normal tension or pressure.   Thus in a circular
cylinder under torsion the strain consists of equal stretch and squeeze
in lines inclined at 45° to the axis in planes orthogonal to perpen-
diculars on the axis, and Sir W. Thomson assumes, as the result of his
experiments in fields below the Villari point, an increased susceptibility
in the direction of the stretch and a diminished susceptibility in the
direction of the squeeze.   This view had been already propounded by
Maxwell, *Electricity and Magnetism*, Vol. II., Art. 447.   Oxford, 1873.
Sir William Thomson thence argues that when the torsion of a wire is
very small the magnetic susceptibility in the direction of its length is

unaltered, and if finite torsions produce a change in susceptibility, it "must ultimately (for very small torsions) vary inversely [? directly] as the square of the amount of torsion" He apparently considers this explanatory of the results of Matteucci, Wiedemann and Wertheim (see our Arts. 703, 712 and 813 (ii)), viz. that in the cyclic state magnetisation is diminished by torsion in either direction and increased by detorsion.

[1735.] Pp. 67–72 and plates 10–12 (*M. P.*, Vol. II., pp. 374–80, plates XIII.–XIX.), describe the effects of torsion on a soft iron wire (22 B. W. G.) exposed to longitudinal pull of various amounts. The wire, whose length was 81 cm., passed through the cycle of twist[1] $\theta°$, $+320°$, $0°$, $-200°$, $\theta°$, where the $+$ sign refers to the direction of the first twist, the angles referring to the twisted end of the wire. Readings were taken for every $20°$ of twist, the magnetizing force being simply the earth's vertical component. The general character of the results was always the same, viz. that with torsion in either direction there was a loss, and with detorsion a recovery of magnetisation. The effect of the torsion varied but little as the longitudinal pull was raised from 10 to 20 lbs., but as the load was further increased the effect of torsion fell off rapidly. The wire was not in a cyclic state, there being always a fall in the magnetisation as the result of the torsion cycle, but in some of the later experiments the result of a second torsion cycle is given. Attention is drawn, p. 72 (*M. P.*, Vol. II., p. 379), to a "lagging of quality", or what Ewing has since called *Hysteresis*.

The necessity for a more exhaustive enquiry, distinguishing between the cyclic and non-cyclic effects, and varying the magnetic field and the torsion cycle, is abundantly shewn by the experiments of G. Wiedemann (*Annalen der Physik*, Bd. 27, S. 376–403. Leipzig, 1886). With a torsion cycle $0°$, $210°$, $0°$ in soft iron wire he found in the cyclic state that the curve whose abscissae give the twists, and ordinates the changes of induced magnetisation, was nearly symmetrical about a maximum ordinate answering to the mean twist (see also our Art. 813, (iii)).

Recent experiments by Nagaoka (*Philosophical Magazine*, Vol. XXVII., pp. 117–132. London, 1889) having shewn that the phenomena which accompany the application of twist to loaded magnetised nickel wires completely alter in character as the load and field are varied, the sign even of the magnetisation being sometimes reversed, fresh experiments were undertaken by Bottomley and Tanakadaté (*Philosophical Magazine*, Vol. XXVII., p. 138. London, 1889) on a piece of the iron wire used by Sir W. Thomson. They tried whether in a very weak field and with a heavy load the effect of twist would change in character—as suggested by what happens with nickel—, but they found no such change, their results being of the same character as Sir W. Thomson's. They do not, however, profess to regard the question as finally settled.

[1] θ the reading on the torsion circle when the torsion was nil seems in general to have been about $+40°$.

Pp. 73–4 (*M. P.*, Vol. II., pp. 381–2) refer to the discovery by Wiedemann (see his *Lehre von der Elektricität*, Bd. III., S. 680) of the production through torsion of longitudinal magnetisation in a wire magnetised by an axial current. Sir W. Thomson refers to his theory of "aeolotropic susceptibility", which gives results according with Wiedemann's if we assume the magnetisation below the Villari point. He believes, however, that explanation to fail, as he supposed Wiedemann's currents so strong as to have given a field above the critical, and in a footnote he adds that experiments he had made with very strong currents gave effects the same as Wiedemann's. A possible explanation has been suggested by Ewing (see his *Magnetic Induction in Iron and other Metals*, pp. 223–4 and footnote. London, 1891).

[1736.]   Pp. 74–9 (*M. P.*, Vol. II., pp. 382–7) describe experiments by the magnetometric method on the effects of longitudinal pull on the magnetisation of bars of nickel and cobalt magnetised by the earth's vertical component, and compare the effects with those in a tolerably soft iron bar similarly situated. In the nickel bar the non-cyclic effect of pull was as in iron to increase the magnetisation, the only difference being the much greater proportional change in the nickel; but when the cyclic state was reached the effect of pull was the exact opposite of that in iron, *i.e.* in nickel the magnetisation was least when the load was on.

In cobalt the same phenomena were observed as in nickel, but the bar broke at an early stage of the proceedings, and no experiments were made in higher fields. Subsequent experiments have confirmed these conclusions for cobalt in weak fields. A critical field however ensues, much higher than the Villari field usually is in iron, and in stronger fields the effect of stress is the same as in iron below the Villari field (see Chree, *Phil. Trans.*, Vol. CLXXXI., A, pp. 329–387. London, 1891, and Ewing, *Magnetic Induction in Iron*...p. 92 and pp. 210–2). Pp. 79–83 (*M. P.*, Vol. II., pp. 388–93) describe further experiments on nickel with higher fields. With cycles of load the residual magnetisation always shewed a distinct minimum when the load was on, and the cyclic change of magnetisation after strong fields shewed no tendency to diminish but seemed to tend to an asymptotic limit. With the induced magnetism there was unmistakeably the same effect in weak fields, but as the field was raised the cyclic change passed through a maximum and then decreased.

An attempt was made to reach a Villari critical field with a second smaller nickel bar, and this seems at first to have been thought successful; but a note dated June 4, 1879, says the result had not been confirmed by later experiments. There remains, however, on p. 83 (*M. P.*, Vol. II., p. 393) an uncontradicted statement that a Villari critical field and a distinct reversal of the effects of pull were obtained by altering the magnetometer, originally opposite an end of the bar, so as to bring it more nearly opposite the centre (cf. our Art. 1733). If this can be trusted, a Villari field actually exists. But Ewing, who has experimented with nickel under both tension and pressure (*Phil.*

Trans., Vol. CLXXIX., A, pp. 325–32 and 333–7.   London, 1889),
while confirming Sir W. Thomson's conclusions as to the opposite
behaviour of nickel and iron in weak fields, has found no trace of a
Villari field in nickel.   He appears, it is true, from his p. 331 and
footnote to have looked for a Villari point in low fields, so his
experiments are perhaps hardly conclusive.   His results and those of
Sir W. Thomson refer to the *total* magnetisation.   For the *temporary*
magnetisation—*i.e.* the magnetisation which disappears on the removal
of the magnetising force—a Villari field has since been found by
H. Tomlinson (*Philosophical Magazine*, Vol. XXIX., pp. 394–400.
London, 1890).   The reader should also consult the conclusions reached
by Shelford Bidwell (*Proceedings of the Royal Society*, Vol. XLVII., pp.
478–9.   London, 1890).   Pp. 84–5 (*M. P.*, Vol. II., pp. 393–5) describe
some experiments by the magnetometric method on the effects of pull
on very soft iron wire.   The results are in agreement with those obtained by the ballistic method.

[1737.]   *Note on the Direction of the Induced Longitudinal
Current in Iron and Nickel Wires by Twist when under Longitudinal
Magnetizing Force.   Philosophical Magazine*, Vol. 29, pp. 132–3.
London, 1890.   A statement is here given of how the direction of
these currents may be specified by reference to the directions of
twist and magnetisation.   A specification had been given for iron
by Matteucci (see our Art. 701).   In nickel under similar conditions
the longitudinal current is opposite in direction to that in iron.
The rule so far as is known applies for all intensities of magnetisation; for though the longitudinal currents diminish in intensity
when the field is sufficiently raised, a reversal in sign has not yet
been observed (see Nagaoka, *Philosophical Magazine*, Vol. XXIX.,
pp. 123–132.   London, 1889, or Ewing, *Magnetic Induction in
Iron*, pp. 225–8).

In tracing the complicated relationships between mechanical
strain and magnetisation the reader will derive much assistance
from a study of pp. 47–72 of J. J. Thomson's *Applications of
Dynamics to Physics and Chemistry* (London, 1888), but a complete
explanation of some of these relations will probably require account
to be taken of possible permanent differences of elastic (and
magnetic) quality in different directions, more especially in the case
of the magnetic phenomena accompanying torsion in wires, as the
strain is then frequently much above the elastic limit.

[1738.]   *The Rigidity of the Earth.   Nature*, Vol. v., pp. 223–4.
London, 1872.   This consists mainly of extracts from the memoir of the

same title published in 1862, and from the *Treatise on Natural Philoso-phy*: see our Arts. 1663 and 1719–25.

The Internal Fluidity of the Earth. A letter to Mr G. Poulett Scrope. Nature, Vol. v., pp. 257–9. London, 1872. This letter brings arguments against the internal fluidity of the earth : see our Art. 1665. Certain arguments introduced into this letter based upon the effects on precession of the elastic yielding of the Earth's surface were withdrawn by Sir William Thomson in 1876 : see *Mathematical and Physical Papers*, Vol. III., p. 321.

[1739.] *The Internal Condition of the Earth, as to Temperature, Fluidity and Rigidity. Transactions of the Geological Society of Glasgow*, Vol. VI. (1876–80), pp. 38–49. Glasgow, 1882. This paper is really a *résumé* without mathematical analysis of work by Sir William Thomson, which so far as it relates to elasticity has been already sufficiently dealt with in our *History* : see, especially for the arguments relating to the rigidity of the earth, to tides, to precession and nutation, our Arts. 1663–5 and 1719–25.

[1740.] *On a new method for discovering and measuring Aeolotropy of Electric Resistance produced by Aeolotropic Stress in a Solid.* A paper read before the Physical Society[1] ; Abstract, *Nature*, Vol. XVIII., pp. 180–1. London, 1878.

A diminution of electric conductivity is produced by stretching metallic wires : see our Art. 1647. Now the torsion of a wire produces slide, which may by Saint-Venant's Theorem (see our Art. 1570*) be resolved into a stretch and a squeeze in the principal axes of the slide, or in directions making angles of very nearly 45° with the axis of the wire. Thus the electricity in a wire would tend to flow in spirals, or have a component of flow *round* the wire. The external effect of this flow would be sensible near the terminals, or *inside* the twisted tube. Evidence of its existence was demonstrated by M'Farlane and Bottomley.

[1741.] *Elasticity.* This is an article contributed to Vol. VII. (pp. 796–825) of the Ninth Edition of the *Encyclopaedia Britannica.* London and Edinburgh, 1878. United to the article on *Heat* contributed to the same work, it afterwards appeared as an off-print (Edinburgh, 1880). Finally it was reprinted on pp. 1–112 of Vol. III. of the *Mathematical and Physical Papers* (Cambridge, 1890). The article incorporates two important memoirs by the author namely : *Elements of a Mathematical Theory of Elasticity. Philosophical Transactions*, Vol. CLXVI., pp. 481–98, London, 1856, and *On the Elasticity and Viscosity of Metals. Proceedings of the Royal Society*, Vol. XIV., pp. 289–97. London,

[1] The title only is printed in Vol. III. of the Society's *Proceedings.*

1865. These memoirs have accordingly not been separately dealt with in their proper chronological order. The article forms one of the chief elementary accounts of the physics of elasticity in the English tongue. All we can do here is to notice individual points in connection with it, especially where the author's definitions differ or his conclusions add to those already adopted or recorded in this *History*.

[1742.] The first 36 sections deal with the definitions of elasticity and treat of the limits of elasticity, of viscosity, etc., etc.

(a) The following definition is given of *perfect elasticity* in § 1 :

The elasticity is said to be perfect, when the body always requires the same force to keep it at rest, in the same bulk and shape and at the same temperature, through whatever variations of bulk, shape and temperature it be brought.

This definition clearly covers the whole range between the usual "limits of elasticity", but this need not necessarily mean the proportionality of stress and strain: see our Arts. 929*, 299 and Vol. I., pp. 891–3. Thus this definition of 'perfect elasticity' covers more than what the mathematicians include in their treatises on "the mathematical theory of elastic solids". The 'perfect' in the one refers in the first place to a physical conception, and in the other to a simplified set of formulae—*i.e.* linearity of the stress-strain relations. Hodgkinson's "defect of elasticity" (see our Vol. I., p. 891) would be covered by Sir William Thomson's definition of 'perfect elasticity'. In § 37 we read :

But now must be invoked minutely accurate experimental measurement to find how nearly the law of simple proportionality holds through finite ranges of contraction and elongation. The answer happily for mathematicians and engineers is *that Hooke's law is fulfilled, as accurately as any experiments hitherto made can tell*, for all metals and hard solids each through the whole range within its limits of elasticity; and for woods, cork, india-rubber, jellies, when the elongation is not more than two or three per cent., or the angular distortion not more than a few hundredths of the radian (or not more than about two or three degrees).

In the light of the researches recorded in the volumes of our *History*, it is impossible to identify generally the range between the elastic limits with proportionality of stress and strain (see our Vol. I., p. 891–3). Sir William Thomson himself adds that a small deviation from Hooke's law has been found by M'Farlane for steel pianoforte wire under combined torsional and tensile strain. The exceptions are wider than this isolated example might lead the reader to infer and occur even for simple tensile tests.

(*b*)   As in Thomson and Tait's *Treatise* (see our Art. 1709 (*d*)) the important distinction between elasticity of bulk and elasticity of shape is emphasised.   Homogeneous solids such as crystals and glasses are stated (§ 3) to probably possess elasticity of bulk to perfection—*i.e.* no amount of *compression* would produce set in them.   It is clear of course that the compressive test is practically the only one to which we can readily subject such bodies, but theoretically it must be considered a very doubtful question whether such bodies would exhibit elasticity of bulk to perfection could we submit them to a uniform surface traction of any arbitrary amount.   To assume that it is so, is to reject *a priori* the maximum stretch-limit to safe-loading.   Such an assumption leaves us indeed in a very vague position as to what the limit of elasticity really means when we are dealing with diverse types of strain, or how we are to apply the results obtained from a tensile test to more complex systems of strain.   Some of the interesting points connected with this subject are noted in §§ 8 and 21[1].   On the whole the treatment of the elastic limits in these sections requires modifying in the light of the splendid researches of Bauschinger and others, to bring the statements quite up to the present state of knowledge[2].   The paper by James Thomson incorporated in §§ 10–20 and to which we have already referred does not, I think, fully represent the state of our existing knowledge on the alteration of the elastic limits: see our Arts. 1379*–81*, 709–10 and 767.

(*c*)   The following definitions of brittle and ductile solids may be compared with those of Rankine (see our Art. 466):

If the first notable dereliction from perfectness of elasticity is a breakage, the body is called brittle,—if a permanent bend [more generally a set?], plastic or malleable or ductile (§ 7).

(*d*)   In § 23 the elastic limit for slide or change of angle appears to be deduced from the elastic limit for stretch.   If a bar be pulled longitudinally till it reaches its elastic limit $\bar{s}$, then on the supposition of isotropy there is a slide in planes at 45° to the axis of the bar of magnitude $\bar{s}(1 + \eta)$, but the converse does not hold, namely, that when there is a slide of this magnitude then there will necessarily be a stretch of magnitude $\bar{s}$.   Indeed a *pure* slide of this magnitude would have for its components a stretch and squeeze each of magnitude $\frac{1}{2}\bar{s}(1 + \eta)$, and would not therefore on our theory correspond to the elastic limit.   As we have shown in the course of our work, an elastic limit for stretch $\bar{s}$ corresponds to an elastic limit for slide $\bar{\sigma} = 2\bar{s}$ and not $= \bar{s}(1 + \eta)$.

[1] The hypothesis of the maximum stretch-limit, supposing the elastic limit to coincide with the limit to linear elasticity, has perfectly definite answers to the questions asked in § 21, and there does not seem to me any *a priori* reason to doubt the physical correctness of the answers it gives.

[2] A correction should be made in § 9 where by a slip it is stated that in the case of the flexure of a bar of any shape of cross-section by opposite bending couples applied at the ends *one-half* the substance is stretched the *other half* shortened;—the *amount* of the substance stretched or squeezed depends on the shape of the section.

Thus for the numerical case taken by Sir William Thomson the limit to angular distortion would, on the maximum stretch theory, be $\frac{1}{43}$ and not $\frac{1}{68}$ of a radian.

[1743.] §§ 29–36 are occupied with a discussion on *viscosity*. We have already referred to the sense in which Sir William Thomson uses the word 'viscosity'. In our *History* a material is termed 'viscous', when a shear, however small, if applied for a sufficiently long period produces set. On the other hand a material is termed 'plastic', if a shear above a certain magnitude is required to produce set. The word shear is here used instead of stress generally, merely to mark that a uniform surface pressure would not give a test of either viscosity or plasticity. The dynamical equations for viscous and plastic materials differ very considerably: see our Arts. 246, and 250. Further it is difficult to associate the phenomena of "after-strain" with anything of the nature of either viscous or plastic action in our senses of these words: see our Arts. 708* ftn. and 1718, (b). The viscosity of fluids may be represented by a force of resistance directly proportional to the velocity of change of shape. Hence the small effects can be superposed. This superposition does not seem to be true for elastic after-strain (see our Art. 717*). Weber, Kupffer and Sir William Thomson himself[1] appear to attribute the diminution of the amplitude of vibrations to elastic after-strain. Lord Rayleigh in his *Theory of Sound* introduces a *Dissipative Function* into his treatment of the vibrations of elastic bodies, which corresponds to true viscous terms (*i.e.* a resistance proportional to the velocity of the strain). Without venturing an opinion as to whether the subsidence of vibrations is due to true fluid viscosity or to elastic after-strain, it seems to me most important to keep the two notions distinct until their real nature and possible relationship have been clearly ascertained. The 'creeping back' in elastic after-strain seems to distinguish it fundamentally from molecular friction or viscosity: see our Art. 750.

[1744.] At the same time Sir William Thomson does not

---

[1] "It was in fact as it would be if the result wore wholly or partially due to imperfect elasticity, or 'elastische nachwirkung'—elastic after-working—as the Germans call it (§ 36) " We may remark that "imperfect elasticity" may mean either an elastic stress-strain relation which is not linear, or a stress accompanied by a set strain. In neither case does it correspond to elastic after-strain, *i.e.* introduce a *time* element.

suppose like Lord Rayleigh that the resistance which he terms viscosity in solids is simply proportional to the velocity of change of shape, he only suggests that this molecular friction is some function of this velocity of change of shape.

After dismissing the thermodynamic dissipation of energy, which occurs with every strain in an elastic solid, as in many cases too small to account for the loss of energy observed (§ 31), he continues:

The *frictional resistance* against change of shape must in every solid be infinitely small when the change of shape is made at an infinitely slow rate, since if it were finite for an infinitely slow change of shape, there would be infinite rigidity, which we may be sure does not exist in nature. Hence there is in elastic solids a *molecular friction* which may be properly called *viscosity of solids*, because, as being an internal resistance to change of shape depending on the rapidity of the change, it must be classed with fluid molecular friction, which by general consent is called *viscosity of fluids* (§ 32)[1].

Sir William Thomson's experiments were made upon the torsional vibrations of round wires supporting different vibrators and his first conclusion § 34 (a) runs:

It was found that the loss of energy in a single vibration through one range was greater the greater the velocity (within the limits of the experiments); but the difference between the losses at low and high speeds was *much less* than it would have been had the resistance been, as Stokes[2] has proved it to be in fluid friction, approximately as the rapidity of the change of shape.

The experiments were not however sufficient to determine any simple law of relation between viscous resistance and strain-velocity.

[1745.] Sir William Thomson's second series of experiments relate to the alteration of the torsional viscosity of wires owing to increase in the longitudinal traction. They may be compared with Kupffer's results cited in our Arts. 735 (iii) and 751 (d). It is not quite clear how far Sir William Thomson's vibrators were sufficiently heavy to produce by

---

[1] If like Sir William Thomson and others (see our Arts. 928*, 192 (a) and 299) we apply Maclaurin's Theorem to deduce Green's expression for the strain-energy, there seems precisely as much, or as little, reason for applying it to the problem of viscosity. If we do apply it, however, we only reach Lord Rayleigh's Dissipative Function, or fluid-viscosity.

[2] The reference is, I suppose, to Stokes' memoir of 1845. Poisson in 1831 and Saint-Venant in 1843 had arrived at the like conclusion, the latter by a method which appears to be as satisfactory as Stokes'.

mere tension either sensible elastic after-strain or set in the wires used. He found that when the weight of the vibrator was increased the viscosity of the vibrator was always at first much increased, but that it diminished day by day and ultimately became as small in amount as it had been with the lighter vibrator (§ 34 (b)). Here again no general law was ascertained.

[1746.] The third series of experiments relate to the subsidence of vibrations in aluminium wires. Sir William Thomson found that the number of vibrations during the subsidence from a higher to a lower ampli tude (say 20 to 10) was less when the vibrator was started at 40, and allowed before counting to sink first to 20, than if it were started at 20 itself (§ 34, (c)). The author does not appear to have noticed the effect remarked on by Kupffer that the period of vibration was a function of the amplitude (see our Arts. 735 (iii), 709 and 751 (d)), nor is it clear whether the drag of the air on the vibrator was allowed for: see our Art. 735 (i). The remark as to the air-resistance on a spring in § 31 does not seem to entirely cover this difficulty. Possibly it was tested and found to be negligible. Kupffer, however, endeavoured to measure and then eliminate it.

[1747:] The third series of experiments seemed to indicate that 'viscous' action depends on previous molecular condition, namely on whether the wire is started from rest, or has immediately beforehand been subjected to still larger repeated strains. A fourth series of experiments was accordingly instituted in which two equal and similar wires with equal and similar vibrators were dealt with,—one being kept in as far as possible a continual state of vibration, the other being vibrated only for the sake of one daily experiment. It was found in the case of two copper wires that the quiescent one subsided through the same range of amplitudes only after longer time and more vibrations with a shorter mean period than the frequently vibrated one[1] (§ 34 (d)).

[1748.] Finally series of experiments with much smaller maximum distortions were made in order to determine : (i) the law of subsidence of range in any single series of undisturbed oscillations, and (ii) the relation between the laws of subsidence for two sets of oscillations with the same elastic body performing oscillations of different periods, owing not to a change of weight, but to a change of the moment of inertia in the suspended vibrator (§ 35). The answer to the first question "so far as the irregularities depending on previous conditions of the elastic substance allowed any simple law to be indicated" was that :

The differences of the logarithms of the ranges were proportional to the intervals of time (§ 36).

[1] Thus while the amplitude was reduced a half the quiescent wire made 98 vibrations with a mean period 2·4 secs., while the frequently vibrated one made 59 vibrations with a mean period 2·45 secs. A possible reduction in the period with the change of amplitude is not referred to.

This held for distortion much smaller than the palpable elastic limit. The result resembles that due to a true fluid viscosity, or that produced by the drag of the air on a vibrator.

The only approach found to an answer to the second question was that:

the proportionate losses of amplitude in the different cases are not such as they would be if the molecular resistance were simply proportional to the velocity of change of shape in the different cases (§ 36).

Here again it seems as if Kupffer's experiments and results might have been found suggestive. On the whole the experiments, especially the last three series, appear to suggest the influence of that 'creeping back' which is peculiar to after-strain and seems quite masked under the term *viscosity*. Sir William Thomson speaks of these later results as shewing a very remarkable "fatigue of elasticity" (§ 30). It would be interesting to know whether this fatigue was only of kinetic or also of static elasticity, and further whether the distortions being below the elastic limit, the elastic limit and even the absolute strength were affected by it. The term fatigue although appropriate has been used by engineers in such a definite sense, namely the lowering of the *absolute strength* of a material by repeated strain below the rupture strain, that it is perhaps unadvisable to give it a new meaning in reference to elasticity. It is clear that Sir William Thomson's 'fatigue' is a phenomenon differing from that dealt with by Braithwaite or Wöhler: see our Arts. 970 and 997–1003.

Sir William concludes his remarks on viscosity by suggesting an elastic vesicular solid, the vesicles being filled with a viscous fluid like oil. Such a model solid would, he holds, suffice to elucidate some, but far from all, of the properties noted in the above series of experiments (§ 36).

[1749.] §§ 37–72 reproduce matter from the author's memoirs or from the *Treatise on Natural Philosophy* which has already been amply dealt with in our *History*. The arguments in favour of bi-constant isotropy from the action of cork and jellies are again referred to. We have already pointed out that they will only become valid when it has been demonstrated experimentally that cork and jelly are true isotropic elastic solids, *i.e.* can have *all* relations between stress and strain expressed by aid of two constants: see our Art 192 (*b*). In this matter we must bear in mind what Sir William Thomson himself (§ 37) says of such "elastic or semi-elastic 'soft' solids" as cork, india-rubber or jellies:

The exceedingly imperfect elasticity of all these solids, and the want of definiteness of the substance of many of them, renders accurate experimenting unavailable for obtaining any very definite or consistent numerical results.

In fact the elastic action of cork on the one hand and of gelatinous substances on the other would probably be best exemplified theoretically by treating them as porous elastic solids, the pores containing air and liquid respectively. The argument used in § 48 for multiconstancy

based on a jointed bar mechanism we shall deal with in our Arts. 1771–2.
It is really an appeal to the principle of "modified action." The
articles on *Resilience*[1], §§ 52–56, are reproduced from the *Treatise*.
They conclude in the reprint in the *Mathematical Papers* (Vol. III.,
p. 47) with a table of the elastic resiliences and the slide- and stretch-
moduli of a variety of wires. This table is based on experiments carried
out in the Physical Laboratory of Glasgow University. In § 62 (1) we
note that Sir William Thomson adopts the Bresse- Saint-Venant mode
of dealing with the flexure of beams when the stretch-modulus varies
from point to point of the cross-section: see our Arts. 169 (*e*)—(*f*)
and 515.

[1750.] §§ 73–6 deal with the thermo-elastic relations, and of
course draw largely on the memoir of 1855: see our Art. 1631.
Turning to Equation (vi) of our Art. 1633, or :

$$H = -\frac{t}{J}\frac{dw}{dt},$$

where t is measured in the absolute scale, let τ be the increase of
temperature due to the sudden application of a stress S corresponding to a strain $-s$, χ the strain produced by an elevation of
temperature of one degree when the body is kept under constant
stress,—this strain being measured in the opposite sense to that
of the constant stress[2], K the specific heat of the substance per
unit mass under constant stress, ρ the density, and J Joule's
equivalent, then :

$$H = K\rho\tau, \quad \frac{dw}{dt} = -S\frac{ds}{dt}, \quad \text{and } \frac{ds}{dt} = \chi,$$

whence we deduce

$$\tau = \frac{t\chi S}{JK\rho}\dots\dots\dots\dots\dots\dots\dots(i).$$

With regard to this formula Sir William Thomson remarks :

The constant stress for which K and χ are reckoned ought to be the
mean of the stresses which the body experiences with S and without S.
Mathematically speaking, S is to be infinitesimal, but practically it may
be of any magnitude moderate enough not to give any sensible difference

---

[1] The historical statement that Lewis Gordon first introduced the word *resilience*
to denote the work done by a spring or other elastic body returning to the unstrained
state from some strained limit is erroneous. He only adopted the word from Young:
see our Vol. I., p. 875.

[2] That is: χ must be an expansion if S denotes a pressure uniform in all
directions, or χ must be a stretch if S denotes a longitudinal compression, etc.

in the value of either K or χ, whether the "constant stress" be with S or without S, or with the mean of the two (§ 74).

[1751.] § 75 deals with the important distinction between static and kinetic elastic moduli. This distinction appears first to have been pointed out in a clear *scientific* manner by Sir William Thomson himself:

When change of temperature, whether in a solid or a fluid is produced by the application of a stress, the corresponding modulus of elasticity will be greater in virtue of the change of temperature than what may be called the static modulus defined as above, on the understanding that the temperature if changed by the stress is brought back to its primitive degree before the measurement of the strain is performed. The modulus calculated on the supposition that the body, neither losing nor gaining heat during the application of the stress and the measurement of its effect, retains the whole change of temperature due to the stress, will be called for want of a better name the kinetic modulus, because it is this which must (as in Laplace's celebrated correction of Newton's calculation of the velocity of sound) be used in reckoning the elastic forces concerned in waves and vibrations in almost all practical cases.

Let M be a static, M' the corresponding kinetic modulus. Clearly, if a body is not allowed to either lose or gain heat, then the strain will be, since there is a change of temperature , equal to

$$\frac{S}{M} - \chi^\tau,$$

but this equals S/M'; equating the two we have by using (i):

$$\frac{M'}{M} = \frac{1}{1 - \dfrac{t\chi^2 M}{JK\rho}} \quad\text{......................... (ii).}$$

Further if K and K' denote thermal capacities of a given quantity of the substance under constant stress and constant strain respectively then :

$$\frac{M'}{M} = \frac{K}{K'} \quad\text{......................... (iii).}$$

The values of the ratios M'/M or K/K' are tabulated in two "Thermo-dynamic Tables [1]" for a temperature of 15° C, the quantities J, ρ, K, M and χ being the experimental data. Thus Sir William Thomson gives for the ratios :

---

[1] In the first Table we find "$J = 42400$ centimetres" and the slip is repeated in the reprints. Here is a chance for the foot-pound, that unhappy "no-system" to have its revenge!

Dilatation-Modulus F'/F	Stretch-Modulus E'/E
Glass (flint)........1·004	Zinc ....1·0080
Brass (drawn)......1·028	Tin ....1·00362
Iron ....1·019	Silver............1·00315
Copper...............1·043	Copper...........1·00325
	Lead ....1·00310
	Glass.............1·00060
Water................1·004	Iron ....1·00259
Ether................1·577	Platinum........1·00129

We have tabulated these values here because they throw considerable light on a point often referred to in our *History*, namely the difference between the kinetic and static moduli. As Sir William Thomson points out, the difference between the values obtained by Wertheim for these moduli cannot be explained by thermal influence, they must be due to errors of observation. A similar opinion had been expressed by Clausius: see our Arts. 1297*, 1350*, and 1403*.

[1752.] In § 18, Tables V., VI. and VII., will be found recorded a number of results for the dilatation-modulus, stretch-modulus, slide modulus, tenacity, elastic stretch, and resilience of a variety of materials. These results are taken from the memoirs or tables of Wertheim, Rankine, Everett, Gray, and others . They are here conveniently brought together and reduced to common units. At the same time such results are only roughly approximate. The elastic moduli and limits are physical quantities which vary very widely with the *form*, exact process of manufacture and individual working of each test piece of a given type of material, and as it is of course impossible in tables of this kind to give information with regard to the actual specimen of each material to which the results refer, the data given cannot be of very great service in accurate physical investigations. It must ever be remembered that the elastic properties of a body are characteristic and peculiar to the preparation of the specimen itself, and are not solely determined by the material of which it is made.

[1753.] §§ 78–81 deal with the important problem of the effect of working, or of permanent molecular changes, on the elastic moduli of a body. They cite the results of experiments made by D. M'Farlane and A. and T. Gray. Sir William Thomson refers to Wertheim and others who have investigated this problem, "but solely" he writes "with reference to Young's modulus" (§ 78). The elaborate researches of Kupffer appear to have escaped his notice : see our Arts. 752–6.

The error by which ice is given double the stretch-modulus of any other material is repeated in the *Papers*: see our Art. 372* *ftn.*

(a) In § 78 and Table VIII. we have results of experiments by
M'Farlane on the results of a set-stretch in wires upon their slide-
modulus. The effect of a set-stretch was partly decrease of density with,
as a rule, decrease of the slide-modulus. The results may be compared
with those of Kupffer : see our Arts. 735 and 741, (b).

(b) Results for the change of the stretch-modulus with the temperature were in the earlier issues of the paper cited from Wertheim's
memoir of 1844 (see our Art. 1292*) but they are removed from § 79 of
the reprint in the *Mathematical and Physical Papers* (Vol. III., p. 80)
as "very far wrong". The sole result cited in the latter work is one for
a steel tuning-fork due to Macleod and Clarke[1], from which it would appear that the stretch-modulus for steel diminishes at the rate of $23 \cdot 2 \times 10^{-5}$
of itself per degree centigrade of elevation of temperature.

Results for the influence of temperature on the slide-moduli of iron,
copper and brass are cited from F. Kohlrausch and F. E. Loomis[2].
There is no reference to the results of Kupffer : see our Arts. 754–6.

[1754.] § 80 records some experiments by J. T. Bottomley on soft-
iron wire, from which it appears that the gradual addition of stress
during a long interval increases the ultimate tensile strength. This
point had been previously noticed by several technical elasticians. An
iron bar tested to the beginning of stricture, will after being left
quiescent for a period suffer striction at a different section and a higher
load, and in this manner the ultimate strength may be raised very
considerably. In some of Bottomley's experiments, the increase of
tensile strength amounted to as much as 15 to 26 p.c. : see our Arts. 1503*
and 1125.

[1755.] Finally in § 81 we have the effect of permanent tort on
the elastic nature of wires. Thus it developed æolotropy in the substance of the wire, and altered both the stretch- and slide-moduli. For
example, the slide-modulus of copper permanently torted decreased with
the increase of tort even to 1/6 of its original value, and then slightly
increased again before rupture. Steel pianoforte wire shewed a diminution and then a slight augmentation of the slide-modulus under tort.
Thus it first sunk from 751×10^6 grammes per sq. centimetre to 414×10^6
and then rose to 430×10^6. Iron wire shewed a diminution of 14 p.c. of
the original value before rupture.

In copper wire the stretch-modulus on the other hand was *increased*
10 p.c. by a permanent tort. In steel wire no sensible alteration due
to tort was noticed in the stretch-modulus.

There is no reference to the experiments of G. Wiedemann on the
subject of tort : see our Arts. 708 and 714.

---

[1] *Phil. Trans.* Vol. CLXXI., Part I., pp. 1–14. London, 1881.
[2] *Annalen der Physik*, Bd. CXLI., pp. 481–503. Leipzig, 1870.

[1756.] As an appendix to the article we have the mathematical theory of elasticity to which reference has been made in our Art. 1648. All but the last Chapter, *i.e.* XVII., appeared in the *Phil. Trans.* for 1856. Several important points in this memoir must be noticed.

Chapter I. Def. I. *A stress is an equilibrating application of force to a body.*

This definition of stress *appears* to identify it rather with load or body-force than with stress in the sense of this *History.* It does not readily suggest the idea of "stress across a plane in the material". The vagueness of this use of the word is, I think, exemplified by Def. I. of *Chapter* II.:

A stress is said to be homogeneous throughout a body when equal and similar portions of the body, with corresponding lines parallel, experience equal and parallel pressures or tensions on corresponding elements of their surfaces.

If a cylindrical shell or part of a spherical shell were turned inside out, it could hardly be described in customary language as having in its new state an application of force, but it is very clearly in a state of stress. It seems better to preserve the primitive use of the word stress, as adopted by Rankine and sanctioned in the *Treatise on Natural Philosophy.*

Chapter III. Cor. 3. Here the following ellipsoid is introduced:

$$(1 - 2eT_1) x^2 + (1 - 2eT_2) y^2 + (1 - 2eT_3) z^2 = 1,$$

where the axes are the principal axes of the stress, T_1, T_2, T_3 are the principal tractions (see our Art. 603*), and e any indefinitely small quantity. This represents the stress in the following manner:

From any point P in the surface of the ellipsoid draw a line in the tangent plane, half-way towards the point where this plane is cut by a perpendicular to it through the centre; and from the end of the first-mentioned line draw a radial line to meet the surface of a sphere of unit radius concentric with the ellipsoid. The tension at this point of the surface of a sphere of the solid is in the line from it to the point P; and its amount per unit of surface is equal to the length of that infinitely small line, divided by e.

The construction does not seem so simple as that of the usual stress-quadric; and, given the direction of any plane, it is not clear how we should find from the above construction except by a tentative process the direction and magnitude of the stress across it.

Chapter IV   Prop. 3.   An ellipsoid of the following type is given :

$$(1 - 2s_1)\, x^2 + (1 - 2s_2)\, y^2 + (1 - 2s_3)\, z^2 = 1,$$

where the axes are the principal axes of the strain (or, as is well-known, of the stress : see our Art. 614*) and s_1, s_2, s_3 are the principal stretches.

...the position, on the surface of this ellipsoid, attained by any particular point of the solid, is such that if a line be drawn in the tangent plane, half-way to the point of intersection of this plane with a perpendicular from the centre, a radial line drawn through its extremity cuts the primitive spherical surface in the primitive position of that point.

[1757.]   We now reach on the basis of the preceding ellipsoids the following definition (*Chapter* IV., Prop. 3, Cor. 1 and Def. 2) :

For every stress, there is a certain infinitely small strain, and conversely, for every infinitely small strain, there is a certain stress, so related that if, while the strain is being acquired, the centre and the strain-normals [=principal axes of strain] through it are unmoved, the absolute displacements of particles belonging to a spherical surface of the solid represent, in intensity (according to a definite convention as to units for the representation of force by lines) and in direction, the force (reckoned as to intensity, in amount per unit of area) experienced by the enclosed sphere of the solid, at the different parts of its surface, when subjected to the stress.

Such a stress and the infinitely small strain related to it are termed *of the same type*.

This type requires *five* quantities to define it, two ratios between principal tractions (or principal stretches) and three angular directions defining the position of the principal axes.

Further definitions of what is meant by orthogonal stresses and strains are given in *Chapter* VI., Def. 1–3 :

A stress is said . . . . to be orthogonal to a strain if work is neither done upon nor by the body in virtue of the action of the stress upon it while it is acquiring the strain.

Two stresses [or strains] are said to be orthogonal when either coincides in direction with a strain [or stress] orthogonal to the other.

[1758.]   *Chapter* VIII. is entitled : *Specification of Strains and Stresses by their Components according to chosen Types.*

Six stresses or six strains of six distinct arbitrarily chosen types may be determined to fulfil the condition of having a given stress or a given strain for their resultant, provided these six types are so chosen that a strain belonging to any one of them cannot be the resultant of any strains whatever belonging to the others.

This follows from the fact that six independent parameters are required to specify any stress or strain whatever. The six arbitrarily chosen types of stresses or strains are termed *types of reference*.

Definition.   An *orthogonal system* of types of reference is one in

which the six strain or stress components are all six mutually orthogonal (*Chapter* IX.). When the types of reference expressing the
strain constitute an orthogonal system then the component stresses may
be expressed by the differentials of the strain-energy with regard to the
six component strains.

This principle is deduced in *Chapters* XI. and XIII. by a consideration of what is defined as *concurrence* between stress and strain.

[1759.] We now turn to the contents of *Chapters* XIV –
XVI. which form perhaps the most important portion of the paper
under consideration.

Let $\xi_1, \xi_2, \xi_3, \xi_4, \xi_5, \xi_6$ specify a strain by means of one system
of types of reference, and $\zeta_1, \zeta_2, \zeta_3, \zeta_4, \zeta_5, \zeta_6$ the same strain by
means of another system. Then any strain ξ_1 will be a linear
function of the ζ-system and the relation will contain six constants.
In general there will be 30 constants connecting the ξ- and
ζ-systems. Now the strain-energy is a quadratic function of the
strain-components and involves 21 constants. We can accordingly
always make use of 15 out of our 30 disposable constants to
eliminate the product terms of the strain-energy by a linear
transformation. Thus in an infinite variety of ways the strain-
energy can be expressed in the form :

$$w = \tfrac{1}{2}(A_1\zeta_1^2 + A_2\zeta_2^2 + A_3\zeta_3^2 + A_4\zeta_4^2 + A_5\zeta_5^2 + A_6\zeta_6^2).$$

In this case a strain of any one of the ζ types, if impressed on the
solid will be accompanied by a stress orthogonal to the five others
of the same system. The stress will be proportional but not generally equal to $dw/d\zeta$.

[1760.] The investigation of the previous article has left us
with 15 disposable constants and we can employ these to make
the six strain types ζ mutually orthogonal ; for the condition that
two strain types shall be mutually orthogonal involves only one
relation and there are just 15 pairs in 6 things. This follows
from the algebraic theory of the linear transformation of quadratic
functions, associated with the condition for orthogonality : see
Chapter X., Cor. 1 and 2.

Thus we reach the following important proposition :

...a single system of six mutually orthogonal types may be determined
for any homogeneous elastic solid, so that its potential energy when
homogeneously strained in any way, is expressed by the sum of the

products of the squares of the components of the strain, according to those types, respectively multiplied by six determinate coefficients (*Chapter* XV. Prop. 1).

Definition. The six strain-types thus determined are called the Six Principal Strain-Types of the body.

[1761.] If ζ_1, ζ_2, ζ_3, ζ_4, ζ_5, ζ_6 denote the six principal strain types, and S_1, S_2, S_3, S_4, S_5, S_6 the corresponding stresses we have the strain-energy of the form :

$$w = \tfrac{1}{2}(A_1\zeta_1^2 + A_2\zeta_2^2 + A_3\zeta_3^2 + A_4\zeta_4^2 + A_5\zeta_5^2 + A_6\zeta_6^2),$$

and generally $S = dw/d\zeta = A\zeta$.

It follows that the stress required to maintain a given amount of strain is a maximum-minimum if it be one of the six principal types (Prop. 4).

We can now return to § 41 of the article on *Elasticity* for the following definitions :

A modulus of Elasticity is the number obtained by dividing the number expressing a stress by the number expressing the strain which it produces. A modulus is called a principal modulus when the stress is such that it produces a strain of its own type.

An æolotropic solid has in general six principal *elasticities*, namely, the A-coefficients of the above value for the strain-energy. Sir William Thomson appears in § 41, (6) of the article on *Elasticity* to identify the six principal *elasticities* with six principal *moduli*. I am not certain how far this is consistent with the definition that a modulus is the ratio of the *number* expressing stress to the *number* expressing the strain which it produces. My point of difficulty is whether a ' principal stress type' is always capable of being expressed by a single numerical stress, or whether it will not often consist of a system of stresses. Thus the bulk-modulus in Sir William Thomson's sense (see our Art. 1776 and footnote) might be a principal elasticity, but, as it corresponds in some cases to a system of stresses, is it always a principal modulus ?

[1762.] Sir William Thomson gives in *Chapter* XV. Prop. 2 the following examples of principal elasticities :

(*a*) For *cubical æolotropy* (see our Arts. 450, (v) and 1639) :
Modulus of compressibility, the rigidity against diagonal distortion in any of the principal planes (three equal elasticities), and the rigidity against rectangular distortions of a cube of symmetry (two equal elasticities).
In the notation of our Arts. 1203 (*d*) and 1206 these moduli would be $\tfrac{1}{3}(a + 2f')$, d and $\tfrac{1}{2}(a - f')$ respectively.

(b)   For *perfect isotropy:*

Modulus of compressibility and the rigidity (five equal elasticities). In our notation these moduli are $\frac{1}{3}(3\lambda + 2\mu)$ and μ.

Further statements as to principal moduli will be found in § 41 of the article on Elasticity, but I do not clearly comprehend their meaning; thus it is said that a crystal of the rectangular parallelepiped (or "tesseral") class has six distinct principal moduli—"three, of the three (generally unequal) compressibilities along the three axes; and three, of the three rigidities (no doubt generally unequal) relatively to the three simple distortions of the parallelepiped...." I do not follow what is meant by the "three compressibilities along the three axes"—they cannot refer to the three stretch-moduli as these are not *principal* moduli.

The whole discussion would have been much clearer if the strain-energy, for a tesseral crystal say, had been written down in terms of the principal moduli and the six principal strain-types, these principal moduli being then given as functions of the usual nine elastic coefficients and the principal strain-types in terms of the usual stretch- and slide-components of strain. I have not succeeded in accomplishing this. I am indeed in doubt as to how to apply the condition for "orthogonality of strains";—nor if a dilatation can be a principal strain am I at all clear what is the corresponding principal stress; it certainly cannot be like most stresses a *directed* quantity.

[1763.]   In *Chapter* XV., Prop. 6, Sir William Thomson remarks that:

A homogeneous elastic solid, crystalline or non-crystalline, subject to magnetic force or free from magnetic force, has neither right-handed nor left-handed, nor any dipolar properties dependent on elastic forces simply proportional to strains.

Hence he argues that the elastic forces concerned in optical phenomena such as occur in quartz or tartaric acid cannot depend on the magnitude, but can solely depend on the heterogeneousness of the strain in the portion of the medium through which the wave passes. Polar properties of crystals whether crystallographic, optical or electrical, can have no corresponding characteristic in elastic forces which are simply proportional to the strain.

[1764.]   *Chapter* XVII. is entitled : *Plane Waves in a Homogeneous Æolotropic Solid.* It does not go further than demonstrating that in general *three* pairs of plane waves are possible in such a medium—in the case of an incompressible solid reducing to two pairs in which the motion is parallel to the wave-front. The three velocities of these three pairs of waves are determined neither in terms of the 21 elastic con-

---

[1] It is easy, Sir William Thomson tells us, to investigate the principal strain-type and principal elasticities for a crystal of the tesseral class (*Chapter* XVI., Cor.).

stants, nor of the direction of the wave-front. The problem had been previously discussed by Blanchet (see our Arts. 1166*–78*) and has been exhaustively dealt with by Christoffel: see *Annali di Matematica*, T. VIII., pp. 193–243. Milano, 1877, and Love: *Treatise on the mathematical Theory of Elasticity*, Vol. I., pp. 134–40. Cambridge, 1892.

[1765.] *Notes of Lectures on Molecular Dynamics and the Wave Theory of Light. Delivered at the Johns Hopkins University, Baltimore. Stenographically reported by A. S. Hathaway.* Baltimore, 1884. This is a shorthand report reproduced by papyrograph of what Sir William Thomson said in twenty lectures delivered at Baltimore before a distinguished audience of physicists and mathematicians in 1884. The preface to Vol. III. of the *Mathematical and Physical Papers* announces that Vol. IV. will contain a printed edition of these lectures. That volume not having yet appeared, our references will be to the pages of the papyrograph (pp. 1–328 + Index). The report was not revised by the lecturer, owing to his departure from America.

We shall put on one side the large portion of these lectures devoted to molecular theories, treating only of those points which relate to the theory of elasticity, and briefly of some problems in which that theory is applied to the luminiferous ether.

[1766.] *Lecture I.* (pp. 1–20) is chiefly historical and introductory. The position of the lecturer at that time is indicated in the following words:

In the first place we must not listen to any suggestion that we must look upon the luminiferous ether as an ideal way of putting the thing. A real matter between us and the remotest stars I believe there is, and that light consists of real motions of that matter, motions just such as are described by Fresnel and Young, motions in the way of transverse vibrations. If I knew what the magnetic theory of light is, I might be able to think of it in relation to the fundamental principles of the wave theory of light. But it seems to me that it is rather a backward step from an absolutely definite mechanical motion that is put before us by Fresnel and his followers to take up the so-called electro-magnetic theory of light in the way it has been taken up by several writers of late. In passing, I may say that the one thing about it that seems intelligible to me, I scarcely think is admissible. What I mean is, that there should be an electric displacement perpendicular to the line of propagation and a magnetic disturbance perpendicular to both. It seems to me that when we have an electro-magnetic theory of light, we shall see electric displacement as in the direction of propagation—simple vibrations as described by Fresnel with lines of vibration perpendicular to the line of propagation—for the motion actually constituting light. I merely say that in passing, as perhaps some apology is necessary for my insisting upon the plain matter of fact dynamics and the true elastic solid as giving what seems to me the only tenable foundation of the wave theory of light in the present state of our knowledge.

The luminiferous ether we must imagine to be a substance which so far as luminiferous vibrations are concerned moves as if it were an elastic solid. I do not say that it is an elastic solid. That it moves as if it were an elastic solid in respect to the luminiferous vibrations, is the fundamental assumption of the wave theory of light (pp. 5–6).

In the last eight years Sir William Thomson has without doubt modified his view as to the respective merits of an elastic solid and an electro-magnetic theory of light: see in particular his papers referred to in our Arts. 1806–16. But the emphasis laid on the "real matter" and "real motion" of the luminiferous ether seems to the Editor of this *History* a grave danger in this method of speaking of the ether. The ideal nature of geometry involves the ideal nature of kinematics and ultimately of mechanism, and the "luminiferous ether" is only an intellectual mode of briefly summarizing certain wide groups of sensations. The advantage of the electro-magnetic over the elastic solid theory of light appears to lie in the wider range of phenomena it enables us to epitomise under one conception.

The difficulty of the passage of the stellar bodies through the ether is explained by aid of the principle first indicated by Sir G. G. Stokes (see our Art. 1266*), *i.e.* that as in the case of cobblers' wax, which vibrates to rapidly alternating forces, long continued but very small forces suffice to produce permanent change of shape[1].

Whether infinitesimally small forces produce change of shape or not we do not know; but very small forces suffice to produce change of shape. All we have got with respect to the luminiferous ether is that the exceedingly small forces required to be brought into play in the luminiferous vibrations do not, in the times during which they act suffice to produce any sensibly permanent distortion. The come and go effects taking place in the period of the luminiferous vibrations do not give rise to the consumption of any large amount of energy,

---

[1] Glycerine is also suggested as an example illustrating the ether on p. 119, and Maxwell's experiment in which the sudden turn of a stick in Canada Balsam gave the medium a double refractive power, which gradually disappeared, is referred to on pp. 119—20.

not large enough an amount to cause the light to be wholly absorbed in say its propagation from the remotest visible star to the earth (p. 8).

[1767.]   *Lecture II.* (pp. 20–5) opens with a brief elementary theory of elasticity containing, however, nothing beyond what is given in the *Encyclopaedia* article on *Elasticity*: see our Art. 1741. *Lecture III.* (pp. 31–3) indicates the general solution of the equations of vibration for a homogeneous isotropic solid. *Lecture IV.* (pp. 38–48) develops this solution, chiefly in reference to the sound vibrations represented by an equation of the type:

$$\rho \frac{d^2\phi}{dt^2} = (\lambda + 2\mu)\,\nabla^2\phi.$$

Lecture VI. (pp. 57–66) continues the discussion of these sound vibrations.

[1768.]   *Lecture VIII.* (pp. 77–91) deals with distortional waves, or those for which the dilatation $\theta = 0$. Consider the function:

$$\phi = \frac{C}{r} \sin \frac{2\pi}{l}\left(r - \sqrt{\frac{\mu}{\rho}}\,t\right),$$

which satisfies the equation:

$$\rho \frac{d^2\phi}{dt^2} = \mu\nabla^2\phi,$$

r being the distance from the origin, C and l being constants.

(*a*)   A solution of the body-shift equations, subject to $\theta = 0$, is given by:

$$u = 0, \quad v = \frac{d\phi}{dz}, \quad w = \frac{d\phi}{dy}.$$

At a considerable distance from the origin the solution takes the approximate form:

$$u = 0, \quad v = -C\frac{2\pi}{l}\frac{z}{r^2}\cos q, \quad w = C\frac{2\pi}{l}\frac{y}{r^2}\cos q,$$

where q is written for $\dfrac{2\pi}{l}\left(r - \sqrt{\dfrac{\mu}{\rho}}\,t\right)$.

Further the twists at a considerable distance are given by:

$$\tau_{yz} = C\frac{4\pi^2}{l^2}\left(\frac{x^2}{r^3} - \frac{1}{r}\right)\sin q, \quad \tau_{zx} = C\frac{4\pi^2}{l^2}\frac{xy}{r^3}\sin q, \quad \tau_{xy} = C\frac{4\pi^2}{l^2}\frac{xz}{r^3}\sin q.$$

Thus there are rotations proportional to $-(\sin q)/r$ round the axis of x, and to $(\sin q)\,x/r^2$ round the radius-vector.

If you think out the nature of the thing, you will see that it is this: a globe, or a small body at the origin, set to oscillating about Ox as an axis. You will have turning vibrations everywhere; and the light will be everywhere polarized in planes through Ox. The vibrations will be everywhere perpendicular to the radial plane through Ox (p. 79).

(*b*) Besides this solution for a torsional vibration, Sir William Thomson gives (p. 84) the solution for a small to-and-fro motion in the axis of x, viz.

$$u = \frac{4\pi^2}{l^2}\,\phi + \frac{d^2\phi}{dx^2}, \quad v = \frac{d^2\phi}{dy\,dx}, \quad w = \frac{d^2\phi}{dz\,dx},$$

ϕ having still the value

$$\frac{C}{r}\sin\frac{2\pi}{l}\left(r - \sqrt{\frac{\mu}{\rho}}\,t\right).$$

At a considerable distance from the origin we have approximately:

$$u = C\,\frac{4\pi^2}{l^2}\,\frac{r^2 - x^2}{r^3}\sin q, \quad v = -\,C\,\frac{4\pi^2}{l^2}\,\frac{xy}{r^3}\sin q, \quad w = -\,C\,\frac{4\pi^2}{l^2}\,\frac{xz}{r^3}\sin q,$$

where q has the same value as above, and clearly the resultant of these shifts is perpendicular to the radius-vector. Further at a great distance there is no appreciable shift at points in the axis of x at all. In the plane of yz we have $v = w = 0$, or the shift is perpendicular to this plane, *i.e.* light would be polarised in this plane (p. 86).

Sir William Thomson refers with regard to this solution to Sir G. G. Stokes' theory of the blue light of the sky. He further deals at considerable length with models of vibrators which would produce vibrations corresponding to either of the above cases. It is clear that the solutions given by Sir William Thomson are special cases of those due to Voigt and afterward dealt with by Kirchhoff: see our Arts. 1309–10.

[1769.] While *Case* (*b*) of the preceding article deals with the to-and-fro motion in the axis of x of a single small body at the origin, *Lecture IX.* (pp. 92–4) considers the case of a doublet of such motions at the origin. Such a motion might be considered as given by discs attached to the two ends of a tuning-fork, neglecting the prongs, or by two small balls connected by a spring and pulled asunder so as to vibrate in and out (p. 94). The expressions for the shifts may be found from those given in *Case* (*b*) above by simply differentiating them[1] with regard to x and introducing a new constant into ϕ. Thus, at a considerable distance from the vibrator the shifts will be approximately of the forms:

$$u = C'\frac{x^2 - r^2}{r^4}\,x\cos q, \quad v = C'\frac{x^2 y}{r^4}\cos q, \quad w = C'\frac{x^2 z}{r^4}\cos q.$$

It is easy to prove that the complete solution represents a distortional vibration ($\theta = 0$), and that the radial component of shift at a considerable distance is zero. There is zero shift in the plane of yz and along the axis of x. Further treating the motion as that of light, we see that light

---

The papyrograph has a slip at this point, it speaks of du/dx, dv/dy and dw/dz as the shifts (p. 93).

would be "polarized in the plane through the radius of the point considered and perpendicular to the radial plane through Ox" (p. 93).

Sir William Thomson holds that : " This is the simplest set of vibrations that we can consider as proceeding from any natural source of
light " (p. 94)[1].

Much of the remainder of this *Lecture*, dealing with the simplest
conceivable form of elementary vibrator in the case of light, is of great
interest, but it would lead us beyond our legitimate subject to discuss
the lecturer's suggestions here.

[1770.]   *Lecture XI.* (pp. 124–37) treats of æolotropic
elastic solids. The first nine pages (pp. 124–32) deal with the
constant' controversy. After referring to the meaning of the
term *æolotropic*, and " the somewhat cloud-land molecular beginning" of the theory of elasticity, Sir William Thomson remarks
that :

...we have long passed away from the stage in which Father Boscovich
is accepted as being the originator of a correct representation of the
ultimate nature of matter and force. Still, there is a never-ending
interest in the definite mathematical problem of the equilibrium or
motion of a set of points endowed with inertia and mutually acting
upon one another with any given force. We cannot but be conscious
of the one grand application of that problem to what used to be called
physical astronomy but which is more properly called dynamical astronomy, or the motions of the heavenly bodies. We have cases in which
we have these motions instead of the approximate equilibriums or infinitesimal motions which form the subject of the special molecular
dynamics that I am now alluding to (pp. 125–6).

It is then pointed out that those who have treated the theory
of elasticity from the standpoint that :

matter consists of particles acting upon one another with mutual forces,
and that the elasticity of a solid is the manifestation of the force required
to hold the particles displaced infinitesimally from the position in which
the mutual forces will balance (p. 126),

have been led to rari-constant equations. This statement should,
I think, be modified by the addition to " mutual forces" of the
words "which act in the line joining the particles and are functions

---

[1] This statement is modified in *Lecture XII.* (p. 145), where the lecturer points
out that the condition for the centroid of a molecule remaining stationary while
the molecule acts as a vibrator, would be satisfied not only by the double to-and-fro
motion of our Art. 1769 but also by *Case* (b) of our Art. 1768, if the vibrator were a
Thomson "shell-spring" molecule,—*i.e.* one with a massive nucleus carrying an
external shell-surface of extremely small mass by means of connecting springs.

only of the mutual distances." The statement of the *Lectures* does not exclude the hypotheses of modified action and of aspect, either of which being admitted lead to multi-constant equations : see our Arts. 276 and 302–6.

Sir William cites Sir G. G. Stokes as having first called attention to "the viciousness of this conclusion (*i.e.* uni-constancy) as a practical matter in respect to the realities of elastic solids." Jelly and india-rubber, our old friends, are referred to as examples of elastic solids which do not fulfil the uni-constant condition, but no attempt is made to complete the validity of the argument by demonstrating that they are true elastic solids at all, *i.e.* that two elastic moduli will suffice to determine absolutely the relations between all types of small stresses and strains in these materials. For example, in the case of these materials are the stretch and squeeze-moduli practically the same, and if the slide-modulus and the dilatation-coefficient (λ, see Vol. I., p. 884–5) be determined from torsion and pure traction experiments, are the values of the dilatation-modulus ($\lambda + \frac{2}{3}\mu$), the spread-modulus $[\mu(3\lambda + 2\mu)/(\lambda + 2\mu)]$ and the plate-modulus $[4\mu(\mu + \lambda)/(\lambda + 2\mu)]$ calculated from these results in agreement with experiment? These points require very careful consideration before the argument from jelly and india-rubber can be recognised as conclusive : see our Arts. 1636 and 1749.

[1771.] Sir William Thomson now raises a more interesting argument against rari-constancy. He introduces it with the following remark :

Stokes also referred to a promise that I made, I think it was in the year 1856, to the effect that out of matter fulfilling Poisson's condition [*i.e.* rari-constant matter] a model may be made of an elastic solid, which when the scale of parts is sufficiently reduced will be a homogeneous elastic solid not fulfilling Poisson's condition. Stokes refers to that promise of mine which was made very nearly 30 years ago. I propose this moment to fulfil it never having done so before. It is a very simple affair (p. 127).

The following is the model suggested.

Take a geometrical right six-face as our element and suppose 8 particles at its angles. These may be connected by the 12 edges, the four internal diagonals and the 12 face diagonals. Each edge will however belong to four such right six-faces, and each face diagonal to two right six-faces ; hence we are left with only 13 disposable links for each element. Suppose these links replaced by 13 springs of different elasticities.

This gives us 13 arbitrary constants. Two further constants come
from the ratios of the three edges, and three from the arbitrary
directions which we may take for our coordinate axes of reference.
Thus we have at present 18 arbitrary constants. To get three more
constants Sir William Thomson places bell-cranks at each corner and
connects them by pieces of wire, so that the wire, thought of for
the moment as continuous through the bell-cranks, passes *twice*
round the edges of the right six-face. This can be done in a variety
of ways. These pieces of wire connecting the bell-cranks can be
taken of different elasticities in the directions of the three principal
axes, and we thus have three more disposable constants, or 21 in all.
Sir William Thomson speaks of this arrangement as "a model of a solid
having the 21 independent coefficients of Green's theory." He draws
attention to the fact that for the case of an isotropic solid if the bell-
crank wires are inelastic, the right six-face can suffer no dilatation. In
fact, we might place smooth rings at the corners and take a continuous
inextensible string twice round the edges; for small strains the solid
would then be inextensible (if not incompressible)[1]

[1772.] Now there seems to me to be grave difficulties about this
model. It consists really of a space framework with a considerable
number of supernumerary bars, besides a binding of wire and bell-cranks.
These involve 18 disposable constants. But why stop at 18? We cannot,
indeed, put in any more *straight* supernumerary bars, but there is nothing,
I think, to hinder us running wire and bell-cranks round the diagonal
bracing bars in a great variety of ways. I see no reason why the disposable constants should stop at 18. Yet no one will assert that because
we can build up a frame with supernumerary bars, bell-cranks and wires
which has 24 or perhaps 30 disposable constants, that therefore we can
have an elastic solid with 24 or 30 disposable coefficients. Clearly there
is a portion of the argument which is very far from completed by the
lecturer. Out of material obeying rari-constant conditions, we can build
up a frame with 18 (or possibly 80 disposable constants), but it has yet
to be proved that the relations between stress and strain for such a frame
will contain *the same number of independent coefficients.* The complexity
of the supernumerary bars in Sir William Thomson's model framework
renders it difficult, if not impossible, to work out the relations between
the elasticities of the various members and the elastic coefficients of the
corresponding elastic solid. Till that is done, however, we have no evidence that certain inter-constant relations may not after all hold for this
model[2]

---

[1] An inextensible string alone would not answer the purpose in the case of an
aeolotropic medium, for if a, b, c be the edges of the right six-face, the condition of
inextensibility gives $\delta a + \delta b + \delta c = 0$, but that of incompressibility $\dfrac{\delta a}{a} + \dfrac{\delta b}{b} + \dfrac{\delta c}{c} = 0$.
The further conditions: $a = b = c$, are necessary and sufficient.

[2] Sir William Thomson remarks on p. 131: "We have 18 available quantities,
which will make by solution of linear equations the required 18 moduluses." This

Even the particular case of isotropy is by no means easy of analysis in the model, we have of course straight off only one edge elasticity, one face diagonal elasticity, one internal diagonal elasticity and further one elasticity of the binding wire, four constants in all. Sir William Thomson tells us (p. 129) that without the binding wire the three other elasticities for isotropy reduce to a single one and that "an isotropic solid made up in this way will have an absolutely definite compressibility; we cannot make the compressibility what we please." It would be an interesting, but I fear complicated piece of analysis to ascertain even in this case the relations between the elasticities of the three bars and to determine whether the stretch-modulus is or is not $\frac{5}{2}$ of the slide-modulus.

[1773.] Since Sir William Thomson introduces *supernumerary bars* into his frame, it is clear that the action between any two particles depends on the action between other pairs, for a strain in one bar produces strain in all the others, which strains of course influence the stress in the first bar. Thus he is really constructing a model which introduces the hypothesis of modified action. This hypothesis is expressly excluded by the assumptions of Navier and Poisson, and we have already recognised that it may lead to multi-constancy. Whether it leads in the case of the model described in this lecture to *complete* multi-constancy, I do not think we have evidence enough to determine. Clearly the model does not carry us further than, if indeed as far as, the statement, that modified action leads to multi-constancy: see our Arts. 1529* (and ftn.), 276 and 305. The remainder of the *Lecture* (pp. 132–7) is devoted to a discussion of wave motion in an aeolotropic medium and covers practically the same ground as the *Encyclopaedia* article on *Elasticity*: see our Art 1764.

[1774.] *Lecture XII.* (pp. 137–43) discusses the differences between aeolotropic and isotropic solids in the matter of wave motion. It indicates rather by suggestion than analysis what is the probable solution for waves in the former case, and also the nature of the conditions which must hold in order that condensational may be separated from distortional waves. As to indications of the former wave Sir William Thomson says:

The want of indication of any such actions is sufficient to prove that if there are any in nature, they must be exceedingly small. But that there are such waves I believe, and I believe that the velocity of propagation of electrostatic force is the unknown condensational velocity that we are speaking of......I do not mean that I believe this as a matter of religious faith, but rather as a matter of strong scientific probability (p. 143).

[1775.] *Lecture XIII.* (pp. 154–62) contains some rather disconnected but still suggestive remarks on aeolotropy and wave motion in aeolotropic

does not I think mean that the 18 coefficients are linear functions of the 18 moduluses. They come out, I think, very complicated functions of the 16 elasticities and the two length ratios, but I do not see à priori why these functions *must* be independent.

solids. On p. 156 the form of the equations for wave motion in an *incompressible* isotropic solid is generally indicated, and the method of obtaining those for an incompressible aeolotropic solid is suggested. Franz and Carl Neumann had dealt previously at some length with these problems: see our Arts. 1215 *et seq.* The lecturer then turns to Rankine's nomenclature and deals especially with cyboïd or cubic aeolotropy: see our Arts. 443–52, especially Art. 450 (v). He points out that Rankine had remarked that according to Sir David Brewster this sort of variation from isotropy was to be found in analcime[1]. He then quotes Sir G. G. Stokes to the effect that no optical phenomenon observed in cubic crystals gives any evidence in favour of the existence of this sort of aeolotropy, and that not even Brewster's experiment is a true instance. Thus we are thrown back on physical elasticity rather than on optics for examples of cyboïd aeolotropy, and the lecturer illustrates it from woven material and basket work, where the elasticity may be the same in the direction of the two (or three) principal axes, but the resistance to shear may vary widely with the direction of the shear. He refers on p. 159 to the error of Rankine noticed in our Art. 421.

Starting from cyboïd aeolotropy, Sir William Thomson, supposing incompressibility and annulling the "difference of rigidities for the principal distortions in each of the three principal planes," reaches an elastic solid with three principal moduli and giving Fresnel's wave surface. For a fuller discussion of the details of this investigation, which is only indicated in the briefest manner in the *Lectures*, we may refer the reader to the memoirs cited in our Arts. 917*–18*, 148–50 and 1214–15. As in Neumann's investigation the shifts lie *in* the plane of polarization (pp. 161–2).

[1776.] *Lecture XIV.* (pp. 173–78) has some interesting remarks and results bearing on various features of aeolotropy.

(a) The first elastic problem is to find the bulk-modulus, *i.e.* the dilatation-modulus for an aeolotropic solid (p. 174). We take the bulk-modulus[2] to be the elastic-constant by which uniform pressure on the surface of any portion of a homogeneous aeolotropic solid must be divided

---

[1] See Herschel's *Light.* Art. 1133. *Encyclopaedia Metropolitana.* London, 1854.

[2] Sir William Thomson here defines the bulk-modulus to be the mean normal pressure divided by the compression when the solid is compressed equally in all directions, *i.e.* when the strain denotes a pure change of size. This, however, does not give the relation between pressure and dilatation for the case which we can actually experiment on, namely: a uniform surface pressure. Further making an aeolotropic body incompressible for the stress which produces pure change of size, does not insure that the body is really incompressible for every form of stress. The bulk-moduli in Sir William Thomson's sense and in our sense of the word coincide only for cubical crystals and isotropic bodies. In other cases it is difficult to see how this modulus in Sir William Thomson's sense satisfies his definition of a modulus of elasticity: see our Art. 1761. It is the ratio not of an actual, but of an *average* stress to the dilatation. This bulk-modulus cannot be ascertained by any simple experiment, and no arrangement of load capable of being practically applied would produce such a pure change of size in an aeolotropic body.

in order to obtain the compression per unit volume of the solid. If this be so, the proper method of procedure seems to be to equate the three tractions in their most general form to the pressure with its sign changed $(-p)$, and further to put the three shears zero. From the six equations so obtained the slides must be eliminated and the three stretches found. The sum of the three stretches then gives the dilatation (*i.e.* the compression) in terms of the pressure, and so determines the dilatation-modulus. The answer can be at once written down in the form of determinants, but to expand them for the most general case of an aeolotropic solid is very laborious. For the case of three planes of elastic symmetry, we find for the dilatation-modulus F in the notation of Art. 117:

$$F = \frac{abc + 2d'e'f' - ad'^2 - be'^2 - cf'^2}{bc - d'^2 + ca - e'^2 + ab - f'^2 + 2(d'e' - cf') + 2(f'd' - be') + 2(e'f' - ad')}.$$

This agrees with Neumann's result in our Art. 1205 for a special case and also with the value of F for isotropy.

Sir William Thomson, with the definition of the footnote to our Art. 1776 deduces from the expression for the strain-energy that (p. 168):

$$F = \tfrac{1}{9}\{|xxxx| + |yyyy| + |zzzz| + 2(|yyzz| + |zzxx| + |xxyy|)\}$$

for the general case of aeolotropy, the notation being that of our Art. 116, ftn. This result does not involve like the previous one the direct slide coefficients nor those of asymmetrical elasticity.

For the case of three planes of elastic symmetry it becomes

$$F = \tfrac{1}{9}\{a + b + c + 2(d' + e' + f')\},$$

which differs from the result given above. It agrees with that result and with the usual value $(\lambda + \tfrac{2}{3}\mu)$ in the case of isotropy.

This value of F given on pp. 168 and 174 leads me to believe that the second problem treated by Sir William Thomson, namely the value of the strain-energy for an incompressible aeolotropic elastic solid, is erroneously worked out[1].

(b) The third problem is entitled : *To annul skewnesses relatively to* Ox, Oy, Oz. This amounts to equating to zero the coefficients of asymmetrical elasticity : see our footnote p. 77.

(c) The fourth problem is : *To annul weblike aeolotropy, the skewnesses being annulled* (pp. 175–8). By "annulling the weblike aeolotropy" Sir William Thomson understands introducing a condition of the following kind :

Take a plane perpendicular to any one of the axes, say that of Ox, and suppose lines in the direction Oy to receive a stretch $\tfrac{1}{2}s$, and lines in

---

[1] I have used the word 'erroneous' here although the matter is rather one of definition. We are dealing with two bulk-moduli (tasinomic and thlipsinomic) differently defined. But it seems to me impossible to consistently define a solid, in which some systems of loading do produce compression, as incompressible.

the direction Oz a squeeze $-\frac{1}{2}s$, then the work done in this strain is to be equal to the work done in giving a face perpendicular to y a slide parallel to z of magnitude s. Geometrically the slide and the stretch and squeeze are equivalent, and Sir William Thomson introduces an isotropy with regard to slide in the planes perpendicular to each of the coordinate axes. Thus the condition is :

to express that there is such a deviation from aeolotropy as would be produced if we were to annul the differences of rigidity relatively to a shear[1] produced by pulling out one diagonal and shortening the other compared with the shear of sliding one face past the other (p. 177).

The strain-energy ϕ for an aeolotropic solid in which the "skewnesses are annulled" is easily seen to be:

$$2\phi = as_x^2 + bs_y^2 + cs_z^2 + 2d's_ys_z + 2e's_zs_x + 2f's_xs_y + d\sigma_{yz}^2 + e\sigma_{zx}^2 + f\sigma_{xy}^2.$$

Hence for : $s_y = \frac{1}{2}s$, $s_z = -\frac{1}{2}s$, we have, all the other strains being zero :

$$2\phi = \tfrac{1}{4}(b + c - 2d')\,s^2 ;$$

and for $\sigma_{yz} = s$, and all the other strains zero :

$$2\phi = ds^2.$$

Thus the condition for annulling weblike aeolotropy is $d = \frac{1}{4}(b + c - 2d')$, or

$$\tfrac{1}{2}(b + c) = (2d + d').$$

Similarly :

$$\tfrac{1}{2}(c + a) = (2e + e'),$$

$$\tfrac{1}{2}(a + b) = (2f + f').$$

Now these are precisely Saint-Venant's ellipsoidal conditions of the second kind (see our Art. 230), or "weblike aeolotropy" is not consistent with the aeolotropy produced by permanently straining an isotropic elastic solid so as to have three planes of elastic symmetry: see our Art. 231.

[1777.] There is another way of looking at the results of the preceding article, which is not without instructiveness. Consider a strain confined to the plane xy, and defined by the three strain-components s_x, s_y, and σ_{xy}. Let r' be a line in this plane which makes an angle θ with the axis of x and let r be a line perpendicular to it. Then we easily find for a solid with the skewnesses annulled :

$$\sigma_{rr'} = (s_y - s_x)\sin 2\theta + \sigma_{xy}\cos 2\theta,$$

and $\widehat{rr'} = \frac{1}{2}\{s_x(f' - a) + s_y(b - f')\}\sin 2\theta + f\sigma_{xy}\cos 2\theta.$

See (vi) and (viii) of our Art. 133.

The second result holds in the case of weblike aeolotropy. Now give

---

[1] *Shear* is here used for *strain*, in the sense of our *slide*.

any uniplanar strain without dilatation, or such that $s_x + s_y = 0$. Then, if there is to be isotropy of slide, we must have:

$$\widehat{rr'} = f\sigma_{rr'},$$

or $\qquad\qquad \tfrac{1}{2}\{(b - f') - (f' - a)\} = 2f,$

i.e. $\qquad\qquad \tfrac{1}{2}(a + b) = 2f + f'$

or the same condition as before. This method of obtaining the result brings out more clearly that absolute isotropy of slide does really exist, when weblike aeolotropy is annulled in each principal plane for strains without dilatation in that plane.

[1778.] Sir William Thomson uses (pp. 169, 177–8) the results of the previous article to obtain an expression for the strain-energy ϕ of an elastic solid without either 'skewnesses', or 'web-like aeolotropy' and strained without dilatation. He takes to insure the latter condition

$$s_x = \tfrac{1}{2}(\beta - \gamma), \qquad s_y = \tfrac{1}{2}(\gamma - a), \qquad s_z = \tfrac{1}{2}(a - \beta).$$

We then find[1]:

$$\phi = \tfrac{1}{2}d\,(a^2 + \sigma_{yz}^2) + \tfrac{1}{2}e\,(\beta^2 + \sigma_{zx}^2) + \tfrac{1}{2}f\,(\gamma^2 + \sigma_{xy}^2)$$
$$- \tfrac{1}{2}(e + f - d)\,\beta\gamma - \tfrac{1}{2}(f + d - e)\,\gamma a - \tfrac{1}{2}(d + e - f)\,a\beta.$$

Thus we have the strain-energy expressed in terms of the three slide-moduli alone, and so in a form suitable for discussing waves of distortion.

[1779.] In *Lecture XV.* pp. 182–93 are devoted to the subject of elasticity and the elastic theory of light. The remarks on pp. 182–3 as to the conditions for incompressibility seem to me doubtful, owing to the use of the particular value of the dilatation-modulus before referred to: see our Art. 1776. Sir William then passes to the thlipsinomic coefficients. Adopting the notation for these coefficients suggested in our Art. 448, we have as types:

$$s_x = (aaaa)\,\widehat{xx} + (aabb)\,\widehat{yy} + (aacc)\,\widehat{zz} + (aabc)\,\widehat{yz} + (aaca)\,\widehat{zx} + (aaab)\,\widehat{xy},$$

$$\sigma_{yz} = (bcaa)\,\widehat{xx} + (bcbb)\,\widehat{yy} + (bccc)\,\widehat{zz} + (bcbc)\,\widehat{yz} + (bcca)\,\widehat{zx} + (bcab)\,\widehat{xy}.$$

Clearly we must then write

$$\theta =$$

$(aaaa)$	$\widehat{xx} + (aabb)$	$\widehat{yy} + (aacc)$	$\widehat{zz} + (aabc)$	$\widehat{yz} + (aaca)$	$\widehat{zx} + (aaab)$	$\widehat{xy}.$
$+ (aabb)$	$+ (bbbb)$	$+ (bbcc)$	$+ (bbbc)$	$+ (bbca)$	$+ (bbab)$	
$+ (aacc)$	$+ (ccbb)$	$+ (cccc)$	$+ (ccbc)$	$+ (ccca)$	$+ (ccab)$	

---

[1] The papyrograph (p. 169) appears to have the factor 2 instead of $\tfrac{1}{2}$ in the three last terms.

Hence we have the following six conditions for complete incompressibility under all forms of stress :

$$(aaaa) + (aabb) + (aacc) = 0, \qquad (aabb) + (bbbb) + (ccbb) = 0,$$

$$(aacc) + (bbcc) + (cccc) = 0, \qquad (aabc) + (bbbc) + (ccbc) = 0,$$

$$(aaca) + (bbca) + (ccca) = 0, \qquad (aaab) + (bbab) + (ccab) = 0.$$

Sir William Thomson remarks (p. 184) :

It is startling to think of six equations to express incompressibility, I have not really noticed it before, but it is quite right...

In thlipsinomic coefficients it is clear that the conditions of incompressibility can only be expressed by the above six relations; but it is not so clear that in the case of tasinomic coefficients six relations will be necessary.

For example, we have, in the case of three planes of elastic symmetry[1], when the plagiothliptic coefficients, *i.e.* those of unsymmetrical pliability, vanish (see our Art. 448):

$$(aaaa) = \frac{bc - d'^2}{\Delta}, \quad (aabb) = \frac{d'e' - f'c}{\Delta}, \quad (aacc) = \frac{f'd' - e'b}{\Delta},$$

$$(bbbb) = \frac{ca - e'^2}{\Delta}, \quad (bbcc) = \frac{e'f' - d'a}{\Delta}, \quad (cccc) = \frac{ab - f'^2}{\Delta},$$

where $\qquad \Delta = abc + 2d'e'f' - ad'^2 - be'^2 - cf'^2.$

Hence the conditions for incompressibility reduce to :

$$\frac{bc - d'^2 + d'e' - f'c + f'd' - e'b}{\Delta} = 0, \qquad \frac{ca - e'^2 + e'f' - d'a + d'e' - f'c}{\Delta} = 0,$$

$$\frac{ab - f'^2 + f'd' - e'b + e'f' - d'a}{\Delta} = 0.$$

Now we can satisfy these by making all three numerators zero, which does not involve any of the tasinomic coefficients being infinite (and certainly not all six, a, b, c, d', e', f' infinite, as seems to be suggested by the lecturer on p. 175), or we can take $\Delta = \infty$, without making the numerators infinite. For example, if we take a and d' infinite, or b and e' infinite, or c and f' infinite, the conditions of incompressibility will be satisfied. To judge from this special case the general rule seems to be the following which is not in complete agreement with that stated by Sir William Thomson. For incompressibility it suffices that six relations be satisfied among the tasinomic coefficients, none of them becoming infinite; but in special cases the becoming infinite of a number

---

[1] The reader must carefully distinguish between the a, b, c of the symbols for the thlipsinomic coefficients which denote *merely directions*, and the a, b, c which are the direct stretch coefficients (tasinomic constants) of an elastic solid with three planes of elastic symmetry.

less than six of the tasinomic coefficients will suffice to ensure incompressibility.

The further condition for the vanishing of the "skewnesses" in thlipsinomic coefficients is discussed on pp. 186–7. It is of course merely the vanishing of Rankine's plagiothliptic coefficients. Sir William Thomson speaks of them here as well as of the plagiotatic coefficients as "side-long coefficients." They are the coefficients such as $|xyyz|$, $|xxxz|$, $(abbc)$, $(aaac)$ etc., which contain an odd number of any subscript letter.

The remainder of *Lecture XV.* (pp. 187–93) and the first part of *Lecture XVII.* (pp. 209–13) contain a criticism of Green's "extraneous pressures." The criticism misses, I venture to think, the real point of what Cauchy, Green and Saint-Venant denote by these "extraneous pressures," or by what we have by preference in our *History* termed *initial stresses*: see our Arts. 616*, 1210* and Vol. I., p. 883. It is of the very essence of such initial stresses that the principle of the superposition of small strains does not apply. Compare our Arts. 129 and 1445–6 with Sir William's remarks on p. 192, noting, however, his p. 212. The footnote p. 189 together with the addition on p. 213 must, I think, be taken as probably marking a withdrawal after further consideration from the standpoint of the lectures: see also our Art. 1789.

[1780.] The only other part of *Lecture XVII.* (which is mainly occupied with considerations as to the reflection and refraction of light at the interface of two media, and as to the plane of polarisation) relating closely to our subject is the further discussion of aeolotropy on pp. 213–6. Sir William Thomson refers in particular to 'web-like asymmetry' and refers to braced structures having only one set of diagonal bracing bars as representing something analogous in framework. He refers also to the probability that crystals of the cubic class possess it, and suggests the importance of experiments. Clearly were we to annul 'web-like asymmetry' in regular crystals, they would become isotropic elastic bodies[1], and they would cease to be crystals from the elastic standpoint. Klang as early as 1881 and Voigt in a series of memoirs have determined the constants a, f' and d for regular crystals, and shewn that the rari-constant relation $a = 2d + f'$ is very far from holding: see our Arts. 1203 (d), and 1212. How far their experiments inspire full confidence will be discussed later.

[1781.] *Lecture XVIII.* (pp. 227–49) deals with the reflection and refraction of light at the interface of two media on the elastic solid theory. The method adopted is very close to Lord Rayleigh's treatment of Green's theory: see the *Philosophical Magazine*, Vol. XLII., pp. 81–970, London, 1871. The discussion is very suggestive on a number of points, but they belong rather to the theory of light than to that of

---

[1] This follows at once if we introduce the annulling condition or $a = 2d + f'$ into the stress-strain relations of our Art. 1203 (d).

elasticity. The first suggestion, I have come across, of using an elastic medium loaded with gyrostatic molecules, as a mode of explaining the rotation of the plane of polarisation by quartz, etc. is given on pp. 242–5.

Lecture XIX. (pp. 256–69) so far as it concerns elasticity deals further with the subject of the reflection and refraction of light at an interface. It discusses chiefly from Lord Rayleigh's standpoint the "condensational wave." The language used (p. 267) as to Neumann's work—especially if we consider the latest form of his researches—seems to me both in the present and previous lectures too severe.

Lecture XX. (pp. 270–88) concludes the body of the work. It deals principally with the theory of light, but one or two points are sufficiently close to our subject to be noted here.

(*a*) Sir William Thomson refers on p. 270 to Rankine[1] as the originator of the idea of "aeolotropy of density" in the medium which transfers light in a crystal. This idea was deduced by Rankine from his hypothesis of "molecular vortices": see our Arts. 424 and 440. Speaking of this hypothesis the lecturer says (p. 270):

I do not think I would like to suggest that Rankine's molecular hypothesis is of very great importance. The title is of more importance than anything else in the work. Rankine was that kind of genius that the names were of enormous suggestiveness; but we cannot say that always of the substance. We cannot find a foundation for a great deal of his mathematical writings, and there is no explanation of his kind of matter. I never satisfy myself until I can make a mechanical model of a thing.

The hypothesis of "aeolotropy of density" has been further investigated by Lord Rayleigh : see the *Philosophical Magazine*, Vol. XLI., pp. 519—28. London, 1871.

It leads to equations practically identical with those adopted by Sarrau and Boussinesq to explain double refraction : see our Arts. 1476, 1480 and 1483. The hypothesis itself is rejected by Sir William on the ground of a paper by Stokes in the *Proceedings of the Royal Society*, Vol. XX., pp. 443–4. London, 1872. Stokes had verified Huyghens' construction as the true law of double refraction for Iceland spar within the limits of errors of observation and had remarked :

This result is sufficient *absolutely to disprove* the law resulting from the theory which makes double refraction depend on a difference of inertia in different directions (p. 444).

(*b*) Some further considerations on the difficulty of the motion of molecules through the ether occur on pp. 277–80 : see our Art. 1766.

---

[1] *Philosophical Magazine*, Vol. I., pp. 444–45. London, 1851.

Here we have the particles going with a velocity of half or a quarter of a kilometer per second in the kinetic theory of gases, and yet we have the molecules creating waves of light by vibrations of a velocity which may not be more than one kilometer per second, and cannot probably be as much as a thousand kilometers per second (pp. 277–8).

Sir William, however, falls back on the analogy of glycerine; namely that it is not the velocity of the vibrations, but the shortness of their period which enables the ether to act as an elastic solid.

Why does a collision between molecules in the kinetic theory of gases give rise to velocities of one or two kilometers per second, or change the velocity one or two kilometers per second? Answer, because the whole time of collision is enormously greater than the four hundred million millionth of a second or than the slowest of vibrations that Langley has found......The medium's being perfectly elastic for the to-and-fro recoverances of motions in the 20 million millionth of a second is perfectly consistent, it seems to me, with its being like a perfect fluid in respect to forces acting perhaps for one millionth of a second (p. 279).

See our Arts. 930* and 444.

(c) On pp. 288–9 will be found: *The Lament of the 21 Coefficients;* this deserves, perhaps, a passing reference here as the one occasion in the history of our subject on which a poet (Professor G. Forbes) has condescended to touch such a serious theme as elasticity.

[1782.] Certain appendices to this volume of lectures may be briefly referred to here.

(a) On pp. 290–3, 320–327 and 328 will be found an Appendix entitled : *Improved Gyrostatic Molecule.* This Appendix not only discusses the dynamics of two types of gyrostatic molecule, but applies the theory of an elastic medium in which an infinitely great number of such molecules are imbedded to explain the rotational effect of certain media on the plane of polarisation of transmitted light. To discuss the details would lead us beyond our proper sphere ; the subject has been very fully treated by J. Larmor in a paper entitled: *The equations of propagation of disturbances in gyrostatically loaded media. Proceedings of the London Mathematical Society,* Vol. XXIII., pp. 127–35. London, 1891.

(b) The second Appendix deals with *Metallic Reflection* and occupies pp. 294–313. It starts with a development of the Green-Rayleigh theory of the reflection and refraction of waves at the interface of two elastic media, and endeavours to apply the results to metallic refraction by making the square of the index of refraction negative. The little chromatic dispersion in reflection at metallic surfaces forms a difficulty in the theory :

We are thus forced to admit that our dynamical theory of metallic reflection is a failure for the present, but it is not unsuggestive and it may possibly help to the true dynamical explanation which is so much desired. That it does indeed contain part of the essence of the true dynamical theory, can scarcely be doubted after we have considered the next two subjects on which we are going to try it: the translucency of thin metallic films, and the effect of magnetism on polarised light incident on polished magnetic poles, or traversing thin films of magnetised iron, nickel or cobalt (p. 313).

(c) The third Appendix entitled: *Translucency of Thin Metallic Films*, occupies pp. 314–9. Here we require an application of the Green-Rayleigh conditions at each of the two faces of the plate or film. Sir William again puts the square of the refractive index negative, and obtains an expression for the intensity of the wave transmitted through the film and for the advance of the phase in the two cases of vibrations in and perpendicular to the plane of the incident and transmitted rays. The results although suggestive are not in accordance with the experiments of Quincke (p. 317). The theory explains Kerr's results for the *normal* reflection of polarised light from magnetic poles, but not Kundt's for the transmission of polarised light through thin magnetised iron sheets. Being unable to abandon a pure imaginary value of the refractive index for metals, Sir William hopes:

that extinctivity on a true dynamical foundation in connection with our molecular theory[1], which it must be remembered is due originally to Sellmeyer, may serve to solve the numerous difficulties in connection with metallic reflection and transmission, which give us so much anxiety (p. 319).

As a last remark on the elastic theory of light we may cite the remaining words of this Appendix:

Extinctivity, however, cannot help to solve the great difficulty as to reflection at the interface between two transparent mediums, in the case of vibrations in the plane of the three rays. Green's attempt to explain this difficulty by gradualness in the transition of physical quality from one medium to another seems to me most unpromising if not utterly hopeless. There remains Green's other suggestion of "extraneous force," by which as we have seen he opened a door for explaining how the velocity of light in a crystal can depend on the direction of the line of vibration irrespectively of the line of propagation. If this suggestion becomes realised it must modify the circumstances at the interface which determine the reflection. Is it possible that it can lead to the true law for reflection of waves consisting of vibrations in the plane of the three rays? (p. 319).

[1783.] Sir William Thomson's *Baltimore Lectures* are undoubtedly a most suggestive and interesting study—such a study as brings the reader into the creative workshop of a great scientist. But they are a study which should be undertaken after rather than before the perusal of what other leading physicists—Green, Neumann, Lord Rayleigh, Sarrau, Boussinesq etc.—have achieved in the same field. This seems to me the sole method of fairly weighing the strength of the author's criticisms and of duly appreciating the importance of his ideas. A careful study of this kind would go a long way to convince the student that the elastic theory of light cannot in the form of "the mathematical theory of perfectly elastic solids" (what-

[1] See pp. 246–7 of the *Lectures* for considerations on the storing of luminiferous energy by the attached molecules, especially in relation to anomalous dispersion.

ever be their degree of aeolotropy) prove serviceable as a dynamical explanation of optical phenomena.

[1784.] *Elasticity viewed as possibly a Mode of Motion. Proceedings of the Royal Institution of Great Britain*, Vol. IX., pp. 520–1. London, 1882. *Popular Lectures and Addresses*, Vol. I., 1st Edn., pp. 142–6. This is a brief résumé of a lecture given on March 4, 1881. Numerous examples are cited,—spinning-tops, hoops, bicycles, chains, etc., in motion—where a stiff elastic-like firmness is produced by motion. The lecturer suggested that the elasticity of every ultimate atom of matter might be thus explained.

But this kinetic theory of matter is a dream, and can be nothing else, until it can explain chemical affinity, electricity, magnetism, gravitation, and the inertia of masses (that is, crowds) of vortices.

[1785.] (a) *Oscillations and Waves in an Adynamic Gyrostatic System* (1883).

(b) *On Gyrostatics*[1] (1883).

The titles of these papers only are given in *Proceedings of the Royal Society of Edinburgh*, Vol. XII., p. 128. Edinburgh, 1884. Their contents relate probably to ' elasticity as a mode of motion.' Some slight account of them will be found in *Nature*, Vol. XXVII., p. 548.

[1786.] *Steps towards a Kinetic Theory of Matter. Report of the British Association* (Montreal Meeting, 1884), pp. 613–22. London, 1885. (*Nature*, Vol. XXX., pp. 417–21; *Popular Lectures and Addresses*, Vol. I., 1st Edn., pp. 218–52.) This paper still further develops the gyrostatic theory of elasticity, *i.e.* elasticity as a mode of motion. In particular the author indicates how a model spring balance might theoretically be constructed from a four-link frame, each link carrying a gyrostat so that the axis of rotation of the fly-wheel is in the axis of the link which carries it (pp. 618–9). He further extends the conception to the constitution of elastic solids and to the model of a solid which would present the magneto-optic rotation of the plane of polarised light (pp. 619 20). The paper concludes by shewing that perforated solids with fluid

[1] On the general theory of gyrostatics : see Arts. 319, Example (G), and 345^{vi}—345^{xxviii} of Thomson and Tait's *Natural Philosophy*, Part I.   Cambridge, 1879.

circulating through them might, if linked together, be made to replace a system of linked gyrostats.

[1787.] *On the Reflection and Refraction of Light. Philosophical Magazine*, Vol. xxvi., pp. 414–25. London, 1888.

The expression for the work of an isotropic elastic medium is given by the integral :

$$W = \tfrac{1}{2} \iiint \{\lambda \theta^2 + 2\mu \left(s_x^2 + s_y^2 + s_z^2\right) + \mu \left(\sigma_{yz}^2 + \sigma_{zx}^2 + \sigma_{xy}^2\right)\} \, dxdydz.$$

If τ be the resultant twist this is easily thrown into the form :

$$W = \tfrac{1}{2} \iiint \left[(\lambda + 2\mu) \, \theta^2 + 4\mu\tau^2 + 4\mu \left\{ \left(\frac{dw}{dy}\frac{dv}{dz} - s_y s_z \right) + \left(\frac{du}{dz}\frac{dw}{dx} - s_z s_x \right) \right. \right.$$
$$\left. \left. + \left(\frac{dv}{dx}\frac{du}{dy} - s_x s_y \right) \right\} \right] dxdydz.$$

Integrating the term in curled brackets by parts we have

$$W = \tfrac{1}{2} \iiint \{(\lambda + 2\mu) \, \theta^2 + 4\mu\tau^2\} \, dxdydz$$
$$+ 4\mu \iint w \left(m \frac{dv}{dz} - n \frac{dv}{dy} \right) + u \left(n \frac{dw}{dx} - l \frac{dw}{dz} \right) + v \left(l \frac{du}{dy} - m \frac{du}{dx} \right) dS,$$

where l, m, n are the direction-cosines of the normal drawn outwards from the element dS of the bounding surfaces. Now if the medium be rigidly fixed at the bounding surfaces (*i.e.* $u = v = w = 0$ there), then the surface-integrals vanish. Further, the medium may change its density at any surface, provided that at this surface u, v, w are functions of the same function of x, y, z and t (Glazebrook : *Philosophical Magazine*, Vol. xxvi., p. 523. London, 1888), and lastly there be equality of μ on both sides of the surface[1]. Subject to these conditions, if there be a fixed boundary or boundaries, W will always reduce to

$$W = \tfrac{1}{2} \iiint \{(\lambda + 2\mu) \, \theta^2 + 4\mu\tau^2\} \, dxdydz.$$

Sir William Thomson now notes that this expression for the work will be positive if $\lambda + 2\mu$ is positive, or even zero, provided μ be positive. Thus the medium as a whole will be stable. According to our Vol. I. p. 885, the dilatation-modulus $= \tfrac{1}{3} (3\lambda + 2\mu)$; hence if $\lambda = -2\mu$, this dilatation-modulus is negative, or the medium would collapse if not

---

[1] The interfaces between two media being either closed surfaces or extending to infinity, the surface-integrals may be thrown into the form :

$$4\mu \iint \{lu (s_x - \theta) + mv (s_y - \theta) + nw (s_z - \theta)\} dS$$
$$= 2 \iint (lu\widehat{xx} + mv\widehat{yy} + nw\widehat{zz}) \, dS - 2 (\lambda + 2\mu) \iint \theta (lu + mv + nw) \, dS.$$

Hence for media for which $\lambda + 2\mu = 0$, we must have at an interface $lu\widehat{xx} + mv\widehat{yy} + nw\widehat{zz}$ the same for both, if these surface terms are to disappear. Sufficient conditions would be : (a) u, v, w, the same and the tractions $\widehat{xx}$, $\widehat{yy}$, $\widehat{zz}$ the same for both media, or (b) u, v, w the same, μ the same and the stretches s_x, s_y, s_z, the same for both media. Case (a) does not appear to involve the sameness of μ.

fixed to rigid boundaries. As an example of this kind of medium, Sir William Thomson cites "homogeneous air-less foam held from collapse by adhesion to a containing vessel, which may be infinitely distant all round" (p. 414). Such a medium " exactly fulfils the condition of zero velocity for the condensational-rarefactional wave; while it has a definite rigidity and elasticity of form, and a definite velocity of distortional wave, which can easily be calculated with a fair approximation to absolute accuracy " (p. 415).

[1788.] Unlike Green, who made his ether absolutely incompressible, Sir William Thomson suggests a "contractile ether," for which $\lambda + 2\mu = 0$, fixed to an infinitely distant containing vessel. He, then, in a manner very similar to Green's, investigates the intensities of the reflected and refracted rays at the interface of two media, and finds Fresnel's sine-law for vibrations perpendicular to the plane of incidence and his tangent-law for vibrations in the plane of incidence (pp. 421 and 425).

In the paper itself the author takes μ the same for both media with a view of simplifying his results. In a *Note* added on pp. 500–1 of the same volume of the *Philosophical Magazine*, Sir William Thomson states that Glazebrook had pointed out to him that the equality of μ for both sides of the interface of two media for which $\lambda + 2\mu = 0$, is needful for stability. Glazebrook himself extends Sir William Thomson's hypothesis of a contractile ether to double refraction, dispersion, etc. in a paper which will be found in the same volume of the *Philosophical Magazine*, pp. 521–40.

[1789.] *On Cauchy's and Green's Doctrine of Extraneous Force to explain dynamically Fresnel's Kinematics of Double Refraction. Proceedings of the Royal Society of Edinburgh*, Vol. xv., pp. 21–33. Edinburgh, 1889. This paper was read on December 5, 1887. It is also printed in the *Philosophical Magazine*, Vol. xxv., pp. 116–28. London, 1888. Our references will be to the pages of the latter journal.

This is an important paper in that it gives an expression for the energy of an incompressible elastic medium initially isotropic, but subjected to a finite homogeneous strain, when a small uniform slide is given to it in any direction. It then applies this result to the elastic theory of light, the ether in crystals being supposed incompressible but subjected to a surface stress which produces a homogeneous strain throughout the interior.

[1790.] Let $S_1 - 1$, $S_2 - 1$, $S_3 - 1$ be the principal stretches of the homogeneous initial strain, and let a slide σ, whose cube may be neglected, be given to the material, so that the plane with direction-cosines l', m', n' receives a slide in the direction l, m, n. Let the directions of the initial principal stretches be taken as axes of x, y, z. Then the point, whose coordinates are before initial strain x, y, z, after the initial strain and the slide is given by the coordinates x', y', z' where:

$$x' = xS_1 + \sigma pl, \; y' = yS_2 + \sigma pm, \left.\right\} \quad \text{.............. (i)},$$
$$z' = zS_3 + \sigma pn$$

where $\quad p = l'xS_1 + m'yS_2 + n'zS_3$, and $ll' + mm' + nn' = 0$.

Let the principal stretches after the slide σ be $S_1 + \delta S_1 - 1$, $S_2 + \delta S_2 - 1$, $S_3 + \delta S_3 - 1$, then they are to be found by making $x'^2 + y'^2 + z'^2$ a maximum or minimum for variations of x, y, z, subject to the condition that

$$x^2 + y^2 + z^2 = 1.$$

As typical result we find, neglecting σ^3:

$$\left(1 + \frac{\delta S_1}{S_1}\right)^2 = 1 + 2\sigma ll' + \sigma^2 \left\{ l'^2 - \frac{S_3^2}{S_3^2 - S_1^2}(nl' + ln')^2 - \frac{S_2^2}{S_2^2 - S_1^2}(lm' + ml')^2 \right\}.$$

Whence:

$$\frac{\delta S_1}{S_1} = \sigma ll' + \tfrac{1}{2}\sigma^2 \left\{ l'^2 - l^2l'^2 - \frac{S_3^2}{S_3^2 - S_1^2}(nl' + ln')^2 - \frac{S_2^2}{S_2^2 - S_1^2}(lm' + ml')^2 \right\} \dots \text{(ii)},$$

with similar values for $\delta S_2/S_2$ and $\delta S_3/S_3$.

Now let $E + \delta E$ be the strain-energy in the condition $S_1 + \delta S_1$, $S_2 + \delta S_2$, $S_3 + \delta S_3$, then it must be a function of $S_1 + \delta S_1$, $S_2 + \delta S_2$, $S_3 + \delta S_3$, or, (neglecting cubes of the small quantities δS_1, δS_2, δS_3) δE must by Taylor's theorem be of the form:

$$\delta E = A \frac{\delta S_1}{S_1} + B \frac{\delta S_2}{S_2} + C \frac{\delta S_3}{S_3} + a_1 \frac{(\delta S_1)^2}{S_1^2} + b_1 \frac{(\delta S_2)^2}{S_2^2} + c_1 \frac{(\delta S_3)^2}{S_3^2}$$
$$+ a_2 \frac{\delta S_2 \delta S_3}{S_2 S_3} + b_2 \frac{\delta S_3 \delta S_1}{S_3 S_1} + c_2 \frac{\delta S_1 \delta S_2}{S_1 S_2} \dots \dots \text{(iii)},$$

where the quantities A, B, C, a_1, b_1, c_1, a_2, b_2, c_2 are functions of S_1, S_2, S_3.

Now since the medium is incompressible:

$$S_1 S_2 S_3 = 1,$$

and therefore:

$$\frac{\delta S_1}{S_1} + \frac{\delta S_2}{S_2} + \frac{\delta S_3}{S_3} + \frac{\delta S_2 \delta S_3}{S_2 S_3} + \frac{\delta S_3 \delta S_1}{S_3 S_1} + \frac{\delta S_1 \delta S_2}{S_1 S_2} = 0.$$

Hence still neglecting cubes we have relations of the type:

$$2\frac{\delta S_2 \delta S_3}{S_2 S_3} = \frac{(\delta S_1)^2}{S_1^2} - \frac{(\delta S_2)^2}{S_2^2} - \frac{(\delta S_3)^2}{S_3^2},$$

which enable us to throw (iii) into the form:

$$\delta E = A \frac{\delta S_1}{S_1} + B \frac{\delta S_2}{S_2} + C \frac{\delta S_3}{S_3} + G_1 \frac{(\delta S_1)^2}{S_1^2} + H_1 \frac{(\delta S_2)^2}{S_2^2} + I_1 \frac{(\delta S_3)^2}{S_3^2} \dots \text{(iv)},$$

where A, B, C, G_1, H_1, I_1 are functions of the initial strains S_1, S_2, S_3.

[1791.]   Noting that $(mn' + nm')^2 = 1 - l^2 - l'^2 + 2 (l^2 l'^2 - m^2 m'^2 - n^2 n'^2)$, (since $ll' + mm' + nn' = 0$, $l^2 + m^2 + n^2 = 1$ and $l'^2 + m'^2 + n'^2 = 1$) with similar relations for $(nl' + ln')^2$ and $(lm' + ml')^2$, we find by transforming (ii), substituting in (iv) and neglecting σ^3, that :

$$\delta E = \sigma \left(All' + Bmm' + Cnn' \right) + \tfrac{1}{2}\sigma^2 \{ L + M + N - Ll^2 - Mm^2 - Nn^2$$
$$+ (A - L) \, l'^2 + (B - M) \, m'^2 + (C - N) \, n'^2 + 2 \left(G_1 + L - M - N - \tfrac{1}{2}A \right) l^2 l'^2$$
$$+ 2 \left(H_1 + M - N - L - \tfrac{1}{2}B \right) m^2 m'^2 + 2 \left(I_1 + N - L - M - \tfrac{1}{2}C \right) n^2 n'^2 \} \dots \text{ (v)},$$

where $L = \dfrac{BS_3^2 - CS_2^2}{S_2^2 - S_3^2}$, $M = \dfrac{CS_1^2 - AS_3^2}{S_3^2 - S_1^2}$, $N = \dfrac{AS_2^2 - BS_1^2}{S_1^2 - S_2^2}$.

This result agrees with Sir William Thomson's on p. 124, if we put $2G$, $2H$, $2I$ respectively for our $G_1 - \tfrac{1}{2}A$, $H_1 - \tfrac{1}{2}B$, $I_1 - \tfrac{1}{2}C$.

[1792.]   A physical meaning can be found for the constants A, B, C. The work done per unit volume in producing a change δS_1, δS_2, δS_3 of infinitesimal magnitude in S_1, S_2, S_3 may, if T_1, T_2, T_3 are the normal forces per unit area in the directions of S_1, S_2, S_3, be written :

$$T_1 S_2 S_3 \delta S_1 + T_2 S_3 S_1 \delta S_2 + T_3 S_1 S_2 \delta S_3 = T_1 \frac{\delta S_1}{S_1} + T_2 \frac{\delta S_2}{S_2} + T_3 \frac{\delta S_3}{S_3},$$

since $S_1 S_2 S_3 = 1$.   Hence by (iii) clearly T_1, T_2, T_3 are equal to A, B, C, or the latter are the initial principal stresses.   Clearly since the material is incompressible

$$A + B + C = 0.$$

[1793.]   Sir William Thomson now supposes a finite plate of the medium of thickness h and very large area a to be displaced by the shear σ, the medium being initially in a state of strain given by S_1, S_2, S_3. The bounding faces of this plate are supposed unmoved and all the solid exterior to the plate undisturbed by σ except some slight strain round its edge.   If σ be given as some function of p the distance from one face of the plate, $= f(p)$ say, then clearly :

$$\int_0^h \sigma dp = 0 \dots\dots\dots\dots\dots\dots\dots\dots \text{ (vi)}.$$

Further neglecting, since the area of the plate is very great, the work done at the edge of the plate as small compared with the strain-energy due to slides, we have for the total strain-energy of the plate :

$$W = a \int_0^h dp \, \delta E$$
$$= \tfrac{1}{2} \{ L + M + N - Ll^2 - Mm^2 - Nn^2$$
$$+ (A - L) \, l'^2 + (B - M) \, m'^2 + (C - N) \, n'^2$$
$$+ 2 \left(G_1 + L - M - N - \tfrac{1}{2}A \right) l^2 l'^2 + 2 \left(H_1 + M - N - L - \tfrac{1}{2}B \right) m^2 m'^2$$
$$+ 2 \left(I_1 + N - L - M - \tfrac{1}{2}C \right) n^2 n'^2 \} \int_0^h \sigma^2 dp \dots \text{(vii)}.$$

By wave-theory the problem is now to find the values of l, m, n which make the coefficient of $\int_0^h \sigma^2 dp$ a maximum or minimum. This reduces to finding the principal diameters of the section in which the ellipsoid

$$\{2\,(G_1 + L - M - N - \tfrac{1}{2}A)\,l'^2 - L\}\,x^2 + \{2\,(H_1 + M - N - L - \tfrac{1}{2}B)\,m'^2 - M\}\,y^2$$
$$+ \{2\,(I_1 + N - L - M - \tfrac{1}{2}C)\,n'^2 - N\}\,z^2 = \text{const.}$$

is cut by the plane

$$l'x + m'y + n'z = 0.$$

These two directions of l, m, n are those for which the force of restitution and the shift coincide in direction. The magnitude of the velocity V of the two simple waves with fronts perpendicular to l', m', n' is then given by

$$V^2 = \{J\}/\rho \dots\dots\dots\dots\dots\dots\dots \text{(viii)},$$

where ρ is the density of the medium and $\{J\}$[1] is the maximum or minimum value of the factor in curled brackets on the right of (vii), such value being obtained from the values of l, m, n found for the principal axes of the section of the above ellipsoid (p. 125).

[1794.] Taking the case of a wave-front perpendicular to the principal plane yz, we have $l' = 0$ and the factor in curled brackets in (vii) will then be a maximum or minimum (p. 125) either for

$$l = 1, \; m = n = 0,$$

(vibration *perpendicular to* principal plane)

or, for

$$l = 0, \; m = -n', \; n = m'$$

(vibration *in* principal plane).

In the first case:

$$V^2\rho = (M + N) + (B - M)\,m'^2 + (C - N)\,n'^2 \dots\dots\dots \text{(ix)},$$

and in the second case:

$$V^2\rho = L + Bm'^2 + Cn'^2 + 2\,(H_1 + I_1 - 2L - \tfrac{1}{2}B - \tfrac{1}{2}C)\,m'^2 n'^2 \dots \text{(x)}.$$

According to Fresnel's theory $V^2\rho$ in (ix) must be a constant, and the coefficient of $m'^2 n'^2$ in the value of $V^2\rho$ in (x) must vanish. These results, taking into account the symmetrical results for the other principal planes, lead to:

$$A - L = B - M = C - N,$$

$$H_1 + I_1 = 2L - \tfrac{1}{2}A, \quad I_1 + G_1 = 2M - \tfrac{1}{2}B, \quad G_1 + H_1 = 2N - \tfrac{1}{2}C,$$

since

$$A + B + C = 0.$$

---

[1] $\{J\}$ is clearly the elastic modulus for the strain when the shift and the force of restitution are concurrent.

If μ' be a function of S_1, S_2, S_3 we find from these equations in the manner indicated by Sir William Thomson on p. 126 that:

$$A = \mu' \left(\frac{1}{S^2} - \frac{1}{S_1^2} \right), \quad L = \mu' \left(\frac{2}{S^2} - \frac{1}{S_1^2} \right), \quad G_1 = \tfrac{1}{2}\mu' \left(\frac{3}{S_1^2} - \frac{1}{S^2} \right) \dots \text{(xi)},$$

where

$$\frac{1}{S^2} = \tfrac{1}{3} \left(\frac{1}{S_1^2} + \frac{1}{S_2^2} + \frac{1}{S_3^2} \right),$$

and B, C, M, N, H_1, I_1, are given by proper interchanges.

[1795.] Substitute (xi) in (iv) and we find:

$$\delta E = - \mu' \left\{ \frac{\delta S_1}{S_1^3} + \frac{\delta S_2}{S_2^3} + \frac{\delta S_3}{S_3^3} - \frac{3}{2} \left(\frac{(\delta S_1)^2}{S_1^4} + \frac{(\delta S_2)^2}{S_2^4} + \frac{(\delta S_3)^2}{S_3^4} \right) \right\}$$
$$+ \frac{\mu'}{S^2} \left\{ \frac{\delta S_1}{S_1} + \frac{\delta S_2}{S_2} + \frac{\delta S_3}{S_3} - \frac{1}{2} \left(\frac{(\delta S_1)^2}{S_1^2} + \frac{(\delta S_2)^2}{S_2^2} + \frac{(\delta S_3)^2}{S_3^2} \right) \right\}.$$

To terms of the third order the coefficient of μ'/S^2 vanishes owing to the considerations stated in Art. 1790 above. To the same order the coefficient of μ' is equal to

$$\tfrac{1}{2}\delta \left(\frac{1}{S_1^2} + \frac{1}{S_2^2} + \frac{1}{S_3^2} \right),$$

or to

$$\tfrac{3}{2}\delta \left(\frac{1}{S^2} \right).$$

Thus we have

$$\delta E = \tfrac{1}{2}\mu'\delta \left(\frac{1}{S_1^2} + \frac{1}{S_2^2} + \frac{1}{S_3^2} \right) \dots\dots\dots\dots\dots\dots \text{(xii)}.$$

Thus, if μ' be constant, we have (p. 127),

$$E = \tfrac{1}{2}\mu' \left\{ \frac{1}{S_1^2} + \frac{1}{S_2^2} + \frac{1}{S_3^2} - 3 \right\}.$$

If in the value (iv) of δE we put $S_1 = S_2 = S_3 = 1$, $\delta S_1 = 0$, we find, since $A = B = C = 0$, and $H_1 = I_1 = \mu'$ by (xi):

$$\delta E = \mu' \left\{ (\delta S_2)^2 + (\delta S_3)^2 \right\}.$$

Now for a pure sliding strain $(S_2 + \delta S_2)(S_3 + \delta S_3) = 1$, whence it may easily be shewn that neglecting terms of the cubic order, the slide σ is given by

$$\sigma^2 = 2 \left\{ (\delta S_2)^2 + (\delta S_3)^2 \right\}.$$

Thus:

$$\delta E = \tfrac{1}{2}\mu'\sigma^2,$$

or, if μ' be considered as a constant, we see that it is the slide-modulus μ of the isotropic material before initial strain.

[1796.] If the value of $\{J\}$ in (viii) be calculated by aid of (xi) we have (p. 128):

$$V^2\rho = \mu \left(\frac{l^2}{S_1^2} + \frac{m^2}{S_2^2} + \frac{n^2}{S_3^2} \right) \dots\dots\dots\dots\dots \text{(xiii)}.$$

Clearly the velocity of a wave for vibrations parallel to any one of three directions of initial principal stretch may be found by dividing the velocity of transverse vibrations in the isotropic material by the corresponding ratio of elongation. Sir William Thomson indicates that the results are entirely in agreement with Fresnel's Kinematics of Double Refraction, and therefore of course with the view that the vibration is perpendicular to the plane of polarisation. If we take the vibration *in* the plane of polarisation, V in (x) must be constant, for this would now be the ordinary ray. But this involves $A = B = C = 0$, or perfect isotropy without of course double refraction.

[1797.] The general method indicated in this memoir of calculating the strain-energy when there are initial strains seems of great value. So far as it relates to the ether the assumptions made are that in a crystal (i) the ether is incompressible, (ii) is in a state of homogeneous initial strain, and (iii) that the quantity μ' of our Arts. 1794-5 is a constant for all values of the initial strains. The investigation seems in several important respects superior to that of Green : see our Arts. 917* and 1779 (p. 465).

[1798.] *Molecular Constitution of Matter. Proceedings of the Royal Society of Edinburgh*, Vol. XVI., pp. 693-724. Edinburgh, 1890. *M. P.*, Vol. III., pp. 395-427. This paper although of very great interest only explicitly touches on the topic of our *History* at one or two definite points and then, alas! without the mathematical analysis which "must be deferred for a future communication": see the final sentence of the memoir. One of the chief results of the memoir is that Sir William Thomson withdraws the reproach he had previously cast on Boscovich's theory : see our Art. 924*. He remarks :

Without accepting Boscovich's fundamental doctrine that the ultimate atoms of matter are points endowed each with inertia and with mutual attractions or repulsions dependent on mutual distances, and that all the properties of matter are due to equilibrium of these forces, and to motions, or changes of motion, produced by them when they are not balanced; we can learn something towards an understanding of the real molecular structure of matter, and of some of its thermodynamic properties, by consideration of the static and kinetic problems which it suggests. Hooke's exhibition of the forms of crystals by piles of globes, Navier's and Poisson's theory of the elasticity of solids, Maxwell's and Clausius' work in the kinetic theory of gases, and Tait's more recent work on the same subject—all developments of Boscovich's theory pure and simple—amply justify this statement (§ 14).

Sir William Thomson's increased respect for Boscovich's theory may possibly have arisen from his discovery that it will suffice to explain multi-constancy. We shall consider below the conditions by which he attains this result, while avoiding the limitations of Cauchy and Poisson.

[1799.] The memoir opens with some introductory remarks which belong so essentially to our subject that they may be quoted here:

The scientific world is practically unanimous in believing that all tangible or palpable matter, molar matter as we may call it, consists of groups of mutually interacting atoms or molecules[1]. This molecular constitution of matter is essentially a deviation from homogeneousness of substance, and apparent homogeneousness of molar matter can only be homogeneousness in the aggregate. "A body is called homogeneous when any two equal and similar parts of it, with corresponding lines parallel and turned towards the same parts, are undistinguishable from one another by any difference in quality" [*Treatise on Natural Philosophy*, Part II., §§ 675–8]. I now add that unless the "part" of the body referred to consists of an enormously great number of molecules, this statement is essentially the definition of crystalline structure. It is, indeed, very difficult to imagine equilibrium, static or kinetic, in an irregular random crowd of molecules. Such a crowd might be a liquid,—I can scarcely see how it could be a solid. It seems, therefore, that a homogeneous isotropic solid is but an isotropically macled crystal; that is to say, a solid composed of crystalline portions having their crystalline axes or lines of symmetry distributed with random equality in all directions. The proved highly perfect optical isotropy of the glass of object-glasses of great refracting telescopes, and of good glass prisms, seems to demonstrate that the ultimate molecular structure is fine-grained enough to let there be homogeneous crystalline portions, which contain very large numbers of molecules while their extent throughout space is very small in comparison with the wave length of light (§ 1).

Sir William Thomson's remarks as to the "isotropically macled crystal" seem to suggest Saint-Venant's *amorphic bodies*.

These bodies (see our Arts. 231 and 308) have elastic constants satisfying relations either of the type: $2d + d' = \sqrt{bc}$, or of the type: $2d + d' = \frac{1}{2}(b + c)$. In both cases *isotropic* "amorphic bodies" have a single interconstant relation $2d + d' = a$, which reduces their stress-relations to the types:

$$\widehat{xx} = (2d + d')\, s_x + d'\,(s_y + s_z), \qquad \widehat{yz} = d\sigma_{yz},$$

---

[1] The Editor of this *History* can hardly pass this sentence without a word of respectful protest. What science seems to him to have achieved is the description (in some respects very accurate) of the sequences of the perceptual world (or world of sense-impressions) by aid of a conceptual model of atoms and molecules—which corpuscles have not necessarily equivalents in the material universe.

or, to the usual bi-constant types. On the assumption of rari-constancy we should further have $d = d'$. On both these hypotheses therefore there is *no* distinction in the elastic constants between an absolutely homogeneous isotropic solid and an isotropic amorphic body (*i.e.* an isotropically macled crystal). The reason for this apparent paradox seems to lie in the fact that the elements, the action between which we consider in our elastic theories, are supposed to contain an enormously great number of the individual crystals, and so are dealt with as if they were essentially homogeneous. If the element does not contain this great number, then, I think, the above stress-strain relations must not be considered as holding for the stress across any individual element but only for the mean of the stresses across a great number of individual elements subjected to the like strain. I think this idea might be used to throw some more light on the question of bi-constant isotropy. Such bi-constant isotropy may be physically due to amorphism, such amorphism not being so fine-grained as to admit practically of the application of that principle of absolutely homogeneous distribution to which the rari-constant elasticians appeal in calculating the stresses from their molecular hypothesis.

[1800.] §§ 3–13 deal with *Space-Periodic Partitioning* and homogeneous distributions of assemblages of points. To consider these matters would lead us beyond our limits. They are still further discussed in §§ 45–61, which contain a *Summary of Bravais' Doctrine of a Homogeneous Assemblage of Bodies*, and deal generally with what Sir William Thomson calls the "molecular tactics" of crystals. Attention may be drawn to the explanation given of H. Baumhauer's discovery of the artificial twinning[1] of Iceland spar by means of a knife in §§ 58–61. The structure of Iceland spar is here built up as suggested by Huyghens (see our Art. 836 (*a*)) of oblate ellipsoids of revolution, and the twinning is described on either of two hypotheses by aid of the turning and sliding of these oblates, accompanied by a shrinkage and an elongation of their figures. The explanation is thus based on a geometrical change in certain rather artificial elements of which Iceland spar is assumed to be built up, and it presents to my mind the old difficulty as to what is the exact physical equivalent of these closely packed geometrical globes and ellipsoids.

[1801]. §§ 14–44 entitled: *On Boscovich's Theory*, and §§ 62–71, *On the Equilibrium of a Homogeneous Assemblage of mutually Attracting*

---

[1] The subject of the artificial twinning of crystals is treated with ample reference to the original memoirs of Baumhauer and others in Th. Liebisch: *Physikalische Krystallographie*, S. 104–18. Leipzig, 1891.

ADDENDUM to Arts. 1801–5.

On June 15, 1893, Lord Kelvin communicated a paper to the Royal Society entitled: *On the Elasticity of a Crystal according to Boscovich*. I owe to the courtesy of the author the sight of a brief abstract of a portion of this paper. Its contents refer to the following topics: (i) Demonstration that the simplest Boscovichian system leads to rari-constancy, (ii) Demonstration that a homogeneous group of double points enables us to give any arbitrarily assigned value to each of the twenty-one coefficients by assigning very simple laws of variation to the forces between points, (iii) Determination of the values to be assigned to the twenty-one coefficients so as to render the medium incompressible. The discussion seems based on action between nearest neighbours. The paper may remove some or all of the difficulties felt by the Editor in the paper of 1890, and the reader is accordingly requested to consider our Arts. 1801–5 in conjunction with this new paper. Models illustrating the "molecular tactics' of crystals discussed in our Arts. 1798–1805 were exhibited at the annual *soirée* of the Royal Society, June 7, 1893. A brief account of them (as well as of a model of an incompressible elastic crystal with twelve arbitrarily given rigidity moduli) will be found in *Nature*, Vol. 48, p. 159. London, 1893.

To face p. 479 of Part II.

Points deal more closely with our subject. They begin by describing the construction of various homogeneous assemblages of points. A homogeneous assemblage of points having been defined as " an assemblage which presents the same aspect and the same absolute orientation when viewed from different points of the assemblage "—such an assemblage as the centres of equal globes piled homogeneously (§§ 21 and 45), Sir William Thomson tells us that he has investigated the moduli of elasticity produced by a homogeneous strain in such an assemblage. He finds that the solid so constituted is not elastically isotropic if we deal only with forces *between nearest neighbours,* and suppose, as on Boscovich's theory, that the forces act in the lines between pairs of points and are functions only of the distances between individual pairs of points (*i.e.* admit no *modified action*). The solid possesses in fact the properties of a cubic crystal, *i.e.* its stress-strain relations may be expressed in terms of the three moduli :—dilatation-modulus F, slide-modulus for a face μ_1, and for a diagonal plane μ_2.[1]

Extending the investigation to include forces between next nearest neighbours, the moduli still remain unequal, but can be equalised by certain hypotheses as to the forces between points. If they are equalised then we find uni-constancy results :—

...it will no doubt be found that this restriction is valid for any single equilibrated homogeneous distribution of points, with mutual forces according to Boscovich, and sphere of influence not limited to nearest and next-nearest neighbours, but extending to any large, not infinite, number of times the distance between nearest neighbours (§ 65).

[1802.] In § 27 Sir William Thomson seems to indicate that for any single homogeneous assemblage of Boscovichian atoms he finds the relation :

$$3F = 3\mu_1 + 2\mu_2.$$

If this be so then in § 65 the *cubical isotropy* of which he speaks— *i.e.* the elasticity of a cubical crystal—is not the cubical isotropy of multi-constancy, but as we see from the footnote to the previous article, it involves $d = f'$, or $|xyxy| = |xxyy|$ the rari-constant condition : see our p. 77, ftn. Hence the single homogeneous assemblage always leads to *rari-constancy*, whether or not we cause it to lead to uni-constancy by taking $\mu_1 = \mu_2$. Thus Poisson's restriction is essential to such a system *apart from the question of isotropy.*

[1803.] Sir William Thomson tells us (§ 28) that the uni-constant relation is not obligatory when :

the elastic solid consists of a homogeneous assemblage of double, or triple, or multiple Boscovich atoms. On the contrary, any arbitrarily chosen values may be given to the bulk-modulus and to the rigidity, by

---

[1] With the notation of our Arts. 1203 (*d*) and $F = \frac{1}{3}(a + 2f')$, $\mu_2 = \frac{1}{2}(a - f')$, $\mu_1 = d$.

proper adjustment of the law of force, even though we take nothing more complex than the homogeneous assemblage of double Boscovich atoms above described.

The two-atom system here referred to consists of two simple homogeneous assemblages of points:

reds and blues, as we shall call them for brevity; so placed that each blue is in the centre of a tetrahedron of reds and each red in the centre of a tetrahedron of blues (§ 69).

Such an assemblage "the next-to-the-simplest-possible mode of arranging an assemblage of points"—Sir William tells us—produces an elastic solid realising Green's ideal, and is of course much easier to conceive than the model of the Baltimore Lectures: see our Arts. 146 and 1771. Unfortunately the mathematical analysis is not as yet published, so that it is difficult to realise whether the statement made depends in any manner on: (1) a difference between the forces between two blues, two reds and a red and a blue, or on (2) the extent of the sphere of intermolecular action. That a very great number of intermolecular actions should go to make up the stress across any elementary plane in an elastic solid and that these actions should be distributed practically uniformly in all directions seems essential to our notion of a practically isotropic elastic solid. It is certainly involved in the principles from which Poisson and Cauchy deduced rari-constant elasticity on the basis of Boscovich's theory. When the condition that a very great number of intermolecular actions cross an elementary plane is not satisfied, then it is difficult to treat the assemblage of points as situated in a like manner with regard to every elementary plane of section, and we thus lose the notion of an isotropic medium.

[1804.] Failing the mathematics of the multi-constant Boscovichian system, we are thrown back on a mechanical model, described by Sir William in §§ 67–8, as a means of elucidating the double-atom homogeneous assemblage.

Suppose six equal and similar bent bows taken and freely jointed together so as to form a tetrahedron. Take four equal bars and joint them to a boss to be placed at the centre of this tetrahedron, and let the bars connect the boss with the angles of the tetrahedron. If the bars are just the distance from the centre to the angles in the unstressed condition of the tetrahedron the rigidities (i.e. the two slide-moduli as in our Art. 1802) remain unaltered by the insertion of the bars.

If the tie-struts are shorter than this, their effect is clearly to augment the rigidities ; if longer, to diminish the rigidities.   The mathematical investigation proves that it diminishes the greater of the rigidities more than it diminishes the less, and that before it annuls the less it equalises the greater to it (§ 67).

Looked at from the standpoint of a Boscovichian system it would seem that the forces between the points at the tetrahedron angles might thus be of a different sign to the forces between the point at the centre and those at the angles in the case where the two rigidities are equal or there is isotropy.   The model is evidently a framework with supernumerary bars (see our Arts. 1772–3).   What would be the nature of the force between two centre-points in the Boscovichian system is hardly suggested by the mechanical model, nor does the model include actions other than those of *nearest* neighbours.

[1805.]   § 71 concludes the memoir as follows :

Leaving mechanism now, return to the purely ideal mutually attracting points of Boscovich ; and, as a simple example suppose mutual forces to be zero at all distances exceeding something between ζ and $\zeta \sqrt{2}$.

Let the group be placed at rest in simple equilateral homogeneous distribution :—shortest distance ζ.   It will be in stable equilibrium, constituting a solid with the compressibility, and the two rigidities referred to in § 27·above [*i.e.* those noted in our Art. 1802].   Condense it to a certain degree to be found by measurements made on the Boscovich curve[1], and it may become unstable.   Let there be some means of consuming energy, or carrying away energy ; and it will fall into a stable allotropic condition.   The Boscovich curve may be such that this condition is the configuration of absolute minimum energy ; and may be such that this configuration is the double homogeneous assemblage of reds and blues described above.   Though marked red and blue, to avoid circumlocutions, these points are equal and similar in all qualities.

According to the above statement it would almost appear as if uni-constant isotropy were the normal condition and bi-constant isotropy a special allotropic condition which might be produced in uni-constant substances by a process of condensation.   At any rate it seems marked by intermolecular force being attractive between certain molecules and repulsive between others.   Until we have before us the promised mathematical investigation it will be impossible to fully realise the nature of the arrangement by

---

[1] The curve which connects intermolecular force with intermolecular distance, and which is marked according to Boscovich by numerous transitions from attraction to repulsion.

which Sir William Thomson has deduced bi-constant isotropy
from a Boscovichian system of points, nor till then can we clearly
recognise the features in which the homogeneity of this system,
and the extent of its sphere of intermolecular action differ from
those of the systems from which Poisson and Cauchy start their
investigations.

[1806.] *On a Mechanism for the Constitution of Ether. Proceedings of the Royal Society of Edinburgh*, Vol. XVII., pp. 127–32.
Edinburgh, 1890. The author describes a model consisting of
telescopic rods connecting " an equilateral homogeneous assemblage
of points" and further of rigid frames built up of three mutually
rectangular bars each of which carries four "liquid gyrostats" and
rests on a pair of the telescopic rods which go to form a tetrahedron
of the equilateral homogeneous assemblage. Such a model :

has no intrinsic rigidity, that is to say, no elastic resistance to change of
shape ; but it has a *quasi*-rigidity, depending on an inherent quasi-
elastic resistance to absolute rotation. It is absolutely non-resistant
against change of volume and against any irrotational change of shape.
Or it is absolutely incompressible (p. 131).

A homogeneous assemblage of points with gyrostatic quasi-rigidity
conferred upon it in the manner described...would, if constructed on a
sufficiently small scale, transmit vibrations of light exactly as does the
ether of nature. And it would be incapable of transmitting condensational-rarefactional waves, because it is absolutely devoid of resistance
to condensation and rarefaction (pp. 131–2).

[1807.] This paper is reprinted as §§ 7–15 of *Article C*,
Vol. III., pp. 467–72 of the *Mathematical and Physical Papers*.
§§ 1–6 of this *Article* (pp. 466–7) contain the translation of a
Note from the *Comptes Rendus*, T. CIX., pp. 453–5. Paris, 1889.
This Note describes a gyrostatic model of the ether. It is built up
by bars terminating in little cups resting on a system of spheres,
these bars carrying gyrostats. Before the gyrostats are "energised"
the model represents a perfectly incompressible quasi-liquid.
When they are "energised" the model possesses a rigidity not
like that of ordinary elastic media, but which depends directly on
the absolute rotations of the bars. This relation between quasi-
elastic forces and absolute rotation is akin to what we require for
the ether as it offers resistance to "irrotational distortion." It is not
however such a complete representation of the ether as the model

referred to in the preceding article, for the irrotational distortion of the structure requires a "balancing forcive" or system of force.

[1808.]   *Motion of a Viscous Liquid; Equilibrium or Motion of an Elastic Solid; Equilibrium or Motion of an Ideal Substance called for brevity* ETHER; *Mechanical Representation of Magnetic Force.* This paper was published for the first time, May, 1890, in the *Mathematical and Physical Papers*, Vol. III., pp. 436–65. Cambridge, 1890.

It compares the analytical expressions in the form of equations, which represent the physical properties of viscous liquid, elastic solid and ether.

[1809.]   §§ 1–11 are devoted to the viscous liquid.   Assuming that stress is proportional to speed of strain (see our Arts. 1264* and 1744) the stresses are of the following type:

$$\widehat{xx} = -p + 2\mu \frac{du}{dx}, \qquad \widehat{yz} = \mu \left(\frac{dw}{dy} + \frac{dv}{dz} \right) \dots\dots\dots\dots\dots\text{(i)},$$

where μ is a constant termed the 'viscosity,' p is the mean pressure, and u, v, w are the speed-components of the point x, y, z of the fluid.   If ρ be the density, and X, Y, Z the body-forces per unit mass, then the type of the equations of motion is:

$$\rho \left(\frac{du}{dt} + u \frac{du}{dx} + v \frac{du}{dy} + w \frac{du}{dz} \right) = \mu \nabla^2 u + \rho X - \frac{dp}{dx} \dots\dots \text{(ii)},$$

where if the motion be slow we need retain only $\rho \dfrac{du}{dt}$ on the left-hand side.

[1810.]   §§ 12–13 deal with the equilibrium or motion of an iso-tropic elastic solid.

The stress-equations are now of the form:

$$\widehat{xx} = -\frac{3\lambda}{3\lambda + 2\mu} p + 2\mu \frac{du}{dx}, \quad \widehat{yz} = \mu \left(\frac{dw}{dy} + \frac{dv}{dz} \right) \dots\dots \text{(i) } bis,$$

where λ is the dilatation-coefficient, μ is now the rigidity and u, v, w the shifts and not the speeds.   The 'pressure' p will be given by:

$$p = -F \left(\frac{du}{dx} + \frac{dv}{dy} + \frac{dw}{dz} \right) \dots\dots\dots\dots\dots\text{(iii)},$$

where F is the dilatation-modulus or bulk-modulus.

Finally the shift-equations will be of the type:

$$\rho \frac{d^2 u}{dt^2} = \mu \nabla^2 u + \rho X - \frac{3(\lambda + \mu)}{3\lambda + 2\mu} \frac{dp}{dx} \dots\dots\dots\dots \text{(iv)}.$$

When $\lambda/\mu = \infty$ the stresses in (i) *bis* become identical in form with (i), and the body-shift-equations are of the type :

$$\rho\,\frac{d^2u}{dt^2} = \mu\nabla^2u + \rho X - \frac{dp}{dx}\ldots\ldots\ldots\ldots\ldots(iv)\ bis.$$

Sir William Thomson speaks of (iv) *bis* as being true for any elastic solid (§ 13) : This is certainly not the case if the constant in (iii) be the bulk-modulus and p the 'pressure' as he supposes. The constants of the pressure in the stress-relations (i) *bis* and in the body-shift-equations (iv) are only equal to unity as in the case of a viscous fluid if $\lambda/\mu = \infty$.

Our author draws attention to the case of $F = \infty$,

or
$$\frac{du}{dx} + \frac{dv}{dy} + \frac{dw}{dz} = 0\ \ldots\ldots\ldots\ldots\ldots\ldots\ (v),$$

i.e. that of an incompressible elastic solid, spoken of as a *jelly*. μ being finite, equations of the types (i) and (iv) *bis* now hold.

We have then always four equations to find u, v, w and p. Their solution for the case of equilibrium is easily written down :

The three equations of type (iv) by aid of (iii) lead to

$$\rho\left(\frac{dX}{dx} + \frac{dY}{dy} + \frac{dZ}{dz}\right) = \frac{\lambda + 2\mu}{F}\,\nabla^2p.$$

Thus p is the potential due to an ideal distribution of matter of density

$$-\frac{F}{\lambda + 2\mu}\,\rho\left(\frac{dX}{dx} + \frac{dY}{dy} + \frac{dZ}{dz}\right)\bigg|\,4\pi,$$

and by aid of (iv) u, v, w are also at once expressible as the potentials due to certain distributions of matter : see our Arts. 1653 and 1715–6.

Sir William Thomson terms that distribution of body-force on matter continuously occupying space for which :

$$\frac{dX}{dx} + \frac{dY}{dy} + \frac{dZ}{dz} = 0,$$

a *circuital forcive*, and says that in this case $p = 0$ and u, v, w are the potentials due to distributions of matter of densities :

$$\rho X/4\pi\mu,\quad \rho Y/4\pi\mu,\quad \rho Z/4\pi\mu,$$

since $\mu\nabla^2u + \rho X = 0,\quad \mu\nabla^2v + \rho Y = 0,\quad \mu\nabla^2w + \rho Z = 0\ldots\ldots(vi).$

Thus if the forcive be circuital, the shifts will be the same whatever the degree of incompressibility of the infinite body may be (§§ 36–8).

[1811.] §§ 14–20 deal with the *Equilibrium or motion of an ideal substance called for brevity, Ether.*

This ether is described as follows :

What I am for the present calling *ether*, is an ideal substance useful for extending the " Mechanical representation of electric, magnetic and galvanic forces "......[see our Art. 1627]......For the present I suppose it absolutely incompressible. It has no intrinsic rigidity (elastic resistance to change of shape); but it has a *quasi*-rigidity depending on an inherent *quasi*-elastic resistance to absolute rotation. This *quasi*-rigidity may be called simply rigidity for brevity; but when it is to be distinguished from the known natural rigidity of an elastic solid it will be called gyrostatic rigidity (§ 14).

Sir William Thomson accordingly introduces shears proportional to the twists (see our Vol. I., p. 882), besides which there may be something of the nature of fluid pressure. He thus has the following system of stresses:

$$\widehat{xx} = \widehat{yy} = \widehat{zz} = -p, \quad \widehat{yz} = -\widehat{zy} = \mu\left(\frac{dw}{dy} - \frac{dv}{dz}\right),$$

$$\widehat{zx} = -\widehat{xz} = \mu\left(\frac{du}{dz} - \frac{dw}{dx}\right), \quad \widehat{xy} = -\widehat{yx} = \mu\left(\frac{dv}{dx} - \frac{du}{dy}\right)\ldots\text{(vii)}.$$

These lead us at once to equations *identical* with (iv) *bis* and (v) above for an incompressible elastic solid.

We may then ask what is the difference between this ether and a jelly? Sir William Thomson answers:

No difference whatever in respect to the equilibrium-displacement, or the motion, throughout any portion of homogeneous substance of either kind, if the position and motion of every point in the bounding surface of the portion considered are the same for the two. But in respect to the traction on the bounding surface of a detached portion, and therefore also in respect of the interfacial relation between portions of the substance having different rigidities, there is an essential difference between the two, of vital importance for the inclusion of magnetic induction in our mechanical representation (§ 17).

When there is equality of rigidity on either side of an interface, while there is discontinuity due to a difference of body-force or of density, then all the interfacial conditions are the same for both jelly and ether, and may be best expressed by saying that p and the nine differential coefficients of u, v, w must have equal values for the two media at the interface (§ 20).

[1812.] §§ 21–8 deal with *Energy of stressed jelly or of stressed ether*.

The strain-energy per unit volume of the stressed ether is

$$= 2\mu\tau^2,$$

where τ is the resultant twist.

The strain-energy per unit volume of the jelly

$$= \mu\left(s_x^2 + s_y^2 + s_z^2\right) + \tfrac{1}{2}\mu\left(\sigma_{yz}^2 + \sigma_{zx}^2 + \sigma_{xy}^2\right).$$

If the boundary of a volume V of jelly or ether be *fixed*, then the strain-energy involved in any specified strain of either substance within this volume is the same for both (§ 24): see our Art. 1787.

Examples of the correspondences between jelly strain-energy and ether strain-energy are given in §§ 25–8. Their bearing is, however, rather on electro-magnetism than on elasticity.

[1813.] §§ 29–45 are entitled: *Mechanical representation of the magnetic force of an electro-magnet.*

Imagine a piece of endless cord, in the shape of any closed curve, to be imbedded in a jelly, and a tangential force to be applied to this cord uniformly all along its length. Further let the substance of the cord be exactly the same as that of the jelly. This "tangential drag" on the jelly causes stress and strain throughout the jelly, becoming nil only at infinitely great distances. The twist at any point of the jelly caused by this circuital force is equal "to half the magnetic force at the corresponding point in the neighbourhood of a conducting wire, taking the place of our tangentially applied force and having an electric current steadily maintained through it" (§ 30).

This is Sir William Thomson's "mechanical representation" of electro-magnetic force due to a closed circuit, and completes what he had reserved for a future paper in 1847: see our Art. 1627. Various special cases are illustrated; thus a circular circuit in §§ 31–2; equal and opposite currents in straight parallel conductors in §§ 33–5. More general cases are referred to in §§ 36–40.

Thus if X, Y, Z denote components of electric current, and F, G, H the components of magnetic force due to the current, the mechanical representation of an electro-magnetic field consisting of any distribution of closed electric currents may be obtained from the jelly as follows:

Take the components of magnetic force equal to twice the twists, then by means of (v) we find:

$$\frac{dH}{dy} - \frac{dG}{dz} = -\nabla^2 u, \qquad \frac{dF}{dz} - \frac{dH}{dx} = -\nabla^2 v, \qquad \frac{dG}{dx} - \frac{dF}{dy} = -\nabla^2 w,$$

whence we see by (vi) that:

the components X, Y, Z of electric current, *i.e.* $dH/dy - dG/dz$, $dF/dz - dH/dx$, $dG/dx - dF/dy$ divided by 4π, are proportional to the body-forces ρX, ρY, ρZ of the previous investigation for the jelly or for any unlimited elastic solid in the case of a "circuital forcive": see our Art. 1810. We see further that if an infinite homogeneous elastic solid be acted upon in some parts by circuital forcives, then at points unaffected by force:

$$F = \frac{d\chi}{dx}; \qquad G = \frac{d\chi}{dy}, \qquad H = \frac{d\chi}{dz},$$

or the twist-components are the differentials of a single function χ. The electro-magnetic analogue to χ is a quantity differing only by a constant factor from the magnetic potential at x, y, z of the electric current system X, Y, Z (§ 39)[1].

[1814.] §§ 41–3 deal with the *Synthesis of a circuital forcive from*

---

[1] The reduction of the strain-energy to a single term proportional to the square of the twist in the case of a jelly with rigidly fixed boundaries was pointed out by the Editor of this *History* in a *Note on Twists in an infinite elastic solid; Messenger of Mathematics*, Vol. XIII., pp. 84–5. Cambridge, 1884. A somewhat different elastic analogue to the electro-magnetic field is given in the same paper.

*a single force applied through a space comprised within an infinitely
small distance from a point in an incompressible elastic solid (jelly).*

Let q denote the tangential force per unit length of the circuit, and
l, m, n the direction-cosines of its element ds, distant r from the point
of which u, v, w are the shifts, then Sir William Thomson deduces
from his solution of the elastic equations given in our Art. 1810 that:

$$u = \frac{q}{4\pi\mu} \int \frac{l}{r} \, ds, \qquad v = \frac{q}{4\pi\mu} \int \frac{m}{r} \, ds, \qquad w = \frac{q}{4\pi\mu} \int \frac{n}{r} \, ds,$$

while

$$\chi = \frac{q\Omega}{4\pi\mu},$$

Ω being the solid angle subtended by the circuit at the point for which
χ is ascertained, and the integrations for s extending round the circuit.

The analogue to the magnetic potential is here obvious.

[1815.]   In conclusion Sir William Thomson asks why, the analogies being so complete, we cannot be satisfied with the jelly for a
mechanical representation of electro-magnetism? The answer lies in the
difference of conditions at the interface between two jellies and between
two substances of different magnetic permeabilities in a magnetic field.

The magnetic force being in our analogy the rotation of the jelly, or ether,
we see...that the proper interfacial condition[1] between substances of different
rigidity (μ) is not fulfilled by the jelly, and is fulfilled by the ether (§ 44).

Referring to his 'ether' Sir William Thomson draws attention to
the fact that it :

whether extending to infinity in all directions, and having vesicular or
tubular hollows, or a finite portion of it given with a boundary of any shape,
provided that only normal pressure act on the boundary, takes precisely the
same motion for any given motion of the boundary as does a frictionless
incompressible liquid in the same space, shewing the same motion of boundary
(§ 46).

The importance of this, *e.g.* in the length it goes towards explaining
Sir G. G. Stokes' theory of aberration, is pointed out (§ 46), but at the
same time Sir William indicates how very obscure still remains our
knowledge of the real relations between ether, electricity and ponderable matter (§ 47).

[1816.]   *Ether, Electricity, and Ponderable Matter.   Mathematical
and Physical Papers*, Vol. III., pp. 484–515.   Cambridge, 1890.   This
paper constituted part of the Presidential Address to the Institution
of Electrical Engineers, delivered on January 10, 1889.

It contains some references to the elastic solid analogies of the

---

[1] Equality of normal components of magnetic force and proportionality of
tangential components to the magnetic permeabilities on either side the interface.

ether and a description of a gyrostatically loaded network (§§ 21–6), which would serve in some respects as a model for the ether. The whole is more fully developed in the memoirs referred to in our Arts. 1806–7. The address concludes with words of hope in future knowledge following on a confession of present ignorance—*i.e.* the inadequacy of existing theories to represent the relation between ether, electricity and ponderable matter.

[1817.] The third volume of the *Mathematical and Physical Papers* closes with two papers (Arts. CIII. and CIV.), which may be looked upon as appendices to the Encyclopaedia article on *Elasticity*. The first deals with Tait's experimental results for the compressibilities of water, mercury and glass, and the second gives *inter alia* (p. 522) the velocity of elastic waves (distortional, pressural in an infinite solid, longitudinal in rod) in iron, copper, brass and glass, as well as the moduli for the same four materials.

[1818.] *Summary.* It is a very difficult task to preserve an accurate historical stand-point with regard to a physicist—or, as we ought to call him, a naturalist (*M. P.* Vol. III., p. 318)—so close both in time and country to ourselves as Sir William Thomson, now Lord Kelvin. We can hardly see him in the same perspective as we see Saint-Venant or Franz Neumann. At the same time the function of this *History* would hardly be fulfilled did its Editor leave this chapter without some slight summary of its contents. To the future must be left any real test of his critical accuracy.

A distinguished biologist once stated to the Editor of the present work that he had for many years given up endeavouring to ascertain what others had done or were doing in his subject. To follow the great mass of contemporary work meant to expend his time in historical investigations rather than in original research. When he devoted his energies to the latter, he was fairly certain that fifty per cent. of his published results would be new contributions to scientific knowledge. A man of Sir William Thomson's surprising productivity—covering almost every field of physical science—must perforce be occasionally content with the rediscovery of known laws. The reader of our present chapter will have marked instances of this in the researches on the elasticity of springs, in those on light in the Baltimore Lectures and more particularly in the investigations on the relations of stress and electro-magnetic properties. The repetitions are, however, small as compared with the new material, or with the fertile conceptions, which abound even in the treatment of old themes.

In two points a further criticism will also probably be raised in the future, a paucity of experimental demonstration, which occasionally accompanies the statement of an important physical law—compare for example the elaborate experiments of Wiedemann, Ewing or Bauschinger with Sir William Thomson's in similar fields,—and further the absence of mathematical analysis at points where the less gifted are liable to stumble, and may feel compelled either to reserve their judgment or to accept on faith—compare for example the molecular discussions of Saint-Venant and the investigations on crystals of Franz Neumann with those of Sir William Thomson's *Molecular Constitution of Matter* or his *Mathematical Theory of Elasticity.* But this occasional paucity of experiment or analysis is largely due to our author's eagerness to reach the physical law as the all-important goal; he rightly recognises experiment and analysis as only means and not ends in themselves. He is in this a pleasing contrast to those mathematical elasticians who are far more desirous of obtaining a complete solution, whatever be its physical value, than reaching any approximation, however important its bearing on natural facts.

Of the great advances in our subject which will always be associated with the name of Sir William Thomson we must mention especially the accurate foundation of the science of thermo-elasticity, the suggestion that the principles of elasticity ought to be applied to the earth itself, and the first consideration following upon this suggestion of tides in the solid earth. Equally fruitful of results—if indeed they are largely negative results—have been his researches on the elastic theory of light, leading as they have done to the rejection of the old elastic theories. Here it is that he has suggested with his gyrostatically loaded medium a new kind of elasticity or *quasi-elasticity*, which bids fair to open up an entirely new field of investigation, and which may in the end make elasticity the predominant physical science.

Not only in the border-land of optics, electro-magnetism and molecular physics have Sir William Thomson's researches widened our knowledge of the possibilities of elastic theory, but in conjunction with P. G. Tait his geometry of strain and his treatment of rods and plates have largely contributed to our appreciation of pure elastic problems, and further have rendered the discussion of them accessible to British students.   In these latter cases, as well as in his more recondite researches, there is that fertility of idea, and that mark of genius which have made Sir William Thomson the leader and characteristic representative of physical science in our own country to-day.

INDEX.

The numbers refer to the articles of this volume and not to the pages unless preceded by (i) p. , or (ii) p. , where the Roman numerals refer to the parts of this volume.

C. et A. = Corrigenda and Addenda to Volume I. attached to Part ii. of Volume II.

ftn. = footnote.

The reader is reminded that the Index to Volume I. is not incorporated with this index, and consequently the absence of an author's name or of any special topic from this index, does not preclude its having been dealt with in Volume I.

Abacs, use of, in strength of materials, 921 and ftn.

Aberration, of light, Boussinesq's theory of, 1449, 1478, 1482; Sir William Thomson on, 1815

Adams, W. A., on railway waggon springs, 969 (*a*)

Adularia, hardness of, 836 (*d*); optic axes of, change with temperature, 1218, ftn.

Aeolotropy, defined by Sir W. Thomson, 1770; Rankine obtains a 16-constant, 429; wave-motion in aeolotropic solid, 1764, 1773—5; dilatation moduli for, 1776 (*a*); weblike annulled, 1776 (*c*), 1777; strain-energy for aeolotropic solid when 'skewnesses' and weblike aeolotropy are annulled and there is no dilatation, 1778; discussed by Sir W. Thomson, 1780; conditions for, incompressible, 1776 (*a*), 1779

Aeolotropy of Density, suggested by Rankine to explain double refraction, considered by Lord Rayleigh, rejected by Sir W. Thomson on ground of Sir G. G. Stokes' experiments on Iceland spar, 1781 (*a*)

After-strain, general remarks on, 748—9; distinguished from frictional action, 750 (*a*); not a pure frictional resistance and masked by term viscosity, 1718 (*b*), 1743; is not proportional to load, 750 (*b*); chief cause in producing subsidence of oscillations according to Seebeck, 474 (*c*)—(*d*); in silk and spider filaments, 697 (*b*); in glass and silk threads, (i) p. 514, ftn.; experimentally discovered in metals by Kupffer, 726, his discussion of, by torsional vibrations, 734; its effect in subsidence of torsional vibrations, 1744—8; effect of working on, 750 (*b*); how influenced by change of temperature, 756; neglected by Wertheim in torsion experiments on metals, 803, his erroneous statement as to, for metals and glass, 819; in guns, 1038 (*g*), 1081; in razors, 1718 (*b*), ftn.; in caoutchouc springs, 851; is proportional to load in caoutchouc, 1161; in organic tissues, 828—35; Weber, E. on, for muscle, 828; stress-strain relation for final load, linear, Wundt's form of after-strain curves, 830; Wertheim's hyperbolic form of

August, E. F., on simple experiments to demonstrate Taylor's law for vibrating strings, (i) p. 573, ftn.

Autenheimer, von, on torsion (1856), 581

Axes, feathered, strength of, 177 (c)

Axes, 'optic' and 'optical' distinguished, 1218, ftn.; dispersion of optic axes, 1218—9 and ftn.; optic, 1476, 1483; different sets of rectangular systems exist in crystal for distribution of different physical properties, 683—7, 1218—20, 1637; Neumann's theory of distinction between optical and elastic axes, 1216—8, 1220; elastic axes do not coincide with optical for alum, 788; optical, thermal and elastic not coincident for gypsum, etc., 1218—9 and ftn.; diamagnetic, electrical and other properties distributed about different axes, 1219

Axes of Elasticity, 135, 137 (iii), 137 (vi), (i) p. 96, ftn., 443—51; defined, 444; orthotatic and heterotatic, 445; euthytatic, 446; metatatic, 446, 137 (vi)

Axles, how affected by prolonged service and vibration, 881 (b) and ftn., 970; strength, 905; of railway rolling stock, calculation of dimensions, 957—9; McConnell's hollow railway axles, experiments on, 988—9; flexure of railway axles under static load, 990; resistance to impact of cast steel, 995, of ordnance, 996; fatigue under repeated flexure of railway, 998, 1000—3, under repeated torsion, 999, 1000—3

Babbage, C., hardness of diamond varies with direction, 836 (d)

Babinet, his proof of velocity of pressural, or sound wave, 219

Baden-Powell, influence of torsion on magnetisation, 811

Baensch, on simple beams and braced girders (1857), 1006

Baker, Sir B., on the actual lateral pressure of earthwork, 1606; his rule for breadth of supporting walls, 1607

Bancalari, R. P., on law of molecular force, 866

Bar, heavy tension bar of equal strength, 1386 (a): see *Rod, Beam, Flexure, Impact,* etc.

Barilari, on statically indeterminate reactions, (i) p. 411, ftn.

Barlow, P., formula for hydraulic press, 901, 1044 (h), 1069, 1076—7; experiments on wrought-iron beams, 937 (c); on combined girder- and suspension-bridges, 1025

Barlow, W. H., attempts to explain 'beam-paradox' by a theory of lateral adhesion (1855—7), 930—8, 1016

Barnes, cuts steel with rapidly rotating soft-iron disc, 836 (h)

Barton, J., on wrought-iron beams, 1016

Basset, A. B., on thin cylindrical and spherical shells, 1296 *bis,* 1234

Baudrimont, A., researches on vibrations of aeolotropic bodies (1851), 821

Baumeister, his experiments on stretch-squeeze ratio referred to, 1201 (e)

Baumgarten, on flexure of solids of equal resistance, 929; on stretch-modulus of calcspar, 1210

Baumhauer, on twinning of Iceland spar, 1800

Bauschinger, his results partially anticipated by Wiedemann, 709—10; on elastic limits referred to, 1742 (b)

Beam, lines of stress in, Rankine, 468, Kopytowski, 556, Scheffler, 652; uniplanar stress in, 582 (c); slide introduced into theory, Bresse, 535, Jouravski, 939, Scheffler, 652, Winkler, 661—2, 665, Airy, 666; strength of, increased by building-in terminals, 571—7, 942—5; strength of, given by graphical tables, 921 and ftn.; for various forms of cross-section, 927; transverse vibrations of, when suddenly loaded, 539; live load on, 540—1; formulae for statical deflection when loaded, 760—2; strength of 'split' beams, 928; general treatment of, 1006; Thomson and Tait on, 1696; of variable cross section, flexure of, 929; small beams relatively stronger than large ones, 936 (iii); central line of, under transverse load, really stretched, 941; supports of beam under transverse load really subjected to side pull, 940; cast-iron beam of strongest cross-section, 176, 177 (b), 951, 1023; strength of various forms of cast-iron beams, and Barlow's attempt to explain paradox, 930—8; paradox neglected by Morin, 881 (a); proper proportion of web and flanges in wrought-iron beams, 1016; formulae for stress-strain relation when stretch- and squeeze-moduli are unequal, 178; rupture of, deduced from empirical stress-strain relation, formulae of Saint-Venant and Hodgkinson, 178: see also *Continuous Beams, Rods, Rolling Load, Torsion, Flexure, Impact,* etc.

Beam-Engine, stress in beam, 358; danger of certain speeds of fly-wheel, 359

of stress for isotropic bodies in a state
of limiting equilibrium (1874), 1605;
on the lateral pressure of a pulverulent
mass with horizontal talus (1881),
1606—7; on horizontal thrust of pulverulent masses against vertical walls,
etc. (1882—5, diverse memoirs), 1608
—25

Summary of Boussinesq's work,
1626

Bracing Bars, on distorted form of, in
multi-bracing, 1017, 1026, 1028; experiments on buckling of, 1019: see
also *Girders*

Braithwaite, F., on fatigue of metals
(1853—4), 970

Brame, Ch., on planes of cleavage, 849;
experiments on iron-plate, 1106

Brass, elastic flexures increase more
rapidly than loads, 709 (i); flexural
sets or bents, how influenced by
alterations of load, etc. 709; effect of
rolling and hammering on stretch-
modulus, 741 (a); thermo-elastic properties of, 752, 754, 756; after-strain
and temperature, 756; relation of
stretch-modulus to density, 759 (e),
(i) p. 531, 824, 836 (b); ratio of kinetic
and static stretch-moduli, 824, of
kinetic and static dilatation-moduli,
1751; slide-, stretch-, and dilatation-
moduli of, 1817; thermal effect on slide-
modulus, 1753 (b); thermo-electric
properties under strain, 1646; rendered brittle by sudden atmospheric
changes, 1188; nature of rupture,
1667

Bravais, on homogeneous assemblages of
bodies, 1800

Breguet, on velocity of sound in iron,
785

Breithaupt, attempts to introduce a
new scale of hardness, 835 (d)

Bresse, Researches of: memoir on the
flexure of arches (1854), 514—30;
treatise on applied mechanics (1859—
65), 532—42; on elliptic flues, 537;
on solution for long train continuously crossing a bridge, 382, 541

References to: on elastic rods of
double-curvature, 291; his treatment
of elastic rods commended by Saint-
Venant, 153; his formula for beams
of varying stretch-modulus, 169 (e),
515; on approximate treatment of
slide due to flexure, 183 (a), 535; on
the core, C. et A. p. 3, 515; corrects
error of Phillips', 540

Brewster, on double-refraction artificially
produced, 792—3; the principle of

his Teinometer adopted by Wertheim,
794, 797 (e); on production of crystalline structure by stress (1853), 864; on
analcine, 1775

Bri-Brachion (? Sir W. Armstrong) on
the cause and prevention of the de-
terioration of wrought-iron (1860), 1189

Brick, strength of, 880 (b), 1173, 1182

Bridges, deflection of railway viaduct at
Tarascon, 520 (b); transverse vibrations of, 539, 1034—5; effects produced by a rolling load, 372—82,
Bresse corrects error of Phillips', 540;
repeated loading of, 1035; deflections
of Flemish bridges, 1020; treatises
and text-books on bridge-construction,
883, 885—90, 915, 950; historical
account of (1857), 890; "Ritter's
method", 915 (b); minor memoirs
on, 1004—36

Special Bridges: Tarascon, 520 (b),
1109; St Louis, U. S., (i) p. 358, ftn.;
Britannia and Conway, 560, 603, 607;
Hungerford, 579; Manchester, 1007;
Newark Dyke, 1012; Cöln, 1019;
Flemish, 1020; Niagara, 1025; de la
Roche-Bernard, 1033; over the canal
Saint-Denis, 1034

Bridges, Suspension, form of chains, 579;
oscillations of, 612, 883; impact on,
883; iron-wire for, 904; when, where
and by whom first introduced, (i) p.
622, ftn.; girder suspension bridges,
1025

Brill, points out an error in Saint-
Venant's memoir of 1863, 239

Brillouin, on the elasticity, fluidity and
rigidity of bodies, 1464

Briot, Saint-Venant's views on his theory
of light, 265

Brittle, defined, 466 (vii), 1742 (c);
metals not rendered, by cold, 697 (c);
state, 1185, 1188

Brix, on strength of railway-rails, C. et
A. p. 11; on fail-points of uniformly
loaded beams, C. et A. p. 12; erroneous
theory of resistance of cylinder to internal pressure, (i) p. 712, ftn.; on
strength of stone, 1181; on set in
cast-iron due to heating, 1186

Bronze, gun-metal, effect of head on casting, 1038 (f), 1050; rupture of rings
of, 1044; guns of, 1045; physical
properties of, 1063 and ftn.; stress-
strain diagrams for, 1084; torsional
strength of, 1039 (c), 1113, 1166; tensile strength of, 1113, 1166

Brooks, C. H., erroneous theory of resistance of hollow cylinders, 1080

Brown, Captain, introduces iron cables

ous Principles of Mathematical Optics, 1391; on reflection at a spherical surface, 1392—1410; accounts of his life and work, 1390, (ii) p. 107, ftn.

Cobalt, effect of longitudinal stress on magnetisation, Villari critical field, 1736

Coefficient of Optical or Photo-elasticity, 795

Coefficient of Plasticity (K), or plastic-modulus, 247, 249, 259; does it vary? 1568—9, 1586, 1593: see also *Plasticity*

Coefficient of Restitution, or dynamic elasticity, 209, 217, 847; really varies with masses, sizes and shapes of colliding bodies, 1682—3

Coefficients, Elastic, names for in this *History*, (i) p. 77, ftn., Tables, 445, 448; *Homotatic*, 136, 446; of *Pliability* are reciprocals of coefficients of Rigidity, 425; of *Extensibility* (longitudinal and lateral, or direct and cross) and *Compressibility*, 425; *Tasinomic* (euthytatic, platytatic, goniotatic, plagiotatic), 445; *Thlipsinomic* (euthythliptic, platythliptic, goniothliptic, plagiothliptic), 448; transformation of, Rankine's use of surface of fourth order, 432; in any direction expressed symbolically, 133; for various crystals, 1203—5; numerical values, 1212; for a material with three planes of elastic symmetry, 307; for amorphic bodies, 282 (8), 308; for equal transverse elasticity, 308 (a); of wood do not admit of ellipsoidal conditions, 308 (a); for bodies possessing various types of elastic symmetry, 281—2; experimental methods of determining, 283, 1205—11; expressions for, in terms of initial stress, 240; effect of initial stress on stretch-modulus, 241; effect of set on cross-stretch coefficients, 194; Lament of, 1781 (c): see also *Constants* and *Moduli*

Cohesion, Herschel, Séguin and Sir W. Thomson endeavour to explain it by molecules of infinitely great density and infinitely small volume attracting according to Newtonian law, 865, (i) p. 600, ftn., 1650; supposed by Zaborowski to depend on absolute continuity of matter, 867: see *Molecules, Strength*, etc.

Colladon and Darier, cut steel, chalcedony and quartz by iron disc in rapid rotation, 836 (h), 1538, ftn.

Colladon and Sturm, their theory of pie-

zometer referred to by F. Neumann, 1201 (c)

Collet-Meygret, on bridge-structure (1854), 1109—12; cited, 169 (e), (i) p. 368, ftn.

Columns, best form of, discussed by Clausen, 476—80; strength of wooden, 880 (a); cast-iron, tables and curves for strength of, 880 (c), do not obey ordinary elastic theory, 1117 (v), Hodgkinson's later formulae for strength of, 973 (cf. 469, 649—50, 956), relative strength of those with rounded and with bedded ends, 974 (a), loss of relative strength due to removal of external crust, 974 (c); strength of square, triangular and circular cross-sections, 974 (d); on strength of long columns, 978; empirical formulae for steel columns, rounded and bedded ends, 978; ditto, for wrought-iron columns, 978: see also *Struts*

Combes, report on Phillips' memoir on springs, 482

Combination of Strains: see *Strain, Combined*

Compatibility, of given system of strains, conditions for, 112, 190 (c); proved by Boussinesq, 112, 1420, by Kirchhoff, 1279

Compression, difficulty of experiments to determine squeeze-modulus, influence of buckling in long and friction in short specimens, 793

Condenser, spherical glass, strain produced by charge, 1318

Conductivity, electric, of iron and copper, how altered by strain, 1647; rendered aeolotropic by aeolotropic stress, 1740

Cone, very sharp, vibrations, notes and fail-point of, 1306—7; truncated, impact longitudinal on, 223, duration of blow, maximum strain, etc., 1542—4

Conjugate Functions, in torsion problem, 285, 1460, 1710; used to solve uniplanar equations of plasticity, 1562—7, to solve those of pulverulence, 1566, 1570

Connecting Rod, stress produced by vibrations in, 583, by variations of pressure, 681—2

Conservative systems of Force, 1709 (a), 1716 (d)

Constants, Elastic, equality of cross-stretch and direct-slide on rari-constant hypothesis, 73; controversy about, 68, 192, 193, 196, 197, 276, 301; bi-constancy of iron and brass wire, 727; bi-constancy investigated by stretching hollow prisms, 802, 1201

subject, 836 (a)—(k); scales of, for metals, 836 (b), for minerals, 836 (d); varies with direction, 836 (a) and (e); use of metal and diamond scribers, 836 (f) and (g); varies with speed of scratching or tearing substance (rotating soft-iron discs cut hardened steel, chalcedony, and quartz) 836 (h); first scientific sklerometer used by Seebeck, 836 (i); problems to be considered in testing hardness, 836 (j); definition and analysis of, 837—8; ray-curves for hardness in various directions, 839; laws connecting hardness with planes of cleavage in crystals, 839—40; relation to atomic and molecular properties, 841; use of sklerometer, 843; scale of hardness of metals and alloys, 845, 846, (i) p. 707, ftn.; Wade's method of testing by indentation, 1040—2; experiments on cast-iron, wrought-iron and bronze, 1042 —3; relation to density in cast-iron, 1042—3; for hardness of various metals and minerals: see under their titles

Hart, erroneous theory of shrunk-on coils for guns, 1071 (b)

Haughton, discovers tasinomic quartic, 136, orthotatic ellipsoid, 137; discussion on his views as to elastic constants by Saint-Venant, 193; his experiments on impact referred to, 217; cited by Rankine (as to $\widehat{xy} = \widehat{yx}$), 428

Haupt, H., on resistance of vertical plates in tubular bridges, 1015

Hausmaninger, on longitudinal impact of bars, 203

Haily, R. J., scale of hardness (1801), 836 (d)

Hawkes, W., on repeated meltings of cast-iron, 1101

Hawkshaw, J., on absolute strength and deflection of cast-iron girders, 1007

Heat, attempted explanation by translational vibrations of molecules, 68; explanation of its effect in dilating bodies, and the nature of coefficient of dilatation, 268; stretch due to thermal vibration, 268; thermal effect depends on derivatives of second order of function giving intermolecular action, 268; diagram of possible law of intermolecular action, (i) p. 179; phenomena of, accounted for by molecular translational vibration, 271; theory does not appear in accordance with spectral phenomena, 271; deduction of pressure on surrounding envelope from this theory, 273; Saint-

Venant rejects kinetic theory of gases, 273; passage of, produces crystalline structure in metals, 1056: see also *Expansion, Coefficient of*

Heat, Mechanical equivalent of, obscure treatment by Resal, 716, by Vogel, 717, by Kupffer, 724—5, 745—6, 823

Heat, Relation to Elasticity: see *Thermal Effect, Modulus*, etc.

Helix: see *Springs, helical*; principal helices of wire, 1692

Helm, G., Die Lehre von der Energie, cited, (i) p. 501, ftn.

Helmholtz, von, remarks on Kupffer's treatment of mechanical equivalent of heat, (i) p. 501, ftn.; generalises *Huyghens' Principle*, 1312; on change of density and on stress due to magnetisation, 1313, 1315, 1316

Henry, on strength of stone, 1180

Heppel, J. M., on Three Moments Theorem with isolated loads (1859), 607; erroneous treatment of web and flanges of iron girders, 1018

Hermite, reports on Saint-Venant's memoir on transverse impact, 104

Herschel, his explanation of cohesion by gravitating molecules adopted by Séguin and Sir W. Thomson, 865, (i) p. 600, ftn., 1650

Hertz, on the impact of two solid elastic spheres (1882), 1515—7, importance of this investigation, 1140, 1684

Hess, on elastico-kinetic analogy, cited, 1267

Heterotatic Axes, 445

Heterotatic Surface, 445, 137 (v); has no existence for rari-constancy, 137 (v)

Hodgkinson, account of his life, 975; G. H. Love on his work, 895; researches on strength of cast-iron pillars (1857), 972—5; his 'Experimental Researches' translated into French, 1095; on the elasticity of stone and crystalline bodies (1853), 1177; his experiments on stretch-modulus referred to, 169 (e); on Emerson's Paradox, 174; his experiments on beam of strongest cross-section criticised by Saint-Venant, 176, rejected by Moll and Reauleaux, 875; experiment on his beam of strongest cross-section, 927; his beam referred to, 951, 1016, 1023, 1031; his experiments on compression criticised by Wertheim, 793; his formula for cast-iron questioned by Bell, 1118; his experiments on cast-iron beams cited by Barlow, 937 (a) and (d); Morin's graphical and numerical presentation of his results

mode of casting and head on strength, 1049—50, 1060, (i) p. 707, ftn. ; effect of remeltings on transverse strength and ultimate deflections, 1097—99, 1101; strength of 'toughened' cast-iron girders, 1105; comparative strength of various kinds of cast-iron (sets, loads and deflections), 1093; nature of fracture of, (i) p. 707, ftn., 1039 (*e*), of small blocks, 321 (*b*), 3⁰

arches, elliptic of, 1011

beams of, comparative strength of various cross-sections, 927, 936; experiments on, 937; of strongest cross-section, 176, 875, 927, 951, 1016, 1023, 1031; Barlow's attempt to explain 'paradox' as to inequality of tensile and transverse strengths, 930—8; of circular cross-section relatively stronger than square, 1038 (*b*); experiments on beams of triangular cross-section, 971; experiments on rupture of, 1024

columns of, 973—4

girders of, comparative strength of cast- and wrought-iron, 954, of ⊥ and ⊥ sections 1007—8, 1031, of toughened cast-iron, 1105: see also *Girders* and *Beams*

ordnance, 891 (*d*), 1037, etc.: see *Guns*

pipes of, Morin and Love's formulae for strength of, 900; rings of, bursting by wedging, 1044

Iron, Meteoric, very ductile, 1165

Iron Plate, stretch-modulus and density of, with and across fibre, (i) p. 531; strength of, 879 (*c*)—(*d*), 902, 1066; resistance to punching, traction, shearing, crushing, 1104; experiments on strength of, 1106; parallel and perpendicular to rolling, 879 (*d*), 902, 1108, 1126—7; absolute strength and stricture with and across direction of rolling, 1141; strength when impulsively wedged asunder, 1107; strength of riveted iron-plates, 1121, 1127, 1135; effect of temperature on unwrought, 1097, 1127; how affected parallel and perpendicular to rolling by change of temperature, 1115, 1126 —7; pipes and water reservoirs of, 904 girders of, 953: see *Girder*

Iron Rivets, strength of, 879 (*d*), 1103; how changed by temperature, 1116; proper thickness for, 1145 (ii)

Iron Scrap, vagueness of term, 1141

Iron Sheet, anomalous action under torsion, 808

Iron Ships, 907—11

Iron, Soft Bar, elastic resistance increas-

ed by tort, 810; influence of torsion on magnetisation, 812

Iron Stays, strength of, 908

Iron Wire, thermal effect of stress, 689 ; effect of annealing, 1131; not rendered brittle by cold, 697 (*c*); thermo-electric properties in relation to strain, 1642 —6; conductivity under strain, 1647; stretch-squeeze ratio (η) for, 1201 (*a*); effect of tort on moduli of, 1755; stretch-modulus of, (i) p. 531, 824; velocity of sound in, 785; attempt to take account of traction of manufacture, 897

strength of, 902, 1033, (i) p. 753, ftn., effect of galvanisation on, 1096, effect of annealing on, 1131, effect of long-continued electric current on, 1187, effect of long-continued stress, 1754, magnetisation, effect of in producing strain, 688; effect of stretching on magnetic properties, 705; effect of pull on magnetisation of, 1727—8; Villari critical field for soft iron-wire, 1730—1, 1736; effect of temperature on Villari field, 1731; effect of torsion on loaded and magnetised iron wire, 1735; relation of magnetisation and torsional elastic strain and set, 708, 714, 812—6; specification of direction of induced current by twist when wire is under longitudinal magnetising force, 1737

Iron, Wrought, thermal effect of stress, 692, 695; thermo-elastic properties of, 752, 756; after-strain and temperature, 756; effect of annealing, 879 (*f*); effect of frost, 1148; effect of heat and working in determining elastic axes, 1065

hardness of, (i) p. 592, ftn., 836 (*b*), 846

stretch-modulus of, how influenced by hammering and rolling, 741 (*a*); value of when rolled, (i) p. 531; generally obeys elastic theory, 1117; Lüders' stress curves for, 1190 and frontispiece to Part (ii); stress-strain curves for, 879 (*a*), stress-strain relation for, 793, 896; elastic limit of, 951; safe tractions for, 176

molecular state of, how affected by repeated loads, 364, 3⁰, by repeated gradual or impulsive torsions, 992—4, 1185; in state of confused crystallisation if forged in large masses, 1066; fibrous and crystalline states of, 861, 881 (*b*), 970, 1067; fracture whether 'fibrous' or 'crystalline' depends on nature of breaking, 1143; effect of sudden load, 1148 impurities render it less liable to 'crystallise,' 1189; molecular constitution of, 1065, (i) p. 736, ftn.; molecular arrangement rather than metallurgical constitution affects strength and elasticity, 1129; how weakened when converted into massive forgings, 1128; nature of fracture, (i) p. 707, ftn., 902, 1140, 1143 strength of, tensile, 902, 966, 1105, 1113, 1133, 1139—40, of bar, angle and plate with and across 'fibre,' 1150, increased by straining up to rupture, 1125, effect of processes of preparation and working on tenacity, 891 (b) and (d); strength when prepared by Bessemer process, 1114; influence of size, skin, forging on absolute strength, 1141; compressive strength equal to tensile, 1118; torsional strength, 1039 (c), 1113; transverse strength, 1105 stricture of, 902, 1139—40, when prepared by Bessemer process, 1114; magnetisation, longitudinal how effected by transverse force, 1733 Axles of, under repeated loading, 1000—3: see *Axles* Beams of, P. Barlow's experiments on, 937 (c): see *Beams* Columns of, formulae for, 978: see *Columns* Girders of, tubular, 1007, proper proportions of web and flanges, 1016, 1018, 1023: see *Girders* Pipes, bursting of, 983

Iron Commissioners' Report, 344, 371 (i), translated into French, 1094

Isochronism, of spiral watch-springs, theory and experiment, 676

Isostatic Cylinders (=conjugate functions), 1562

Isotropy, defined, 4 (η); its rarity, 4 (ι), 115; very doubtful, if it exists in wires, 1271, 1273

Jackson, on steel, 897

Jee, A. S., on deflection and set of cast-iron girders, 1007

Jelly, 1749; equations of motion and equilibrium of, and their solution, 1810; strain-energy for, 1812; comparison of equations with those of viscous fluid and ideal ether, 1809—12

Johnson, R., on hardness of metals and alloys (1860), 845

Johnson, W. R., on strength, etc., of stone (1851), 1175

Joint, equable elastic rotating, 1697 (a)

Jones, J., table of pressures necessary for punching plate-iron (1853), 1103; empirical formulae for his results, 1104

Joule, J. P., effects of magnetism on dimensions of iron and steel bars (1846), 688, cited 1321, 1727; on the thermo-electricity of ferruginous metals and on thermal effects of stretching bodies (1857), 689; on thermal effects of longitudinal compression (1857), 690; on thermo-dynamic properties of solids (1859), 691—6; on testing steam boilers (1861), 697 (a); after-strain and thermal effects in silk and spider filaments (1869), 697 (b); action of cold in rendering iron and steel brittle (1871), 697 (c)

Jouravski, his approximate method of treating slide due to flexure of beams (1856), 939, 183 (a); on vibrations in lattice and plate girders, 1034

Journals, strength of, for railway axles, 959

Junge, on strength of "split" beams (1855), 928

Kant, Saint-Venant's criticism of his antimony, (i) p. 187, ftn.

Karmarsch, on absolute strength of metal wires (1859), 1131

Kaumann, experiments on flexure of railway axles under static load, 990

Kelvin, Lord: see *Thomson, Sir W.*

Kendall, T., on Barnes' discovery that steel may be cut by rotating disc, 836 (h)

Kennedy, A. B. W., experiments on rupture by pressure, (i) p. 215, ftn.

Kenngott, A., on relation between atomic weight and hardness (1852), 841

Kerr, his results for *normal* reflection of polarised light from magnetic pole deduced from Sir W. Thomson's theory, 1782 (c)

Ketteler, adopts F. Neumann's view of dispersion, 1221

Kinematics, elastic deformations treated kinematically, 294

Kinetic Energy, loss of by impact, 209, 217, 1517, 1684

Kinks, in twisted and bent wire, 1670

Kirchhoff, his memoirs on elasticity, 1231; accounts of his life, (ii) p. 39, ftn.; on the equilibrium and motion of thin elastic plates (1848—50), 1232—43; on the elastic equations when the shifts are not indefinitely small (1852), 1244—50; on the equilibrium and motion of an indefinitely thin rod (1858), 1251—70; on the stretch-squeeze ratio of rods of hard steel (1859), 1271—3; on the reflection and refraction of light at the surface of

memoir, 1; gives expression for slide in any direction, 4 (δ); uses doubtful limit of safety, 5 (c); reports on Saint-Venant's memoir on transverse impact, 104; Saint-Venant on his results for cylindrical boiler, 125; his views on propagation of light cited, 146, 1216—20, 1274; Saint-Venant's views on his theory of light, 265; criticism of his deduction of stress-strain relations, 192 (a); his definition of stress, 225; his contributions to curvilinear coordinates, 544; his views as to vibrations, 546; his contributions to figure of earth, 562—8; his error as to 'direct' potential, 1487 (a)

Lamé's Problem, investigated by Sir W. Thomson, 1651

Lamé and Clapeyron, their condition for rupture, 166; insufficiency of their solution of infinite elastic solid, bounded by plane subjected to load, 1487; their solution for infinite plane plate reached by Sir W. Thomson, 1660

Lang, V von, determination of constants which occur in solution of equation for transverse vibrations of rods (1858 —9), 614, 616

Langer, J., on wooden and iron lattice girders, 1022

Larmor, J., on flaws in torsion bars, 1348 (f); on gyrostatically loaded media, 1782

Latticed Girders: see *Girders*

Laugel, A., on the cleavage of rocks, 850

Lavalley, on steel, 897; on iron plate, 902

Laves, introduces "split" beam or girder (1859), 928

Laws of Motion, how far legitimately applicable to atoms, 276 and ftn., (i) p. 185, ftn., 305

Lead, thermal effect of stretching, 692, 695; thermo-elastic properties of, 752; hardness of, (i) p. 592, ftn., 846, 836 (b); thermo-electric properties under strain, 1645—6; ratio of kinetic and static stretch-moduli, 1751; rupture surface of, 1667; bursting of pipes of, 983

Lefort, report by Saint-Venant, Tresca and Resal upon a memoir by, 266

Lemoyne, obscure theory of transverse impact, 965

Le Roux, on thermal phenomena accompanying vibrations (1860), 827

Lévy, M., pupil of Saint-Venant, 416; on stability of loose earth, 242; form of equilibrium of pulverulent mass studied by him, 1590; defect of his

theory, 1613; establishes general body-stress equations of plasticity, 243, 245, 250; his general equations of plasticity corrected by Saint-Venant, 263 —4; his method of finding deflection of circular plate anyhow loaded, 336; his assumptions in theory of thin plates, 385; his memoir on thin plates (1877) and controversy with Boussinesq as to 'local perturbations and contour conditions,' 394, 1441; his views on thin plates criticised by Saint-Venant and Boussinesq, 394, 397; his treatment of local perturbations discussed, 1522—4; his controversy with Boussinesq as to collapse of belts subjected to external pressure, 1556

Lévy-Lambert, abac for Phillips' spring formulae, 921, ftn.

Liebisch, Th., his treatise on physical crystallography cited, 1800, ftn.

Light, relation to elasticity, 101; propagation of, when ether has initial stresses, 145—6; theory of, Saint-Venant's discussion of views of Cauchy, Green, Briot, Sarrau, Lamé and Boussinesq, 265; Rankine's theory of molecular vortices applied to polarised light, 440, his oscillatory theory of, 441; luminiferous ether requires for its refractive action fewer constants than those of crystalline elastic medium, 452; Fresnel's equations deduced by Ménabréa, 551 (b); reflection and refraction for in crystalline media, 594, 1274; historical treatment of elastic jelly theory, 1213; Cauchy's views, F. Neumann's views (Fresnel's laws with different plane of polarisation), 1214; C. Neumann's incompressible ether, 1215; F. Neumann's researches, 1229 (a)—(e); Kirchhoff, F. Neumann, and MacCullagh on reflection and refraction, 1274; Kirchhoff's elastic theory of, 1301; source of, in elastic ether, 1308—10; Clebsch on circularly-polarising media, 1324; Clebsch starts really from a rari-constant basis, 1391; Boussinesq's theory of luminiferous waves, 1449; he retains second shift-fluxions, 1465; he supposes aeolotropy produced by initial stresses in isotropic medium, 1467; discussion of wave-motion, quasi-transverse and quasi-longitudinal waves, Fresnel's wave surface, etc., 1468—71; assumptions made in Boussinesq's final elastic theory, 1478—80; general remarks on, 1484; Boussinesq and Sarrau's equations for waves in double-refractive medium, obtained for deformed isotropic medium, 1559; on waves in aeolotropic medium, 1764, 1773, when incompressible, 1774—5, 1776 (*a*), 1778; Sir W. Thomson on theory of luminous waves, 1765—83, *passim:* see also *Refraction, Double, Ether, Rotation of Plane of polarised light,* etc.

Limestone: see *Stone*

Limits, elastic, pulverulent and plastic, nature of, 1568—9, 1585—7, 1593—5, 1720: see also *Elastic Limits, Failure, Fail-Point, Strength,* etc.

Line of Pressure, in arches, 518, 1009

Link-Polygon, defined, (i) p. 354, ftn.; used in theory of arches, 518

Links: see *Chains, Links of*

Lippich, erroneous theory of vibrations of light loaded rod, 774

Lissajous, J., on transverse vibrations of bars (1858), 825 (*d*); on the optical study of vibrations (1857), 826

Littmann, his experiments on stretch-squeeze ratio referred to, 1201 (*e*)

Live-Load: see *Rolling-Load*

"*Lloyd's*" experiments on iron plates and rivetting, 1135

Load, equivalent statical systems of, produce same elastic strains, 8, 9, 21, 100; this "principle of elastic equivalence of statically equipollent loads" applied to plates, 1354, 1440, 1522—4, 1714; principle stated and demonstrated for body-forces, effect of equal and opposite forces and of couple, 1521; effect of local load in producing stress in extended elastic solid, 1487 (*c*); sudden, effect on iron and frozen iron, 1148; repeated, how affecting materials, 364: see also *Fatigue, Girder, Axle, Torsion, Flexure,* etc.; distribution of, over base of a prism not adhering to a surface, 516; measured by a scale of colours, 794

Loadpoint, 515—6

Load-Systems, classified, 461

Lohse, on the buckling of the bracing bars of latticed girders, 1019

Lommel, adopts F. Neumann's view of dispersion, 1221

Longridge, J. A., on the construction of artillery (1860), 1076—81

Loomis, his experiments on stretch-squeeze ratio referred to, 1201 (*e*); his experiments on influence of temperature on slide-modulus, 1753 (*b*)

Lorberg, on electro-striction referred to, 1313

Lorgna, on statically indeterminate reactions, (i) p. 411, ftn.

Louvel, graphic tables of resistance of iron bars, 921, ftn.

Love, G. H., his Treatise on strength of Iron and Steel and their use in Construction (1859), 894—905; reduces elasticity to an empirical science, 894—5; on strength of pillars of steel (1861), 978; his account of Hodgkinson, 975; his empirical formulae for Fairbairn's experiments on flues, 987

Love, A. E. H., his treatment of flexure, (i) p. ix., ftn.; on boundary conditions for thin shells, 1234; on thin elastic shells, 1296 *bis*; on Kirchhoff's assumptions in theory of plates, (ii) p. 86, ftn.; his Treatise on Mathematical Theory of Elasticity referred to, 1764

Lüders, W., first drew attention to network of curved lines on surface of bar-iron, cast-steel and tin,—*Lüders' curves*—in wrought-iron I-sectional beams, 1190; in round holes punched in steel plates, etc., (i) p. 761, ftn. and (ii), frontispiece

Luminous Point, in elastic ether, theory of, 1308—10; types of motion started by, 1768—9

Lynde, J. G., experiments on cast-iron girders of Hodgkinson's section, 1031

MacConnell, on hollow railway axles (1853), 988—9

MacCullagh, his theory of light referred to, 1274

McFarlane, on aeolotropic electric resistance produced by aeolotropic strain, 1740; experiments on steel pianoforte wire under combined tractive and torsional stress, 1742 (*a*); as to effect of permanent molecular change on elastic moduli, 1753

Maclaurin's Theorem, doubtful use of in elastic theory, 1636—8, 1744, ftn.: see *Approximation*

Macleod and Clarke, on alteration of stretch-modulus of tuning-fork with temperature, 1753 (*b*)

Macvicar, J. G., metaphysical views on atomic theory (1860), 872

Magnetisation, Permanent, as test for purity of iron, 1189

Magnetisation, Relation to Stress and Strain, historical notices, Réaumur on impulsive stress and magnetisation, (i) p. 564, ftn., Scoresby, bending and twisting, (i) p. 564, ftn., Baden-Powell on torsion, 811. De Haldat, Becquerel, Matteucci, etc., on torsion, 811—2; influence of torsional elastic and set strain on temporary and permanent magnetisation, 814—6; Wertheim does not recognise 'critical twist,' 818; he falls back on a theory of ether vibrations to explain magnetic phenomena, 817; no sensible results in case of torsion applied to diamagnetic bodies, 814 (xii); effect of long repeated torsions on magnetic properties of wrought-iron axles, 994; influence of torsion on electro-magnetism, 701—4, on magnetisation, 703; Matteucci's "bundle of fibres" theory to account for magnetic effect of torsion, 701, 704; influence of torsion on magnetisation of steel bars, 712; influence of magnetisation on torsion of iron and steel wires, 713—4; comparison of magnetic and torsional phenomena, 714 (13)—(16); effect of temperature on magnetisation, 714 (17)—(19); Wiedemann's mechanical theory of magnetisation, 715; twisting of iron-wire, when magnetised in a direction inclined to axis, explanation of Maxwell, reversal observed by Bidwell, 1727; soft-iron wire subjected to longitudinal traction and then twisted under earth's vertical magnetic component, effect on magnetisation, 1735; effect of twist on loaded and magnetised nickel wires, 1735; production by torsion of longitudinal magnetisation in wire magnetised by axial current, 1735; on direction of induced longitudinal current in iron and nickel wires by twist under longitudinal magnetic force, 1737

Joule's experiments as to effects of magnetisation on dimensions of iron and steel bars, free and under tension, 688; no magnetic influence on copper wires, 688; influence of tension on magnetisation, 705; influence of longitudinal load on induced and residual magnetisation of iron and steel wires, 1728; statement of Sir W. Thomson's results for steel pianoforte wire and consideration of how they must be limited in light of Villari critical field, 1728—9; Villari critical field for soft-iron, 1730—1; influence of temperature, 1731—2; effect of longitudinal pull on magnetisation of cobalt and nickel, Villari critical field for cobalt, 1736, as to this field for nickel (?), 1736; effects of transverse stress on longitudinal magnetisation of iron, 1733; develop-

ment of aeolotropic inductive susceptibility by stresses other than pure compression, 1734; diamagnetic power of bismuth increased by compression, 700

magnetic rotation of plane of polarisation in flint and crown glass affected by compression, no sensible rotation in slightly compressed crown glass, 698; rotatory effect on plane of polarisation is influenced and can be annulled by mechanical stress, 797 (d)

Kirchhoff's theory of strain produced by magnetisation, 1313—21; he assumes the coefficient of induced magnetisation constant and neglects square of strain, doubtful character of his results, 1314, 1321; strain in isotropic iron sphere due to uniform magnetisation, 1319—20; neglect of terms connecting intensity of magnetisation with strain, 1321

Magnetism, elastic analogue to magnetic force, 1627, 1630, 1813—5; analogues to electro-magnetic force and to magnetic potential in strained jelly, 1813—4; failure of analogue in the conditions at interface of two jellies and of two substances of different magnetic permeabilities, 1815

Magnus, on thermo-electric currents produced by strain, 1645 (iv)

Mahistre, on stress produced by rotation of wheels (1857), 590; erroneous theory of stress produced by rapidly moving load (1857), 963

Mainardi, on equilibrium of strings (1856), 580

Malfatti, on statically indeterminate reactions, (i) p. 411, ftn.

Malleable, defined, 466 (vi)

Mallet, on deflection of girders due to rapidly moving load, 964; on physical conditions involved in the construction of artillery (1855), 1054—72; on resilience and influence of size of casting, 1128—9; reports on properties of metals (1838—43), (i) p. 707, ftn.

Mallock, his experiments on stretch-squeeze ratio referred to, 1201 (e)

Manger, J., on strength of cements (1859), 1170

Manometer, tested by teinometer, 797 (e)

Mantion, theoretical study of a canal bridge (1860), 1034

Marble: see *Stone*

Marcoux, on axles, (i) p. 610, ftn.

on elastic limit, 878 ; on elasticity of aluminium, 1164

Mortar, strength of, 880 (*b*)

Moseley, graphical construction for line of pressure, 1009

Multi-constancy, remarks on, 4 (ʄ), 192, 193, 196, 197, results from hypothesis of modified action, (i) p. 185 ; model illustrating, 1771—3 ; may be deduced according to Sir W. Thomson from Boscovichian system of atoms, 1798—1805 : see for further references, *Constants, Elastic, Constant Controversy, Rari-constancy*

Muscle, after-strain in, 828, 829—30, 832; stretch-modulus of, 830

Musschenbroek, his method of measuring scale of hardness, 836 (*b*)

Müttrich, his experiments on nodal lines of square plates, C. et A. p. 4

Nagaska, on current in nickel wire under longitudinal magnetizing force, produced by twist, 1735, 1737

Navier, gives formula for value of stretch (s_r) in any direction, 4 (δ); his lectures edited and annotated by Saint-Venant, 160; on summing intermolecular action, 228; on impact of elastic bar, 341; his memoir on rectangular thin plates, 399; on statically indeterminate reactions, (i) p. 411, ftn. ; first applies theory of elasticity to arches, 1009; F. Neumann adopts his methods, 1193, 1195

Nerve, after-strain of, 830; stretch-modulus of, 830

Neumann, C., general theory of elasticity (1860), 667—73, 1195; introduces idea of incompressible ether, 1215 and ftn.; his generalised equations of elasticity, 670, 1250; discussion of his views as to elastic constants by Saint-Venant, 193; his method of finding strain-energy, 229

Neumann, F., his "Lectures on the theory of the elasticity of solids" (delivered, 1857—74, published, 1885), 1192—1228; on the optical properties of hemiprismatic crystals (1835), 1218 and ftn.; on the double-refraction of light (1832), 1229, (1841), 1221; on the reflection and refraction of light (1835), 1229; on Fresnel's formulae for total reflection (1837), 1229; his "Lectures on theoretical optics" (1885), 1229 ; his formula for torsion, 1230

chief features of his researches : thermo-elastic equations, 1196—7 ; general proof of uniqueness of elastic

equations, 1198—9; his discussion of crystals, 1203—12, 1219; first determines stretch-modulus quartic, 151, remarks on this quartic, 309, error in Vol. I. about this quartic corrected, p. 209, ftn., C. et A. p. 3; his treatment of waves and elastic theory of light, 1213—8, 1220—1, 1229; his theory of impact of bars, 203, 1224—5; his definition of the plane of polarisation of light, 1214; gives among the first a true theory of dispersion, 1221; erroneous method of approximation, 1225—6

erroneous identification of crystal-line axes (788*—795*), 684; his theory of influence of traction on torsional vibrations, 735 (iii); his method of finding stretch-squeeze ratio by distortion of cross-sections under flexure, 736 and ftn.; notices that, up to a certain limit, volume of wire increases under traction, 736; his theory of photo-elasticity referred to, 792, 793 (iii); list of his pupils, 1192 and ftn.

Neutral Axis, distinguished from neutral line, (i) p. 114, ftn. ; for flexure under asymmetrical loading, 171; relation to ellipse of inertia and stress-centre, 515; applied to elastic bodies resting on rigid surfaces, 515—6, 602; attempted extension to curved surfaces, 602; does not pass through centroid, if there be any thrust, 922; existence or not of strain at, 1016 ; in cast-iron beams does not pass through centroid (?), 971, 1091, 1117 (iii); erroneously placed by Thomson and Tait, 1689

Neutral Line, distinguished from neutral axis, (i) p. 114, ftn. ; coincides closely with central line if load be transverse, 930

Newton, his experiments on impact, 209 ; his theory of impact criticised, 1682; his proof of velocity of sound, 219; treated intermolecular force as central, 269

Nickel, admixture with cast-iron tends to reduce strength of latter, 1165; prevents crystallisation of iron, 1189; effect of torsion on loaded and magnetised nickel wire, 1735; effect of longitudinal pull on magnetisation of, 1736; as to existence of a Villari critical field, 1736; direction of induced longitudinal current by twist under longitudinal magnetic force, 1737

Nicking, how it produces change in tough iron, 1067

Steel, Bessemer, strength of exaggerated by shape of test piece, 1146
Steel, Cast, stress-strain diagrams for, 1084; Lüders' curves for, 1190; tensile and torsional strengths, stricture of Krupp's, 1113; strength and elasticity when prepared by Uchatius' process, 1114; crushing strength of, 1039 (*e*); *axles of*, resistance to impact, 995, 1000; processes of manufacture, 891 (*e*); *plates of*, strength of, parallel and perpendicular to direction of rolling, 1130; elastic limit and structure of, 1130; annealing only slightly reduces strength, 1130; effect of tempering, annealing, etc., on absolute strength, rupture stretch, stretch-modulus, elastic limit, resilience, 1134: see also *Steel, Plates*
Steel, Columns, experiments and formulae for strength of, 978
Steel, Plates, absolute strength and stricture greater in direction of rolling, if puddled, converse if cast, 1142; if rivetted and hardened in oil, as strong as unrivetted plates, 1145; Lüders' curves in steel plates of dredger buckets, 1190, ftn. and (i) frontispiece
Steel, Puddled, for links of cables, absolute strength of, 1132
Steel Wire, thermo-electric properties under strain and working, 1646; stretch-modulus and density of, (i) p. 531; effect of tort on moduli, 1755; Kirchhoff's determination of stretch-squeeze ratio, 1271—3; absolute strength of, (i) p. 753, ftn.; pianoforte, absolute strength of, 1124
Stefan, J., general equations of a vibrating elastic medium (1857), 594; on transverse vibrations of rods (1859), 616
Stephenson, R., on neutral axis, 1016; experiments on cast-iron, 1093
Stiffness, defined, 466 (v); how it affects note of musical string, 472—3, 1374, 1432; how it affects note of membrane, 1439
Stirling, J. D. M., on transverse and tensile strength of cast- and wrought-iron (1853), 1105
Stokes, Sir G. G., discussion of his views as to elastic constants by Saint-Venant, 193; first calls attention to difficulties of uni-constancy, 1770; on his doctrine of continuity, 196; his results for bridges subjected to rolling load, 372, 378—9; comparison of his solution of Willis' Problem with Boussinesq's, 1553; his experiments on

Iceland spar cited against Rankine's hypothesis of aeolotropy of density in ether, 1781; his solution of equations for vibrations of infinite elastic medium reached, 1526; extension of his results for diffraction, etc., to aeolotropic medium of simple kind, 1560
Stoletow, on coefficient of induced magnetisation for soft-iron, 1314
Stone, stress formulae and elastic constants for, 314; rupture of, 321 (*b*), 1°; empirical law for crushing strength of, 1175; strength in frozen condition, 1176; defect of Hooke's law in, 1177; important influence of manner in which faces of cube of stone are bedded during test, 1175, 1180; strength of, 880 (*b*), 1133, 1153, 1176, 1179; strength and deflection, 1174; crushing strengths and rupture surfaces of granite, limestone and sandstone, 1182; strength and density of sandstone, marble and granite, 1178, 1180; cracking and crushing loads of German stones, 1181; crushing and transverse strength of colonial stones, 1183; crushing strength of Irish Basalt, 909, of American stones, 1175, of colonial stones, 1183, of Italian stones, 1184; *résumé* of English and French experiments, 1175
Stoney, B. B., on strength of long pillars (1864), 977; on lattice girders (1862), 1029—30
Storer, H. R., on bursting of gutta-percha tubes (1856), 1160
Strain, pure, definition of, 1677; appropriated by Rankine to relative displacement (1850), 419; homogeneous, Thomson and Tait on, 1672—80, Kirchhoff's treatment, 1276; resolution of homogeneous strain into stretch, slide and dilatation, 1675, combinations of pure strains, 1678; general analysis of, Saint-Venant, 4, Boussinesq, by simple geometry, 1456 —9, in terms of principal stretches, 1575; components of, might be taken as the stretches in the six edges of a tetrahedron, 1640; Sir W. Thomson's general analysis of stress and strain, types of reference, orthogonal systems, 1756—8, principal strain types, 1760 —1; generalised expressions for components of, when shifts or strains are large, 4 (δ), 1248—50, 1445, 1661; permanent, effect on bodies primitively isotropic, deduction of ellipsoidal distribution on multi-constant lines,

230—1; initial state of, in general equations, 237; error of Saint-Venant's method of dealing with on multi-constant lines, 238—9, 1469

apparatus for recording automatically, 998—9, 1032; directions of maxima and minima rendered visible by applying acid to a planed section, 1143, ftn., 1190; graphically analysed by aid of Lüders' curves, 1190 and ftn.

in spherical condenser, 1318; in isotropic iron-sphere due to magnetic force, 1319—21; effect of strain on thermo-electric properties of metals, 1642—6; thermal effect produced by sudden strain, 689—96, 1638, 1750—2

Strain, Combined, slide, flexure and torsion, 50; of prism of elliptic cross-section, 52; case of two equal stretches, two slides equal and third zero, 53; case of two slides vanishing at fail-point, elasticity asymmetrical, general solution for prism under flexure, traction and torsion, 54; case of non-distorted section subjected to slide and torsion, 55; case of cantilever, 56, Case (iv); influence of length of short rectangular prisms on resistance to flexure and slide, 56, Case (i); prism of circular cross-section subjected to flexure, torsion and traction, 56, Case (iii), of elliptic cross-section, 1283; flexure and torsion in shaft, 56, Case (v); torsion and flexure for prism of rectangular cross-section, 57, Case (vi); special cases of skew loading, 58; flexure and torsion of prism of elliptic cross-section, 59; numerical examples of combined strain, 60; flexure, traction and slide, 180; torsion and flexure, 183; traction and flexure, 1289; traction and shearing in case of axles, 1000

Strain-Ellipsoids, 159, 1194, 1673, 1677; inverse strain ellipsoid, 1676; Sir W. Thomson's strain ellipsoid, 1756

Strain-Energy, first legitimate proof that it depends only on strain and not on manner in which strain is reached, 1641; function only of initial and final configurations if equilibrium of temperature maintained, 1463; as quadratic function of strain-components, 1254, 1277—8, 1709 (c); in terms of orthogonal strain-components, 1759; in terms of principal strain types, 1760—1; in terms of principal stretches, 1235; in terms of stresses, when elasticity is ellipsoidal, 163; expressed symbolically, 134; deduced

from rari-constancy by Lagrange's process, 229, 667; when products of shift-fluxions are not negligible, or shifts are large, 1250, 1444—6; when thermal terms are included, 1200; is of two kinds, elastic and ductile, the sum expressing total resilience of body, 1085, 1088; ductile strain-energy erroneously calculated by Mallet, 1128

of rod, 1261, 1266, 1268, 1283 (b); of plate, 1237, 1296, 1699, 1703; of wire (or thin rod), 1690, 1692; for infinite elastic medium, with zero shifts at infinity, 1787, when incompressible, 1812—3, ftn., when subjected to uniform initial slide and incompressible, 1789—97; for jelly and for ideal ether, 1812

Strehlke, his experimental values of nodal circles of circular plates tested by Kirchhoff's theory, 1242—3; his views on nodal lines of square plates criticised by Müttrich, C. et A. p. 4

Strength, ultimate (=absolute), 466 (i), *Proof*, 466 (ii); *limit to*, a stretch rather than a stress, 5 (c), 321 (a), 321 (d), 1327, 1348 (g)—(h), 1386 (b), 1720; in hard solids, 1667; in plastic solids a shear (? a slide), 236, 247, 1586. 1667; *tensile*, how related to density, 891 (a), 1039 (a), 1086; increased by repeated stress, 1754; increased by straining up to rupture, 1125; measured by resilience, 1128; ought to be measured for iron and steel by breaking stress per unit-area of section of stricture, 1150; tensile and compressive increased by solidification under pressure, 1156; *crushing*, of stone increased by lateral support, 1153, 1180; *tensile*, of wrought-iron cables, 879 (e); tensile and crushing of glass in various conditions and forms, 854—6, 859—60; ratio of *tensile* to *shearing* for iron, 879 (d), 903, 966, for steel, 1145 (ii); *transverse* or *flexural*, 920; of beams under flexure produced by skew-loading, 65; graphical tables in case of beams, 921 and ftn.; strength of materials used in construction, views and theories of Ortmann, With, Grashof and Roffiaen, 922—5 (see on transverse strength, *Beams, paradox in theory of, Iron, Cast, Iron, Wrought*, etc.); *torsional*, with empirical stress-strain relation, prisms of circular and rectangular cross-section, 184 (b) and (c); mutual relations of tensile, transverse, torsional and crushing strengths in cast-iron, 1043, theory of, 1051—2,

action of a punch, 1511, 1602 (*d*); on the elasticity and strength of steel plates, 1134; on the elasticity of aluminium, 1164

Truss, history of, C. et A. p. 5 (i): see *Framework*

Tubes, strength of simple tubes and tubes strengthened by belts, 654—5; collapse of tubes, used as boiler flues, experiments and empirical formulae, 982—4; bursting of, by internal pressure, 983; empirical formulae for collapse of, 986—7; bursting of gutta-percha, 1160, of earthenware, 1171—2: see also *Flues, Pipes*

Tubular Bridges and Girders, 1007, 1015

Twist, geometrical discussed, hodograph for, 1669—71; components of strain, 1679; expression of integral tangential shift in terms of, 1681

Twisting, defined, 466 (*a*)

Uchatius, steel prepared by his process, 1114

Undulatory Theory: see *Light* and *Ether*

Uni-constancy: see *Constants, Stretch-Squeeze Ratio, Rari-constancy*, etc.

Uniqueness, of solution of equations of elasticity, 1198, 1199, 1240, 1255, 1278

Unwin, W. C., assists Fairbairn in experiments on collapse of tubes (1858), 984; his *Testing of Materials of Construction* (1888), 1046

Variations, Calculus of, use of in elastic problems, 229, 667—9

Vector-Polygon, defined, (i) p. 354, ftn.; used in theory of arches, 518

Velocity, of pressural and slide waves proved in elementary manner, 219; of elastic waves of various types in diverse materials, 1817

Vène, on statically indeterminate reactions, (i) p. 411, ftn.

Verdet, bibliography and criticism of Wertheim's researches, 820

Vibrations, mode of counting, 822; Lissajous' mode of rendering visible and of compounding, 826; thermal effect of damping, 827; influence of in changing constitution of metal, 1185, 1189 (see also *Iron, Wrought*); coexistence of longitudinal and transverse vibrations, 825, of torsional and transverse, 825; influence of on magnetisation, 811; general laws of, 578

Vibrations of Elastic Media, isotropic, Rankine's form of solution, 434, Po-

poff's solution, 510, Boussinesq's solution by aid of potentials, 1485, by aid of 'spherical' potentials, 1525; form of, when started by various types of elementary vibrators, 1767—9; about a fixed and rigid spherical surface, 1392—1410; aeolotropic, 1764, when there are three planes of elastic symmetry, 594; of a medium obtained by deformation of an isotropic medium, 1557

Vibrations, Stability of, in case of elastic solids, 1328—30

Vibrations of Special Bodies: of *ellipsoidal shell*, 544—8; of *sphere*, radial, 551 (*i*), 1327; of *plates*, 613, 1241—4, 1296 bis, 1300 (*b*), 1383—4, when aeolotropic, 1415, when *infinite*, 1462; of *membranes*, 551 (*h*), 1223, 1300 (*c*), 1385, when stiff, 1439; of *rods*, deduced from systems of particles, 550—1, transverse, 614—6, 821—2, 825, 1228, 1291, 1372—3, 1431, when loaded, 751 (*c*), 759 (*a*), 769, 774—84, 1431, when cross-section varies, 1302—7, longitudinal, 823—4, 825, 1224, 1291, 1373, 1431, torsional (for prism, rod or wire), 191, 751 (*d*), 1373—4, 809, 1291, 1431, subsidence of, 734, 739, 1744—8; of *curved rods*, Bresse's equation, 534; of *strings*, 617, 1291, 1374, deduced from those of systems of particles, 550—1

Vicat, his experiments on rupture cited by Saint-Venant, 32, by Morin, 880 (*b*); on cohesive power of cements, 1168

Vignoles, on adaptation of suspension-bridges to railway traffic (1857), 1025

Villarceaux, Y., on hydrostatic arch, 468

Villari, on relation of stress to magnetisation (1865), 1729, 1731

Villari Critical Field, for soft-iron, 1730 —1, 1733, for cobalt, 1736, for nickel, 1736

Virgile, his memoir, criticised by Saint-Venant, 122

Virtual Velocities, applied to theory of elasticity, 427—9, 667, 1195

Viscosity of Fluids, equations for, 1744, ftn., 1809

Viscosity of Solids, 734, 748, 750; in Sir W. Thomson's sense, 1666; confusion of after-strain with frictional resistance, 750, 1718 (*b*), 1743; how related to plasticity, 1743; according to Sir W. Thomson no simple law between viscous resistance and strain-velocity, 1744; experiments on subsi-

819; memoir on flexure, 820; commits suicide (1861), 820

References to: on torsion, his experiments and errors, 191; on caoutchouc, 192 (*b*)—(*c*); (and Chevandier), experiments on wood, 169 (*f*), 198 (*e*), 284; Rankine on his hypothesis of $\lambda = 2\mu$, 424, tested by Kirchhoff's theory of thin circular plates, 1242—3; his hollow prism method of finding stretch-squeeze ratio considered by F. Neumann, 1201 (*b*); his views as to set and his values of kinetic moduli criticised by Seebeck, 474; his experiments on influence of temperature on moduli referred to by Kupffer, 723; his value for stretch-modulus of gold, 772; his results for the stress-strain relation in organic tissues confirmed by Volkmann, 831—2

Weyrauch, his contribution to law of intermolecular action, C. et A. p. 1

Wheels, tires of, flexure and stress in, when shrunk on, 584—8; Mahistre's obscure treatment of, 590

Wiebe, H., on strength of rivetted iron plates, 1121

Wiedemann, G., on torsion, flexure and magnetism (1860), 706—15; on axes of electrical conduction in crystals, 1219; his treatise on electricity and magnetism cited, 818; on relation of stress to magnetisation, 1727, 1735

Willis' Problem, of rolling loads on bridges, 344, 372, solved by Boussinesq, 1553

Wilmot, F. E., report on cast-iron ordnance (1858), 1048—51

Wilson, J., on 'grooved' plate springs, 969 (*c*)

Winkler, C., on lattice girders and the distorted form of bracing bars (1859), 1028

Winkler, E., on the strain and strength of links of chains (1858), 618—41; on strength of flues, boilers, and flywheels (1860), 642—7; general equations of stress and application to special case of flexure (1860), 660—5; on continuous beams (1862), 949

Wire, stretch-modulus and density of metals, 773, of brass, iron, steel and copper, (i) p. 531; tables of slide- and stretch-moduli for, 1749; want of isotropy in, 1271—3, 1692, ftn.; drawing, decreases density, 1149, influence of, on strength, 1131; strength absolute, depends partly on first power and partly on square of diameter, 1131;

subsidence of vibrations in, due to 'viscous' action, 1744—8

Wire-Rope, how weakened by heating, splicing, etc., 1136

Wires, Theory of, with central line of double-curvature, Kirchhoff on helical, 1268—9; Thomson and Tait on helical, 1693; kinks in twisted and bent, 1670; Thomson and Tait's general theory, 1687—97; disregard distortion of cross-section and yet require it to reach torsional rigidity, 1687, 1691, really fall back on Bernoulli-Eulerian theory, 1691; strain-energy for, 1690; principal torsion-flexure rigidities, 1692; principal helices, 1692; wire of equal flexibility as equable rotating joint, 1697 (*a*); wire with circular central line and plane of greatest flexibility inclined to plane of central line, bending of, 1697 (*b*); special cases of bending and rotating wires in form of hoops, etc., 1697 (i)—(iii); see also *Rod, Springs,* etc.

With, on strength of materials, 923

Wittstein, on strength of screws, 966

Wöhler, A., initial form of central line, that a bar when loaded may become straight (1853), 919; early experiments on repeated flexure and torsion of railway axles (1858—60), 997—1003; on plate-web and lattice girders (1855), 1017; cited as to effect of alternating load on strength, 407 (1)

Wollaston, on hardness of diamond, 836 (*e*)

Wolters, G., on deflection of Flemish railway bridges (1856), 1020

Wood, thermo-elastic properties of, 694; variation of stretch-modulus of, across trunk of tree, 169 (*f*); elastic constants of, 198 (*e*), 282 (9), 308, 312—3; stress formulae and elastic constants for, 314; safe tractions for, 176; stretch-moduli of various kinds, 1157; strength of, 879 (*e*); tensile and crushing strengths, with and across fibre, deflections of 80 specimens, 1158, of 3000 specimens (including set and elastic strain), 1159; flexure of various kinds, limit of elasticity (oak, beech), large beams have less stretch-modulus than small, 1157; rupture of, 321 (*b*), 2^0; torsion of wooden prism, 186; bars of, under flexure have increased strength, if subjected to traction, 918; best method of cutting beams from tree, 1167; advantageous forms of wooden trusses, 1167; arches of, ex-

CORRIGENDA AND ADDENDA
TO VOLUME I.

CORRIGENDA.

Art. 922.

I have used an expression in this article with regard to Weyrauch's contribution to the problem of rari-constancy which is undoubtedly liable to misinterpretation. It might be supposed from what I have written that Weyrauch had obtained rari-constant equations on the assumption that the intermolecular action although central was any function *whatever*, e.g. a function of aspect' or involving 'modified action terms.' What he really does (*Theorie elastischer Körper*, 1884, p. 132) is to take a central action R between two elements of masses m and m', at distance r of the form :

$$R = mm' \{F(r) - i\} \quad\dots\dots\dots\dots\dots\dots\dots\text{(i)},$$

where, in his own words :

"*mm'i* ganz allgemein eine Function derjenigen Grössen bedeutet, welche neben der Entfernung r auf R Einfluss nehmen."

This of course is something different from taking R of the form :

$$R = mm' F(r, i) \quad\dots\dots\dots\dots\dots\dots\dots\text{(ii)}.$$

Further, if i_0 represents the value of i before strain or at time t_0, and i the value at time t, Weyrauch assumes (p. 134) that $i - i_0$ for the material in the neighbourhood of the element m may be treated as constant and brought outside the sign of summation for elementary actions. This would be impossible, if $i - i_0$ were due to 'modified action,' because the modifying elements (or molecules) would be themselves in the immediate neighbourhood of m, and the modifying action would probably be a function of their distances which are themselves commensurable with the linear dimensions of the "neighbourhood of the element m."

By taking R of the form (i) and not (ii) Weyrauch much limits the generality of his results, and by choosing $i - i_0$ a constant for the neighbourhood of an element, he practically reduces his $(i - i_0)$ to little more than the temperature-effect. But even this may serve to indicate that wider laws of intermolecular action than that in which it is central and a function of the distance only may be found to lead to rari-constant equations.

Art. 959.

The formulae for the buckling load on struts were taken from notes of mine in which $2l$ and *not* l was the length of the strut. This, however, does not apply to the point of maximum traction or other results of this same Article. We have with this correction the following results for a strut of length l:

Buckling force for doubly built-in strut

$$= E\omega \frac{\dfrac{4\pi^2 \kappa^2}{l^2}}{1 + \dfrac{4\pi^2 \kappa^2}{l^2}}.$$

Buckling force for built-in pivoted strut

$$= E\omega \frac{\dfrac{\pi^2 \kappa^2}{l^2}\, 2{\cdot}047}{1 + \dfrac{\pi^2 \kappa^2}{l^2}\, 2{\cdot}047}$$

Buckling force for doubly pivoted strut

$$= E\omega \frac{\dfrac{\pi^2 \kappa^2}{l^2}}{1 + \dfrac{\pi^2 \kappa^2}{l^2}}$$

I much regret that this error should have escaped my attention, and trust all possessors of the first volume will make the above changes in the text.

Arts. 795—6.

I have reproduced an error of Neumann's which I ought to have seen and corrected. The wrong signs are given to all the quantities M, N, P in Art. 796. If these are corrected a negative sign must be inserted in the second table of Art. 795 before all the $1/F$'s. The value of $1/E$ in Art. 799 is then accurate.

Arts. 1392—3.

The word 'copper' should be replaced throughout by 'brass.'

Art. 1467.

The form of the beam section, which is ⊥, has dropped out of the type.

Index, p. 899, Column (ii) and Arts. 813—16.

The title *Bresse* has been inserted between *Bevan* and *Binet*, when it ought to follow *Braun* on p. 900, Column (i). There should also be a reference under *Bresse* to Arts. 813—16. I find that the lithographed course of lectures there referred to is due to this scientist, to whom we thus probably owe the first theory of the 'core.'

ADDENDA.

Arts. 352, 353, 354—5, 745—6.

A paper by A. Müttrich on Chladni's figures for square-plates appeared in 1837 in the *Geschichte des altstädtischen Gymnasiums. Dreizehntes Stück*, Königsberg. It is entitled: *Beitrag zur Lehre von den Schwingungen der Flächen*, and contains 8 pages and a plate of figures. Pp. 1—5 suggest practical methods of supporting the plates, of setting them vibrating, and of keeping their surfaces dry and clean. Pp. 6—8 give Müttrich's conclusions and the grounds on which he bases them. Two of them are opposed to Strehlke's views of 1825 as given in our Vol. I., Art. 354, namely Müttrich holds:

(i.) Straight lines are possible forms for the nodal lines of plates with free edges.

(ii.) Nodal lines can intersect one another.

The experimental proof of these results lies in the demonstration of a *gradual* transition from one system of nodal lines to another, when intermediate stages are necessarily intersecting straight lines.

Müttrich's third conclusion is that the nodal lines themselves are in a state of vibration and that only their nodal points are true nodes for the plate. It seems to me possible that this oscillation of the nodal lines results from longitudinal vibrations in the plate which again are due to its sensible thickness, or to the mode of support and excitation.

Art. 937.

A copy of Ardant's work which was printed as a separate publication by "order of the minister of war" has reached me since the printing of Vol. I. The title is: *Études théoriques et expéri-mentales sur l'établissement des charpentes à grande portée*, Metz, 1840. It contains *Avertissement* pp. i—v; the report referred to in our Vol. I., Art. 937, pp. vi—xvii; the text of the work pp. 1—94; *Appendice* pp. 95—122, and concludes with five pages (123—127) of contents and twenty-nine plates of figures. It is obvious that the work is one of considerable size, and as it possesses some importance, I give here a résumé of its contents.

[i.] Chapter I. (pp. 1—11) briefly describes the origin and history of wooden trusses designed to cross considerable spans, more especially roof-trusses. These range from the 4th century roof of the Basilica of Saint-Paul's, through the frame 'à la Palladio,' the arched truss of Philibert de l'Orme, and the Gothic roof to the English truss with iron tie-bars, and to the arched forms common in France in 1840. Ardant gives at the end of the chapter a summary of the conclusions he has formed upon the comparative merits of arched timber trusses and trusses built up of straight pieces of timber. He believes the former to be very inferior to the latter in both economy and strength; while the latter can be easily made to present as pleasing an artistic effect. He holds the adoption of the former to have arisen partly from the mistaken notion that a semi-circular arch produced little or no thrust on the abutments, partly from an unreasoning extension of the theory of stone arches to wood and iron:

Dans la première de ces constructions, on utilise la pesanteur, la rigidité et l'inflexibilité relatives des pierres; dans les secondes, c'est l'élasticité et la cohésion des parties qui sont les qualités essentielles (p. 10).

Chapter II. gives an account of the fifteen arches and frames (with spans so large as 12·12 metres and rise so large as 5·41 metres), upon which experiments were made, as well as the apparatus with which they were made.

[ii.] Chapters III., IV. and V. cite the theoretical results of the Appendix for the thrust in terms of the load in the cases of circular arches and of a simple roof-truss of straight timbers. The thrust for the latter is not materially greater than that for the former. Hence no gain is obtained by combining the two, which appears to have been frequently done in practice :

On tirera de cette comparaison une conclusion assez opposée à l'opinion de la plupart des constructeurs, savoir ;

Que dans les cas ordinaires de la pratique, un cintre demi-circulaire exerce autant de poussée que la ferme droite sans tirant, à laquelle on le réunit pour composer une charpente en arc ; et que, par conséquent, on pourrait, en augmentant l'équarrissage de cette ferme, supprimer le cintre sans qu'il en résultât sur les appuis, une action horizontale plus considérable (p. 25).

These chapters then compare the experimental measure of the thrust with that given by theory. The comparison gives an accordance fairly within the limits of experimental error. Unfortunately Ardant did not make a sufficiently wide range of observations for the results to be quite conclusive. He cites an experiment of Emy which led the latter to believe that circular arches had no thrust. He then considers experiments made by Reibell at Lorient. These appear to be the only other important experiments which had been made on large circular wooden arches. An account of them was published in the *Annales maritimes et coloniales* 22ᵉ année, 2ᵉ série, T. XI., p. 1009. Reibell did not get rid of the friction at the terminals of the arch, but allowing for this Ardant finds the corrected values of the thrusts agree well with his formulae (pp. 32—33). From this double set of experiments he draws the following conclusions :

(*a*) The thrust of a semi-circular arch due to an isolated central load never exceeds $\frac{1}{5}$ of the load.

(*b*) Whatever be the manner in which a continuous load is distributed along the arch, the thrust for a semi-circular arch never exceeds $\frac{1}{4}$ to $\frac{1}{3}$ of the total load.

(*e*) That flatter arches produce thrusts which are to those which arise in the case of a semi-circular arch in the ratio of the half span to the rise.

(*f*) That the thrust is independent of the particular mode of construction of the arch, when its figure, dimensions and the load-distribution are the same.

Chapter V. shews that the thrust-formula obtained in the Appendix for the truss with straight timbers, and without a tie, is confirmed by experiment.

[iii.] Chapter VI. begins with some general discussion on elasticity, the elastic constants and the coefficients of rupture. Ardant then cites a formula of the following kind for the deflection, f, of a circular arch at the summit, the terminals being both pivoted:

$$f = K \frac{P Y^2 X}{E \omega \kappa^2},$$

where $2X$ is the span, Y the rise, E the stretch-modulus, $\omega \kappa^2$ the moment of inertia of the cross-section, P the total load and K a constant depending on the distribution of the load etc. Here the arch is supposed to be of continuous homogeneous material and of uniform cross-section. Ardant now applies this formula to the deflections he has found by experiment for his arches built up of curved pieces or planks pinned or bound together. The results given in Chapter VII. he holds to satisfy this formula, provided E be given values depending on the nature of the structure, from $\frac{3}{8}$ to $\frac{3}{50}$ of its value for a continuous arch or beam of the same material. The experiments even on the same arch seem to me to give such divergent values for E, that I think this method of exhibiting the deflection can only be looked upon as an expression of experimental results for practical purposes. With certain assumptions Ardant also obtains an expression for the deflection of a roof truss without tie, built up of straight beams (pp. 48—49). I do not consider this expression to be theoretically or experimentally justified. Ardant proceeds at the end of Chapter VII. (pp. 61—68) to determine the resistance to rupture of his arches. Here he applies to rupture a formula deduced from the theory of continuous arches on the hypothesis that linear elasticity holds up to rupture. At best the theory could only apply to the *fail-point* (i.e. failure of linear elasticity) of continuous arches. A like treatment of rupture leads to absurd results in the case of the flexure of beams, so it can hardly be expected to give better results in the case of arches: see our Vol. I. Art. 1491 and Vol. II. Art. 178. Thus, as we might naturally expect, his "coefficient of rupture" varies from arch to arch, and its ratio in each case to the "coefficient of rupture" for a continuous arch is equally variable. The results however, of his experiments resumed on (pp. 67—8) are suggestive for the practical design of such arches and roof-trusses as he has experimented on.

[iv.] In Chapter VIII., it is sufficient to notice here Ardant's conclusion that the truss built up of straight beams is for the same amount of material stronger than the built-up wooden arch :

Il semble d'après cela que si les charpentes en arc conservent quelque avantage sur les fermes droites, c'est uniquement celui d'avoir une forme plus gracieuse, et que sous les rapports importants de la solidité et de l'économie, les premières sont très-inférieures aux autres (p. 75).

Chapter IX. gives methods of calculating suitable cross-sections for the various parts of arches of the types on which Ardant has experimented. It also gives some attention (pp. 77—80) to the thickness and height of the masonry which will stand the thrust of a given roof-truss. It concludes with two numerical examples of the application of the formulae of the appendix to the calculation of the dimensions of metal arches.

[v.] We now reach the *Appendice*, which is entitled : *Théorie de la flexion des corps prismatiques dont l'axe moyen est une droite ou une courbe plane* (pp. 95—122). This contains the first theory of circular arches which attains to anything like completeness (see our Vol. I. Arts. 100, 278, 914), and it anticipates Bresse's later work on this subject : see our Vol. I. Arts. 1457—8, and Vol. II. Chapter XI. for an account of the book referred to in these Articles. We note a few points with regard to this Appendix.

(a) Pp. 95—100 give the ordinary Bernoulli-Eulerian theory of flexure. On p. 98 Ardant speaks of the product of the stretch-modulus and moment of inertia of the cross-section (namely $E\omega\kappa^2$ in our notation) as improperly termed the *moment d'élasticité*. It is the *moment de roideur* of Euler (Ek^2 in his notation : see our Vol. I. Art. 65) or the 'moment of stiffness.' This 'moment of stiffness,' $E\omega\kappa^2$, occurs so frequently that we have ventured to term it the 'rigidity' of a beam. It follows from this definition that the product of the rigidity and curvature is equal to the bending-moment. Thus for the same value of the bending-moment the curvatures of a series of beams vary inversely as their rigidities.

(b) Pp. 100—103 deal with rupture on the old lines, i.e. as if linear elasticity lasted up to rupture. The results obtained are thus only of value when we treat the 'coefficient of rupture' R which occurs in them as the 'fail-limit.' Accordingly the Tables on p. 103 for rupture-stresses are meaningless when applied to the previous flexure formulae. On pp. 99 and 101 we have the rigidity and fail-moment (here called *moment de rupture*) calculated for 'skew-loading' or for the case when the load-plane does not pass through a principal axis of inertia of each cross-section : see our Vol. I. Arts. 811, 1581, Vol. II. Arts. 14, 171. To judge by Ardant's reference to Persy's lithographed *Cours*, the latter possibly did more for the theory of skew-loading than I judged from an examination of only one edition of that *Cours*: see Vol. I. Art. 811. The value given by Ardant on p. 101 for the fail-moment of a beam of *rectangular* cross-section under skew-loading is incorrect, it applies only to the case of *square* cross-section. The true value is given in our Vol. II. Art. 14.

(c) Pp. 104—115 are occupied with a consideration of the elastic line under various systems of loading in the case of straight beams, besides a discussion of combined strain. The results obtained are afterwards applied to various types of simple roof or bridge trusses, in which the members are supposed mortised and not merely pinned at the joints. Ardant's treatment of these trusses seems to me from the theoretical standpoint extremely doubtful, and I should hesitate before applying his results even to the practical calculation of dimensions. The remark in § 34, p. 107, on the sign to be given to a certain quantity is, I think, erroneous. The fail-point of a beam is not necessarily where the *stress* is greatest, as Ardant like Weisbach (see Vol. I. Art. 1378) holds. It will be at the point of *maximum stretch*, and this will be at the side of the cross-section in tension or compression according as the load-point is outside or inside the whorl of the cross-section : see Vol. I. p. 879.

(*d*) Pp. 115—121 contain the theory of flexure of circular ribs or arches. Ardant's work here was up to his date the most complete treatment of the subject, and his Table on p. 45 for thrust and deflection based upon this theory may even now be of practical service. He obtains the thrust and deflection for circular ribs with an isolated load, or with uniform loading distributed along either the span or rib, when the terminals of the rib are *pivoted.* He finds also for a complete semi-circle, that the points of maximum horizontal shift are about 63° from the vertical. He throws all his results into very simple approximate forms, which he holds accurate enough for practice. I refrain from quoting these theoretical results, because they have been worked out with greater generality and accuracy by Bresse in a work with which I shall deal fully in Chapter XI. At the same time Ardant's researches must be remembered as an important historical link between those of Navier and Bresse. That the latter had studied them may be seen from our Vol. I. Art. 1459.

What I have noted in Ardant's memoir will probably be sufficient to mark its importance. Experiments on such large wooden arches and frames have I believe not been repeated and it seems improbable that they ever will be. The results obtained will therefore remain of value, so far as roof-structures of the types with which Ardant dealt are concerned. In addition to the experimental data of the memoir I may mark Ardant's conclusion, that the same theoretical formulae hold for an arch of continuous material and one built-up of bent pieces of wood or planks bolted or bound together, provided we reduce the stretch-modulus in a certain proportion. Finally I have already noted the historical value of the memoir as a step in the theory of circular arches or ribs.

Art. 974.

Poncelet. *Cours de mécanique industrielle, fait aux artistes et ouvriers messins, pendant les hivers de* 1827 *à* 1828, *et de* 1828 *à* 1829. *Première partie. Préliminaires et applications.* Metz, 1829. I have procured a copy of this work since the publication of Vol. I. It contains xvi pages of prefatory matter, 240 pages of text, and 8 pages of contents at the end. The first preliminary 145 sections agree with those in the third edition by Kretz (1870). In the *Applications* the Metz edition agrees fairly with Kretz's up to section 197; after this it deals with the resistance and motion of fluids, thus containing nothing concerning the resistance of solids to which the *Deuxième Partie* of the 3rd edition is devoted. The few paragraphs on the *Élasticité des corps*, pp. 17—20, are thus all it contributes to our subject: see our Vol. I. Art. 975. The chief interest of the work is the place it takes in the origin of modern technical instruction.

Art. 1249.

A further memoir by Brix which had escaped my attention may be referred to here: *Ueber die Tragfähigkeit aus Eisenbahnschienen zusammengesetzter horizontaler Träger.* This is an offprint from the *Verhandlungen des Vereins zur Beförderung des Gewerbfleisses in Preussen*, Berlin, 1848, 16 pages and a plate.

Owing to some peculiar local conditions at a Berlin mill it was necessary to build bridges, of which the girder-depth had to be very small, over the mill-races. For this purpose pairs of railway rails with flat bases ('*sogenannte Vignolsche*') were placed base to base and used as girders. The bases were riveted together at short intervals. Experiments were made on the flexure and ultimate strength of two such girders; in the one the bases were riveted close together, in the other there were placed at the rivets small intervening blocks of cast-iron. The first part of the paper (pp. 1—6) is occupied with an account of the experiments made upon these two girders, for the details of which—too individual to be of much general use—I must refer to the paper itself. The rupture, by shearing of the rivets, only seems to shew that the area of the riveting was very insufficient, as the load required to produce failure in a bar under flexure by longitudinal shearing is immensely greater than that required to produce failure by stretch in the 'fibres,' the order of the ratio of these loads being practically that of the length to the diameter of the bar.

The second part of the paper—that specially due to Brix—deals with the theory of the flexure of a beam (a) with both terminals supported, (b) with one terminal supported and one built-in, (c) with both terminals built-in—the load in all cases being partially uniform and continuous and partially isolated and central. The treatment of these problems by the Bernoulli-Eulerian theory presents no difficulties, but it has long been known that the *absolute strength* of beams under flexure calculated by this theory is very far from according with experiment (see our Vol. II. Art. 178). Hence there does not seem much value in the numerical results given on pp. 12—16 and based on the preceding experiments. Two points in Brix's work may be noticed. He *assumes* the maximum curvature (which gives the maximum stretch and so the fail-point) to be either at the built-in end or the centre of the beam in case (b), but this is by no means obvious, it requires an investigation similar to that given by Grashof in Arts. 58—9 of his *Theorie der Elasticität*, 1878. Secondly, he shews, I believe for the first time, that the fail-point for a *uniformly* loaded beam, either doubly-built-in or built-in and supported, is at the built-in end; in the former case the bending-moment at the centre is only half its value at the built-in ends.

Arts. 1180 and 1402, ftn.

A copy of Seebeck's paper in the *Programm* of the Dresden Technical School (1846) has reached me. It contains a good deal of valuable matter, and I have taken the opportunity of referring to it with other papers of Seebeck's in the course of Vol. II. Art. 474.

CAMBRIDGE : PRINTED BY C. J. CLAY, M.A. AND SONS, AT THE UNIVERSITY PRESS.

Printed in the United States
By Bookmasters